AF386389

The Frontiers of Science

The Frontiers of Science

By

Benjamin Lewin

Hardback ISBN: 978-1-83767-967-6
EPUB ISBN: 978-1-83767-968-3
PDF ISBN: 978-1-83767-969-0

A catalogue record for this book is available from the British Library

The Royal Society of Chemistry is a charity, registered in England and Wales, Number 207890, and a company incorporated in England by Royal Charter (Registered No. RC000524), registered office: Burlington House, Piccadilly, London W1J 0BA, UK, Telephone: +44 (0) 20 7437 8656.

For further information see our website at www.rsc.org

For general enquiries, please contact books@rsc.org

For EU product safety enquiries, please email books@rsc.org or contact Royal Society of Chemistry Worldwide (Germany) GmbH, Römischer Hof, Unter den Linden 10, 10117 Berlin.

Printed in the United Kingdom by CPI Group (UK) Ltd, Croydon, CR0 4YY, UK

For Ann

Scientific progress on a broad front results from the free play of free intellects, working on subjects of their own choice, in the manner dictated by their curiosity for exploration of the unknown. Freedom of inquiry must be preserved...

Vannevar Bush, *Science, the Endless Frontier*, 1944.

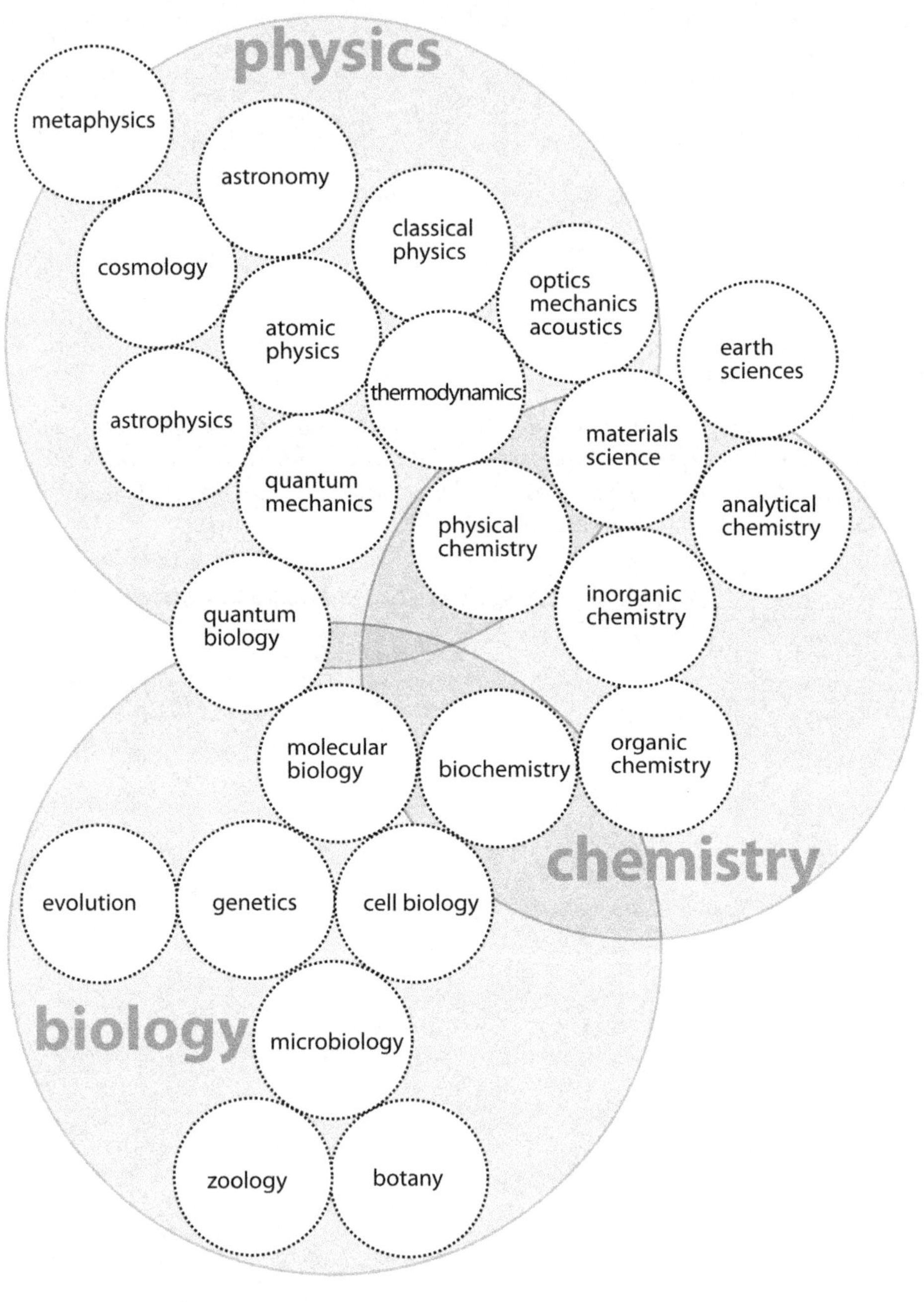

physics
chemistry
biology
metaphysics
astronomy
cosmology
classical physics
optics
mechanics
acoustics
atomic physics
earth sciences
thermodynamics
astrophysics
materials science
analytical chemistry
quantum mechanics
physical chemistry
inorganic chemistry
quantum biology
molecular biology
biochemistry
organic chemistry
evolution
genetics
cell biology
microbiology
zoology
botany

Preface

How did science come to seize the commanding heights of the intellect? For make no mistake about it, science now dominates intellectual activity in a way nothing has done since religion in the Middle Ages. Its influence is equally strong for those who accept science and for those who battle against it. The question is how and why science came to occupy this role (and whether it deserves it).

Science is a relatively modern concept, a self-contained system in which theories depend on facts, and change when new facts emerge. Science has never been more influential, but it is also under attack from those who deny its legitimacy. Sociologists are wont to claim that science is nothing more than a fiction of the society that creates it, that it does not really have any special character. This makes it all the more important to explain the basis for the claim that science is distinguished from other intellectual activities by its objectivity.

An apple falls to the ground by gravity no matter where the tree is on the planet. Gravity is a *fact*, independent of time or place or who first noted the phenomenon. The independence of science rests upon the idea that facts are universal. Theories are accepted or rejected according to whether they agree with the facts.

The humanities are more sensitive to societal context, and more inclined to the position that all views can be equally valid.

The Frontiers of Science
By Benjamin Lewin
© Benjamin Lewin 2026
Published by the Royal Society of Chemistry, www.rsc.org

Is a clash of cultures between science and the humanities just a failure of one to understand the other, or does it go further into an irreconcilable difference of attitudes?

The politically correct approach to the history of science compounds the problem by placing societal influence above scientific discovery. One recent historian of science admits, "History of science, far from serving as a bridge between the arts and the sciences, nowadays offers the scientists a picture of themselves that most of them cannot recognize. It has become part of the problem, not part of the solution."[1]

As a scientist rather than a historian, I take a different approach, focusing on science as an enterprise in its own right. Especially in the context of explaining science to the public, it is important for the narrative to be one that scientists might recognize. My aim is to relate history to science as it is practiced today.

I show how science has morphed from sharing a common approach with other intellectual activities to a distinctive system with its own precepts. The major challenge is to understand how and why the scientific ideas of today have replaced those of the past. However, it is just as important to understand false turns, why inadequate theories sometimes persist longer than they should, or why there can be resistance to new theories.

Historians today reject the vantage point of the present, emphasizing the need to understand scientists and their discoveries exclusively in the context of their times. They reject the 'great men' theory of science. (Ironically, restricting analysis to the terms of the then-contemporaneous society emphasizes even more the remarkable leaps of inspiration that enabled the 'great men' to break out of their constraints.)

To be sure, to understand the mindset of researchers historically, we have to wind back the clock to the worldview of the past. But to assess the contribution of historic discoveries to our understanding of science today, and to its impact upon society, it is at least equally important to assess discoveries with respect to current knowledge.

I want to understand how and why the modern mindset developed. That said, this is not simply a record of how we got from there to here, because you can be wrong for the right reasons, or right for the wrong reasons. The reality (or otherwise) of the facts, and the logic of the argument, are equally important.

It's a fair question how far science today may be limited by our current worldview, which leads down the slippery slope to relativism, believing that 'facts' are valid only in the context of current society. To be sure, social and cultural context can impact the conduct of science. Indeed, science cannot function in isolation from society. It needs the support of society to be a successful endeavor. But science stands as an independent activity in the way that theories are ultimately validated or rejected within the body of science.

A key difference between science and history is that science obeys laws, but history is changeable. Historian Hugh Trevor-Roper said, "History is not merely what happened: it is what happened in the context of what might have happened. Therefore it must incorporate, as a necessary element, the might-have-beens."[2] By contrast, the laws of physics have been unchanging since very early in the evolution of the universe. You cannot defeat the law of gravity. The apple will always fall down to the ground. But history could easily have had another outcome. This is an essential difference with science.

Viewing the past in its own terms can certainly explain the dominance of a theory that we now hold to be erroneous. However, it's less successful in explaining the contribution of a new theory to our present understanding. As Nobel Prize-winning physicist Steven Weinberg pointed out, a difference between science and other intellectual activities is that "in scientific history we really can say who was right."[3]

The history of science is a history of discoveries that changed our worldview. Why did everything seem to revolve around the Earth before we realized the Sun is at the center of the solar system? How did we turn from the absolutism of Newton's law of gravity to Einstein's relativity? What forced the transition from believing in protein as the hereditary material to the centrality of DNA? Examples such as these allow us to follow the evolution of science as an intellectual enterprise. I want to give the flavor of discoveries at the time as well as to explain how they contribute to modern science.

I am concerned here with the nature of scientific thought, with the development of scientific attitude, with what causes science to advance (or impedes it, internally or externally). This is not a history of science as such, but discoveries are the framework for assessing the great enterprise of science.

Knowledge of the details of science is not required to follow this book. I have concentrated on principles, not details. I want to demystify science without losing sight of its essential object-ives. (In the discussion of contemporary physics, for example, there are no equations—ignoring $E = mc^2$, which, I feel, as the most famous equation of all time, is an exception.)

I present science as advancing to its present state through a series of revolutions.[4] They are not necessarily revolutions in the popular sense of sudden discrete moments when everything changes. Often enough, a discovery was not recognized imme-diately, but in retrospect marked a dramatic change in under-standing. Most scientific revolutions have actually played out over a protracted period, sometimes over several decades. But comparing the end with the beginning, they are revolutions nonetheless.

These revolutions were almost certainly inevitable. In science, facts are the facts, and there comes a time when the accumu-lation of knowledge requires a revolution in theory. Yet many of the revolutions described in this book were driven by certain individuals, or are very much associated in the public mind with them. Emphasizing that science is a human activity, I have named the chapters for the geniuses who sparked the revo-lutions. Each of their names is, as it were, shorthand not only for a discovery, but for a whole way of thought associated with it.

Each chapter starts with a timeline outlining its scope, but it's also useful to look out of the period, either forward or backward. To sharpen the focus on the transition between periods, the timelines are followed by (imaginary) dialogues between a major personality of the period and a historic figure from a different era.

Four key ideas whose ramifications underpin the ascent of science are:

- the Big Bang—the origin of the universe;
- the atom—the unity of matter;
- the double helix—the unity of all life;
- artificial intelligence—the future?

The diversity, extent (and specialized terminology) of science can seem overwhelming, but together these ideas define the grip of the natural sciences (physics, chemistry, and biology) on modern life.

Even the enemies of science—perhaps this is a bit strong but the vehemence with which science is attacked makes it seem appropriate—admit that it is a dominant intellectual force in today's society. Perhaps that is why they attack it so fiercely, fearing it as an alien way of thought. So it is all the more important to understand how it came to occupy its prominent position. Will this continue in the context of today's society, and how may the 'anti-science' movement impact science?

In my previous book, *Inside Science*, I looked at the internal workings of science and asked how the scientific process works. Now I ask how science took its current form. I want to trace the line—or more realistically, zigzag—from the ancient belief in deduction based on common sense observations (which can be deceptive) to the modern belief in data based on experiments (which should be objectively independent of time and place). Indeed, the question of objectivity is at the heart of what scientists believe science is about and the nature of critics' objections to it.

I differ in several critical aspects with the perspective from which the humanities tend to view science today. Philosophers may deny there is any demarcation between science and humanities, viewing science as an intellectual activity no different in principle from any other.[5] Historians of science seem dominated by the view that it is inadmissible to analyze the development of science from a modern perspective.[6] They take the view that the *only* legitimate perspective is that of society in the appropriate historical period.[7] (Of course, the notion that only one way of thought is permissible is completely contrary to the very basis of science itself.) Many sociologists see science as an artificial construct reflecting only the society in which it is based.[8]

These may be interesting views that broaden the overall perspective from which to view science. But taken to extremes, as they often are, they present a distorted view of science that scientists cannot recognize, and that the public cannot penetrate. I hold science to a higher standard, that of verification and reproducibility, and I believe that it is different and distinctive from other intellectual pursuits. My aim in this book is to speak for science.

Benjamin Lewin

NOTES AND REFERENCES

1. D. Wootton, *The Invention of Science: A New History of the Scientific Revolution*, Harper, New York, 2015, p. 16.

2. H. Trevor-Roper, *History and Imagination*, Clarendon Press, Oxford, 1980, p. 15.

3. For full quotation see the section on The Culture of Science.

4. I discuss Thomas Kuhn's view of how revolutions overthrow paradigms in Chapter 11.

5. Relativism is not unique to the sociology of science, of course. In 1987, Allan Bloom famously complained that, "There is one thing a professor can be absolutely sure of: almost every student entering the university believes, or says he believes, that truth is relative… It's something with which they have been indoctrinated." A. Bloom, *The Closing of the American Mind*, Simon & Schuster, New York, 1987, p. 25.

6. The perspective of looking back from the present was called the 'whig view of history' by Butterfield, *The Whig Interpretation of History*, Bell, London, 1931.

7. Well before the fashion of the modern era for dismissing thoughts with which you disagree as unacceptable "-isms," *presentism* was used to describe the interpretation of history in terms of present-day values as opposed to the values prevailing in the period. For the past century, presentism has been used exclusively in a pejorative sense; it has even been called a fallacy. See A. Walsham, Introduction: Past and … Presentism, *Past Present*, 2017, **234**, 213–217.

8. In my view, the relativistic view is a caricature of what science is about. This would be a good example. "In black-and-white versions of the world, science is set apart, as though it were a unique type of intellectual activity yielding unassailable truth. Yet what counts as a scientific fact depends not only on the natural world, but also on who is doing the research—and where and when." P. Fara, *Science. A Four Thousand Year History*, Oxford University Press, Oxford, 2009, p. xvii. My position is that this is complete nonsense.

Acknowledgements

It's a real pleasure to thank people who read some or all of the manuscript and suggested many helpful improvements, especially Philip Feil, Tony Freeth, Stuart Gould, Ron McKay, Hugo Rienhoff, Larry Stern, and Larry Walker. In a book of this scope, especially one extending beyond the author's original expertise, there will inevitably be some errors: they, of course, are mine alone.

The Frontiers of Science
By Benjamin Lewin
© Benjamin Lewin 2026
Published by the Royal Society of Chemistry, www.rsc.org

Contents

The Frontiers of Science
By Benjamin Lewin
© Benjamin Lewin 2026
Published by the Royal Society of Chemistry, www.rsc.org

Becoming Science

The Nature of Science

Physical Sciences

Prologue: Believing in Science

"BLACK HOLES ARE STRANGER THAN ANYTHING DREAMT UP by science fiction writers," said Stephen Hawking.[1] Photographing a black hole may seem like a contradiction in terms. Produced by the collapse of a star to form a body so dense and concentrated that the force of gravity prevents any light from escaping, it is exactly what its name suggests: totally black. But its effects on surrounding space produce a ring of light around the horizon of the black body. In fact, the dimensions of the ring are an important factor in analyzing a black hole. Its details are obscured by the limitations of technology, such as the resolution of the telescope. This means that the procedures used to analyze the picture of the black body have a crucial effect on our conclusions. The extension of those procedures by using AI (artificial intelligence) may make a great change in the nature of science. Because the way AI works is not entirely understood, its use sharpens the dilemma of modern science: what confidence can we place in our conclusions given the intricacy of the analysis?

The Frontiers of Science
By Benjamin Lewin
© Benjamin Lewin 2026
Published by the Royal Society of Chemistry, www.rsc.org

Timeline for Views on Changing Beliefs and Knowledge

BCE	Thales of Miletus questions nature of reality
	Buddha emphasizes experience over intellectual belief
	Socrates says "I know that I know nothing"
400	Plato argues reality is "Forms" not experience
	Aristotle argues reality depends on perception
200	
CE	
200	Ptolemy considers mathematical models are not reality
400	
	Saint Augustine discusses faith versus reason
600	
800	
1000	Avicenna explores nature of knowledge and belief in Islam
1200	
	Thomas Aquinas says "philosophy is the handmaiden of theology"
1400	
	Copernicus says heliocentricity is a calculation
1600	Galileo insists astronomy reflects reality
	Descartes subjects all beliefs to skepticism: "Cogito ergo sum"
	Locke claims knowledge comes from experience
1800	
	Maxwell & Helmholtz view thermodynamics as analogy for real world
	Mach asks whether atoms are reality or useful fiction
2000	Planck & Einstein view quanta as mathematical solutions
	Bohr views quanta as physical reality
	Sociologists question whether science reflects reality

A Dialog between an Ancient Greek Philosopher and a Modern Physicist

A Greek philosopher's view of space beyond the Earth would be based on deductions from observations made with the naked eye, aided by some primitive instruments for alignment.

A modern astrophysicist would actually be more likely to use a radio telescope, or a space telescope, than optical instruments, which have diameters up to 10 m.

Reproduced from https://commons. wikimedia.org/wiki/File:100_Hours_ of_Astronomy-_Observing_the_sky_ %28iau0904a%29.jpg under the terms of the CC BY 4.0 license, https:// creativecommons.org/licenses/by/4.0/ deed.en.

Philosopher: I see a thinker from another era. Greetings, esteemed Physicist. My eyes send light to illuminate you.

Physicist: Salutations, venerable Philosopher. It's an honor to exchange ideas with someone whose contemplations have influenced humanity for millennia. Indeed, our vision depends on light, but light comes from an amalgam of colors that enter the eye.

Philosopher: In my time, we pondered the cosmos deeply. I see the spirits in the heavens, I see the constellations of stars revolving around the Earth, I see the Sun rising and falling around the Earth. I conclude the Earth must be at the center of the universe.

Physicist: We view the cosmos differently today. The movement of any two bodies around one another is relative, but we know that only the 'fixed' stars are far away. The others, which today we call planets, like the Earth, revolve around the Sun.

Philosopher: We thought deeply about the aether, a quintessence permeating the universe, serving as the medium for propagating light and celestial motion. There must be an aether to enable objects to influence one another.

Physicist: After much debate, and many attempts to detect the aether, we have concluded in the modern era that it does not exist. Space is a vacuum. Objects can influence one another at a distance through the force of gravity.

Philosopher: How can a vacuum exist? If there is no resistance to motion, an object could move at infinite speed. Clearly this is impossible. Perhaps your tools for detecting the aether are inadequate?

Physicist: It's true there are limitations on our knowledge. Matter, as we would both recognize it, accounts for only 5% of the universe. So far, we have been unable to identify the nature of the other 95%.

Philosopher: We had greater certainty in my day. Common sense observations created a rigorous and harmonious understanding that set every object in its place. We had confidence that our knowledge showed the reality of the universe. Your advanced techniques seem to have created a state of uncertainty about the reality of your knowledge.

Physicist: Your certainty in ancient times was based on the inability to test theories. As we developed a framework for theoretical analysis in physics, the question arose as to whether it's merely a useful description with valuable predictive properties, or whether it actually represents physical reality. The uncertainties of modern physics leave this as an open question. It's crucial that we define the boundaries and limitations of our knowledge. I rest my case.

BELIEVING IN SCIENCE

> *I conclude that, while it is true that science cannot decide questions of value, that is because they cannot be intellectually decided at all, and lie outside the realm of truth and falsehood. Whatever knowledge is attainable, must be attained by scientific methods; and what science cannot discover, mankind cannot know,*[2] Bertrand Russell, 1935.

Science is about knowledge, and how we know it. When Galileo looked through his telescope and saw stars, planets, and moons, the data were irrefutable. You could argue about the significance of the observations, but at the end of the day, anybody could make the same observations with the same equipment. Here is the basic principle of science: its transparency allows anyone, irrespective of time or place, to reproduce observations.[3]

This principle remained true for centuries, but became more involved in the present era as equipment became increasingly complicated and methods of analysis became more sophisticated. In the 21st century, performing science on a scale that extends beyond the analytical capacity of individuals challenges that principle by eliminating the tradition that every participant in a research project fully understands it. And the development of artificial intelligence (AI) pushes it to breaking point by obscuring the basis for interpreting observations.

The concept of distinguishing knowledge from belief was explicit in the Latin *scientia*, first used in the first century BCE. This gave rise to *science,* first in Old French, and then in English in the Middle Ages. The question for science becomes what we knew in each era, and how it was distinguished from belief. Moving from the classical period to the present, as science took over from philosophy, the question turned towards asking whether observations necessarily reflect reality.

A theme that's repeated through the ages is whether our models are calculations that make useful predictions—you might even say a fiction—or whether they represent physical reality. This was an issue in astronomy from Plato to Galileo. It has been a hot topic in particle physics since the construction of quantum theory in the 20th century.[4] The question takes an even more pointed form with the introduction of techniques of AI.

To be part of science, observations and analyses have to be reported in a way that allows them to be questioned and repeated. The challenge here is posed by the very nature of AI. Even if the source code for a program is provided, even if the methods of training the software are specified, an AI program is a black box whose inner workings are impenetrable. There's no simple explanation of how it arrives at its results. So how can we question the results, let alone reproduce them?

Every era has had limitations on knowledge. They may be intrinsic, resulting from available technology, or extrinsic, resulting from constraints imposed by society. The combination of dependence on available technology, the weight of history, and the context of society have influenced the development of science at every turn. Astronomy was restricted by the need to rely on the naked eye until the telescope was invented. It was unthinkable that the earth might revolve around the sun when religion required it to be at the center of the universe.

New observations (or experiments) may cause a break with the past and (sometimes) society. Until that happens, we are stuck in a paradigm. Great advances involve what philosopher Thomas Kuhn called a paradigm shift. This book is about the series of paradigm shifts that led us to our present state of knowledge.

But our present state of knowledge is, of course, a paradigm in itself. We need to ask: how reliable is that paradigm? What do we really know?

If there is any single assumption underlying the paradigm, it's that knowledge rests upon data that can be verified and reproduced by anyone who has access to the appropriate equipment. And we need not only to assess the data and their reliability, but also the methods used for analysis.

As science has evolved, analytical methods have become increasingly more complex. Ancient astronomers analyzed their observations with simple mathematical tools. The invention of the telescope improved the data, but analysis continued to use methods that any competent mathematician could understand. Newton's introduction of differential calculus made for more complicated analysis, but it still used pencil and paper.

Even through the first part of the 20th century, scientists in all fields made their own equipment and analyzed the results by manual methods. A major change came with the digital era, with equipment that internally corrected results, the production of images that could be manipulated, and the move to science on a much larger scale.

To be sure, science has never relied on simply accepting all the data sight unseen. There has always been selection of data to dismiss outliers, or normalization of data to compensate for interfering effects. But it has been possible in principle to maintain transparency by describing the basis for any adjustments.

It has always been a question as to how much manipulation of data is appropriate. That question moves into another dimension with the application of AI. This may be the biggest single flux in the evolution of science to date. The analysis of black holes is a perfect metaphor for the situation of science today.

Because it absorbs all light, its very nature makes a black hole impossible to see directly. But because it bends light or distorts the orbits of other stars, a black hole can be visualized indirectly by its effect on the vicinity. The first photograph of a black hole was obtained in 2017.[5] A second black hole was photographed in 2022.[6]

The original photograph of the black hole has a blurry black center surrounded by an arc of light in a fuzzy ring. A few years later, with AI programs now in vogue, it was much more sharply defined by applying software that had been trained on simulations of black holes. In fact, this image is even sharper than the potential maximum resolution of the telescope. The surrounding ring becomes somewhat fuzzier when the image is downgraded to that resolution.[7]

So our view of the black hole, and its implications for cosmology, depend very much on whether we can trust the AI program. Science has yet to come to grips with the full implications of using AI to analyze results (or to design experiments!).[8] How do we allow for possible bias in the AI software? How do we maintain the principle that results must be verifiable and reproducible when we do not fully understand the software? (And, of course, the expertise needed to approach an AI program is completely different from the expertise involved in designing experiments in any area of science.)

The big question is whether AI is compatible with the traditional view of science or whether we are forced to change our criteria for how science is done in order to accommodate it?

Is the standard for judging AI different from the historic rule? If an astrophysicist said, "my experience and intuition say that

Will the real black hole light up, please? Top image shows the original photograph of the M87 black hole taken by the Event Horizon Telescope in 2017. Center shows the image reconstructed by the PRIMO AI program. Bottom shows the PRIMO image blurred to the resolution of the EHT array.

the boundaries of the black hole should be sharper, and the ring of light should be finer," they would be ridiculed, if not accused of fraud. But if an AI program—trained on simulations!—refines the data, the conclusions are accepted.

We are in a new realm of knowledge here. Even its creators do not know exactly how an AI program works. Experience with AI programs shows that the way they are 'trained' can influence the results. Using AI raises the question of whether we can maintain the principle that science rests on transparency. So can we trust results that are analyzed using the latest technology? How does this affect our belief in science?

Paradigm shifts in science are driven by discoveries that are inconsistent with the prior view of the world and that require a new way of looking at things. Often they are driven by developments in technology. Galileo could not have destroyed the classical view of the relationship between Earth and the solar system without the invention of the telescope. Cajal could not have demonstrated the cellular nature of the neuron without the discovery of Golgi stain. But AI may be different. It's changing the paradigm of scientific discovery itself.

Our view of the black hole owes more to AI's perspective than to peering up a telescope. This is a dramatic example, but the effects of AI are not confined to physics. They will soon pervade all science. In biology, the AlphaFold AI program can analyze protein structure more effectively than physical equipment can. (I discuss AI more fully in Chapter 23.) Indeed, the transition to a new way of science was recognized by the award in 2024 of Nobel Prizes in both physics and chemistry for developments driven by AI. AI itself has become the paradigm.

NOTES AND REFERENCES

1. In a lecture at Harvard in 2008. Online at www.hawking.org.uk/in-words/lectures/into-a-black-hole.
2. B. Russell, *Religion and Science*, Thornton Butterworth, London, 1935, p. 243.
3. As Bertrand Russell pointed out, this principle is specific for science as opposed to other intellectual activities.
4. The dilemma between fiction and reality was a concern of philosophers of science in the early 20th century. Pierre Duhem related it to the concept of 'save the phenomena,' meaning that observations were adjusted to fit the model, and he pointed out that the practice of science was constricted by taking the view that models must represent reality. P. Duhem, *To Save the Phenomena. An Essay on the Idea of Physical Theory From Plato to Galileo*,

1908, translated by E. Dolan and C. Maschler, Chicago University Press, Chicago, 2015.

5. Visualizing the bending of light requires a large-scale effort, using satellite telescopes or apparatus all over Earth. The first black hole was detected by the Event Horizon Telescope, a 'virtual telescope' consisting of a global network of synchronized radio observatories.

6. The first black hole photographed is at the core of the M87 galaxy, 53 million light years away. The second, Sagittarius A, is 27 000 light years away at the heart of the Milky Way. More than 50 black holes have now been identified.

7. L. Medeiros, *et al.*, The Image of the M87 Black Hole Reconstructed with PRIMO, *Astrophys. J. Lett.*, 2023, **947**, L7.

8. The Royal Society issued a report expressing concern that AI may produce 'transformational changes' in science. It included recommendations that AI should be used in the context of 'open science,' but could not come to grips with the central issue that the intrinsic nature of AI makes it impossible to completely describe the basis for analysis. See *Science in the Age of AI*, Royal Society, London, 2024.

Proto-Science

The lover of mind will not allow that there are any prime causes other than the rational and invisible ones. . . As I said at first, all things were originally a chaos in which there was no order or proportion. The elements of this chaos were arranged by the Creator, and out of them he made the world, Plato, c. 360 BCE

CHAPTER 1

Prequel:
Aristotle's Doctrines: 600–300 BC

"SCIENCE LEADS TO KNOWLEDGE, BUT OPINION BREEDS ignorance," said Hippocrates. It may seem irrelevant to start an account of modern science with the philosophy of the ancient Greeks, but so many of their ideas were the driving force as science began to develop in Western Europe. The Sun revolves around the Earth. Everything in the world is made of air, water, earth and fire. Living organisms can arise by spontaneous generation when they obtain a "vital spark" from the air. These ideas came to us from Aristotle. Preceded by Socrates and Plato, Aristotle was the last of the great trio of ancient Greek philosophers. With interests ranging from astronomy to biology, and from linguistics to politics, Aristotle's doctrines dominated Western thinking for the next thousand years, to the point at which he was simply called "The Philosopher." After studying at Plato's academy, he left Athens, and later became tutor to the future Alexander the Great. He returned to Athens for the last decade of his life, and started the Lyceum school. Most of his writings date from this period, but only a small part has survived. His legacy is immense.

The Frontiers of Science
By Benjamin Lewin
© Benjamin Lewin 2026
Published by the Royal Society of Chemistry, www.rsc.org

Timeline for Key Events in Astronomy and Mathematics in Ancient Greece

600 BCE

Thales predicts solar eclipse (perhaps)

Anaximander proposes Earth floats in space

550

Xenophanes questions form of the Gods
School of Pythagoras develops mathematics

500

Anaxagoras explains eclipses and describes Sun as fiery mass

450

Democritus proposes atomic theory

400

Plato asks why Earth does not fall

Eudoxus proposes circular orbits

350 Aristotle places spherical Earth at center of universe

300 Euclid develops geometry and publishes *Elements*

Aristarchus of Samos proposes heliocentric system

250 Eratosthenes calculates circumference of Earth

Archimedes develops theory of buoyancy

200

Hipparchus discovers precession of the equinoxes and maps 1000 stars

150

A Dialog between Anaximander and Aristotle

Anaximander of Miletus (610–546 BCE) was the first philosopher to replace the supernatural with explanations based on natural forces. As an astronomer, he developed models for the Earth and the Sun.

Aristotle (384–322 BCE) was the supreme philosopher of ancient Greece, developing a worldview that encompassed everything from the universe to the origins of animal life and the nature of humanity.

Anaximander: Greetings, Aristotle. I believe you have become *"The Philosopher,"* but I too have explored the origins and underlying elements of the cosmos. I believe there is a fundamental substance, the "apeiron," from which all things arise. It's neither water nor any of the so-called elements, but has a different nature from them. It is infinite.

Aristotle: Greetings, Anaximander. You are delving into the fundamental principles of existence. Your concept challenges the composition of elements. I assert that everything in existence is comprised of earth, water, air, and fire. We can only understand the world in terms of these entities. But let us consider matters closer to home, the Earth and the Sun.

Anaximander: I have shown that the Earth is cylindrical, and only the upper part is inhabited. It floats unsupported in the infinite.

Aristotle: I beg to differ. In eclipses, the outline is always curved. Since it is the interposition of the Earth that makes the eclipse, the form of this line will be caused by the form of the Earth's surface, which must therefore be spherical.

Anaximander: The Earth floats in the center of infinite space and stays in the same place because of its indifference. The Sun is equal in size to the Earth, but the circle by which it is borne through the heavens is 27 times as large as the Earth.

Aristotle: This is ingenious, Anaximander, but the infinite cannot have a center. Eudoxus worked out the patterns of movement of celestial bodies. I deduce that the Moon and Sun move on crystalline spheres around the Earth. It's clear that the Earth does not move. It can only lie at the center of the finite universe. I rest my case.

ARISTOTLE'S DOCTRINES

There was a time, called the Big Bang, when the universe was infinitesimally small and infinitely dense... Time had a beginning at the Big Bang, in the sense that earlier times simply would not be defined... An expanding universe does not preclude a creator, but it does place limits on when he might have carried out his job![1] Stephen Hawking, 1990.

The relationship between the Sun and the Earth is at the heart of what science is about. When you look at the sky, it seems that the Sun revolves around the Earth. Theories of how this might be drove religion and philosophy from ancient times until the late Middle Ages. Placing the Sun at the center, against resistance (especially on religious grounds) marked the beginning of science: accepting theories based on data extending beyond personal evidence. Seeing is not believing.

The ancient world viewed natural phenomena as expressions of the thoughts or moods of the Gods. The first glimmerings of a rational, you might almost say a scientific, approach, came in ancient Greece. Philosophers began to interpret natural phenomena in terms of observations of the real world. Lasting from 600–400 BCE, this is sometimes called the Intellectual Revolution. The philosophers are called the pre-Socratic philosophers because the period ended when Socrates (*c.* 470–399 BCE) moved philosophy from cosmology towards politics and ethics.[2]

The philosophers addressed a series of issues anticipating contemporary astrophysics. Democritus (*c.* 460–370 BCE) considered the nature of matter. "Nothing exists except atoms and empty space," he said. Leucippus (*c.* 480–420 BCE) considered the origin of the universe. "Worlds are formed when atoms fall into the void and are entangled with one another. The substance of the stars arises from their motion as they increase in bulk." Anaximander (*c.* 610–547 BCE) even considered the possibility of multiple universes. "There are many worlds and many systems of Universes existing all at the same time." Even earlier, a common theme in ancient mythology was that everything started with chaos. Aristotle (384–322 BCE) summarized this as, "In the infinite chaos there can have been neither above nor below."

The starting point for the Intellectual Revolution has traditionally been equated with the story that Thales of Miletus (*c.* 622–546 BCE) predicted the solar eclipse of 585 BCE. It's not important whether Thales really predicted the eclipse. It's doubtful whether the means to do so existed at the time. But the very concept that it was possible to predict an eclipse undercut the idea that astronomical phenomena were due to whims of the Gods.[3] This dates the beginning of rational analysis as opposed to superstition.[4]

The Intellectual Revolution is sometimes called the Ionian Enlightenment because it started in Ionia, a small region in the Mediterranean on the west coast of what is now Turkey. Miletus was one of the most prosperous cities. We don't know why it became a major intellectual center, but Thales is thought to have been one of the founding intellects. He was followed by Anaximander, who attempted to show how meteorological phenomena could be explained by natural forces. Anaximander has been called the 'first scientist.'[5]

Miletus was a major focus of intellectual activity from 625–425 BCE. In astronomy, Thales may have predicted a solar eclipse in 585 BCE, and Anaximander conceived the idea that the universe could be infinite. Located at the coast of Western Turkey, there are ruins of the ancient Greek theater, dating from the middle of the 5th century BCE.

Anaximander's ideas were breakthroughs in their time. They included the concepts that the Earth floats in space (circled by the Moon, Sun, and stars), and that it was originally covered by water, where life originated. Anaximander's importance does not lie in how closely he approached reality, but in the fact that he introduced analysis of data as a means of understanding the world.

The philosophers of Miletus were much concerned with identifying the 'first principle' as the basis of all matter. They conceived this as a single unit that could be transformed in multiple ways. Thales proposed water, Anaximander proposed 'apeiron,' which has been interpreted as the infinite, and Anaximenes proposed air. It doesn't matter that in modern terms these ideas are incredible. Their importance was the concept that there could be a basic building block for matter.

A century later in Athens, Leucippus, followed by Democritus, developed the theory of atomism. 'Atomos' means indivisible. Atoms were seen as the basic building blocks of matter, uniform and infinite in number. They were impossible to divide into smaller components. The argument was basically that if you keep dividing an object in half, eventually you come to a point when it cannot be divided any further: the atom. Atomism shows the development of the view that deductions could be based on observations devoid of divine influence.[6]

Aristotle was responsible for rejecting atomism. He called the Miletan philosophers *physikoi*, from *physis,* meaning nature, to distinguish them from *theologoi*, who believed natural phenomena came from the Gods. But he did not believe that the first principle could exist in a void. He proposed that matter is a continuum consisting of four elements: air, fire, water, and earth (see Chapter 7). Atomism faded from view (partly because it was violently opposed by the medieval Church) and remained obscure until it became a forerunner of the mechanistic theory of the 17th century (see Chapter 5).

One reason why intellectual activity flourished in this period may have been the absence of any dominant religious or ruling caste that would have felt its privileges challenged by new thinking. Indeed, history since medieval times suggests that science flourishes only in an open society where it's possible to question authority. A clash between religion (or other authoritarian systems) and science is a running theme in the history of science.

Another philosopher from Miletus, Xenophanes (570–480 BCE), drew a distinction between knowledge and faith. The question as to what is *knowable* resonated through the Intellectual Revolution.[7] This was formalized in the distinction between mythos (stories of the gods) and logos (reasoned thought). That distinction was the beginning of the scientific attitude.

The most direct skepticism about the form of the Gods came from Xenophanes. "Mortals consider Gods to be born, and to have their clothing, voice and form... The Ethiopians say that their Gods are snout-nosed and black, the Thracians that they are blue-eyed and red-haired... If cows and horses, or lions, had hands, or were able to draw with their feet and produce the

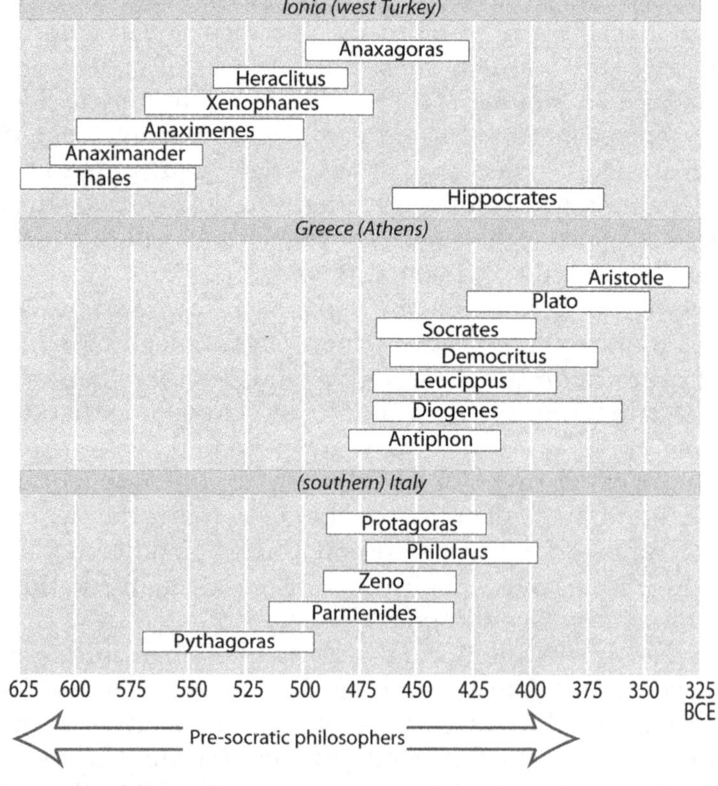

Pre-Socratic philosophers were centered in three areas. There were probably ~100 altogether. The most important are shown here.

works which men do, horses would depict the forms of Gods like horses, and cows like cows, and they would make the Gods' bodies the same shape as their own."[8]

That clear-eyed view was lost for more than two millennia. It was not until Europe emerged from the medieval era after the Scientific Revolution, and entered the age of the Enlightenment in the 18th century, that theism could be questioned. The Enlightenment philosopher, Montesquieu (1689–1755), famously expressed much the same sentiment in 1721. "If triangles were to make a God they would give him three sides."[9]

The basic impetus for the Intellectual Revolution was the idea that the world could be understood if it was properly observed, but the emphasis was on observation rather than experiments.[10] There were exceptions, most notably in biology, where Aristotle undertook a systematic approach to identifying a range of animals, often describing their anatomy on the basis of dissection. He believed in close observation, but he did not often resort to what we would today call experimentation. Yet he expressed the concept, most famously in his work on salt water. "Salt water when it turns into vapor becomes sweet, and the vapor does not form salt water when it condenses again. This I know by experiment."

One feature resembling modern science is that philosophers would challenge one another's assumptions and conclusions. Socrates was even known to argue with himself. The nearest modern parallel might be the 'thought experiments' that were popular, for example, when Einstein and his circle imagined various scenarios, often involving paradoxes, to try to resolve the best way forward for future experiments.[11] It would be a mistake, however, to draw too close a parallel with the methodology of modern science.

Natural philosophy was not science, in the sense of performing experiments to test and develop theories, but it was perhaps proto-science in using observations to formulate models. However, it could be limited by the constraints of society, as witness the trial and subsequent death sentence on Socrates for impiety. Socrates was prosecuted essentially on the grounds that he was an atheist, because he rejected the Gods of Athens. This is a theme that will resonate for centuries.[12]

Socrates is famous for introducing the technique now called Socratic dialog, in which pretended ignorance forces a series of questions by following each thought to its logical conclusion.

This is not unlike the way science is supposed to proceed: testing to see whether the prediction of a hypothesis is correct, and then testing the logical conclusion of the result, ad infinitum. It follows from Socratic method that nothing is sacred.

The importance of the Intellectual Revolution is now sometimes questioned because it was supplanted, first by Socrates, and then by Plato (428–348 BCE) and Aristotle (384–322 BCE). It petered out under its own steam, as it were. But it marks the introduction of the idea that natural phenomena have rational causes. In the following period, Plato and Aristotle turned the argument more towards the nature of reality.

A significant difference in their positions was that Plato aspired to understand an ideal realm that was represented imperfectly in our reality. A 'Form' was the ideal abstract case of an object or property. It was, in a sense, more real than any individual example, although it could be visualized only by the mind, not experienced in any tangible way. Actual objects were only imperfect realizations of the Form.[13]

By contrast, Aristotle believed that reality depends on experience of concrete objects or observations. It was the physical objects that represent reality. At the risk of oversimplifying, the difference foreshadows the issue that runs through natural philosophy into science. Do theories of the construction of the world reflect reality or are they useful fictions that produce predictable results? (I consider the issue of reality in physics in Chapter 16.)

Aristotle moved towards a more teleological approach to philosophy, meaning that things were analyzed in terms of the ends or purposes that they serve. ('Telos' means 'end.')[14] The general emphasis on identifying *causes* of things meant that experiments would be regarded as an artifice, not necessarily representing the situation in Nature.[15] The distinction between 'natural' and 'artificial' situations encouraged observation rather than experimentation. This led to a question that persisted through the Middle Ages. Can a compound (or situation) created by human efforts be the same as one found naturally, and therefore reveal the workings of Nature? (see also Chapter 6).[16]

Mathematics was viewed as a link to reality, and was central to the philosophical approach in this period. It may be apocryphal that "Let no one ignorant of geometry enter here," was engraved over the entrance to Plato's Academy, but it summarizes the belief that the universe was constructed in terms of mathematical principles.[17] Plato's concept of the universe as a rational construction, working on predictable principles, might be viewed as the start of an attitude that makes science possible.

Pythagoras (*c.* 570–495 BCE) may have been the first to express the philosophical importance of numbers when he said, "all things are numbers."[18] The full glory of Greek mathematics emerged later with Euclid (325–265 BCE). As Max Born, awarded a Nobel Prize in 1954 for his work on quantum physics said: "If you are a modern mathematician you can of course look at geometry as a product of pure thinking... But the Greek philosophers... believed [their geometry] was dealing with properties of real things. The fact that the predictions of their theories were confirmed by experience in all cases led to the conviction that the axioms of Euclidean geometry contain final truth."[19] The view that everything could be learned by deduction bypassed any need for experiments.

The focus on theory carried over to the view of paradoxes. The Greek philosophers were good at finding paradoxes, for example, Zeno's famous paradox to show that motion was impossible, or the fable of Achilles and the hare, but they were not very interested in investigating why the paradox did not accord with reality.[20] In modern science, by contrast, such a paradox, for example, Maxwell's demon, which seems to contradict the second law of thermodynamics, provokes intense discussion (as discussed in Chapter 9).

Aristotle believed that natural philosophy should draw conclusions from undisputed premises. The system failed as it became clear that its premises were not, in fact, indisputable. Aristotle's dogma dominated Western thought for more than the next thousand years. It was only after scientific enquiry diverged from philosophy that it became more directly connected with data based on experiments.

The pre-Socratic philosophers certainly showed a spirit of enquiry with scientific elements, but they achieved an intellectual revolution, not a scientific revolution. There was a major difference between natural philosophy, as practiced in ancient Greece, and science, as it developed later in Europe. Philosophy confined itself to using logic to draw deductions from observations, whereas science subordinated that logic to empirical observations or experiments.

When did science begin, in the sense of testing theories based on observations? Perhaps during the first part of the Hellenistic Age (323–30 BCE), which succeeded the classical period. The starting point was the death of Alexander the Great. Greek influence extended all around the Mediterranean from then until Rome conquered Egypt in 30 BCE.

Alexander the Great had founded Alexandria (in Egypt) in 332 BCE. Largely due to its library, it was a focus for intellectual activity. Constructed as the new capital of the Empire, Alexandria became the center of civilization. Its lighthouse, an impressive building about 90 m tall (equivalent roughly to a 20-storey building), constructed around 280 BCE, dominated the harbor and was considered one of the wonders of the world. Founded around 300 BCE, the library was not far from the harbor.

Under Pharaohs Ptolemy I, II, and III, the library had an active policy of acquiring all knowledge, sending for books from Athens and elsewhere. Legend has it that ships docking at Alexandria were required to deposit any books they carried at the library so they could be copied. The collection grew so large that an extension, the Serapeum, was built. The librarians were major figures in ancient scholarship. The library appears to have had tens of thousands of papyrus rolls.[21]

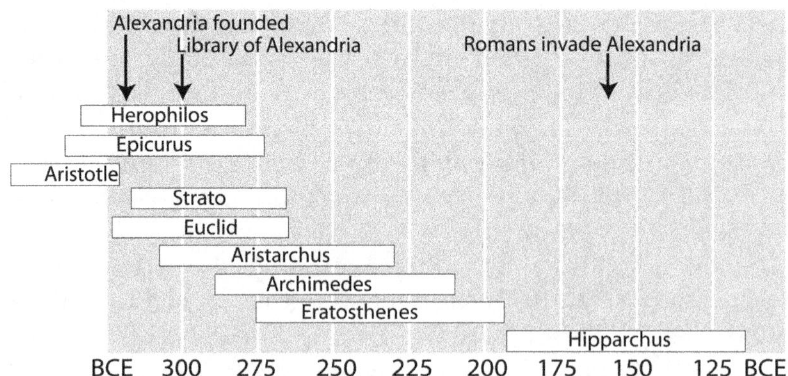

Major philosophers and political events at the start of the Hellenistic Age. The philosophers could claim to include the first anatomist (Herophilos), mathematician (Euclid), physicist (Archimedes) and astronomer (Hipparchus).

Alexandria became a center for science and mathematics. Two millennia before the Scientific Revolution of the 17th century, there was a similar period of ferment, with significant activity in mathematics, physics, astronomy, and even anatomy. It lasted from around 300–200 BCE, and continued later at a slower pace.

The role call of philosophers and mathematicians who worked in Alexandria is as impressive as the list of philosophers from Miletus or Athens in the preceding centuries:

- Herophilos (335–280 BCE) developed anatomy based on dissecting human cadavers.
- Euclid (325–265 BCE) was one of the most important mathematicians of the ancient world.
- Archimedes (287–212 BCE) may have spent time in Alexandria around 269 BCE. Later, he corresponded with mathematicians in the city.
- Eratosthenes (*c.* 276–194 BCE), the third Head Librarian, was an astronomer who famously made one of the first calculations of the size of the Earth.
- Ptolemy (*c.* 100–170 CE) (no relation with the Pharaohs) was the most important astronomer between the ancient Greeks and Copernicus.
- Hypatia (355–415 CE) was notable for being the first woman in mathematics, and the last great mathematician of Alexandria.

When the torch passed from Athens to Alexandria, the nature of enquiry changed. There was a quasi-scientific approach with data used to construct theories that were subjected to observations or experiments (albeit not in the full sense of those terms as used today). Scientific activity in Alexandria ended *c.* 145 BCE when the Greek population of the city was persecuted. Then scientific research resumed through the first and second centuries CE, for example with Ptolemy in astronomy.

A vast amount of learning from the ancient world was lost when the library at Alexandria was finally destroyed, probably during an invasion around 275 CE, although it was already in decline.[22] The murder of Hypatia in 415 CE is usually taken as defining the end of ancient science.[23]

Knowledge did not reach the level of Alexandria again for several hundred years, until it was discovered or rediscovered in Arabia and Europe. The loss of the knowledge gained in Alexandria over the following centuries was so complete that it has even been called an "erasure."[24] The decline started with the Romans' disdain for theory in the first centuries CE, and then accelerated. There was virtually no scientific activity in Europe for the next thousand years.

NOTES AND REFERENCES

1. S. Hawking, *A Brief History of Time*, Bantam, New York, 1998, p. 9.
2. Pre-Socratic is a bit of a misnomer because some of the philosophers included in the school were in fact contemporaries of Socrates; it is sometimes instead referred to as Early Greek Philosophy. The terminology (pre-Socratic) dates from the nineteenth century but even though it's inaccurate, the term has stuck.
3. The claim that Thales predicted the eclipse is attributed to Herodotus around 430 BCE.
4. G. E. R. Lloyd, *Early Greek Science – Thales to Aristotle*, Chatto & Windus, London, 1970, pp. 16–23.
5. C. Rovelli, *Anaximander: And the Birth of Science*, Riverhead Books, New York, 2023.
6. The works of Leucippus have been lost, and only fragments remain of Democritus. Our main source of information is Aristotle's description of atomism, when he rejected the theory. He said that, "[Democritus argued that] problems result from supposing that any body whatever of any size is everywhere divisible... Since magnitudes cannot be composed of contacts or points, it is necessary for there to be indivisible bodies and magnitudes."
7. E. R. Dodds, *The Greeks and the Irrational*, University of California Press, Berkeley, 1951, p. 181.
8. J. H. Leser, *Xenophanes of Colophon*, University of Toronto Press, Toronto, 1992.
9. "Si les triangles faisaient un Dieu, ils lui donneraient trois côtés." Charles de Secondat Baron de Montesquieu (1721) Lettres Persanes (1721) no. 59 (trans, J. Ozell, 1722).
10. Greek philosophy was not devoid of experimentation, although it was rarely thought out well enough to be useful in the physical sciences. See G. E. R. Lloyd, Experiment in Early Greek Philosophy and Medicine, *Proc. Cambridge Philological Soc.*, 1964, **190**, 50–72.
11. The types and legitimacy of thought experiments are controversial. See R. A. Sorensen, *Thought Experiments*, Oxford University Press, Oxford, 1992.
12. Of course, while Socrates was actually condemned for blasphemy, there was a back-story in that he had taken an unsuccessful (anti-democratic)

political position, and his brusque manner towards discussion had not rendered him popular.

13. It has sometimes been argued that this philosophy is antagonistic to science, and that Plato therefore held back the development of science. On the other hand, Plato's argument for the importance of mathematics, and advocacy for analyzing the abstract laws behind empirical phenomena, are quasi-scientific. See G. E. R. Lloyd, *Early Greek Science – Thales to Aristotle*, Chatto & Windus, London, 1970, pp. 66–79, 84–85.

14. A phrase that Aristotle often used was '*coming to be for the sake of*,' applied as he moved into biology (for example, to the purpose of an organ of the body).

15. T. S. Kuhn, Mathematical *vs.* Experimental Traditions in the Development of Physical Science, *J Interdisc Hist*, 1976, 7, 1–31. Reprinted in T. S. Kuhn, *The Essential Tension. Selected Studies in Scientific Tradition and Change*, University of Chicago Press, Chicago, 1977.

16. W. R. Newman, *Promethean Ambitions: Alchemy and the Quest to Perfect Nature*, University of Chicago Press, Chicago, 2004, pp. 238–289.

17. Plato represents Socrates in the Republic (Book 7, section 530) as arguing that astronomy should follow the same principles as geometry.

18. This is quoted in various related forms. I am following the usage of Bertrand Russell, *The History of Western Philosophy*, Simon & Schuster, New York, 1945, p. 35. The concept may have originated in the discovery of harmonic scales in music, where the notes are related by a geometric progression, leading to the concept of harmony.

19. M. Born, *Experiment and Theory in Physics*, Cambridge University Press, Cambridge, 1943, p. 3.

20. The arrow paradox says that at any instant (duration-less) of time, the arrow is at one position: it cannot move because no time has passed. The paradox of Achilles and the hare says that each time Achilles runs some distance, the hare will have moved on, so Achilles can never catch him up. Basically the arrow paradox divides time into points, and the Achilles paradox divides distance into multiple parts. Calculus is needed to derive a proper solution, but the short answer is that an infinite series can have a finite total.

21. R. MacLeod, *The Library of Alexandria: Centre of Learning in the Ancient World*, I. B. Tauris, London, 2004.

22. The library had been declining for at least two centuries. It was damaged by Julius Caesar in 48 BCE and subsequently became less important, but the exact details of its destruction are unclear.

23. L. Russo, *The Forgotten Revolution: How Science Was Born in 300 BC and Why It Had to Be Reborn*, Springer, Berlin, 2003, pp. 11–15, 194–196.

24. L. Russo, *The Forgotten Revolution: How Science Was Born in 300 BC and Why It Had to Be Reborn*, Springer, Berlin, 2003, pp. 1–9.

CHAPTER 2

Saint Augustine's Legacy: 500–1300

"THE AGE OF IGNORANCE COMMENCED WITH THE CHRISTIAN system," said Thomas Paine in *The Age of Reason* in 1807.[1] Christianity became the religion of the Roman Empire in 380 CE. For the following millennium, intellectual activity was dominated by the theological philosophy of Saint Augustine (354–430 CE). The Carolingian Renaissance under Charlemagne was a short-lived (secular) beacon of light in the Medieval Era from 789–814 CE. It's hard to find much original intellectual activity in Western Europe outside of theology until the dark began to lift at the end of the 11th century with the foundation of the first universities. Glimmerings of interest in science came from religious figures. The most prominent were Robert Grosseteste, Bishop of Lincoln, and Roger Bacon, a Franciscan friar, in the 13th century. Augustine's legacy was created by the extent of his output, the breadth of his arguments, and the elegance with which he wrote (in Latin). It's not really an exaggeration to say that he believed all intellectual activity should be judged in the light of theology. He remains a significant figure even today.

The Frontiers of Science
By Benjamin Lewin
© Benjamin Lewin 2026
Published by the Royal Society of Chemistry, www.rsc.org

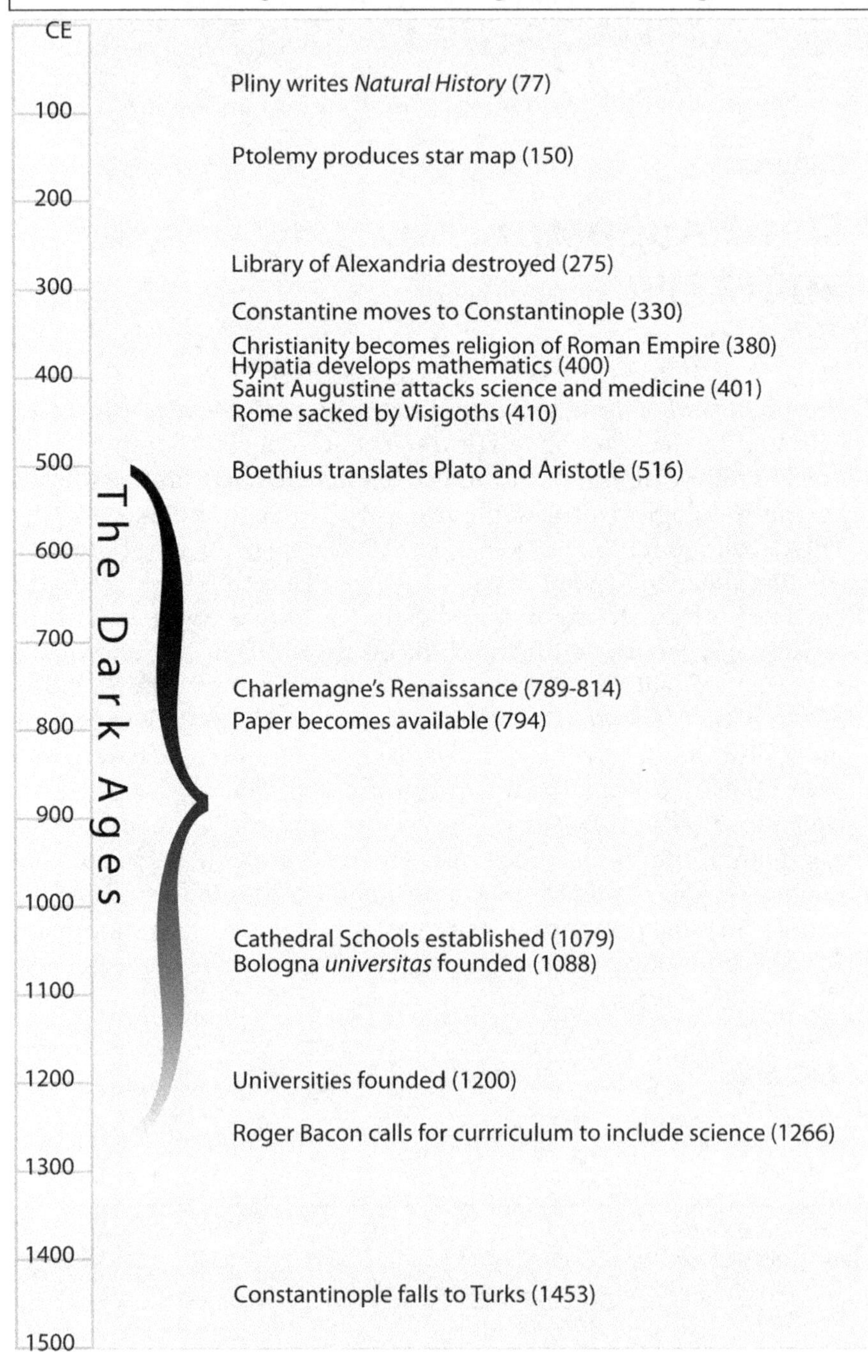

Timeline for Events Leading to the Dark Ages
CE
100
200
300
400
500
600
700
800
900
1000
1100
1200
1300
1400
1500
The Dark Ages
Pliny writes Natural History (77)
Ptolemy produces star map (150)
Library of Alexandria destroyed (275)
Constantine moves to Constantinople (330)
Christianity becomes religion of Roman Empire (380)
Hypatia develops mathematics (400)
Saint Augustine attacks science and medicine (401)
Rome sacked by Visigoths (410)
Boethius translates Plato and Aristotle (516)
Charlemagne's Renaissance (789-814)
Paper becomes available (794)
Cathedral Schools established (1079)
Bologna universitas founded (1088)
Universities founded (1200)
Roger Bacon calls for currriculum to include science (1266)
Constantinople falls to Turks (1453)

A Dialog between Saint Augustine and Roger Bacon

Saint Augustine (354–430) was the most important theologian of the millennium. As a bishop in North Africa, he set the tone for the Dark Ages by arguing that all knowledge should be subordinated to its role in supporting theology.

© Ann Ronan Pictures/Print Collector/Getty Images.

Roger Bacon (1214–1292) became a Franciscan friar whose enquiries placing knowledge on a scientific basis set the tone for the transition out of the Middle Ages, although his work was suppressed by the Franciscan order.

© Oxford Science Archive/Print Collector/Getty Images.

Saint Augustine: Ah, Brother Bacon, welcome. I believe I have established the precepts for basing reason on faith. What brings you to this discourse on matters of philosophy?

Roger Bacon: Greetings, Saint Augustine. It's a pleasure to meet such a distinguished theologian and philosopher. I believe that the pursuit of knowledge, grounded in empirical observation, can enhance our understanding of the divine.

Saint Augustine: An interesting proposition, Brother Bacon. However, is not faith the foundation of true knowledge? The scriptures reveal the divine truth. Through faith, we gain insight into the mysteries of God.

Roger Bacon: I do not dispute the importance of faith, Augustine, but I argue that reason can complement and enrich our understanding. The study of nature and the natural world can be a path to a deeper appreciation of the Creator's wisdom.

Saint Augustine: But, my dear brother, is there not a danger in relying too heavily on human reason? Curiosity is a disease. Trying to discover secrets of nature that we should not know can only divert us from the path of true faith.

Roger Bacon: I acknowledge the limitations of human reason, Augustine. But Boethius says, "I warn the man who spurns the paths of knowledge that he cannot philosophize correctly." The strongest arguments prove nothing so long as the conclusions are not verified by experience.

Saint Augustine: Experience can be misleading. False analyses may lead down a path away from the divine. Man should follow the path of faith.

Roger Bacon: Faith should be complemented by employing mathematics. The knowledge of mathematical things is almost innate in us. Proper understanding of philosophy and theology cannot be achieved without mathematics.

Saint Augustine: (*frowning*) The good Christian should beware the mathematician and all those who make empty prophecies. The danger already exists that the mathematicians have made a covenant with the devil to darken the spirit and to confine man in the bonds of hell.

Roger Bacon: The truth is that there has never been so great ignorance and such deep error. It's easier for a man to burn down his own house than to get rid of his prejudices. Experience is to experiments as faith is to reason. I rest my case.

SAINT AUGUSTINE'S LEGACY

> *In the revolution of ten centuries, not a single discovery was made to exalt the dignity or promote the happiness of mankind. Not a single idea had been added to the speculative systems of antiquity, and a succession of patient disciples became in their turn the dogmatic teachers of the next servile generation,*[2] Edward Gibbon, 1777.

Imagine a world without science. Well, to be more exact, imagine a world in which all intellectual activity excluded science or never even considered its existence. What sort of questions would be asked to explain the natural world?

Europe following the fall of Rome offers insight into such a state. In *The Decline and Fall of The Roman Empire* (published 1776–1789),[2] Edward Gibbon propounded the concept of the Dark Ages, a thousand years of barbarity and depravity resulting from the fall of Rome, enhanced by the intolerance of the Church.[3] This view held sway for at least a century, but a revisionist view has now taken over that the Dark Ages were not so dark. It holds that the Church was not necessarily responsible for the collapse of intellectual activity.

"No serious historian believes any more in the concept of the Dark Ages," is a constant refrain in books on the history of the medieval period. Be that as it may, we should ask what happened to science during this period—and what was responsible for advancing or suppressing science?

In the first half of the 2nd century BCE, roughly from 200–150 BCE, the Romans conquered all of Greece. As Horace (65–8 BCE), the leading poet in Rome, famously pointed out, the Romans won the military war, but lost the intellectual peace.[4] While the Romans absorbed much of Greek knowledge, they did not extend it.

There was a stark contrast between attitudes in Greece and Rome. The Greeks believed in theory and placed less emphasis on technology. The Romans had little interest in abstractions and were concerned with practical results.[5] Archimedes (287–212 BCE), who at the end of the day accomplished some of the greatest engineering feats in Greece, supposedly repudiated "as sordid and ignoble the whole trade of engineering."[6] Cicero (106–43 BCE) commented that "Geometry was in high esteem [with the Greeks], therefore none were more honorable than mathematicians. But we have confined this art to bare measuring and calculating."

Roman engineering was formidable, but required little more than rudimentary knowledge of mathematics. 'Roman science' is all but a contradiction in terms. Our major source for early Roman science is the series of books, *Natural History,* written by Pliny (24–79 CE). They are encyclopedic, but derivative. Essentially they copy and index the works of the Greek philosophers.[7] This was the trend of the era: cataloguing rather than analyzing.

The success of the Roman Empire is only one example of the ability to form an advanced society without relying on science. China developed some technologies centuries ahead of the West, including gunpowder, the magnetic compass, using cast iron, and the printing press. Yet with (perhaps) the exception of astronomy there was little basic understanding of science. Centuries earlier, Egyptian society was extremely sophisticated, but never developed a scientific outlook.

The Roman Empire crumbled over a protracted period. In reality, there was no single point marking a change of era from the Roman to the Medieval. The sack of Rome by the Visigoths in 410 CE is a symbolic point for the collapse. With the center destroyed, there was no hope that the benefits of Roman civilization could survive. But just what were those benefits in the scientific arena?

Following the fall of Rome, Greek cosmology, particularly the works of Plato and Aristotle, survived in the West in the form of bastardized translations.[8] Pythagorean arithmetic survived, but Euclidian geometry was rather mangled. The Latin-speaking West was cut off from the Greek-speaking East in the 4th–5th centuries CE.[9] This led to the loss of the knowledge of the ancient Greeks.

Scientific knowledge survived better in the East than the West, although it was derivative rather than original. It inherited, or at least perpetuated, learning from Athens and Alexandria.[10] The Eastern Empire became known as the Byzantine Empire after 330 CE, when Emperor Constantine moved the capital to Byzantium, which he renamed Constantinople (Istanbul today).

Even after the library was destroyed around 275 CE, Alexandria remained a major intellectual center. Constantinople fell into

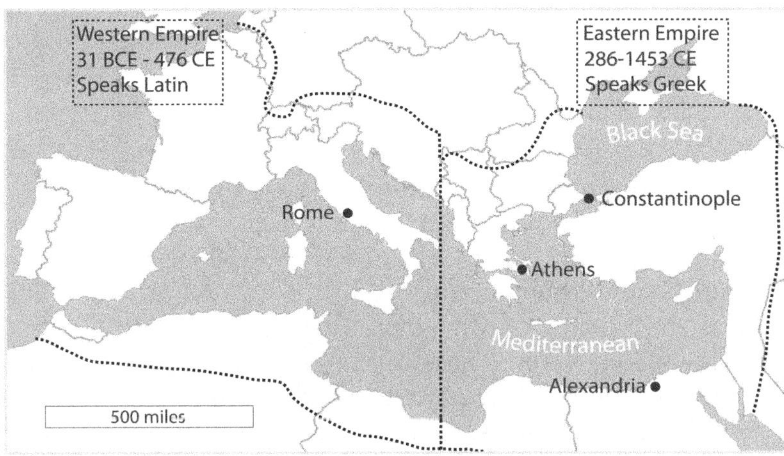

The Imperial period in Rome dates from the first Emperor in 31 BCE. The empire reached its maximum size in 117 CE. It split in 286 CE into the Western Empire (ruled from Rome) and the Eastern (or Byzantine) Empire, ruled from Constantinople. The Western Empire suffered successive waves of attacks by the Visigoths, from 410 CE until the Empire ended in 476 CE. (The Holy Roman Empire formed in a smaller area across Italy-France-Germany in the Middle Ages.) The Eastern Empire fell when the Turks took Constantinople in 1453.

disarray after 1204 through the ravages of the Crusades, but became an intellectual center again after it was restored to the Byzantine Empire in 1261. Its importance persisted until it fell to the Turks in 1453. The fall led scholars to migrate to the West, taking Greek manuscripts with them.[11]

It says a great deal about the nature of the medieval era that for several hundred years the dominant intellectual influences were the great figures of antiquity: Aristotle, Ptolemy, Euclid, and Galen. Almost all analytical activity was devoted to explaining or commenting on their works. The predominant belief was that all knowledge was contained in the deductions of Aristotle and other Greek philosophers.

The state of astronomy refutes any belief that the medieval period was not really dark for science. Catalogs of star positions had been produced in ancient Greece. Ptolemy used them in Alexandria in 150 CE to produce a map that remained the basis for astronomy for the next thousand years. (I discuss the underlying astronomy in Chapter 4.)

∞ ∞ ∞

500 (CE) is a nice round number that is sometimes taken to mark the transition from Antiquity into the Medieval era. By then, Christian theology had become the dominant religious influence in Europe. Saint Augustine (354–430 CE) was the most influential of the theologians.

"Make me good, O Lord—but not just yet" is no doubt the saying for which Saint Augustine is best known. Coming from his autobiographical *Confessions*, it reflects his early change of lifestyle from a (stable) sexual relationship into celibate Christianity. Berber in origin, around 396 CE he became bishop of Hippo Regius in North Africa. His legacy consists of his writings, which were prodigious, consisting of about 100 books, 300 letters, and 500 sermons.[12] The general drift was to adapt classical philosophy, particularly Platonism, to Christian theology, especially in his book *The City of God* (written from 413–416). This interprets and extends the Bible as a prescription for human behavior.

St. Augustine inveighed against mathematics, science, and medicine. Taking the view that the only way to true knowledge was through understanding God's purpose placed limits on intellectual curiosity. "There is another form of temptation, even more fraught with danger. This is the disease of curiosity. . . It's this which drives us to try and discover the secrets of nature, those secrets which are beyond our understanding, which can avail us nothing and which man should not wish to learn."[13]

The early Middle Ages were dark in the West, where there was no center of learning. Literacy was more or less confined to the Church. Insofar as there was any study of natural philosophy, it was focused on practical matters, such as the need to determine the movements of the Sun and the Moon in order to calculate the date of Easter. The first light came with the Carolingian Renaissance under Charlemagne, who ruled the West from 768 to 814.

This mostly took place around the Court in Aachen (Germany). Charlemagne recruited scholars from around the Empire. Medieval Latin was developed as a common language for scholarship, and many of the old Latin texts were retrieved. But the focus was in the humanities, arts, architecture—there were no new

The first page of Exodus from an early 9th century copy, with a handsomely illustrated first letter, followed by conventional upper case, and then the main text in Carolingian miniscule (which distinguishes lower case and upper case).

developments in science.[14] All the same, the Carolingian Renaissance was a rare break with the restriction of intellectual activity to the Church. It's an unusual example of a secular driving force.

The practical issue in addressing activity from, say, 500–1000 CE, is the difficulty of proving a negative. Lack of activity means there are no sources. As Sherlock Holmes remarked about the dog that [didn't] bark in the night, the absence of information is in itself suggestive. Christianity continued to spread throughout the period. The Church became a major institution. Theological arguments developed into the schism between the Western and Eastern Churches in 1054. There may have been art and music, there may have been artisanal development of crafts—but there was precious little science in Europe.

∞ ∞ ∞

It's a challenge to find much in the way of technological development in the period before 1000 CE.[15] As the Dark Ages began to peter out over the following three centuries, technology made its way from the East into Western Europe. It included the use of paper (1056), the compass (1190), and gunpowder (1270). Later there were some original developments, such as the introduction of spectacles (1286) or the mechanical clock (1280). But after Ptolemy, there had been no original scientific thinking for a thousand years.

The first widespread revival came with the establishment of universities in the 11th–13th centuries. They originated with the transformation of cathedral schools that Pope Gregory VII established in 1079.[16] The original intention was to educate the clergy, but when the torch passed to the universities, education became more general.

Bologna was the first, where the term *universitas* was coined, founded in 1088, followed by Paris (1200), Oxford (1200) and Cambridge (1209), and others in Spain, Italy, and France. By the end of the 14th century, there were around 20 universities. By the end of the 15th century, there were up to 100 universities all over the West.[17] Their origins varied from religious to secular. Bologna was effectively run by students focused on the law, Paris was based in theology, and Oxford and Cambridge operated under Royal charters.[18]

But what was the curriculum? The Universities of Paris and Oxford, for example, had faculties of Arts, Medicine, Law, and Theology. Religion, as a separate discipline, still ranked well above science (or natural philosophy, as it was then), which was included in the Arts, and sometimes regarded as a useful adjunct to theology. Natural philosophy was mostly viewed as an opportunity to understand God's purpose. This was captured by the famous expression of Thomas Aquinas (1225–1274), following Saint Augustine, that philosophy was a 'handmaid' to theology.

The university curriculum was based on reading Aristotle for philosophy, and Hippocrates (from Greece) and Galen (from Rome) for medicine, after their works were translated into Latin in the 12th–13th centuries.[19] There was not much new thinking. There were certainly no new data to change views of natural phenomena. Based on reconciling Greek philosophy with Christian theology, the so-called scholastic movement dominated the university curriculum until the 15th century.[20]

In the strict sense of defining science as constructing theories based on observations or experiments, *science* remained dormant for a thousand years. There is little basis really for current revisionist views that there was lively intellectual activity (at least in science) through the Dark Ages. Of course, the concept of the Dark Ages applies specifically to (Western) Europe. In Arabia, India, and China there was flourishing activity in science and technology (see Chapter 3).

∞ ∞ ∞

A thousand years apart, the contrast between the relationship of science and religion in the 5th century BCE and the 5th century CE is striking. Natural philosophy was the glory of the Intellectual Revolution in Athens and elsewhere in Greece. Theology was the major game in town in the remnants of the Roman Empire in the West. Could this be connected with the Pagan basis for religion in the period BCE and the dominance of monotheism in the period CE?[21]

The existence of many gods allows for multiple paths to the truth, so natural philosophy can be accepted as just another path (with some sporadic exceptions).[22] Monotheism assumes that there is only one true path. The idea that there can be objective truths independent of that path is such a challenge that it must be suppressed. What would have happened had Christianity failed to take over, and the Western Roman Empire had remained Pagan?

The great intellectual tradition of Greece did not just simply fade away. The Church took a view of 'learning' that was incompatible with dissemination of other knowledge. The purpose of knowledge was defined as gaining insight into divine will.[23] Of course, the Church was far from a monolithic organization. But if the Church could even ban the Bible—to be more precise, translations into the vernacular were banned at various times after 1229 —how could there be any expectation of tolerance for independent investigations of natural philosophy?[24] Considering St Augustine's views on science and medicine, it's hard to exempt the Church from responsibility for suppressing intellectual activity.

Language was more of an issue in the Latin-speaking West than in the East, where at least the language was the same as the classic Greek texts.[25] But one of the means by which religion suppressed alternative knowledge was to weaponize language. Confining all written texts to Latin was part of the view that a little knowledge is a dangerous thing, best kept from the masses. (Part of the loss of knowledge in the medieval era was due to the fact that Latin was not a sufficiently sophisticated language to express the subtleties of Greek philosophy. And mistakes accumulated in copying the texts.)[26]

I hold no brief for any religion, or animus against one, but in the sense of this book, religion can be counter-revolutionary. In the ancient world, natural phenomena were viewed as messages from the Gods. For example, disease was ascribed to invasion of the body by evil spirits. To break out from this necessarily implied questioning the role of the God(s). For at least a thousand years, there was a yin and yang in which free thinkers followed alternative trains of thought, and the religious authorities found this so threatening they periodically tried to ban it.[27,28]

Revisionist historians have argued that, far from being part of the cause of the Dark Ages, the Church was if anything an ameliorating influence. They argue that it preserved literature

Attempts by Religion to Suppress Rational Thought

Year	Religion	Action
450 BCE	Pagan	Anaxagoras imprisoned in Athens for impiety (asserting that the Sun was not a deity).
399 BCE	Pagan	Socrates condemned to death in Athens for impiety (questioning the Gods).
1190	Jewish	Maimonides Guide for the Perplexed banned (in some communities).
1277	Christian	Bishop Tempier of Paris prohibits teaching of 219 philosophical and theological theses.
1455	Islam	Caliph bans printing material in Arabic.
1515	Christian	Pope Leo X bans wide range of books (in fact, the majority of printed books), followed in 1599 by Index of Forbidden Books (*Index Librorum Prohibitorum*).
1544	Christian	Gerard Mercator imprisoned for producing maps contradicting religious doctrine.
1553	Christian	Michael Servetus burned at stake for heresy (including rejecting Galen's view of blood).
1580	Islam	Observatory at Istanbul destroyed because studying heavens was 'blasphemous.'
1600	Christian	Giordano Bruno, a committed Copernican and atomist, burned at stake.
1616	Christian	Church bans Copernicus' theory of Earth's motion.
1633	Christian	Galileo condemned for heresy.
1928	Christian	Teillhard de Chardin forbidden to publish on evolution.

and literacy, elevated the status of women, and made technical and scientific advances.[29] The argument has even been taken to the extreme of arguing that the rise of science and reason in the West could have happened only in the environment of Christianity.[30] But the problem with arguing that the Church was responsible for the rise of science is that its stranglehold on literacy and learning made any alternative impossible. The view of the Church as a beacon of light contrasts with the reality of an era in which (literally) millions of people were killed in religious wars. It seems superfluous to remark that the very concept of heresy is antithetical to the ideals of science.

I would concede that the Church, in the form of various monastic movements, was responsible for maintaining and improving knowledge of winemaking. The monks at Clos Vougeot in Burgundy, at Haut Brion in Bordeaux, and at Kloster Eberbach in Germany, were at the peak of their craft. Otherwise, literacy was for the clergy, literature was ecclesiastical (it was during the Dark Ages that the bulk of ancient literature was lost),[31] and the role of women was best described as "get thee to a nunnery."

∞ ∞ ∞

By the 13th century, the rediscovery of Aristotle in Europe led to attempts to reconcile his doctrines with Christian thinking. Thomas Aquinas (*c* 1225–1274) was famous for bringing together Aristotelian philosophy with Christian theology.[32] His contemporary Roger Bacon (1214–1292) was more concerned with bringing science into the theological orbit.

The life of Roger Bacon symbolizes the transition from what used to be called the Dark Ages into the coming Scientific

Issues for the Medieval Church in reconciling Aristotle with Christianity

Aristotle	Christianity
• The universe is eternal	• God created the universe
• The universe is self-sufficient	• God intervenes with miracles
• No distinction between mind and body	• The soul exists independently of the body
• Man is born with power of reason	• Man is born with original sin

Renaissance. A crowd including representatives from the Vatican and from the University of Paris, as well as Oxford University, assembled in June 1914 for the installation of a statue in Oxford commemorating the 7th centenary of his birth. The statue was imaginative. There is no known portrait of Roger Bacon, but the statue was described as "giving an impression of alertness, shrewdness, and pugnacity."[33] Bacon was dressed in the traditional garb of a Franciscan friar. It's a sign of changing assessments of the period that at the start of the 20th century Bacon was regarded as a major figure worthy of commemoration, but there was no comparable event for the 8th centenary.

Doubts about the year of his birth—variously given as 1214 or 1220—typify the uncertainties about Roger Bacon.[34] Was he a polymath who introduced the concept of experimental science or just a medieval thinker in the scholastic tradition? He stands very well as an illustration for the ambiguous state of knowledge in the late Middle Ages.

Born into a wealthy family in England, he went to the University of Paris, where he lectured in a conventional enough way on Aristotle. Exactly when he went to Paris is uncertain, perhaps around 1237, but around 1247 he returned to Oxford.[35] For the next decade, Bacon became involved in the sciences—optics, astronomy, mathematics, and alchemy. At least until recently, he was regarded as an early advocate of experimental work to the point that he was often viewed as a forerunner of modern science.

The story goes that he was repeating some of Aristotle's observations when he realized they could not be correct. This led him to question the basis for knowledge and to develop a belief in the importance of experimentation. At the least, he developed an interest in optics and studied the Greek and Arabic works. There was a drastic change in lifestyle when Bacon became a Franciscan Friar in 1257. No one knows why. One possibility is that the change was connected with the impoverishment of his family. Things did not go well at the Franciscans. He became frustrated by rules of the Order that prevented him from publishing.[36]

Through a connection with Pope Clement IV, he was able to obtain a Papal commission to assess the state of knowledge (albeit sworn to secrecy about it). His *Opus Majus*, completed in 1267 or 1268, proposed a reform of the university curriculum that would strengthen theology by adding studies in language and science. It was in effect a manifesto advocating the production of an encyclopedia of scientific knowledge. A treatise of 878 pages, it considers the nature of knowledge and the relationship between philosophy and theology, includes a study of Biblical languages, and covers mathematics, optics, and a range of experimental sciences from alchemy to astronomy.[37] Bacon proposed a revision of the calendar that anticipated the reforms eventually introduced in 1582. His work was centuries ahead of its time in speculating on inventions such as microscopes, telescopes, and machinery that could operate independent of human intervention.

Opus Majus was sent to the Pope. Pope Clement died in 1268, however, and then in 1277, the Church banned the teaching of certain philosophical doctrines. The Church may have imprisoned Bacon in the 1280s for "suspected novelties."[38] There is no contemporaneous evidence to bear on the question, although Bacon does not seem to have been active during this decade, when he did not write any new works.[39]

The statue of Roger Bacon is in the Oxford University Museum of Natural Science.

Bacon's death in Oxford in 1292 is about the only certain part of the chronology. It's unclear what reputation he had during his lifetime. After his death, there were many myths about him, including his reputation as a magician—he was often referred to as Doctor Mirabilis. (Science fiction writer James Blish wrote a novel in 1964, called *Doctor Mirabilis*, which imagined Roger Bacon's life as a struggle to develop a 'Universal Science.')[40] It's possible that Bacon's interest in alchemy led to his reputation for magic, although this would be ironic given his belief in science and his skepticism about medieval magic.

He stands as a metaphor for his times. It does not seem that his religious faith ever wavered. All his efforts in science were with the intention of strengthening theology. In that regard, he was thoroughly medieval. Yet he shows the first signs of breaking with the scholastic tradition—that all knowledge was known by the Greek philosophers—into questioning some of Aristotle's conclusions and advocating the use of experiments. It's unclear whether he actually performed experiments himself. He accepted the strictures of his era, but there are the first glimmers of an approach that will lead to alternative views and (possibly) a reaction of the authorities against them.

NOTES AND REFERENCES

1. T. Paine, *The Age of Reason*, Fellows, New York, 1807, p. 59.
2. E. Gibbon, *The History of the Decline and Fall of the Roman Empire*, Strahan & Cadell, London, 1776–1788, vol. I–VI. Modern reprint: Everyman's Library, 2010, Online at www.gutenberg.org/ebooks/25717.
3. The concept of a Dark Age originated in 1343 with Petrarch. "When the darkness has been dispersed, our descendants can come again in the former pure radiance." The phrase "Dark Age" was introduced by Caesar Baronius in 1602, in his history of the Church (C. Baronius, *Annales Ecclesiastici*, Vol. X. Roma, 1602, p. 647) specifically to describe the period between the end of the Carolingian Empire in 888 and the first reforms under Pope Clement II in 1046. The term then became used more broadly. Today the Dark Ages (if used at all) more often refers specifically to the first half of the period known as the early Middle Ages, from 500 until 1000–1100 CE.
4. "Captive Greece captured, in turn, her uncivilized Conquerors, and brought the arts to rustic Latium." Horace, *Epistles II.1.156*, 14 BCE.
5. W. H. Stahl, *Roman Science: Origins, Development, and Influence To The Later Middle Ages*, University Of Wisconsin Press, Madison, WI, 1962, p. 5. Warfare was a notable exception. Russo makes the case that during the

Hellenistic Age, the Greeks also developed more technology in other areas than they are usually given credit for. See L. Russo, *The Forgotten Revolution: How Science Was Born in 300 BC and Why it Had to Be Reborn*, Springer, Berlin, 2003, pp. 95–142.

6. This widely quoted view is based upon a single comment written 300 years later by Marcellus (*c.* 100 CE), but it captures the spirit of Greek interest.

7. W. H. Stahl, *Roman Science: Origins, Development, and Influence to the Later Middle Ages*, University of Wisconsin Press, Madison, WI, 1962, pp. 101–119.

8. The main project of Boethius (480–524 CE), the leading intellectual of the day in Rome, was to translate Aristotle and Plato.

9. D. C. Lindberg and M. H. Shank, *The Cambridge History Of Science: Volume 2, Medieval Science*, Cambridge University Press, Cambridge, 2013, p. 7.

10. A. Tihon, Science In The Byzantine Empire, in *The Cambridge History Of Science: Volume 2, Medieval Science*, ed. D. C. Lindberg and M. H. Shank, Cambridge University Press, Cambridge, 2013, pp. 190–206.

11. For example, Euclid's works became available in printed form in Latin translation in 1482, in Italian in 1543, and in English in 1570. See D. Wootton, *The Invention Of Science: A New History Of The Scientific Revolution*, Harper, New York, 2015, p. 187.

12. Near the end of his life, he produced an annotated catalog of all his works.

13. He continued on to argue against medicine. "With a cruel zeal for science, some medical men, who are called anatomists, have dissected the bodies of the dead… and have inhumanly pried into the secrets of the human body."

14. Although he rejects the concept of the Dark Ages, Lindberg, says, "Nothing in… Carolingian achievements in astronomy and other scientific topics represents a revolutionary leap forward. The importance of Charlemagne, his reforms, and the scientific achievements of the Carolingian period is to be found not in novelty, but in the recovery and preservation of important portions of the classical tradition." D. C. Lindberg, *The Beginnings of Western Science: The European Scientific Tradition in Philosophical, Religious, and Institutional Context, Prehistory to A.D. 1450*, University of Chicago Press, Chicago, 2007, p. 197.

15. While there was no single cause for the Dark Ages, both P. Watson, *Ideas. A History of Thought and Invention From Fire to Freud*, HarperCollins, New York, 2005 and C. Freeman, *The Closing Of The Western Mind: The Rise Of Faith and The Fall Of Reason*, Vintage Books, New York, 2003, put a major responsibility on Christianity, and its opposition to any learning that was not theological. Watson, p. 249, shortens the period: "the true dark ages…extend from the middle of the sixth century to the middle of the ninth." Both Watson and Freeman are clear about the reality of the Dark Ages as a period devoid of intellectual development. By contrast, many current historians indignantly refute the very concept of the Dark Ages, although I have yet to find one who can list intellectual accomplishments in the period. Many of the revisionists who argue that the Dark Ages were in fact a time of intellectual activity point to technological developments that actually occurred from 1000–1300 CE. The intellectual dishonesty is

breathtaking. For example: J. Hannam, *The Genesis of Science: How the Christian Middle Ages Launched the Scientific Revolution*, Regnery Publishing, Washington DC, 2011 or P. Rossi, *The Birth of Modern Science*, Blackwell, Oxford, 2000. S. Falk, *The Light Ages: The Surprising Story of Medieval Science*, W. W. Norton & Co., New York, 2020 actually deals exclusively with the period around 1400.

16. A Papal Decree ordered the establishment of cathedral schools.

17. Freeman makes the point that the universities lost their impact after the 14th century, more universities closed than opened in the 18th century, and they recovered only in the 19th century. See C. Freeman, *The Reopening Of The Western Mind: The Resurgence Of Intellectual Life From The End Of Antiquity To The Dawn Of The Enlightenment*, Knopf, New York, 2023, pp. 160–161.

18. One estimate is that 1% of the population was educated by the universities. See O. Gal, *The Origins Of Modern Science*, Cambridge University Press, Cambridge, 2021, p. 122.

19. The classical curriculum, inherited from Greek philosophy, was divided into two parts. The trivium included grammar, rhetoric, and logic and lasted two years; those who passed could go on to the quadrivium, which included arithmetic, geometry, astronomy (including astrology), and music. This slowly evolved into separate disciplines.

20. The major resources were Aristotle's *Organon* for cosmology, Ptolemy's *Almagest* for astronomy, Euclid's *Elements* for mathematics, and Galen's *Galenic Corpus* for medicine.

21. Schrödinger argued for the importance of the lack of powerful forces such as the state or religion, that might have an interest in suppressing freethinking. See E. Schrödinger, *Nature and The Greeks*, Cambridge University Press, Cambridge, 1954, pp. 55–56.

22. Anaxagoras had to flee Athens in 434 BCE because he argued that the Sun was not a God but merely a fiery object; there were the tribulations of Socrates in 399 BCE; and later Aristotle's escape from a possible similar trial in 323 BCE.

23. Freeman makes a powerful case that "the Greek intellectual tradition was suppressed rather than simply faded away." See C. Freeman, *The Closing of the Western Mind: The Rise of Faith and the Fall of Reason*, Vintage Books, New York, 2003, p. 340.

24. William Tyndale was burned at the stake in 1530 for translating the Bible into English.

25. "By the sixth century virtually no western scholar was able to understand Greek." P. Watson, *Ideas: A History of Thought and Invention, from Fire to Freud*, HarperCollins, New York, 2005, p. 247.

26. C. Freeman, *The Reopening Of The Western Mind: The Resurgence Of Intellectual Life From The End Of Antiquity To The Dawn Of The Enlightenment*, Knopf, New York, 2023, p. 726.

27. The strength of the focus on religion is shown by the effort devoted to constructing the great gothic cathedrals of France. A casual estimate is that a quarter of the GNP (Gross National Product) may have been devoted to

constructing about 80 cathedrals (mostly in northern France), and another 500 or so large churches, in the 12th–13th centuries. See E. Strosberg, *Art and Science*, Abbeville Press, New York, 2015, p. 56.

28. Religion was even more directly responsible for costs of war in the period. The 7th Crusade (1248–1254), led by Louis IX of France, chewed up 6× France's annual income. S. Runciman, *A History of the Crusades*, Cambridge University Press, Cambridge, 1987.

29. P. Campbell, *The Church and the Dark Ages (430–1027): St. Benedict, Charlemagne, and the Rise of Christendom*, Ave Maria Press, Notre Dame, IN, 2021.

30. R. Stark, *The Victory of Reason: How Christianity led to Freedom, Capitalism, and Western Success,* Random House, New York, 2007.

31. C. Freeman, *The Reopening Of The Western Mind: The Resurgence Of Intellectual Life From The End Of Antiquity To The Dawn Of The Enlightenment*, Knopf, New York, 2023, pp. 160–161.

32. For a discussion of how various religions reconciled Aristotle's view that everything has a cause (in effect excluding divine intervention) with monotheistic beliefs, see O. Gal, *The Origins Of Modern Science*, Cambridge University Press, Cambridge, 2021, pp. 139–145.

33. F. A. D., The Commemoration of Roger Bacon at Oxford, *Nature*, 1914, **93**, 405–406.

34. The alternatives are so far apart because they are calculations based on an ambiguous reference Bacon made much later to his early years. The fact is that there is no evidence for either.

35. It's unclear whether he came directly under the influence of Robert Grosseteste (1168–1252), who was at Oxford until 1235, and influential in the revival of Greek learning in the West.

36. "They forced me with unspeakable violence to obey their will." Quoted in S. C. Easton, *Roger Bacon and His Search for Universal Science*, Blackwell, Oxford, 1952, p. 122.

37. Y. Kedar, Roger Bacon, in *Encyclopedia of Medieval Philosophy*, ed. H. Lagerlund, Springer, Dordrecht, 2018, DOI: 10.1007/978-94-024-1151-5_449-2.

38. About 80 years after Bacon's death, a history of the Franciscan order claimed that Bacon was imprisoned for "suspected novelties." The imprisonment could have lasted from around 1279 to any time up to 1290 when there was a general release of Franciscans who had been imprisoned.

39. S. C. Easton, *Roger Bacon and His Search for Universal Science*, Blackwell, Oxford, 1952, pp. 186–205.

40. J. Blish, *Doctor Mirabilis: A Vision*, New English Library, London, 1964.

Science in the East: 750–1453 CE

"THE MOVING FINGER WRITES AND HAVING WRIT MOVES on," said Omar Khayyám in his epic poem, The Rubaiyat. This could stand as an epitaph for the Golden Age of Islam in the 8th to 13th centuries. Science and other cultural activities went through a glorious period, but it was a dead-end for the culture. It's a conundrum how Arabia could have become the world's leading center for intellectual activity across sciences from astronomy to biology and medicine and then collapsed. The flame of knowledge had passed from the Greeks to the Abbasid Caliphate (750–1258) as the result of Arab conquests across Alexander the Great's empire. It incorporated knowledge from all across Asia. Although it was later extinguished in the Arab world, it passed into Western Europe, partly through the Moors during their occupation of Spain (from 711 to 1492). The renaissance of science in the late medieval period in Europe was fueled by Greek texts that had been conserved by translation into Arabic, and were then translated back into Latin. It was another century before science took off as an independent intellectual activity in Europe.

The Frontiers of Science
By Benjamin Lewin
© Benjamin Lewin 2026
Published by the Royal Society of Chemistry, www.rsc.org

Timeline for Key Events in the Rise and Fall of Science in Arabia

Brahmagupta writes *Brāhmasphutasiddhānta* introducing zero (*c.* 628)

750 — Abbasid Caliphate founded in Baghdad (750)

Samra ibn Jundab al-Fazari translates Indian text *Brāhmasphutsiddhānta* into Arabic (770)

800 — House of Wisdom established in Baghdad (815)
al-Khwārizmi introduces algebra (820)
al-Kindī popularizes Greek philosophy; starts cryptography (833-870)

850

900 — al-Rāzī (Rhazes) writes textbooks on medicine (*c.* 903)

950

1000 — al-Zahrawi writes 30-volume *Kitab al-Tasrif* on surgery (*c.* 1000)
al-Biruni measures size of Earth using trigonometry (*c.* 1001)
al-Haytham publishes 7 volumes of *Book of Optics* (1011-1021)
Avicenna (Ibn Sina) writes 5-volume *Canon of Medicine* (1025)

1050 — First dedicated observatory built at Isfahan (in Persia) (1073)
Omar Khayyam writes book on algebra (1078)

1100 — Abelard of Bath translates al-Khwārizmi into Latin (1126)

1150 — al-Idrisi creates *Tabula Rogeriana* map (1154)
Gerard of Cremona translates 87 Arabic works into Latin (1160-1187)
Averroës (Ibn-Rushd) defends Aristotelian philosophy (*c.* 1169)

1200 — Fibonacci publishes *Liber Abbaci* introducing Arabic numerals (1202)

1250 — Abbasid Caliphate ends when Mongols sack Baghdad (1258)
Construction of observatory at Marāgha (in Persia) (1259)

1300 — al-Fārisī experiments on basis for rainbow (1309)

1350

1400 — Construction of Ulugh Beg Observatory in Samarkand (1422)

1450

A Dialog between Al-Haytham and Robert Grosseteste

Al-Haytham (965–1040) was a mathematician, physicist, and astronomer in Cairo who became one of the world's first true scientists, known especially for his experimental and theoretical work in optics.

Reproduced from https://commons. wikimedia.org/wiki/File:Ibn_al-Haytham.png under the terms of the CC BY-SA 4.0 license, https:// creativecommons.org/licenses/by-sa/ 4.0/deed.en.

Robert Grosseteste (1168–1253) was a theologian and philosopher, who lectured at Oxford University and became prominent in the Church. He was interested in the scientific method and is best known for his work in optics.

Al-Haytham: Greetings, Robert Grosseteste. As a philosopher from the Golden Age of Islam, I approach the natural world with a focus on observation and experimentation. My books on optics have been published over a decade starting in 1011. I have heard of your work on optics and the metaphysical implications of light. Tell me, how do you reconcile the study of natural phenomena, such as light, with your theological commitments?

Grosseteste: The pleasure is mine, Al-Haytham. As Bishop of Lincoln from 1235 to 1253, my first allegiance is to theology, but I am known as an enlightened philosopher with an interest in optics. To your question, I see no conflict between the study of the natural world and theology. In fact, I believe that by understanding the principles of nature, we come closer to understanding the divine order.

Al-Haytham: To me, the physical world operates according to definable laws. My work is concerned primarily with uncovering the principles that govern the behavior of light and vision. My studies in optics led me to understand that light travels in straight lines and that our perception of the world is shaped by the way light interacts with our eyes.

Grosseteste: I too believe that the natural world can be understood through reason, but I view light as both a physical phenomenon and a symbol of divine wisdom. It was the first form created in the universe, from which all other forms and matter emanate. What is your approach in trying to understand the natural world?

Al-Haytham: Experimentation is the cornerstone of my method. I have conducted numerous experiments to test the behavior of light, such as observing the way it reflects and refracts. I used a camera obscura to understand how images are formed. How does experimentation fit within your approach to studying nature?

Grosseteste: Experimentation is a valuable tool, but I believe it must be guided by reason and philosophical principles. We must ensure that our interpretations are aligned with a broader understanding of the cosmos and its divine origin. I see experimentation as part of a holistic approach to knowledge, one that integrates observation, mathematics, and metaphysical insight.

Al-Haytham: I focus primarily on the empirical aspects of science. I believe that hypotheses must be tested through careful observation and experimentation, rather than accepted based on authority or tradition. My approach has led to an understanding of light that will not be equaled in Europe for another four centuries. I rest my case.

SCIENCE IN THE EAST

> *The seeker after the truth is not one who studies the writings of the ancients and, following his natural disposition, puts his trust in them, but rather the one who suspects his faith in them and questions what he gathers from them, the one who submits to argument and demonstration, and not to the sayings of a human being whose nature is fraught with all kinds of imperfection and deficiency. Thus the duty of the man who investigates the writings of scientists, if learning the truth is his goal, is to make himself an enemy of all that he reads, and, applying his mind to the core and margins of its content, attack it from every side,*[1] Al-Haytham, c. 1026.

While Western Europe was sliding backward into the medieval era, there was a Golden Age in Islam, achieving new intellectual heights from philosophy to economics.[2] The Golden Age coincided with the Abbasid Caliphate (750–1258 CE), which extended all along the southern shore of the Mediterranean to the border of India. Knowledge from ancient Greece was absorbed and extended, but did not make its way to Western Europe until much later.

If we take a snapshot of the world in 1000 CE, there was intellectual activity in the East, but little in Europe, where the knowledge of the Greeks had been lost following the fall of the Roman Empire. The remnants of Greek knowledge were to be found in Arabia, which rapidly advanced beyond them in astronomy, physics, and medicine. Arabia is a bit of a misnomer because the geniuses of the era (genius is not an exaggeration) were spread across an area from Baghdad through Central Asia.[3] The common feature was that they wrote (mostly) in Arabic.

The most advanced mathematics was being developed in India. (The original texts developing a decimal system were written in Sanskrit.) Astronomy was developing independently in China, reaching a peak under the polymath Su Song during the Song dynasty. (I include China with Arabia for convenience, because in both cases, although their period of development overlapped with the medieval era in Europe, their discoveries did not penetrate the Western world at this time.)

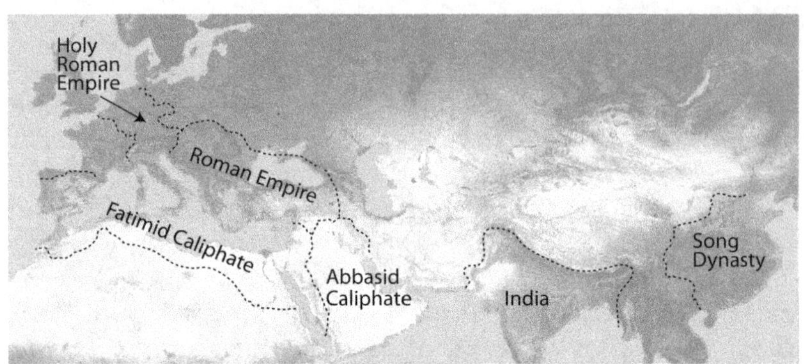

The major centers of learning in year 1000 were in widely separated areas of the Abbasid Caliphate (especially at the House of Wisdom in Baghdad), in India (especially the development of decimal arithmetic), and in the Song Dynasty in China (especially astronomy).

Greek influence had spread across the Mediterranean with the conquests of Alexander the Great (356–323 BCE). Following Alexander's death, Alexandria became the center of learning (see Chapter 2). Alexander's conquests then became part of the Islamic Empire as it expanded rapidly a few years after Mohammed's death in 632 CE. The Abbasid Caliphate founded its capital in Baghdad in 750 CE. Baghdad became a major driving force for intellectual development in the Arab world.

The Caliph set up a library in Baghdad around 815 CE, the House of Wisdom (Bayt al-Hikmah). This was a great stimulus for intellectual activity.[4] During this period, Hunayn ibn Ishaq (809–873 CE) was responsible for translating more than 100 Greek works into Syriac and Arabic. Virtually all of the important Greek works had been translated into Arabic by 1000 CE.[5]

The Golden Age was not confined to science, of course, but extended across a wide intellectual realm. It was much enhanced by the art of papermaking (which the Abbasids learnt from Chinese prisoners captured in 751 CE during a war). They improved the technology and built paper mills. Baghdad became famous for its bookstores.

The Caliphate lasted until 1258 CE, when the Mongols (led by a grandson of Genghis Khan) sacked Baghdad. The river Tigris ran black with ink from the books that were destroyed, or red with blood from philosophers who were killed, according to

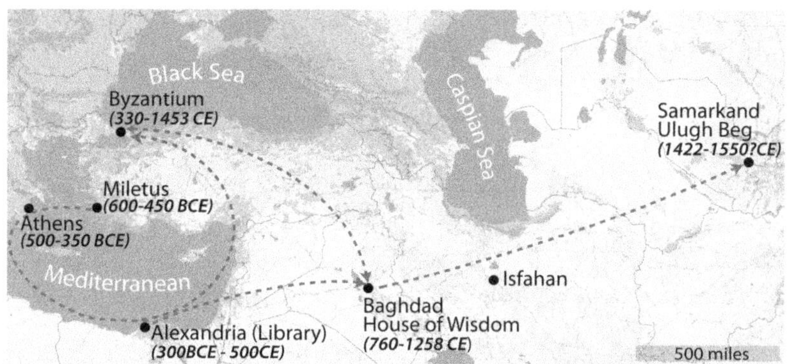

The migration of knowledge from the ancient world into Arabia followed a path from Greece to Alexandria to Byzantium, and across Asia.

various accounts. Knowledge diffused farther to the East, especially with astronomy pursued in Persia and Central Asia.

Transmission through the East became the most important route for preserving the knowledge of the ancient Greeks. The most important works were collected (or written) in Alexandria, may have migrated to Byzantium with the fall of Alexandria, and made their way to the East, most notably to Baghdad, where they were translated into Arabic.[6]

Starting from the work of the Greeks, science in the East developed its own impetus. Education in the Islamic world took place in the madrasas, which focused exclusively on religious matters, including the 'Islamic sciences.' Natural science, as we would define it today, was called 'foreign sciences.'[7] Science was a side matter, taking place privately or in centers supported by a patron. Even so, mathematics and physical sciences flourished far beyond anything practiced in Western Europe after the fall of the Roman Empire. And it was very much a secular activity. Indeed, scientific investigation was at best tolerated by the religious authorities.

Astronomy was highly advanced—there was nothing comparable in Europe until the time of Copernicus (1473–1543).[8] Indeed, the concept of the observatory was developed in the Islamic world. The first dedicated observatory was built at Isfahan in 1073 (now in Iran). Some places became famous as centers of learning, such as the observatories at Marāgha (built in 1259 in Iran) and the Ulugh Beg Observatory at Samarkand (built in 1422), today part of Uzbekistan. They tended to be built as an enthusiasm of the ruler of the day, and to fall into decay after his death, but even so, there was continual progress in astronomy.

Progress in science was driven by an overlapping trio of great scientists, Ibn al-Haytham (965–1040) in Cairo, al-Bīrūnī (973–1048) in Central Asia, and Ibn Sīnā (980–1037), known as Avicenna in the West, who was a physician at various courts across Asia.

One of the greatest astronomers of all time was the polymath al-Bīrūnī, born in Central Asia in what is now Uzbekistan. He worked in the area that is now Afghanistan, ending up in the capital, Ghazni, where he became the court astronomer. He wrote mostly in Arabic, although his major work, *Kitab al-Tafhim*, was written in both Arabic and Persian. His early years were colored by the turmoil of invasions and war. He was forced to flee from one area to another more than once. His surviving works give the impression that he did not suffer fools gladly.

In an acrimonious correspondence with his contemporary, the famous physician Ibn Sīnā, he questioned many of Aristotle's conclusions centuries before there was any comparable criticism in Europe. Although he was officially an astrologer, he wrote a book expressing skepticism about the capacity of astrology to foretell the future. His cartography mapped the known world

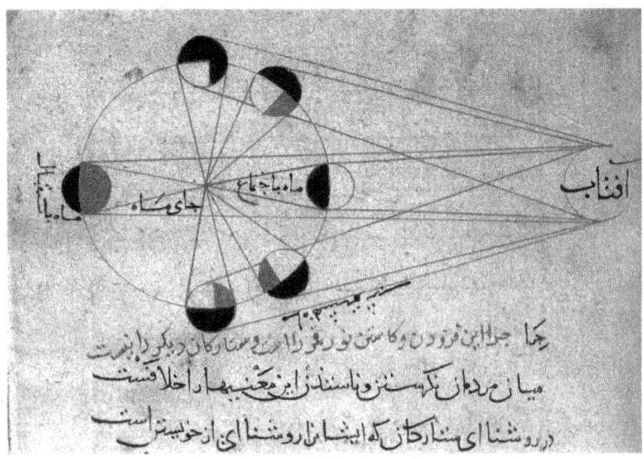

Biruni's explanation of phases of the moon.

and considered the existence of equivalent landmasses in the southern hemisphere. He wrote a book called *Chronology* that converted different calendar systems, and another defending al-Khwarizmi's *Algebra*, called *Disproving the Lie*. He played a role in bringing Indian mathematics and astronomy into the Arabic world with his book *India*, based on travels there in 1022–1024.

In his final work, known as *Masud's Canon* (because it was dedicated to the ruler of the period), which summarized all of his work on astronomy, he even considered the possibility of heliocentricity. "For it is the same whether you take it that the Earth is in motion or the Sky. For, in both the cases, it does not affect the Astronomical Science. It's just for the Physicist to see if it is possible to refute it." In the context of his era, al-Bīrūnī was the ultimate rationalist.[9]

In Alexandria *c.* 150 CE, Claudius Ptolemy had produced a model for the revolution of the Sun around the Earth that employed a complicated series of motions (see Chapter 4). In a book of 1049 with the title, *Doubts about Ptolemy*, Ibn al-Haytham pointed to flaws in Ptolemy's use of epicycles—"this is an absurd impossibility"—although he did not suggest a better model.[10] Nasïr al-Dïn al-Tūsï (1201–1274) produced a model in 1260,[11] now known as the Tūsï couple, that simplified Ptolemy by combining two circular motions.[12]

The Marāgha school of astronomy advanced beyond Ptolemy, culminating in the work of Ibn al-Satire *c.* 1375, to produce models that gave predictions as good as the model later developed by Copernicus (although they remained geocentric).[13] It's controversial whether Copernicus might have known about their results (see Chapter 4).[14] It's a tribute to the astronomers of the time that they developed such strong models. It's a cautionary note with relevance to modern science that models can be built on incorrect theories but produce correct predictions.

Ulugh Beg (1394–1449) was the grandson of Mogul Emperor Tamerlane, and ruler of part of the empire, when he constructed an observatory at Samarkand around 1428. It was part of a complex also containing a mosque, a hospice, and extensive gardens. Within two decades, he was overthrown (and beheaded), but the madrasa remained a center of learning. Destroyed some time after the 16th century, its exact location remained unknown until excavations in the 20th century. These revealed a huge sextant, with a radius of 40 m, which must have enabled observations at an unprecedented level of resolution. (A sextant is an instrument for determining the angle between the horizon and a celestial body. Hand-held versions were used in navigation for measuring latitude.) The observatory produced the first star catalog based on new observations since Hipparchus, 1600 years earlier in ancient Greece. The Zīj-i Sulṭānī (Emperor's Star Table) of 1437 CE catalogued 1018 stars.

Cartography was another area in which the Islamic world was centuries ahead of developments in Europe. Al-Idrisi (1100–1165) was the most gifted cartographer of the era. He was commissioned by King Roger II of Palermo to create a world map in 1154 that was known as the *Tabula Rogeriana*. Taking 15 years to complete, based on extensive travels through North Africa and Europe, the map occupied 70 double-page spreads. It included cultural and economic information as well as descriptions of natural features. A form of the map for the king was engraved on a disk of silver, 2 meters in diameter.[15] Comparable accuracy was achieved in maps produced in Western Europe only in the 16th century.

Because glass making and metallurgy were well developed, there was special interest in optics. Optics provided a notable exception in which Arabia went beyond observations into developing experimental science. Theoretical advances in optics relied on the Arabic strength in geometry. Ibn Sahl (*c.* 940–1000) analyzed the refraction of light and deduced rules similar to the modern law of refraction around 985 CE.

A major breakthrough came with Ibn al-Haytham's studies on optics. He described his work extensively in the *Book of Optics*, published in 7 volumes from 1011–1021 CE.[16] (Translated into Latin around 1250 CE, it remained the definitive work on optics until the 17th century.) Knowledge of optics had practical consequences, for example, in developing primitive treatments for cataracts.

The Greeks, in particular Euclid and Ptolemy, had proposed the 'extra mission' theory of optics, in which the eye transmitted light to illuminate objects. Al-Haytham proposed the 'intromission theory,' in which vision is caused by light entering the eye. He used experiments to test his theories, including the first

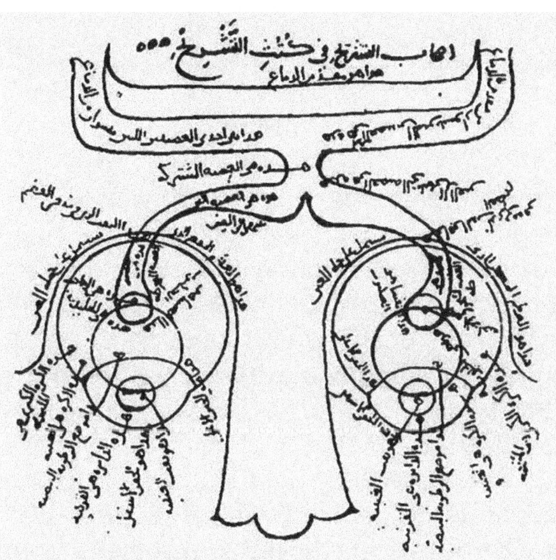

The structure of the human eye according to Ibn al-Haytham.

use of a camera obscura (projecting an image through a small hole). He demonstrated that light travels in straight lines. He analyzed the refraction of light when it passes from one medium to another. Al-Haytham expressed his views on the scientific method and its limitations. He was by any measure a polymath who was one of the world's first true scientists.

Somewhat later, al-Fārisī (1267–1319) tested al-Haytham's theory of the rainbow, which had been based on reflection. Al-Fārisī suggested that water drops refract rays of light from the sun. He tested the theory by using a transparent sphere filled with water.[17] Here was the true scientific approach: questioning the existing theory, and replacing it with a new theory, supported by experiments. This was three centuries before René Descartes published his theory of optics, including an explanation for the rainbow.[18]

Medicine was also more advanced in Arabia than in Europe (see Chapter 21). Abū Bakr al-Rāzī (865–925 CE), known as Rhazes in the West, was head of the hospital in Baghdad, and a famous diagnostician. Al-Zahrawi (936–1013), known as Abulcasis in the West, who was the court physician in Córdoba, was a pioneer in surgery. One section of his 30-volume encyclopedia on medicine, *Kitāb al-Taṣrīf*, became the standard text on surgery in Western Europe after it was translated into Latin. Ibn Sīnā (Avicenna) wrote in both Persian and Arabic. His 5-colume *Canon of Medicine* became hugely influential in the West in Latin translation.

The principal reason why the flowering of science in Arabia came to an end may have been increasing intolerance for anything felt to be inconsistent with religion. (Would science in Arabia have remained ahead of the West had it not been stifled?) It has been a popular view that the decline of science was given an impetus, if not entirely caused, by al-Ghazālī, an influential philosopher (*c.* 1058–1111), whose favorite saying was "May God protect us from useless knowledge." In a book called *The Incoherence of the Philosophers*,[19] he argued that logic is dangerous, because of its reliance on arguments outside religion.[20] Echoes of St. Augustine!

A counter-view limits al-Ghazālī's influence, pointing to significant work in astronomy, optics, and medicine through the 13th and 14th centuries.[21] (The great observatory at Marāgha was built in this period, for example.) However, a blow to science

The great comet of 1577 (star at top right) being observed by an astronomer using a quadrant (right) in Taqī al-Dīn's observatory in Istanbul. Tycho Brahe observed the same comet in Denmark.

came when the printing press was banned in 1455, followed a few years later by a ban on possessing printed material.[22] One view is that the Arab world became inward looking from the 12th century, partly as the result of dealing with foreign invasions, partly from a more restrictive view of monotheism. This led to the eclipse of science.[23,24]

In Istanbul, then part of the Ottoman Empire, an observatory was constructed around 1575 for the astronomer Taqī al-Dīn al-Rasid. But after a comet was observed in 1577, the Sultan was persuaded that studying the heavens was blasphemous. The observatory was destroyed in 1580. It seems to have been comparable in sophistication to the observatory built around the same time in Denmark where Tycho Brahe extended the work of Copernicus that led to the Scientific Revolution in the West (see Chapter 4). The overall cultural problem was that the basic value system held that truth could be achieved only with the aid of divine revelation. Science can succeed only in a society that accepts its precepts (or at least allows that they are valid in the context of scientific enquiry).

The Golden Age of Islam saw the introduction of scientific method into physics, astronomy, and medicine. Mathematics was more advanced in Arabia than anywhere else in the world. It's unclear to what extent this way of thought permeated society. For example, there does not seem to have been any employment of scientists in military affairs, comparable, say, to the involvement of Archimedes or Galileo in their eras. Scientific enquiry declined in the 13th and 14th centuries and collapsed by the 15th century. Even reports of the Scientific Revolution from Europe in the early 17th century did not revive activity.[25] The Golden Age was well and truly over.

∞ ∞ ∞

Of course, Greece and Arabia were not the only places in the ancient world that developed a scientific approach. Greek thought (and its succession in Arabia) had by far the greatest influence on the development of scientific thought centuries later in Western Europe, but other civilizations had significant influence also.

China offers an interesting parallel for the rise and fall of the scientific approach, although here the reason for decline was less religious than societal. The Chinese did not have access to the classic works of Aristotle, Euclid, or Ptolemy. One puzzle about China is that the opportunity to incorporate Arabic mathematics was ignored. Chinese mathematics lacked the trigonometry necessary to resolve astronomical problems, to the point at which an Islamic Bureau of Astronomy was established in Beijing in 1271 CE to compensate.[26]

Although Chinese mathematics relied solely on algebra, it produced some impressive technology. Difficulties with the calendar caused Su Song (1020–1101), a senior bureaucrat in the Empire, to construct an astronomical clock. Traveling on an important diplomatic mission, he had arrived a day late because the calendar was wrong. Su Song designed a water-powered 40 ft high mechanical clock tower in 1088 that followed the movements of the sun, moon, and planets as well as keeping time. He described the details in a book that included 47 illustrations of its construction.[27] (The first astronomical clocks in the West were not built until a century later.)

A polymath by any measure, Su Song also constructed an impressive star map in 1092 and compiled a pharmacopoeia. His contemporary, Shen Gua (1031–1095), another polymath, was

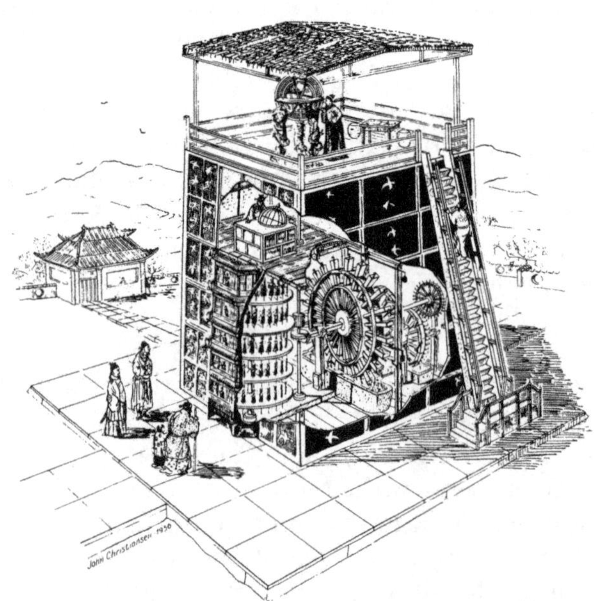

Illustration of Su Song's clock tower.

also driven to use astronomy to improve the prediction of the seasons for agricultural purposes. Indeed, star maps with more than 1400 stars were produced in China from the 4th century BCE.

The Bureau of Astronomy was responsible for calculating the calendar from astronomical observations. Their responsibilities included astrological predictions that were an important part of the culture. By the early 17th century, when the Jesuits were bringing the results of the Scientific Revolution to China, the Bureau was encountering serious problems in maintaining its calendar.[28] It turned to Western systems, but knowledge was restricted to a small circle around the Palace.[29] It never extended into any new theoretical analysis. The whole issue of abandoning traditional Confucian analysis was controversial.

The focus in China was on solving practical problems rather than establishing underlying theories. For example, the printing press was invented in China around 1040 CE, four centuries before Gutenberg. Indeed, because science did not develop from natural philosophy, "Chinese had sciences but no science, no single conception or word for the overarching sum of all of them."[30]

The failure to develop science as such has often been attributed to Confucianism, with its focus on top-down rule from the center, competition to express the conformity of conventional wisdom, and disdain for debate or Socratic dialogue. It placed the emphasis on moral matters, and a search for harmony in nature, rather than reliance on sets of laws (or religious belief).[31] It was also static. The educational syllabus remained unchanged for hundreds of years.

Even though there were considerable differences between the Islamic and the Chinese worlds, one thing they had in common was that science—such as it was—was set aside from society. It did not take place in institutions associated with other forms of learning (such as the universities that were being established in the Western world). This makes the point that science cannot contribute more widely to society in the absence of institutional support.

There was an important difference between the effects on the West of the developments in China and Arabia, respectively. Knowledge from Arabia was a driving force in the revival of intellectual activity in Western Europe. Knowledge from China did not really impact the development of science in the West, although it did affect technological developments, such as papermaking, the introduction of the compass, or the use of gunpowder.[32] (Gunpowder was invented in China in the 9th century CE—possibly as the result of experiments in alchemy—while they were still fighting with swords and armor in the Western world.) However, scientific traffic was in the opposite direction, as Western knowledge began to impinge on China with the arrival of Jesuit missionaries in the 17th century.

There are disagreements on the cause of the decline of science in Arabia and China, just as there is dissension on the cause of its rise in the West. However, the fact remains that science declined in China and Arabia and rose in the West.

∞ ∞ ∞

The development of science required a simple system for numeration. The Babylonians (following the earlier system of the Sumerians) recorded sky movements using a 60-base system. That is why even today we have the apparently illogical system of 60 seconds per minute and 60 minutes per hour. The Greeks were impeded by a system based on letters. This explains the emphasis on geometry, and the failure to progress to other forms of mathematics. The Romans (as the Egyptians before them) used a system based on powers of 10 (but with a cumbersome system of letters).

Indian mathematicians invented the base 10 system between the 1st and 4th centuries CE. They introduced a 'place value' system, meaning that the value of a number is determined by its position. Before then, the meaning of a given numeral was fixed—X means 10 in the Roman system wherever it is placed.

The great Indian mathematician Aryabhata (476–550 CE) used the system to calculate the length of the solar year. Centuries ahead of his time, he proposed that the earth is a sphere rotating on its axis, albeit in a geocentric model. In addition to presenting the decimal symbol set, Indian mathematics introduced zero as a mathematical object. (Zero was necessary to make a place value system work.) The concept of zero first appeared in a work on mathematical astronomy, written in Sanskrit by Brahmagupta *c.* 628 CE.[33]

Sanskrit was by no means a local language, but became dominant across much of Asia. Work in Sanskrit was translated into Arabic when a delegation from the Raja of Sindh (in present-day Pakistan) brought a manuscript of mathematical theory, the *Sindhind*, to Baghdad in 773 CE. It could be translated because, as the result of earlier Arab invasions of India, no less a person than the Grand Vizier came from a family that had converted from Hindu to Moslem. Written traditionally as a rhyming Sanskrit poem, the *Sindhind* contained Brahmagupta's work.[34]

Mathematics became one of the glories of the Golden Age of Islam. By translating Brahmagupta's work into Arabic, al-Fazari (746–806) made possible the later introduction by Musa al-Kwārizmī (*c.* 780–850 CE) of what we now call Arabic numerals.[35] Working at the House of Wisdom, al-Khwārizmī wrote two books on arithmetic and algebra. *Concerning the Hindu Art of Reckoning* introduced the Hindu numeral system in 825 CE. "We have decided to explain Indian calculating techniques using the nine characters and to show how, because of their simplicity and conciseness, these characters are capable of expressing any number." Zero was described as "the tenth figure in the shape of a circle."[36] *Kitab al-Jabr* (The

Brahmi		—	=	≡	+	𝼀	𝼁	𝼂	𝼃	𝼄
Hindu	०	१	२	३	४	५	६	७	८	९
Arabic	٠	١	٢	٣	٤	٥	٦	٧	٨	٩
Modern	0	1	2	3	4	5	6	7	8	9

Modern numerals originated in a system that came from India. The Brahmi system dates from the 3rd century BCE and used 9 symbols, but it was nonpositional and lacked a symbol for zero. The Hindu system was developed in the 1st–4th centuries CE. It was positional (the value of a symbol is determined by its position) and introduced zero. After it entered Arabia, it spread to Muslim Spain and then to Western Europe, where it went through several changes of symbols during the medieval era, taking its present form in the 15th–16th centuries.

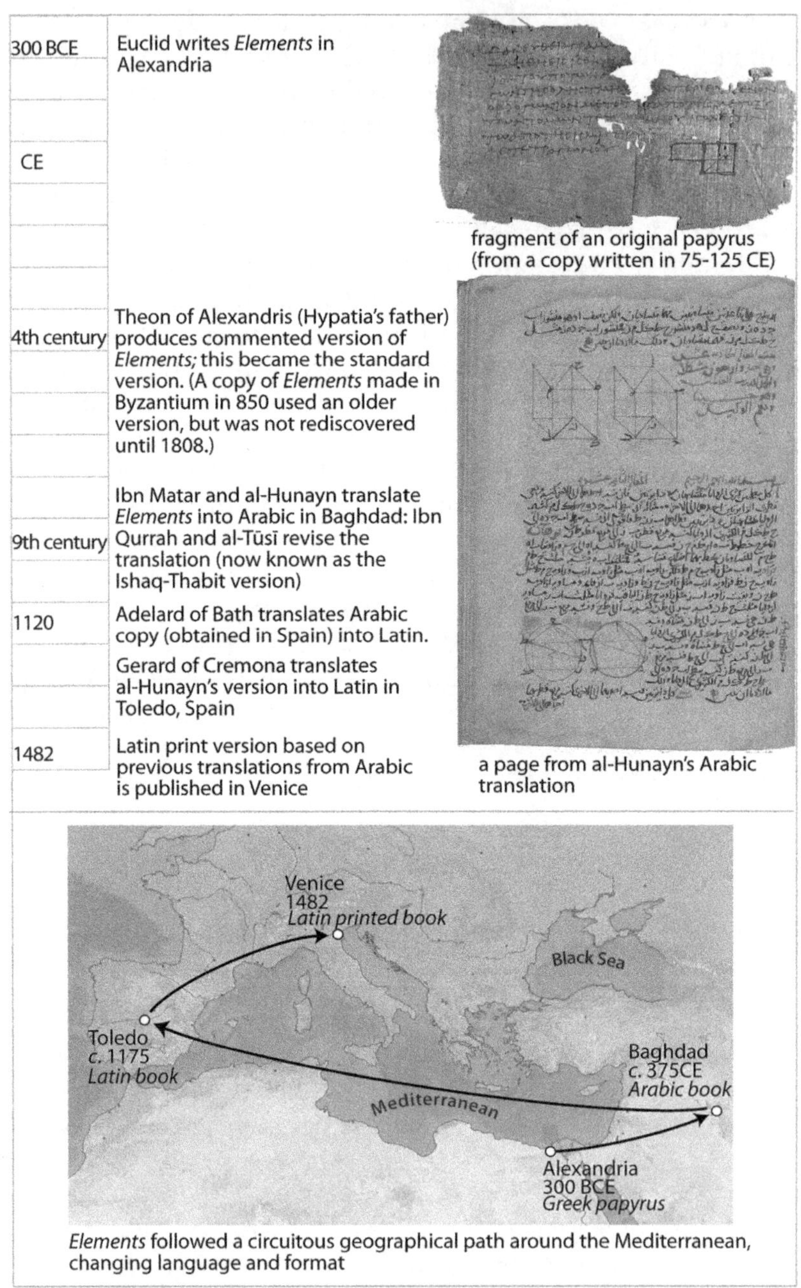

300 BCE	Euclid writes *Elements* in Alexandria
CE	
4th century	Theon of Alexandris (Hypatia's father) produces commented version of *Elements*; this became the standard version. (A copy of *Elements* made in Byzantium in 850 used an older version, but was not rediscovered until 1808.)
9th century	Ibn Matar and al-Hunayn translate *Elements* into Arabic in Baghdad: Ibn Qurrah and al-Tūsī revise the translation (now known as the Ishaq-Thabit version)
1120	Adelard of Bath translates Arabic copy (obtained in Spain) into Latin.
	Gerard of Cremona translates al-Hunayn's version into Latin in Toledo, Spain
1482	Latin print version based on previous translations from Arabic is published in Venice

fragment of an original papyrus (from a copy written in 75-125 CE)

a page from al-Hunayn's Arabic translation

Elements followed a circuitous geographical path around the Mediterranean, changing language and format

Extant copies of Elements *identify a route for the transmission of Euclid's work from Greece to Western Europe that took almost 2000 years.*

Compendius Book on Calculation by Completion and Balancing) extended Greek mathematics by introducing algebra around 830 CE.

The Hindu system became the forerunner of modern numerals after it entered Europe—it was first mentioned in the Codex Vigilanus of 976 CE—and became known as Arabic numerals. They became generally used for arithmetic in the West as the result of a book, *Liber Abaci* (Book of Calculation), written by Leonardo of Pisa, now known as Fibonacci, in 1202. "I was introduced to a wonderful teaching that used the nine figures of the Indias. With the sign 0, any number whatsoever can be written," he said.[37] The book itself became obscure, but was the basis for many simpler descriptions that founded European arithmetic.[38] This was essential for the development of science. For computing (in the electronic sense) to become successful, the binary (base 2) system was necessary, introduced in 1948.

The etymology of the terms for various parts of mathematics shows an amalgam from various cultures. Arithmetic comes from the Greek *arithmos* (number). Geometry comes from the Latin *geometria*, which may have come from the Greek *metria* (measurement). Algebra comes from the Arabic *al-jabra* (reunion of broken parts). Al-Kwārizmī's book on *al-Jabr* was translated into Latin in the 12th century. As *Algebra,* it was a major textbook in European universities until the 16th century. 'Algorithm' is a transmogrification of al-Kwārizmī's name.[39]

∞ ∞ ∞

The philosophy and mathematics of ancient Greece remained unknown in Western Europe until translations from Arabic into Latin began in Spain. (Direct translations from Greek into Latin did not happen until later.) After the Moors lost Toledo to the Christian Reconquista in 1085, it became a center where many works were translated. Gerard of Cremona (1114–1187), who worked in Toledo, remains the best-known translator.[40]

Translations included both the Arabic versions of Greek works and the achievements of the Golden Age of the Abbasid Caliphate. Many of the works became standard textbooks in Europe, especially in medicine.[41] The same route was responsible for the transmission of alchemy from Alexandria to the West. The major players were Jābir ibn-Hayyān (722–815 CE), also known as Geber, who worked on transmutation, and Rhazes and Avicenna (who worked on pharmacology). I discuss their roles in Chapter 7.

The route for recovering ancient knowledge in the West was indirect and protracted.[42] Greek philosophy prior to the 4th century CE came from Alexandria. Indian mathematics from the 5th–7th centuries came from Sindh. They were translated from Greek and Sanskrit into Arabic from the 8th century on in Baghdad. When Arabic texts were translated in Spain after the 11th century, the process was sometimes convoluted by translation into Spanish that was then translated into Latin. Then the translations made their way to the civilized parts of Europe. The lingua franca changed from Greek and Sanskrit to Arabic, then to Latin, and then (much later) to English.

Intellectual activity was at its peak in the East while Europe was in the throes of the Dark Ages. It then declined in the East as it revived in the West. The baton passed to Western Europe, with the fall of Constantinople to the Turks in 1453 precipitating a major movement of scholars and information directly from the East to the West.

NOTES AND REFERENCES

1. Ibn al-Haytham, in *Doubts Concerning Ptolemy (Al-Shukūk 'alā Batlamyūs)*, published 1025–1028, Reproduced in *Ibn al-Haitham*, ed. A. Sabra and N. Shehaby, Dar al-Kutub, Cairo, 1971.
2. M. E. Falagas, E. A. Zarkadoulia and G. Samonis, Arab science in the golden age (750–1258 CE) and today, *FASEB J.*, 2006, **20**, 1581–1586.
3. S. F. Starr, *Lost Enlightenment: Central Asia's Golden Age From the Arab Conquest to Tamerlane*, Princeton University Press, Princeton, 2013.
4. The House of Wisdom has also been ascribed other roles, and variously described as an academy, university, or research center, and as a center for translation, but may actually have been the library of the palace. It was not necessarily the center for translations. See E. Ihsanoglu, *The Abbasid House Of Wisdom. Between Myth and Reality*, Routledge, New York, 2022.
5. D. C. Lindberg, *The Beginnings Of Western Science: The European Scientific Tradition In Philosophical, Religious, and Institutional Context, Prehistory To A.D. 1450*, University of Chicago Press, Chicago, 2007, p. 172.
6. Although Byzantium and the Abbasid Caliphate were often at war.
7. T. E. Huff, *The Rise of Early Modern Science: Islam, China and the West*, Cambridge University Press, Cambridge, 2nd edn, 2003, p. 53.
8. T. E. Huff, *The Rise of Early Modern Science: Islam, China and the West*, Cambridge University Press, Cambridge, 2nd edn, 2003, pp. 55–64.

9. S. F. Starr, *The Genius of Their Age: Ibn Sina, Biruni, and the Lost Enlightenment*, Oxford University Press, Oxford, 2023.

10. G. Saliba, *Islamic Science and the Making of the European Renaissance*, The MIT Press, Cambridge, MA, 2007, pp. 97–108.

11. In his book *Al-Tadhkirah fi'ilm al-hay'ah* (a memoir on the science of astronomy), published in 1260. For a translation see F. J. Ragep, *Nasir al-Din al-Tusi's Memoir on Astronomy*, Springer Verlag, New York, 1993, vol. 1.

12. G. Saliba, *Islamic Science and the Making of the European Renaissance*, The MIT Press, Cambridge, MA, 2007, pp. 109–111.

13. T. E. Huff, *The Rise of Early Modern Science: Islam, China and the West*, Cambridge University Press, Cambridge, 2nd edn, 2003, p. 215; G. Saliba, *Islamic Science and the Making of the European Renaissance*, The MIT Press, Cambridge, MA, 2007, pp. 196–226.

14. T. E. Huff, *The Rise of Early Modern Science: Islam, China and the West*, Cambridge University Press, Cambridge, 2nd edn, 2003, pp. 54–58; F. J. Ragep, Copernicus and His Islamic Predecessors: Some Historical Remarks, *Hist Sci.*, 2007, **45**, DOI: 10.1177/007327530704500103.

15. J. Lyons, *The House of Wisdom: How the Arabs Transformed Western Civilization*, Bloomsbury Press, New York, 2009, pp. 87–99.

16. A. I. Sabra, *The Optics of Ibn al-Haytham, Books I–III, On Direct Vision*, The Warburg Institute, London, 1989.

17. D. C. Lindberg, *The Beginnings Of Western Science: The European Scientific Tradition In Philosophical, Religious, and Institutional Context, Prehistory To A.D. 1450*, University of Chicago Press, Chicago, 2007, p. 184.

18. R. Descartes, *Discours de la Méthode Pour bien conduire sa raison, et chercher la vérité dans les sciences*, 1637.

19. A. H. M. Al-Ghazālī (c. 1091) *The Incoherence of The Philosophers* (Tahāfut al-Falāsifah): A Parallel English-Arabic Text. Trans. Marmura, M E, Brigham Young University Press, Utah, 2002.

20. J. Lent, *The Patterning Instinct: A Cultural History of Humanity's Search for Meaning*, Prometheus Books, Amherst, New York, 2017, p. 322.

21. G. Saliba, *Islamic Science and the Making of the European Renaissance*, The MIT Press, Cambridge, MA, 2007, pp. 244–247.

22. T. E. Huff, *The Rise of Early Modern Science: Islam, China and the West*, Cambridge University Press, Cambridge, 2nd edn, 2003, pp. 231–232.

23. I. B. Cohen, Revolution In Science, Belknap Press, Cambridge, MA, 1987, pp. 64–72.

24. H. Ofek, Why the Arabic World Turned Away from Science, *The New Atlantis*, 2011, **30**, 3–23.

25. T. E. Huff, *Intellectual Curiosity and the Scientific Revolution. A Global Perspective*, Cambridge University Press, Cambridge, 2011.

26. J. Needham, *Science and Civilisation in China, Volume 3: Mathematics and The Sciences Of The Heavens and The Earth*, Cambridge University Press, New York, 1959, p. 49.

27. J. Needham, *Science and Civilisation in China, Volume 4: Physics and Physical Technology, Part 2, Mechanical Engineering*, Cambridge University Press, New York, 1965, p. 446.

28. T. E. Huff, *Intellectual Curiosity and the Scientific Revolution. A Global Perspective*, Cambridge University Press, Cambridge, 2011, pp. 82, 89–91, 95–106.

29. T. E. Huff, *Intellectual Curiosity and the Scientific Revolution. A Global Perspective*, Cambridge University Press, Cambridge, 2011, p. 108.

30. N. Sivin, Why the Scientific Revolution Did Not Take Place in China – or Didn't It?, *Chin. Sci.*, 1982, **5**, 45–66, p. 48.

31. The failure of modern science to develop in China is sometimes called "Needham's grand question." See H. F. Cohen, *The Scientific Revolution: A Historiographical Inquiry*, University of Chicago Press, Chicago, 1994, pp. xv, 16.

32. Papermaking was invented in China around 105 CE, and made its way to the West via the Silk Road, reaching India in 645 CE, Baghdad in 751 CE, and Samarkand in 793 CE. It replaced the use of papyrus by the 10th century CE. It reached the West in Spain by 1150 CE. Paper was used for woodblock printing in China by 600 CE, again almost a millennium before printing started in the West. See T. E. Huff, *The Rise of Early Modern Science: Islam, China and the West*, Cambridge University Press, Cambridge, 2nd edn, 2003, p. 74.

33. *Brāhmasphuasiddhānta* (Correctly Established Doctrine of Brahma) was written in verse with 25 chapters and 1008 stanzas. It introduced the concept of zero, rules for handling positive and negative numbers, methods for solving quadratic equations, and how to calculate square roots, among other formulae.

34. W. Dalrymple, *The Golden Road: How Ancient India Transformed the World*, Bloomsbury Publishing, London, 2024, ch. 9.

35. C. B. Boyer and U. C. Merzbach, *A History of Mathematics*, John Wiley & Sons, New York, 1991, p. 206

36. Quoted in J. Lyons, *The House of Wisdom: How the Arabs Transformed Western Civilization*, Bloomsbury Press, New York, 2009, p. 69.

37. Quoted in J. Lyons, *The House of Wisdom: How the Arabs Transformed Western Civilization*, Bloomsbury Press, New York, 2009, p. 170.

38. K. J. Devlin, *The Man of Numbers: Fibonacci's Arithmetic Revolution*, Bloomsbury, London, 2011.

39. The Latin title of al-Khwarizmi's book *Concerning the Hindu Art of Reckoning* was *Algoritmi de Numero Indorum*.

40. Gerard of Cremona is thought to have translated more than 70 works, including the *Almagest*, Aristotle's *Physics*, Ibn al-Haytham's version of Euclid's *Elements*, works by Ibn Sina, and many others.

41. Books from Arabia that became standard textbooks in late medieval Europe include al-Khwarizmi's *Kitab al-Jabr* (Algebra), published 830, translated 1145; Rhazes' *Kitab al-Hawi* (The Comprehensive Book on Medicine), published 1094, translated 1279; al-Zahrawi's *Kitab-al-Tasrif* (30 volumes) (The Method of Medicine), published 1000, translated 1184; Avicenna's *Kitab al-Qanun fi al-tibb* (5 volumes) (Canon of Medicine), published 1025, translated *c.* 1184.

42. V. Moller, *The Map of Knowledge: A Thousand-year History of how Classical Ideas Were Lost and Found*, Doubleday, New York, 2019.

CHAPTER 4

Copernicus' Revolution: 1543–1609

"IT'S THE DUTY OF AN ASTRONOMER TO COMPOSE THE history of the celestial motions through careful and expert study," said the introduction to Copernicus' great book, *De Revolutionibus*.[1] Regarded in retrospect as a great astronomer who started the revision of the Ptolemaic system, Copernicus was not so well appreciated in his own lifetime. Nominally holding a position in the Church (administrative rather than as a member of the clergy), he began to work on astronomy on his own. He developed his theory over three decades, initially writing it up in a private commentary, and finally publishing it only at his death. The theory did not make a great impact among astronomers of the period. Viewed in its era more as a matter of technical adjustment, Copernicus' move to place the Sun at the center of the universe was only recognized as an epochal turning point in the following century. In the same year that Copernicus published on astronomy, 1543, the great physician Andreas Vesalius published his work in anatomy, *De Humani Corporis Fabrica Libri Septem.* Also that year, Petrus Ramus, a philosopher in Paris, published a book, *Animadversions on Aristotle.* The publications in 1543 set in train the decline of Aristotle's influence. In the following century, the Scientific Revolution made major advances in medicine and biology as well as in astronomy and physics.

The Frontiers of Science
By Benjamin Lewin
© Benjamin Lewin 2026
Published by the Royal Society of Chemistry, www.rsc.org

Timeline of Critical Events Leading to the Scientific Revolution

1440

Gutenberg invents printing press (1448)

1460

1480

Columbus travels to New World (1492)

1500 — Leonardo da Vinci investigates motion & gravity (1500)

Martin Luther begins Protestant Reformation (1517)

1520

1540 — Copernicus publishes *De Revolutionibus* (1543)
Vesailius publishes *De Humani* (1543)
Ramos publishes *Animadversions on Aristotle* (1543)

1560

Mercator introduces map projection in cartography (1569)
Brahe observes supernova (1572)

Brahe observes the Great Comet (1577)

1580

Galileo analyzes falling bodies (1589)

1600 — Kepler publishes *Mysterium Cosmographicum* (1596)
Brahe publishes *Astronomiae* star catalog (1598)
Kepler publishes *Astronomia Nova* (1609)
Galileo publishes *Sidereus Nuncius* (1610)

1620

A Dialog between Ptolemy and Copernicus

Claudius Ptolemy (90–168 CE) was the most important astronomer of ancient times. Living in Alexandria, he wrote the *Almagest,* which developed the geocentric model placing the Earth at the center of the universe.

Nicolaus Copernicus (1473–1543) was a Polish astronomer who developed the heliocentric model placing the Sun at the center of the universe. His book, *De Revolutionibus*, was published only at his death.

Ptolemy: Greetings, Nicolaus Copernicus. It's not often I have the chance to converse with a fellow astronomer. I understand you have proposed a radical shift in our understanding of the cosmos. Your model with the Helios (Sun) at the center has stirred quite a commotion.

Copernicus: Greetings, Claudius Ptolemy. Indeed, I have posited that the Earth revolves around the Sun, challenging the geocentric view that has held sway for centuries. The heliocentric model better explains the observed motions of the planets.

Ptolemy: A heliocentric model? The audacity! For centuries, we have relied on the Ptolemaic system, where the Earth stands at the center of the universe. The celestial bodies move in epicycles to account for their motions.

Copernicus: Claudius, the Ptolemaic system seems increasingly cumbersome as we gather more accurate observations. My model simplifies our understanding, by placing the Sun at the center, and the planets, including Earth, in orbits around it.

Ptolemy: Simplicity is not always the hallmark of truth. Your heliocentric model might explain some phenomena, but does it not fly in the face of our senses and the teachings that have guided astronomers for centuries? My model has stood the test of time for a thousand years. I adjusted it carefully to fit the ancient data of Hipparchus.

Copernicus: I do not dismiss the wisdom of the ancients, Ptolemy. However, my heliocentric model offers a more elegant explanation for the observed retrograde motion of planets.

Ptolemy: Mathematical beauty, you say? The Ptolemaic system, despite its complexity, accurately predicts planetary positions. I question whether astronomers will readily abandon the geocentric view that has served us so well for so long. The heliocentric model introduces unsettling notions. In any case, it requires almost as many additional features of epicycles, equants, and deferents to make predictions as accurate as mine.

Copernicus: Change is never easy, Ptolemy. Yet, I believe that as we refine our instruments and gather more evidence, the heliocentric model will become an accepted truth. It's a shift that aligns with the natural order of the cosmos. Geocentricity is to heliocentricity as papyrus is to the printed book. I rest my case.

COPERNICUS' REVOLUTION

And so we must first know that the movers of the heavens are substances separate from matter, namely Intelligences, which the common people call Angels, Dante, 1306.

I can easily conceive, most Holy Father, that as soon as some people learn that in this book which I have written concerning the revolutions of the heavenly bodies, I ascribe certain motions to the Earth, they will cry out at once that I and my theory should be rejected,[2] Nicolaus Copernicus, 1543.

The remarkable feature of the Copernican Revolution is that it was not a revolution. It did not identify flaws in the data or show that the existing Ptolemaic model did not conform to the data. Rather it was an alternative model whose advantages were extolled as aesthetic or pragmatic. In fact, as a model it was only slightly less complicated than current versions of the Ptolemaic system. Its predictions were only slightly better.[3] It was presented in a work, *De Revolutionibus,* written in such mathematical detail as to render it obscure to all but astronomers.[4] ("Mathematics are for Mathematicians," Copernicus said.) In retrospect, *De Revolutionibus* may mark the transition from antiquity to modernity, but its importance did not become clear immediately. It was the next century before the full effect of moving the center of the universe from the Earth to the Sun was realized.

The Earth was flat before it was round. All civilizations before the Intellectual Revolution in Greece viewed the world as flat.[5] Anaximander (610–547 BCE), a contemporary of Thales at Miletus, proposed the revolutionary idea that the Earth is free-floating.[6] He thought that celestial bodies make circles around it by passing underneath after they disappear over the horizon.[7] This led to a vigorous debate about the shape of the Earth.[8]

The question was settled by Aristotle's argument that the Earth made a circular shadow during an eclipse.[9] This left the question of why the Earth did not fall if it was free-floating. (The ancient world was much concerned with the fear that the heavens might fall.) Aristotle's answer (simplified) was that the Earth is at the center of the cosmos.[10]

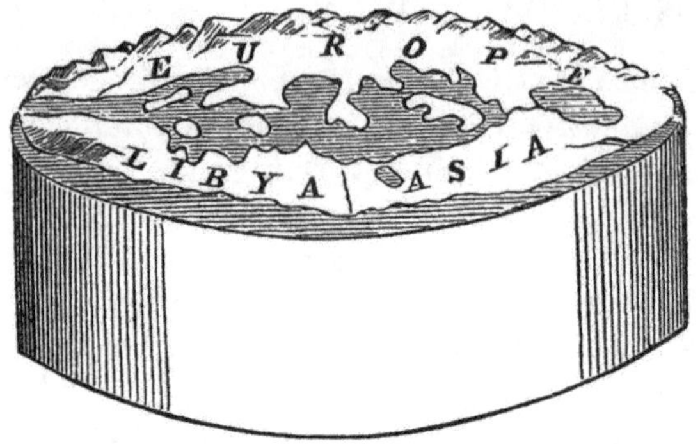

A reconstruction of Anaximander's view of the world, based on a map drawn later by Hecataeus of Miletus (c. 550–476 BCE). The original map was probably on a concave metal surface.

Astronomy has to explain celestial movements. The Sun rises in the East and sets in the West. At night the Moon moves across the sky towards the West. A fixed pattern of stars moves slowly across the sky in the same direction. Five of the 'stars' are not in fact fixed but show movement relative to the fixed stars. These are the planets, which move in the same direction, but occasionally reverse their direction before resuming their trudge to the West.

The obvious explanation was that all the celestial bodies revolve around the Earth. The Sun, Moon, and planets have individual motions, whereas the stars move collectively. As long ago as Babylon, and possibly even earlier, astronomers were able to predict the paths of the celestial bodies.[11] Changing interpretations since then of what we see in the sky give as a clear a view as anything of the transition from ancient philosophy to modern science.

We owe to Aristotle (384–322 BCE), in his work *On the Heavens* (350 BCE), the worldview that prevailed for the next thousand years. Aristotle's view of the universe held that:

- The Earth has existed since eternity.
- It's a stationary sphere, at the center of the universe. All other bodies revolve around it. ("Observations of the stars show not only that the Earth is spherical but that it is of no great size.") It must be stationary because objects fall directly to the ground.

- The sublunar sphere (below the Moon) is made of four basic elements: earth, air, fire, and water.
- Nature abhors a vacuum. ("There is no void existing separately, as some maintain.") Celestial bodies (beyond the Moon) are embedded in concentric spheres composed of ether (also known as quintessence). The celestial realm is unchanging and has a finite size.

The astronomer Eudoxus (*c.* 408–355 BCE), who may have been a student at Plato's academy, introduced the first formal scheme of circular orbits to explain planetary motions. This was probably only a mathematical construction. Aristotle gave the model a physical meaning by viewing the orbits as hard shells,[12] perhaps because he could not conceive how objects could whirl unsupported in space.

Aristotle viewed the Earth as a tiny sphere suspended at the center of a much large sphere.[13] The stars are fixed on the surface of the outer sphere. The movement of the stars across the sky is explained by the rotation of the outer sphere relative to the inner sphere. The space between the inner and outer spheres is not empty, but is filled with crystalline shells. The planets are embedded in the shells. (The Moon and Sun are not distinguished from the planets.)

This geocentric model was widely accepted, although Aristarchus (310–230 BCE) proposed a heliocentric view with the Sun at the center. (Helios is Greek for Sun.) Archimedes (287–212 BCE) was aware of this. "The 'universe' is the name given by most astronomers to the sphere the center of which is the center of the earth... Aristarchus has brought out a book [hypothesizing] that the universe is many times greater than the 'universe' just mentioned. His hypotheses are that the fixed stars and the sun remain unmoved, that the earth revolves about the sun on the circumference of a circle." Aristarchus also proposed that the stars were suns that were very far away.[14]

With no decisive observations to distinguish between models, the geocentric model dominated. The inability to test between two competing hypotheses marks this more as natural philosophy, depending on rational analysis and argument, than 'science,' depending on objective data. At all events, for the Greek philosophers the geocentric model was a conclusion, not the act of faith it became for the Catholic Church a millennium later.

Hipparchus of Rhodes (*c.* 190–120 BCE) was the greatest astronomer of ancient time. He's famous for discovering precession of the equinoxes, a shift in the position of the stars that results from a very slow change in the Earth's angle of rotation.[15] He wrote at least 14 books (only one has survived) and made what was probably the first complete star catalog (with 850 stars). The catalog is mythic in the sense that references to it abound (Pliny referred to it) but the map itself has been lost.

The map was based both on original observations and on data from Babylon—Hipparchus discovered precession by comparing his observed coordinates with those from Babylon (about 400 years earlier). Astronomical data that may have been part of the map were uncovered on a palimpsest found in 2022 in St. Catherine's Monastery in Sinai.[16] After Hipparchus, astronomy in the ancient world petered out. There appears to have been no further advance until the work of Claudius Ptolemy, *c.* 150 CE, in Alexandria.[17]

Because the planets do not appear simply to revolve around the Earth, but sometimes move backward (called retrograde motion),[18] the astronomer Apollonius had introduced epicycles around

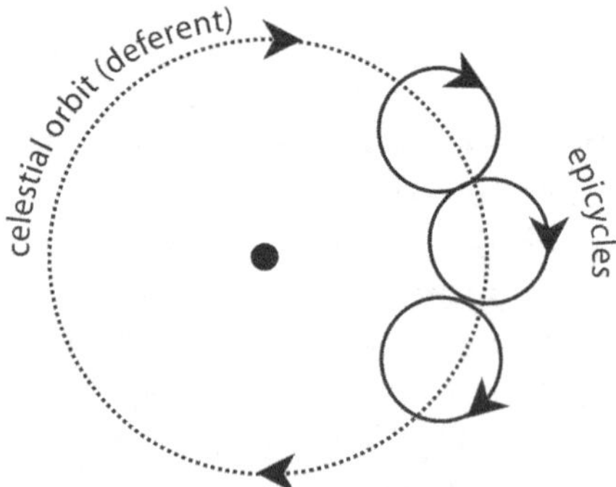

Epicycles are circular sub orbits around the celestial orbit (which is called the deferent). To make the model work, Ptolemy offset the center of the deferent from Earth itself. The center is called the equant. One of the weaknesses of the model is that this is a mathematical conceit, producing predictable results, but not related to the actual basis of motion.

A reconstruction of the Antikythera mechanism shows dials on the front for representing positions of planetary objects, driven by a complex geared mechanism.

Reproduced with permission, Copyright © 2021, Images First Ltd.

200 BCE. These are small circular orbits centered on the celestial orbit, the path of one celestial body around another. Epicycles were used to make calculations of the movements of the planets.

Lack of trigonometry made astronomical calculations difficult, but the Greeks had an apparatus that could make calculations based on a geared mechanism. The Antikythera mechanism was discovered in 1901 as fragments in a shipwreck off the coast of the Greek island Antikythera.[19] Scanning the fragments suggests that it had up to 67 meshing gears. It was a hand-powered device in which the rotation of the gears was used to follow the movements of the Sun and the Moon (allowing eclipses to be predicted).[20] Missing fragments may have extended it to follow the planets as well.[21]

It's controversial when the machine was calibrated. Opinions range from 205 BCE to the date of the shipwreck in 70–75 BCE. Because Hipparchus was involved in creating astronomical instruments, there are speculations he could have been a designer of the Antikythera mechanism.[22] Nothing so complex would be developed in the West for the next millennium.

∞ ∞ ∞

One of the reasons the ancient astronomical model was so successful was that calculations were simplified because all motions were circular. Ptolemy (*c.* 90–168 CE) extended and improved the existing model by adding refinements. Ptolemaic astronomy remained the dominant paradigm for a thousand years. We know Ptolemy through his works, which were written in ancient Greek, but we know little about him as a person.

Ptolemy (no relationship with the Pharaohs of the same name) developed tables to compute the positions of the planets, and prepared a star catalog. It had been thought he simply copied Hipparchus' data[23]—and he has been accused of fraud for presenting the results as though he had observed them himself.[24] But the most recent evidence from some differences in the co-ordinates uncovered on the Hipparchus palimpsest suggests that Ptolemy also had some original data (or obtained coordinates from other sources).[25]

Whatever the source of the data, analysis shows that "his observations are surprisingly bad while his final parameters are amazingly good."[26] This may be an example, not uncommon in physics, even in the 20th century, of believing that a clash between theory and data should be resolved by 'correcting' the data. Indeed, Ptolemy regarded this as his responsibility (see Chapter 6).

Ptolemy was famous not only for his astronomy but also for a system to map the Earth. Published in his work *Geographike Hyphegesis* (Guide to Drawing the Earth), he introduced latitude and longitude. Ptolemy's procedures for projecting spherical coordinates onto a two-dimensional map were lost with other ancient works and rediscovered in West in the 15th century. From 1475, they were used for constructing maps, and they remain the basis for maps even today.[27] Ptolemy also wrote a treatise on astrology, *Tetrabiblos*. For his book *Optics*, he actually performed some experiments.[28]

It's a tribute to mathematical ingenuity that it was possible to construct models making accurate predictions about planetary movements based on a geocentric universe. Other schemes could have given equally good predictions by *ad hoc* tuning, emphasizing the fact that these were mathematical models rather than representations of reality. (One difference between Aristotle's model and Ptolemy's was that Aristotle was concerned to find a mechanical explanation for the movement of the spheres. Ptolemy was more concerned with the mathematics.[29])

Ptolemy's model, especially as elaborated later in Arabia, was surprisingly accurate, but rather complicated, with epicycles piled upon epicycles. (It's a demonstration of where the intellectual action was that Ptolemy's work on astronomy is known as the *Almagest*. This is not its original Greek name, but a derivation of the Arabic title of a 9th century translation into Arabic.[30]) The *Almagest* consisted of an impressive 13 books, covering movements of the Sun, Moon, and planets, and a catalog of 1000 stars. Latin translations were printed in Venice from 1493 and were widely available in the 16th century. An English translation runs to almost 700 pages.[31]

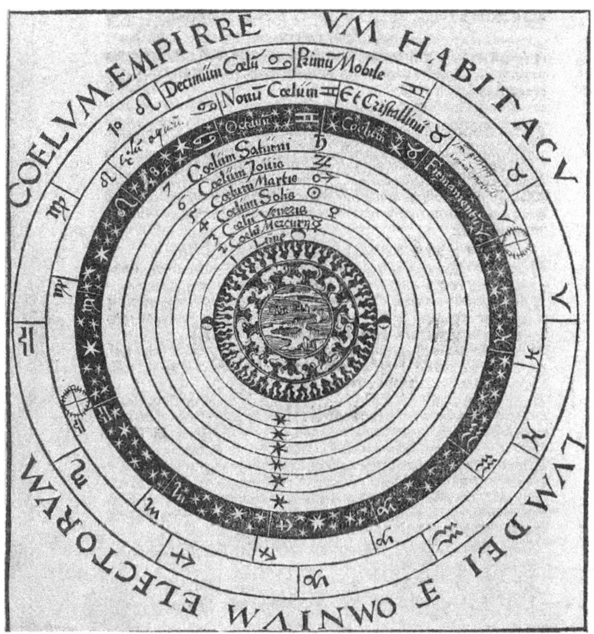

Everything revolves around the Earth. The Ptolemaic view of the universe was pictured in a drawing in Peter Apian's Cosmographicus, *first published in 1524 in Antwerp. The Earth is at the center, surrounded by ten concentric spheres, starting with the Moon, Mercury, Venus, the Sun, Mars, Jupiter, and Saturn. At the exterior is the Primu Mobile, the prime motion, which is responsible for moving all the others.*

The ancient view of the universe is very well summarized by the six hypotheses that start the *Almagest*:

- The heavens move like a sphere.
- The Earth, taken as a whole, is sensibly spherical.
- The Earth is in the middle of the heavens.
- The Earth has the ratio of a point to the heavens.
- The Earth does not have any motion from place to place.
- There are two different primary motions in the heavens.

Ptolemy's model came under increasing pressure over time as discrepancies with the geocentric assumption accumulated. The model was adjusted by introducing more complications (to borrow a term from Swiss watchmakers). Finally, it became apparent that the model needed to be replaced. But this took a thousand years.

∞ ∞ ∞

Aristotle's concept of the universe was more than a paradigm: it was an all-encompassing worldview. It's impressive in its coherence, and difficult to refute if you accept its starting assumptions. However, anyone who wants to challenge the reality of the Dark Ages—at least as a period devoid of intellectual development in Europe—has to explain why those assumptions were never questioned. How did the same (erroneous) theory stand unchallenged for a thousand years? The word of Aristotle, who was often referred to simply as "The Philosopher," was final.

According to Aristotle, the world was a sphere with habitable zones in both the northern and southern hemispheres that excluded the poles (too cold) and the equator (too hot).[32] Ptolemy argued that there must be a landmass in the southern hemisphere to balance the known area of land in the northern hemisphere. But nothing was known of that land until the voyages of Columbus (from 1492) and Vespucci (from 1499). The discovery was a revelation.

Vespucci is supposed to have said, "Pliny did not touch upon a thousandth part of the animals and birds that exist in this region."[33] As exploration (not to say exploitation) continued, vast numbers of new species were identified.[34] The conceptual importance of these discoveries was the direct demonstration that the Ancients had not, in fact, known all knowledge. This no doubt contributed to subsequent openness in questioning conventional wisdom.[35]

The first practical contribution from new knowledge actually came from mathematics. Gerard Mercator (1512–1594) developed projections of the spherical world on to a two-dimensional surface in 1569. This was a major advance in cartography with enormous consequences for navigation. Curiously, Mercator himself, who was a Catholic but seems to have been a dissenter to some extent (as he was imprisoned by the Church for six months), moved from his origins only to a more tolerant city not very far away in Belgium. He never traveled much.[36]

Mercator made his first corrections to Ptolemy's maps with a map of Europe in 1554. He published *Chronologia*, a world history, in 1569—banned by the Church because it included Luther. Later the same year, he published his world map in the form of 18 separate sheets.[37] His objective was to produce a map more suitable for navigation. The project required mastery of many aspects of mathematics.

∞ ∞ ∞

Nicolaus Copernicus (1473–1543) was not a revolutionary.[38] Indeed, he was part of the established Church. He was the fourth child in a prosperous family in Poland (then part of Prussia). He studied at the University of Kraków, which was strong in mathematical astronomy, obtained a position in the Church (although not as a priest), and studied canon law at the University of Bologna. He began his studies in astronomy at Bologna, and continued them at Padua. In 1503, he returned to Poland to assist his uncle, the Bishop of Warmia. When his uncle died, around 1512, he moved to his own house, and it was then he made most of his observations in astronomy.

His 6-volume work *De Revolutionibus Orbium Coelestium*[39] (On the Revolutions of the Heavenly Spheres) was published in 1543—legend has it that the first copy was presented to him on his deathbed. It was preceded by a preface addressed to Pope Paul III that expressed some diffidence. "That I allow the publication of these my studies may surprise your Holiness... I came to dare to conceive such motion of the Earth, contrary to the received opinion of the Mathematicians and indeed contrary to the impression of the senses... by nothing less than the knowledge that the Mathematicians are inconsistent in these investigations. [They] are so unsure of the movements of the Sun and Moon that they cannot even explain or observe the constant length of the seasonal year." Hardly the words of a revolutionary!

In spite of this caution, Copernicus was well aware of the magnitude of the change he was proposing. "It's the common view among authorities that the earth is at rest, and that the contrary view is inconceivable or ridiculous. But if we pay more attention to the matter, it seems that the question is not resolved," he said.[40] Copernicus had developed his ideas at least twenty years earlier. A brief written account, a 20-page pamphlet now called the *Commentariolus*, was circulating by 1514, with all the main conclusions, but was never published, although it was common knowledge. It was lost until copies were discovered in 1878.[41]

Commentariolus was based on a summary of Ptolemy's *Almagest* written by Copernicus' predecessor, Regiomontanus. It describes the basic theory that was expanded in *De Revolutionibus*. Regiomontanus may have known of the work of

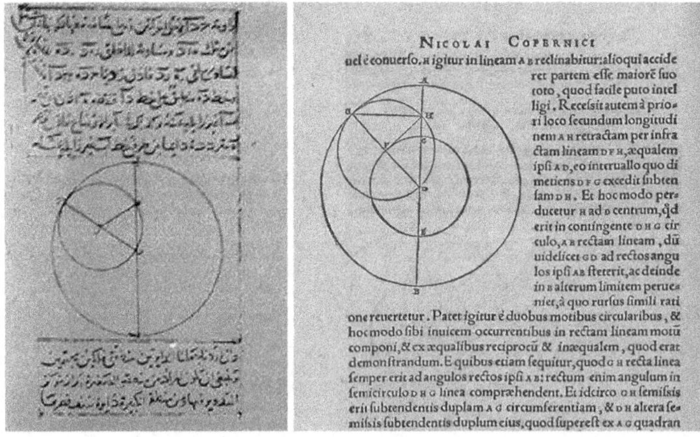

A comparison of Tūsī's book Al-Tadhkirah *of 1260 with* De Revolutionibus *in 1543 shows similarities of analysis. The Tūsī couple worked out the geometry of a circle revolving around the inner circumference of a circle twice its diameter (left). Copernicus used a similar scheme (right) using the same letters (translated from Arabic) as Tūsī.*

the Marāgha School in Arabia. There are parallels between Copernicus' analysis and improvements that Arabian astronomers made to Ptolemy's theory. This has created a controversy as to whether Copernicus might have known of their results. However, the leap to heliocentricity appears to have been all his own.[42]

The key elements in Copernican theory were that:

- The Earth moves in an annual cycle around the Sun.
- The Earth rotates on its own axis.
- The Moon continues to revolve around the Earth.
- The planets occupy a series of orbits at successive distances around the Sun, based on circles and epicycles.
- The universe is finite. The stars remain a fixed distance far away (much farther than the distance from Earth to Sun).

Perhaps because *De Revolutionibus* was obscure to laymen, the Church raised no immediate objection to it. Objections came from astronomers. Most preferred to stay with Ptolemaic models. In fact, all the leading astronomers thought Copernicus was wrong in principle, although in practice his astronomical tables became the standard reference.[43] In the absence of compelling data, it remained a matter of choice as to which theoretical model to believe. It was the best part of a century until opinion changed.

From a perspective in which mathematical calculations do not necessarily represent reality, models did not necessarily clash with religion. This was a theme that would resonate more sharply later in Galileo's clash with the Church. Georg Joachim Rheticus (1514–1574), who was Copernicus' sole pupil, and was instrumental in persuading him to publish, wrote an introduction to *De Revolutionibus* in 1540 called *Narratio Prima* (first account). He quoted the 12th century Islamic philosopher Ibn Rushd (1126–1198), who lived in Cordova in Muslim Spain, known in the West as Averroës, as saying, "Ptolemaic astronomy is nothing, so far as existence is concerned; but it is convenient for computing the nonexistent."[44]

Aside from *Commentariolus*, which had limited circulation, Rheticus' book was the first account of Copernicus' theory. An anonymous preface to *De Revolutionibus* (written by Andreas Osiander, who had taken over from Rheticus the job of getting

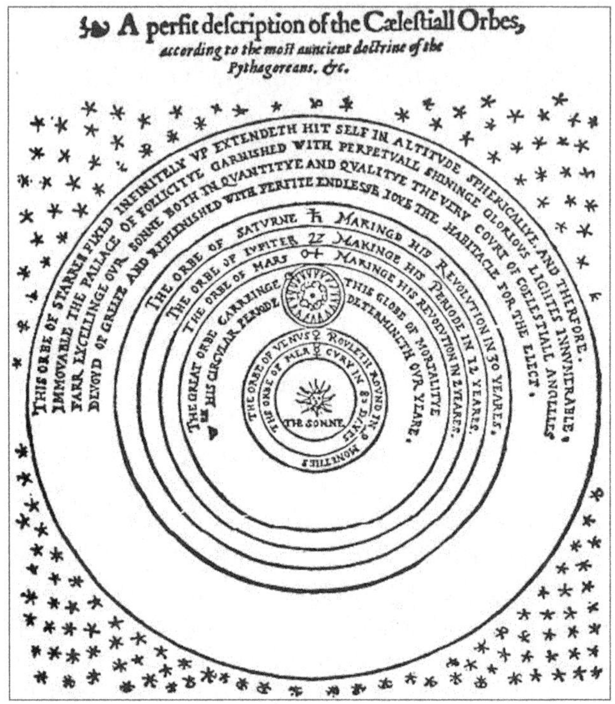

The Copernican universe placed the Sun at the center, as depicted in the Perfect Description of the Celestial Orbs, *published by Thomas Digges in 1576. Digges extended the Copernican model by arguing that the stars are not on the surface of the celestial sphere, but extend beyond it into infinite space.*

De Revolutionibus into print) tried to ward off religious criticism by making the same point as Averroës. "It's not necessary that these hypotheses should be true, or even probably; but it is enough if they provide a calculus which fits the observations."[45]

∞ ∞ ∞

One of Copernicus' problems was the existence of errors in the data he was using. This was partly due to intrinsic difficulties in achieving accurate enough measurements, and partly due to the accumulation of errors in the historic data over time. It was not until the second half of the 16th century that Tycho Brahe (1546–1601) made more precise measurements, making it possible to judge competing models. His construction of an observatory is one mark for the start of the Scientific Revolution in establishing a specialized facility in which to make observations.

Tycho Brahe came from a wealthy family in the Danish nobility. The oldest of twelve children, he was actually brought up by his uncle (an admiral in the Danish navy). After a study tour in Europe, he returned home and built his first observatory. Then under the aegis of King Frederick II, he constructed an observatory in 1576 on the island of Hven (not far from Copenhagen).[46] The observatory extended underground for increased stability. It also included a large alchemical laboratory. (It was a sizeable enterprise, employing around 100 people.) Brahe worked there until 1597, when he was forced into exile by political difficulties, ending up in Prague (at the court of Rudolf II). He was an argumentative character, known for wearing a silver prosthesis to replace the end of his nose, which had been cut off in a youthful duel.

The impetus for constructing the new observatory was Brahe's dissatisfaction with existing data. He had noticed previously, for example, that predictions for an eclipse were one day off. His data achieved a new degree of precision. However, Brahe accepted neither Ptolemaic nor Copernican models. He introduced his own (Tychonic or geoheliocentric) system. The Earth was based at the center, the Moon and Sun moved around the Earth, but other planets moved in epicycles around the Sun. In pragmatic terms, it was a good compromise with high predictive value, but the Earth had not moved yet.

Stellar parallax provides a means to distinguish between the geocentric and heliocentric models. Parallax describes the situation that when an object is observed from two different positions, it appears to be at different angles. Parallax makes nearer objects appear to be displaced relative to distant objects, so stars that are nearer to Earth should change their position relative to stars that are farther away.

Using trigonometry, if you know the distance between the locations where the two measurements were made, and the angle at which the object is observed each time, the distance to the target can be determined. In fact, the first measurement of the distance of an astronomical body using parallax was supposedly

The Stjarneborg Observatory, which Tycho Brahe constructed from 1576–1584 at Uranienborg Castle on the island of Hven, was described in a book containing the astronomical correspondence between Brahe and Wilhelm IV, Landgrave of Hesse (who was himself an astronomer). The legend (in free translation) says: Outside stone columns are arranged to the east and the west to show the Ptolemaic rules. Supporting spheres are arranged to reveal the stars. A round stone table has a portable quadrant and can be used for other small instruments.

when Hipparchus observed a solar eclipse from two locations to calculate the distance of the Moon.

One of Brahe's arguments against heliocentricity was the failure to observe stellar parallax. This meant that the Earth is stationary, or that the stars are very far away, in fact at a distance that was unimaginable in a historic context. By contrast, Galileo argued that the stars indeed were too far away for parallax to be observed with the naked eye or even with available telescopes. (In fact, stellar parallax was not observed until the improvement of telescopes in the early 19th century. The first measurement identified a star 10 light years away from Earth.[47]) The difference of opinion really came down to an assumption as to the plausibility of stars being at vast distances from Earth.[48]

It took dramatic events to demolish the view that astronomical objects were organized in spheres surrounding the Earth. First there was a new star: then there was a comet. Tycho Brahe observed a new star in 1572. A supernova (generated by nuclear fusion), it was extremely bright and could be seen even in daylight.[49] Brahe was able to show that it must be "far above the sphere of the Moon in the very heavens."[50] This refuted the concept that the celestial sphere could not change.[51]

The Great Comet of 1577 was visible all across Europe—in fact, it was the brightest object in the sky—and was recorded in detail by Tycho Brahe (among others). It was also observed at the observatory in Istanbul (see Chapter 3). By measuring parallax for the comet, Brahe was able to demonstrate that the comet's path was about three times farther away than the Moon.

In Aristotle's universe, objects such as comets could occur only in the sublunary area, that is, between Earth and the Moon. Because the comet moved through the spheres of the planets (and also because his system required Mars to cross the sphere of the Sun), Brahe abandoned the crystalline spheres. He proposed instead that the Sun, Moon, and planets float freely in space. He published his observations of the comet and his geoheliocentric theory in a book printed at Uranienborg in 1577.[52] His star catalog, with details of 1000 stars, was published in 1598.

The breakthrough to heliocentricity came with Johannes Kepler (1571–1630), who worked as an assistant to Tycho Brahe during Brahe's last year. Coming from a poor family, and weak as a child, Kepler studied at the University of Tübingen, where he acquired a reputation in mathematics and astrology. He then became a teacher of mathematics in Graz (now in Austria), and published a book, *Mysterium Cosmographicum* in 1596. As the result of corresponding with Brahe (as well as others) about his book, he went to work with Brahe in Prague in 1600. When Brahe died, Kepler was appointed as his successor.

Kepler's approach was partly scientific and partly mystical. He discovered that he could fit 'perfect solids' to fill the intervals between the planets. (A perfect solid is a three-dimensional shape where all the faces are identical. Only 5 types of polygon satisfy the condition. Kepler thought this explained why there were only 6 known planets, as one solid of each type could fill each interval.) It was hard for science to break out of the medieval mindset! On the other hand, Kepler then went on to say, "[If] astronomical determination of the orbits...do not confirm the thesis, then all our previous efforts have doubtless been in vain," in a transition to the scientific approach.[53]

Committed from the start to Copernicanism, Kepler believed that pure mathematical analysis could explain the motion of the planets. "Kepler improved Copernicus' mathematical system by applying strict Copernicanism to it," says Thomas Kuhn.[54] Over a decade, he worked out the orbit of Mars, achieving the crucial realization that it is elliptical rather than circular. He published his laws of planetary motion in his book *Astronomia Nova* in 1609. Crystalline orbs were replaced with 'orbits,' describing the movement of the planets in space.[55]

Astronomia Nova was notable for introducing the concept that physics could be brought to bear on astronomy. Kepler said, "I have mingled celestial physics with astronomy."[56] The book was

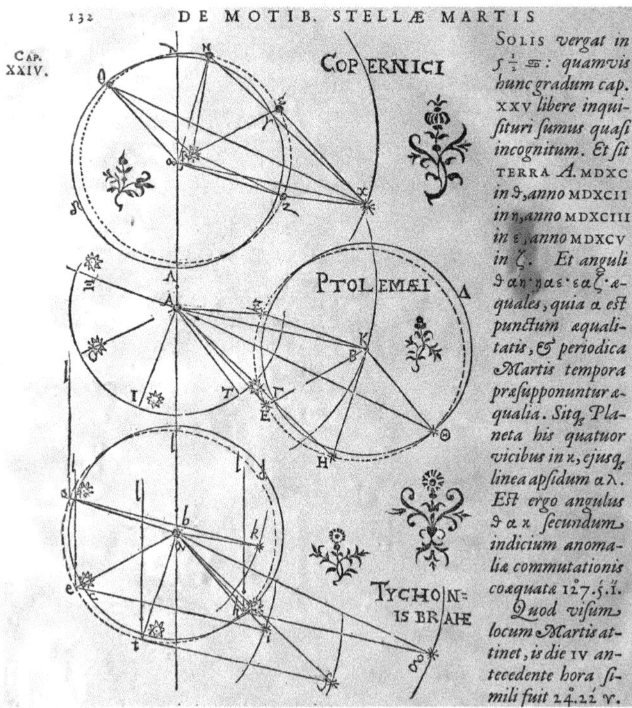

Kepler compares the models of Ptolemy, Copernicus, and Tycho Brahe in the third section of Astronomia Nova.

difficult to follow, because rather than presenting a mathematical explanation of the observations in what is now the style of science, it presented an anecdotal record of Kepler's progress. Kepler was exasperated by Tycho's data. "Tycho's astronomy so shackled me that I nearly went out of my mind," he wrote in a letter in 1601 after his first visit to Brahe.[57] (Kepler also later wrote one of the first science fiction novels, a story about a trip to the Moon, involving a demon and a witch.[58] It drew on events at Uranienborg and on his mother's trial for witchcraft.)

Astronomia Nova was at least a half-ways move towards introducing a scientific method. Kepler's increasingly precise measurements of the positions of the planets and stars culminated in publication of the Rudolphine Tables in 1627. These provided the basis for astronomy for the rest of the century.[59] Based on the tables, Kepler rejected successive models that did not conform well enough to his observations. On the other hand, the elliptical model did not follow inevitably from the data, but required intuition and additional assumptions.[60]

∞ ∞ ∞

Astronomy moved fully into the realm of science with the invention of the telescope. This was an early demonstration that science can be driven as much (if not more) by technology as by theory. Two makers of spectacles independently created telescopes in the Netherlands in 1608. Galileo heard about the invention and then created his own telescope.

"About 10 months ago, a report reached my ears that a certain Fleming had constructed a spyglass by means of which visible objects, though very distant from the eye of the observer, were distinctly seen as if nearby... [This] caused me to apply myself wholeheartedly to investigate means by which I might arrive at the invention of a similar instrument," Galileo said.[61]

Grinding his own lenses, Galileo (1564–1642) produced his first telescope in 1609. In rapid succession, he revised many of the tenets of classical theory. He showed that that the Moon has a distinct surface structure not so different from that of the Earth, and that the planet Venus passes through phases (like the Moon). He also showed that the Milky Way consisted of innumerable separate stars, and that the stars were not so large as they had appeared to the naked eye.

Galileo's first telescope was 32 inches long, magnifying 21×.

A famous exchange about the discovery came when Galileo tried to persuade his friend, the philosopher Cesare Cremonini, to accept the observations of the Moon. Cremonini was a committed Aristotelian, in fact to the point at which the Inquisition persecuted him. (He actually warned Galileo not to leave Padua for the Court at Florence, as this would place him at greater risk from the Catholic authorities.[62]) Invited to view the Moon, Cremonini said, "I do not wish to approve of claims about which I do not have any knowledge, and about things which I have not seen ... and then to observe through those glasses gives me a headache. Enough! I do not want to hear anything more about this."[63]

The phases of Venus provided an argument for heliocentricity. If Venus lay between the Earth and Sun and circles the Earth, it would always present the same appearance. Passing through phases (from crescent to full circle) means that it presents different aspects to Earth depending on its position. It must therefore circle the Sun.[64] (However, it was still possible to argue, as Tycho Brahe did, that Venus orbited the Sun, while the Sun still circled the Earth.[65])

Although Democritus had speculated more than a millennium earlier that the Milky Way consisted of stars, Aristotelian cosmology held that it was the point where the celestial sphere contacted the terrestrial spheres. Classical theory required the Moon to be completely smooth. And most significantly, Galileo's discovery of four moons of Jupiter in 1610 showed that they revolved around the planet in the same way as the Moon around the Earth. The geocentric model was finally dead.

Of course, it did not know that, or to be more precise, the Church did not know that. Until the discovery of the telescope, astronomy had been the preserve of astronomers, but the telescope made it so simple to look beyond Earth that amateurs entered the field in abundance. The breaking point may have been the transition from viewing astronomical theories as mathematical devices that were useful for predicting planetary movements into regarding them as representations of the real world. Once the Earth began to be removed from its position at

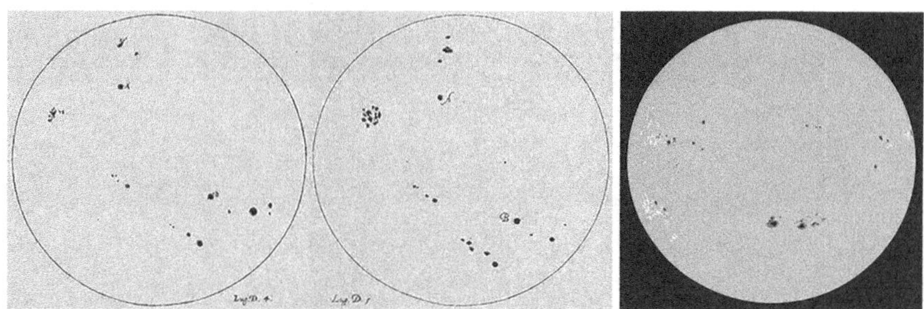

Sunspots observed on July 4, 1613 (left) and July 5 (center) as drawn in Galileo's notebook, showing movement over 24 hours. The accuracy of the scale of his drawing can be seen by comparison with a photograph of sunspots (right), taken by NASA in 2001.

ESA/NASA Solar and Heliospheric Observatory (SOHO).

the center of the universe in popular opinion, the threat to traditional theology became obvious.

One of the first observations Galileo made with his telescope was that sunspots were moving across the surface of the Sun. This challenged the belief that the Sun was "most pure and most lucid." The heavens were no longer unchanging.[66] Galileo was able to publish his observations in 1610 (an *imprimatur* from the Church was required).[67] By 1616, however, the Church declared that the Copernican system was heresy. The works of Copernicus and Kepler were banned. Galileo was warned not to "hold, teach, or defend" the Copernican system.

Reconciling belief in religion with science, Galileo had said previously, in 1615, "I do not feel obliged to believe that that same God who has endowed us with senses, reason, and intellect has intended us to forgo their use," as good a rationale of the situation as might be stated in the period.[68]

In spite of the ban on heliocentric works, Galileo was able to publish his *Dialogo*, now known (in English) as *Dialogue Concerning the Two Chief World Systems,* in Florence in 1632. The original title was *Dialogue on the Ebb and Flow of the Sea,* but had to be simplified because the Inquisition refused to approve the full title. They thought this would constitute acceptance of the theory that tides were caused by the motion of the Earth. Ironically, Galileo was wrong about this anyway.[69]

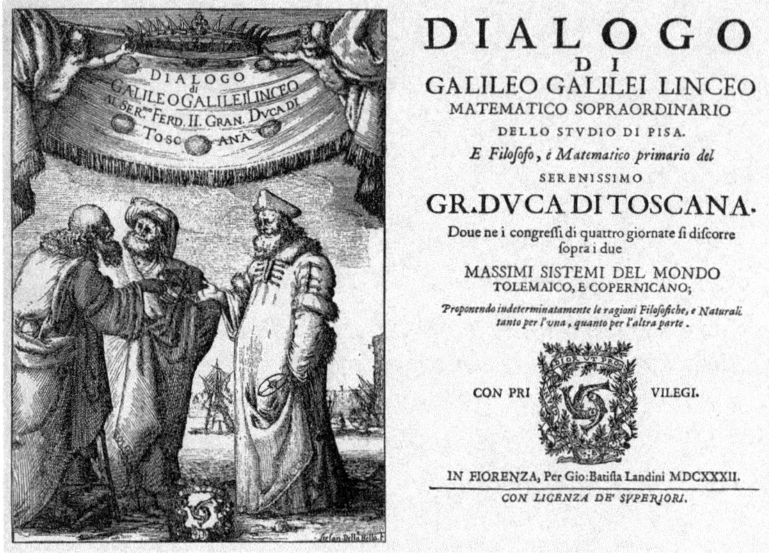

Frontispiece and title page of Galileo's Dialogue, *1632. The three figures are based on Aristotle, Ptolemy (with a turban to indicate his oriental origins), and Copernicus (dressed as a Polish priest).*

The book took the form of a dialogue between characters representing the Copernican and Ptolemaic systems, in which Copernicus eventually trumps Ptolemy. Yet only a year later, in Rome, Galileo was found to be "vehemently suspect of heresy," and the book (and all other works on Copernicus) was banned.[70]

No doubt this is the most famous clash in history between the Church and science, in the end much more damaging to the Church. It's really only the tip of the iceberg of hostility that religion has shown to science through the centuries.[71] (Of course, from the *realpolitik* of the need to protect its position, the Church was quite right. By the end of the 18th century, the rationality of science made atheism respectable.[72])

The argument between Galileo and the Church can be interpreted in terms of the same question that bedevils physics today. Should mathematical constructions be regarded as a useful fiction for describing phenomena, or do they represent actual physical reality? The Church would let the astronomical calculations pass so long as they were clearly presented as no more than a calculating device.[73] But the moment they were seen as

Galileo versus the Church

Galileo: As to the arrangement of the parts of the universe, I hold the Sun to be situated motionless in the center of the revolution of the celestial orbs while the Earth rotates on its axis and revolves about the Sun.

Inquisition: The proposition that the Sun is the center of the world and does not move from its place is absurd and false philosophically and formally heretical, because it is expressly contrary to Holy Scripture.

The proposition that the Earth is not the center of the world and immovable but that it moves, and also with a diurnal motion, is equally absurd and false philosophically and theologically considered at least erroneous in faith.[79]

---

representing reality, they came into conflict with the reading of the Scriptures—which, of course, only the Church had the authority to interpret.[74]

It's a curious situation that the only scientific work undertaken today by the Church is astronomy. Pope Gregory XIII (1502–1585) built an observatory at the Vatican in order to check calculations when the Gregorian Calendar replaced the Julian Calendar (in 1582). It was in use when the Church condemned Galileo. It has always been manned by Jesuits (who were instrumental in bringing Galileo to trial). It's still functional.[75] In fact, the Jesuits confirmed Galileo's data, but nonetheless opposed his conclusions.[76]

The popular view is that Galileo's trial was due to a simple conflict between the Church and science.[77] Without accepting the arguments of Catholic apologists, Koestler (and others) have argued that the clash was not inevitable, but was brought about at least in part by Galileo's arrogance in refusing to show tokenism towards the Church's view. "He is ready to settle for a quiet life by postulating angels who, without difficulty or complications of any kind, direct the known movement of the celestial bodies," Galileo said of Father Riccardi, who was the Chief Censor.[78] But from a modern perspective, either there is free thought or there is no free thought. Of course, it's a reasonable question whether in the present era of political correctness, scientists today practice self-censorship in order to avoid difficulties with funding.

∞ ∞ ∞

It's no exaggeration to say that the creation of science would not have been possible without Johannes Gutenberg (*c.* 1400–1468), who invented the printing press in 1448 in Mainz. Its technology depended on developments from unrelated fields. These included conversion of the winepress for printing, the discovery that linseed oil could make a durable ink, the invention of paper, and the ability to cast metal type to acceptable tolerance.

Before Gutenberg, the only form of scholarly information was handwritten, essentially transcribed with copying errors that multiplied over the centuries. As the scribes were mostly in the monasteries, the Church had a strong influence on the dissemination of information. Gutenberg let Pandora out of the box. Information could be easily and accurately reproduced. It became much more difficult to control circulation.[80] It also stimulated the transition to writing in the vernacular rather than medieval Latin.[81]

It could take a scribe more than a year to copy a manuscript. With a printing press, there was a delay while the press was set up, but then copies could be produced rapidly. The twin advantages of the press were its speed and the increased accuracy—all the copies were identical. Accuracy of reproduction was especially important for the diagrams necessary for cartography and medicine.

Gutenberg sparked a revolution in communications.[82] To be sure, the first result was wider dissemination of the Bible, but printing soon spread to scholarly areas.[83] Ptolemy's *Geographia* went through several editions by 1500, reflecting wide interest in cartography. About 10% of published books dealt with scientific matters, not so much detailed works for the specialist, where the audience was limited, but more as works for other interested parties. (One could hardly say 'the public' given the limitations of literacy in the period.[84])

The analysis of Ptolemy's system produced by Regiomontanus, *Epitome*, was printed in 1496. It's about half the length of Ptolemy's more difficult *Almagest,* which was not published until 1515. This was the beginning of the popularization of science.[85] Copernicus was the first major scientist to benefit from Gutenberg by publishing his results in a print book, followed by Galileo.

It might have been possible to suppress heliocentricity in a previous era, when publication was confined to manuscripts, but the dissemination of printed books was too difficult to stop. *Dialogo* was particularly threatening, because it was written in Italian and was therefore generally accessible (in Italy). It was translated into Latin in 1635 (and smuggled out to be published in Protestant Holland). This was necessary for it to be understood outside of Italy. (The Latin title was *Systema Cosmicum*, clearly implying the existence specifically of one astronomical system.) After *Dialogo* was banned, the price on the black market in Italy increased 10×.[86]

The irony is that the early astronomers were all fervent believers. In fact, their formulation of new theories of astronomy was to some extent held back by their beliefs. Copernicus held a traditional worldview, with beliefs based in religion. His approach followed Ptolemy. Otherwise he might have gone further. Brahe's observational skills were unparalleled for the era, but he could not get beyond the belief that the Earth was at the center.[87] Kepler thought that God had designed the universe, and that mathematics would open a window into the divine creation.[88]

The Copernican revolution was not achieved by a progression of brilliant intellects—or at least, the brilliance of the intellects was dimmed by the society in which they lived. Arthur Koestler even called them 'sleepwalkers.'[89] Copernicus was reluctant to publish his conclusions and took years, even decades, to do so. Kepler had evidence for ellipses relatively early, but kept trying to explain his results in circles and epicycles to conform to the prevailing wisdom: only the perfection of the circle could be permitted for the orbits of heavenly bodies.

The view of the universe was now that:

- The Sun is at the center of the solar system.
- Planets move around the Sun in elliptical orbits.
- Moons orbit the planets.
- The stars are far away and extend into infinite space.

Considering the opprobrium the Catholic Church has had from its suppression of Galileo, it's another irony that the first opposition actually came from the Protestant movement. In

1539, Martin Luther referred to Copernicus as a "fool [who] wishes to reverse the entire science of astronomy".[90] The forces leading to the Enlightenment were not necessarily enlightened towards science![91] When Pope Gregory XIII revised the calendar in 1582, he was pragmatic enough to use tables for calculation based on the Copernican system.

Britain did not change from the Julian calendar to the Gregorian calendar until 1752. This necessitated jumping 11 days.[92] The English calendar riots were characterized by mobs yelling, "Give us our eleven days," under the belief their lifespan had been shortened.[93] Looking at the anti-vaxxer movement of the COVID epidemic, it scarcely seems that we have made any progress towards educating the populace in the past three centuries.

Summarizing, from the enlargement seen in the telescope, Galileo made a series of observations conflicting with the Aristotelian view of the cosmos:

- Venus exhibits phases.
- The Moon is not perfectly spherical.
- Four moons circle Jupiter.
- Sunspots change the appearance of the Sun.
- The Milky Way consists of many stars.

Placing the Sun at the center of the solar system was a challenge that could not be ignored. And the implication that the same laws apply on Earth and in the celestial sphere was a great threat to the religious order. Yet the main movement towards the development of science as an independent way of thought came from physics, with developments in optics and motion, both fields that had been of interest to the Ancients. Here what needed to be overthrown was nothing less than the basis of received wisdom: Aristotle's worldview.

NOTES AND REFERENCES

1. In the anonymous preface (actually written by Andreas Osiander) to N. Copernicus, *De Revolutionibus*, 1543.
2. N. Copernicus, Preface to *De Revolutionibus*, 1543. For a modern translation see *On the Revolutions of Heavenly Spheres*, Prometheus Books, New York, 1995.

3. T. S. Kuhn, *The Copernican Revolution. Planetary Astronomy in the Development of Western Thought*, Harvard University Press, Cambridge, 1957.

4. The book did not sell out its first print run of 400 copies, but most significant astronomers of the period appear to have had a copy. C. Freeman, *The Reopening Of The Western Mind: The Resurgence Of Intellectual Life From The End Of Antiquity To The Dawn Of The Enlightenment*, Knopf, New York, 2023, p. 362.

5. D. L. Couprie, *Heaven and Earth in Ancient Greek Cosmology: From Thales to Heraclides Ponticus*, Springer, New York, 2011.

6. The concept of the free-floating Earth was revolutionary, because previously the cosmos was viewed as a closed system. Bodies such as the Sun, Moon, and planets were placed at different distances from the Earth, another break with the ancient idea of a firmament containing all the bodies at fixed distance.

7. Anaximander viewed the Earth as a cylinder, with its surface like the skin of a drum. "Its shape is rounded, like a column in stone. It has two surfaces, one made of the ground beneath us, and another opposite this." Quoted in C. Rovelli, *Anaximander: And The Birth Of Science*, Riverhead Books, New York, 2023, p. 48.

8. Subsequent pre-Socratic philosophers improved the model, for example, correcting the order of the Sun, Moon, and planets, but accepted the basic idea of the flat Earth. Socrates and Plato considered the possibility of a spherical Earth. It's possible that Pythagoras proposed a spherical Earth a century earlier.

9. D. L. Couprie, *When the Earth was Flat: Studies in Ancient Greek and Chinese Cosmology*, Springer, Switzerland, 2018.

10. "For a thing established in the center, and equally related to the extremes, has no reason to move up or down or to the sides; and since it is impossible for it to move simultaneously in opposite directions, it will necessarily remain where it is."

11. A good account for the general reader of explanations for the movements of the celestial bodies can be found in T. Timberlake and P. Wallace, *Finding Our Place In The Solar System. The Scientific Story Of The Copernican Revolution*, Cambridge University Press, Cambridge, 2019.

12. S. P. Blake, *Astronomy and Astrology in the Islamic World*, Edinburgh University Press, Edinburgh, 2014, p. 6.

13. Aristotle was aware of proposals that the Earth might not lie at the center. "As to its position there is some difference of opinion. Most people—all, in fact, who regard the whole heaven as finite—say it lies at the center." He dismissed alternative views, saying, "They are not... [accounting]... for observed facts, but rather forcing their observations and trying to accommodate them to certain theories and opinions of their own." In *On the Heavens*, Book 2, Part 13. (Translated by J. L. Stocks).

14. Archimedes, *The Sand Reckoner* (Archimedis Syracusani Arenarius & Dimensio Circuli). See T. Heath, *Aristarchus Of Samos*, The Ancient Copernicus, Clarendon Press, Oxford, 1913.

15. It takes 26000 years for the stars to rotate around an axis connecting the Earth to the Sun. This is the *ecliptic axis*, running from the Earth to the Sun

at the equinox. Put another way, precession describes the difference between the time it takes for the Sun to return to the equinox (the twice-yearly moment when the Earth's axis is not tilted relative to the Sun, when hours of daylight and darkness are virtually equal) with the time it takes to return to the same fixed position relative to the stars. It's about 15 minutes per year.

16. A palimpsest is a manuscript in which original writings have been erased to allow reuse. The manuscript from which Hipparchus' coordinates were recovered by multispectral analysis is one of many discovered at Saint Catherine's monastery in Sinai. V. Gysembergh, *et al.*, New evidence for Hipparchus' Star Catalogue revealed by multispectral imaging, *J. Hist. Astron.*, 2022, 53, 383–393.

17. It's unclear why there was little activity in Alexandria between (say) 150 BCE and 150 CE. The most likely explanation is that astronomy was supported by the whim of patronage by the ruler of the day.

18. We now know this is caused by the relative motions of the planets around the Sun. For example, Earth has a greater orbital speed than Mars so can appear to 'overtake' it.

19. Lumps in the remains of a wood case were separated into four main fragments, which in turn were subsequently divided into many smaller parts. The current count is 82 fragments. Presumably the Antikythera mechanism was not unique, but it is the only example of a working model that has been found.

20. T. Freeth, *et al.*, A model of the cosmos in the ancient greek antikythera mechanism, *Sci. Rep.*, 2021, **11**, 5821.

21. The inscriptions on the front and back covers of the device suggest that it followed planetary movements. Discussion with Tony Freeth, October 2024.

22. An argument against this is that the Antikythera Mechanism seems to have included planetary motions (see preceding note) but Hipparchus did not develop a planetary theory. Discussion with Tony Freeth, October 2024.

23. On the basis that his coordinates fit the measurements of Hipparchus by applying an (incorrect) value for precession.

24. R. R. Newton, *The Crime of Claudius Ptolemy*, Johns Hopkins University Press, Baltimore, MD, 1977.

25. V. Gysembergh, *et al.*, New evidence for Hipparchus' Star Catalogue revealed by multispectral imaging, *J. Hist. Astron.*, 2023, **53**, 383–393.

26. O. Gingerich, Was Ptolemy a Fraud?, *Q. J. R. Astron. Soc.*, 1980, **21**, 253–266.

27. *Geographika* was lost and rediscovered in the Western world only in 1295. It was translated in Arabia, in the 9th century, and was used by cartographer Muhammad al-Idrisi to create detailed maps around 1138. The first Latin translation was in 1406.

28. Optics has survived only in an incomplete version. The experiment used a bronze disk, marked in degrees, to compare the angle of refraction between air and water, air and glass, and water and glass. D. C. Lindberg, *The Beginnings Of Western Science: The European Scientific Tradition In Philosophical, Religious, and Institutional Context, Prehistory To A.D. 1450*, University of Chicago Press, Chicago, 2007, pp. 106–109.

29. It was a source of some concern in the 13th and 14th centuries how to reconcile epicycles with Aristotle's model of nested spheres. D. C. Lindberg, *The Beginnings Of Western Science: The European Scientific Tradition In Philosophical, Religious, and Institutional Context, Prehistory To A.D. 1450*, University of Chicago Press, Chicago, 2007, p. 261.

30. The original name was *Mathēmatikē Syntaxis*. *Almagest* is probably an Arabic corruption of *Hē Megistē Syntaxis* (the great work), the name by which it later became known.

31. J. G. Toomer, *Ptolemy's Almagest*, Duckworth, London, 1984.

32. Aristotle argued that it would be impossible to cross the equator because it would be too hot. When Jesuit explorer José de Acosta (1539–1600) crossed the equator in 1570, he recorded. "Then I came to the Equator... I felt so cold that I was forced to go into the sun to warm myself. What could I do then but laugh at Aristotle's *Meteorology* and his philosophy? For in that place and that season [everything] should have been scorched by the heat." J. de Acosta, *The Natural and Moral Historie of the Indies*, ed. E. Grimston and C. R. Markham, Hakluyt Society, 1880, Reprinted Cambridge University Press, Cambridge, 2009, p. 90.

33. In a letter from Amerigo Vespucci to Lorenzo Pietro di Medici on his third voyage, in March 1503. However, there is some doubt as to its authenticity. Online at www.gutenberg.org/cache/epub/36924/pg36924-images.html.

34. In 1577, Francisco Hernández produced 16 volumes identifying 3000 new species of plants and reporting on their medical uses. *The Florentine Codex* of 1578 consisted of 12 books cataloguing plants and animals, and listing Aztec medicine, as well as religion and history.

35. J. Poskett, *Horizons: The Global Origins of Modern Science*, Mariner Books, New York, 2022, pp. 1–45.

36. N. Crane, *Mercator: The Man Who Mapped The Planet*, Henry Holt and Co., New York, 2002.

37. *Nova et Aucta Orbis Terrae Descriptio ad Usum Navigantium Emendate Accommodata* [New and more complete representation of the terrestrial globe properly adapted for use in navigation] Online at www.bl.uk/turning-the-pages/?id=223c7af8-bad6-4282-a684-17bf45bd0311&type=book.

38. In this section I have drawn on T. S. Kuhn, *The Copernican Revolution. Planetary Astronomy in the Development of Western Thought*, Harvard University Press, Cambridge, 1957.

39. The first volume tried to explain the heliocentric theory in general terms. Kuhn describes it as "profoundly unconvincing." The next five volumes present mathematical theories dealing with spherical astronomy, motions of the Sun, the Moon and its orbits, and detailed mathematical explanations. T. S. Kuhn, *The Copernican Revolution. Planetary Astronomy in the Development of Western Thought*, Harvard University Press, Cambridge, 1957, p. 145.

40. N. Copernicus, *De Revolutionibus*, 1543, p. 3.

41. More copies were found in 1880, and then in the 1960s. N. M. Swerdlow, The derivation and first draft of Copernicus's planetary theory: a translation of the commentariolus with commentary, *Proc. Am. Philos. Soc.*, 1973, **117**, 423–512.

42. N. M. Swerdlow and O. Neugebauer, *Mathematical Astronomy in Coperni-cus's De Revolutionibus*, Springer, Berlin, 1984.

43. D. Wootton, *The Invention Of Science: A New History Of The Scientific Revo-lution*, Harper, New York, 2015, p. 145.

44. G. J. Rheticus, *Narratio Prima, in Three Copernican Treatises*, 1540, trans-lated by E. Rosen, Dover, New York, 1959, pp. 194–195.

45. In Osiander's preface to N. Copernicus, *De Revolutionibus*, 1543, p. 3.

46. Brahe was appointed Feudal Lord of Hven, which was not popular with the inhabitants of the island.

47. Distances to the first three stars were published between 1837 and 1840.

48. The failure to observe parallax can be used to calculate a lower limit for the distance of the stars based on the resolution of measurement. This in-creased from 200 astronomical units (200× the distance from the Earth to the Sun) in ancient Greece, to 4000 in Galileo's era, to 20 000 as calculated by Newton, to 700000 (= 10 light years) when the first star's distance was measured by parallax. See D. Hoffleit, The Quest for Stellar Parallax, *Popular Astron.*, 1949, **LVII**, 259–272.

49. The supernova appeared on November 2, 1572, reached its peak brightness by November 16, and remained visible until 1574. It's now known as SN 1572 or Tycho's Supernova. It's one of only 8 supernova that have been visible to the naked eye.

50. Quoted in J. R. Christianson, *On Tycho's Island: Tycho Brahe, Science, and Culture in the Sixteenth Century*, Cambridge University Press, Cambridge, 2000, p. 17.

51. He published the results in a book, *De nova et nullius aevi memoria prius visa stella* (Concerning the Star, new and never before seen in the life or memory of anyone) in 1573. Johannes Kepler was responsible for re-printing it in 1602 and 1610.

52. Volume 2 of *Astronomiae Instauratae Progymnasmata* (Introduction to the New Astronomy). Volume 1 was published only posthumously.

53. J. Kepler, *Mysterium Cosmographicum*, 1596.

54. T. S. Kuhn, *The Copernican Revolution. Planetary Astronomy in the Develop-ment of Western Thought*, Harvard University Press, Cambridge, 1957, p. 210.

55. Following an elliptical path also meant abandoning the notion of crystal-line spheres. Rotation of a solid sphere generates a circular path, but there is no solid body whose rotation could generate an eclipse.

56. Quoted in J. Kepler, *Selections from Kepler's Astronomia Nova*, 1609, trans-lated by W. H. Donahue, Green Lion Press, Santa Fe, New Mexico, 2008, p. 4.

57. Quoted in M. Caspar, *Kepler,* translated by C. D. Hellman, Dover, New York, 1993, pp. 106–107.

58. J. Kepler, *Somnium, seu opus posthumum De astronomia lunari* (The Dream, Or Posthumous Work on Lunar Astronomy), 1634. Published after Kepler's death by his son (Sagani Silesiorum, Frankfurt).

59. Named in memory of Holy Roman Emperor Rudolf II (1552–1612).

60. T. S. Kuhn, *The Copernican Revolution. Planetary Astronomy in the Development of Western Thought*, Harvard University Press, Cambridge, 1957, pp. 215–216.

61. *Sidereus Nuncius* (Starry Messenger), translated in S. Drake, *Discoveries and Opinions of Galileo*, Doubleday, New York, 1957, p. 29. *Sidereus Nuncius* was a pamphlet published (in Latin), March 13, 1610.
62. The university was a protected environment and Padua was a semi-democratic republic, whereas Florence was under absolutist government.
63. Quoted in a letter from Paolo Gualdo to Galileo in 1611 in *Le opere di Galileo Galilei XI*, doc 564 (Edizione Nazionale, 1890), p. 165.
64. The general principle is that inferior planets (lying between Earth and the Sun) show phases, but superior planets (lying beyond Earth) do not. This cannot be reconciled with a geocentric system.
65. M. Sharratt, *Galileo: Decisive Innovator*, Cambridge University Press, Cambridge, 1996, pp. 86–91.
66. Others had observed sunspots previously, in particular Fabricius followed by the Jesuit Christoph Scheiner, who argued they represented objects (planets) moving between Earth and the Sun. See M. Sharratt, *Galileo: Decisive Innovator*, Cambridge University Press, Cambridge, 1996, pp. 98–106.
67. *Sidereus Nuncius* was a pamphlet published (in Latin), March 13, 1610.
68. Letter to Cristina di Lorena, Grand Duchess of Tuscany, 1615, translated in S. Drake, *Discoveries and Opinions of Galileo*, Doubleday, New York, 1957, p. 183. The letter became public only in 1636. It was an expansion of a reply Galileo had written previously in response to a letter from Castelli. It laid out his view of how science could be reconciled with (or at least allowed to proceed in spite of) theology, and might ward off an attack by the Church on Copernicanism. See M. Sharratt, *Galileo: Decisive Innovator*, Cambridge University Press, Cambridge, 1996, pp. 107–136.
69. Galileo was also wrong about the comet of 1604, which he interpreted as an optical effect instead of a celestial body.
70. Dialogo was not removed from the Index of Forbidden books until 1835.
71. It took 400 years before the Church admitted Galileo was right (in 1992).
72. Atheism is usually described as an 18th century phenomenon associated with the Enlightenment. This is not entirely true, at least insofar as it was regarded as threat to the natural order. In the previous century, philosopher Henry More published a diatribe against atheism in 1662: *An Antidote Against Atheism* (William Morden, Cambridge). He also commented that the philosophy of the Royal Society would 'rout' atheism.
73. The device of discussing *ex hypothesi* was used in the universities, where it allowed topics that might otherwise be considered to be heretical to be introduced. Today 'heresy' is defined in terms of political correctness rather than religion.
74. The Church took its authority very literally. "Not only the words but even every comma has been supplied by the Holy Spirit," was the position taken by the Dominican Melchior Cano in 1585, in *de Locis Theologicis*, a treatise intended to set out an authoritative position for the Church. See p. 2.17.
75. The Jesuit astronomers observed the phases of Venus independently of Galileo, although they did not form the same conclusions! See M. Sharratt, *Galileo: Decisive Innovator*, Cambridge University Press, Cambridge, 1996, p. 89.

76. T. E. Huff, *Intellectual Curiosity and The Scientific Revolution. A Global Perspective*, Cambridge University Press, Cambridge, 2011, p. 67–68.

77. The trial took place against the background of the Protestant revolution against the Catholic Church, the ensuing Counter Revolution, the Thirty Years War between Protestants and Catholics, and the realpolitik of the Vatican's relationship with the Hapsburg Empire. See M. A. Finocchiaro, *On Trial For Reason: Science, Religion, and Culture In The Galileo Affair*, Oxford University Press, Oxford, 2019.

78. Quoted in A. Koestler, *The Sleepwalkers*, Hutchinson & Co., London, 1968, p. 480.

79. Papal condemnation of Galileo, signed by 7 cardinals. Quoted in G. de Santillana, *The Crime of Galileo*, University of Chicago Press, Chicago, 1955, p. 306.

80. The economic effects on the intellectual world were considerable. In the century after Gutenberg, the price of a book fell by 90% (J. Dittmar, *The Welfare Impact of a New Good: The Printed Book*, Department of Economics, American University, 2011). This is not very different from the fall in the price of a personal computer in the decades after its introduction, but the collateral effects were greater: The printed book produced a new means of disseminating knowledge, whereas the computer facilitated existing means (N. Ferguson, *The Square and the Tower*, Penguin, New York, 2017, p. 94).

81. Literature moved into the vernacular earlier than technical publications. Chaucer's *Canterbury Tales* was written in Middle English in 1387. *La Chanson de Roland* (the Song of Roland) is a 'chanson de geste,' written in Old French around 1040.

82. Marshall McLuhan—famous for 'the medium is the message'—goes further and argues that Gutenberg was responsible for (or at least enhanced) the split between the humanities and science. M. McLuhan, *The Gutenberg Galaxy: The Making of Typographic Man*, University of Toronto Press, Toronto, 1962, p. 81.

83. When the press was invented in 1448, it used moveable type—blocks of individual letters cast in metal. By 1467 it became possible to include engravings made on copper plates.

84. M. Boaz, *The Scientific Renaissance 1450–1630*, Harper, New York, 1962, pp. 28–29.

85. About 8% of new nonfiction published in the United States each year today deals with science or scientific issues. www.statista.com, 2023.

86. B. Moynahan, *The Faith. A History of Christianity*, Aurum Press, London, 2002, p. 440.

87. T. S. Kuhn, *The Copernican Revolution. Planetary Astronomy in the Development of Western Thought*, Harvard University Press, Cambridge, 1957.

88. "My aim is to show that the machine of the universe is not similar to a divine animated being, but similar to a clock... God ... has fixed the number of heavens, their proportions, and the relations of their movements." Quoted in G. Holton, Johannes Kepler's Universe: Its Physics and Metaphysics, *Am. J. Phys.*, 1956, **24**, 340–351.

89. A. Koestler, *The Sleepwalkers*, Hutchinson & Co., London, 1968, p. 480.

90. Quoted in A. D. White, *A history of the warfare of science with theology in Christendom*, Appleton, New York, 1896, vol. I, p. 126. Online at www.gutenberg.org/files/505/505-h/505-h.htm.
91. The fundamentalism of Lutherans and Calvinists clashed with scientific development.
92. The calendar moved directly from Wednesday 2nd September 1752 to Thursday 14th September 1752.
93. There is a view that the riots were an urban myth, and either never happened or at least were much less serious than supposed.

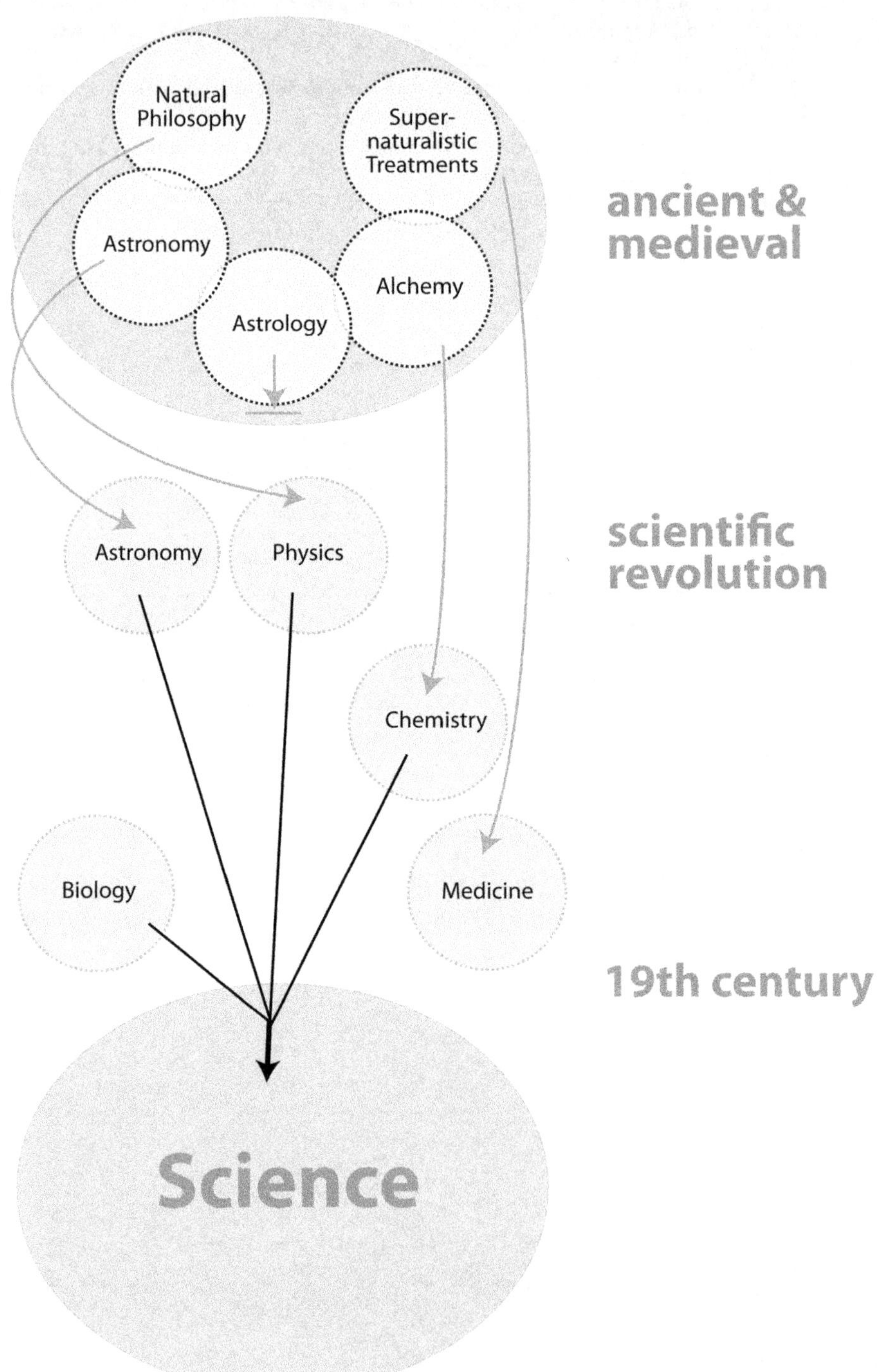

Natural Philosophy
Super-naturalistic Treatments
Astronomy
Astrology
Alchemy
ancient & medieval
Astronomy
Physics
scientific revolution
Chemistry
Biology
Medicine
19th century
Science

Becoming Science

All of the sciences could indeed be represented as an unbroken circle… mathematics, indeed, proceeded from logic, but from mathematics derived mechanics and thereby physics; from physics came chemistry, from chemistry biology, and from that, on the one hand sociology with the human sciences, on the other hand psychology with the theory of awareness; from the latter, finally, the logic developed which closed the circle, Jean Piaget, 1918.

Galileo's Movement: 1610–1659

"EPPUR SI MUOVE"—AND YET IT MOVES—GALILEO SUPPOSEDLY said when the Inquisition released him after the most famous clash in history between religion and science.[1] Galileo was a giant who dominated astronomy and physics at the start of the 17th century. Sometimes regarded as the 'first scientist' for his experimental and deductive approach, he was an essential spark for the Scientific Revolution. If only for his willingness to abandon precedent when demanded by the data, he deserves to have a revolution named after him. His importance goes far beyond demonstrating that the Earth moves, or analyzing the movement of individual objects, into basing scientific conclusions on reproducible observations and experiments. A (typically) arrogant attitude is captured by his saying, "As to the mere authority of ancient and modern philosophers and mathematicians, I say that that has no power at all to establish a knowledge of any physical proposition... Modern observations deprive all former writers of any authority, since if they had seen what we see, they would have judged as we judge."[2] The persecution of Galileo could be a parable for modern times. Refusal to replace Ptolemy with Copernicus represents climate change denial. The insistence of the Church on placing its belief system above the facts represents creationism. And those who refused to look through Galileo's telescope because they might be deceived by appearances stand for anti-vaxxers.

The Frontiers of Science
By Benjamin Lewin
© Benjamin Lewin 2026
Published by the Royal Society of Chemistry, www.rsc.org

Timeline for Key Events Defining the Scientific Revolution of the 17th Century

1600 — Janssen makes first microscope (1600)
Gilbert proposes Earth rotates on its axis (1600)
Bacon publishes *Advancement of Learning* (1605)
Lippershey & Janssen invent telescope (1608)
Kepler defines ellipses (1609)
1610 — Galileo publishes *Sidereus Nuncius* (1610)
Napier introduces logarithms (1614)
Catholic Church bans Copernican doctrine (1616)

1620 —

Kepler publishes *Rudolphine Tables* (1627)
Harvey discovers circulation of the blood (1628)
1630 — Galileo publishes *Dialogo* (1630)
Galileo tried for heresy by Inquisition (1633)

Descartes enunciates *Cogito Ergo Sum* (1637)
1640 — Galileo publishes *Two New Sciences* on motion (1638)

Torricelli demonstrates vacuum (1643)
Descartes publishes *Principia philosophiae* (1644)

Pascal demonstrates vacuum on mountain (1647)

1650 —

von Guericke makes vacuum in Magdeburg hemispheres (1657)
Huygens invents pendulum clock (1657)
Boyle develops air pump (1659)
1660 —

Hooke publishes *Micrographia* (1665)

1670 —

von Leeuwenhoek discovers microorganisms (1676)
Rømer measures speed of light (1676)

1680 —

Newton publishes *Principia* with laws of motion (1687)
1690 —

1700 — Newton publishes *Opticks* (1704)

A Dialog between Aristotle and Galileo

Aristotle's (384–322 BCE) worldview placed the Earth at the center of the universe, excluded the existence of a vacuum, and held that the properties of objects are governed by their intrinsic nature.

Galileo (1564–1642) challenged traditional views on the basis of new observations (in astronomy) and experiments (in physics), which defined motion in terms of velocity, momentum, and acceleration.

https://www.loc.gov/item/ 2021666740/, World Digital Library

Aristotle: Ah, Galileo Galilei! It's an honor to converse with a mind of your reputation. I have laid the groundwork for understanding motion through my philosophy, so I am curious to hear what new insights you have found in Nature.

Galileo: Aristotle, I've been conducting experiments on falling objects and rolling balls. I've observed that all objects, regardless of their mass, fall at the same rate when dropped from the same height. This challenges your idea that heavier objects fall faster.

Aristotle: A bold claim indeed, Galileo. But tell me, how do you justify this conclusion? My philosophy, based on extensive observations of the real world, posits that objects fall at rates proportional to their weights due to their innate tendencies.

Galileo: Aristotle, I propose that the medium through which objects move plays a crucial role. In a vacuum all objects fall at the same rate because they are free from air resistance. It's not about an object's nature. It's about the influence of external factors.

Aristotle: An interesting idea, but Nature abhors a vacuum, so how can one draw conclusions from a scenario so removed from reality?

Galileo: That is a mistaken concern, Aristotle. It's only by eliminating variables, such as resistance to the medium, that we can discern fundamental principles.

Aristotle: This seems an artificial concept compared to motion as I have described it—a reflection of an object's inherent tendencies and its pursuit of its natural place.

Galileo: Aristotle, I propose that motion is not tied to an object's nature, but is a property imposed upon it by the physical world. It's not about objects seeking a natural place. It's about forces that cause them to move.

Aristotle: Galileo! Your perspective challenges the very essence of my philosophy. The only way to understand phenomena is to know their purpose. How can we know that your observation of motion is not an illusion created by the artificial circumstances of your observations?

Galileo: I believe we can refine our understanding of the natural world only through rigorous experimentation and observation. This shows that motion results from the interplay of forces and the environment. Innate tendencies are to propulsion as imagination is to reality. I rest my case.

GALILEO'S MOVEMENT

It's vital today that the world should recognize that 17th-century Europe did not give rise to essentially 'European' or 'Western' science, but to universally valid world science, that is to say, 'modern' science as opposed to the ancient and medieval sciences... Once the basic technique of discovery had itself been discovered, once the full method of scientific investigation of Nature had been understood, the sciences assumed the absolute universality of mathematics, and in their modern form are at home under any meridian, the common light and inheritance of every race and people... This ... communicates ... a body of incontestable scientific truth acceptable to all men everywhere.[3] Joseph Needham, 1959.

I am going to take the publication of Galileo's book *Starry Messenger* (*Sidereus Nuncius*) in 1610 as the true start of the Scientific Revolution. This was the spark for recognizing scientific thinking as a distinct, different type of activity. It was a threat to the established order that the authorities needed to suppress. (Copernicus is an alternative starting point.) Of course, I am anticipating my response to the question of whether there was in fact a Scientific Revolution, and if so, what was its nature?[4]

Not to put too fine a point upon it, the worldview was quite different at the end of the 17th century from the beginning. Religion (and superstition) had scarcely been banished, of course, but scientific thinking, and indeed a rationalist approach, had become accepted. If this was not a revolution, what is?

The term 'Scientific Revolution' is relatively recent, used in the title of a review by French historian Alexandre Koyré in 1943.[5] He took the view that the revolution started with Galileo's work on motion and ended with Newton's definition of gravity. This places the emphasis very much on the development of physics, including optics, magnetism, and vacuums.[6] Yet over the same period there were also major advances in biology, most notably Harvey's proposal for circulation of the blood, von Leeuwenhoek's discovery of microorganisms, and Robert Hooke's microscopy.[7]

The Scientific Revolution is sandwiched between the Renaissance and the Enlightenment. Marking the transition from the medieval to the modern era, the Renaissance is usually considered to have started in Italy in the 14th century. It spread into Europe in the 15th century, and reached England only in the 16th century. The Renaissance encouraged freethinking, which was necessary for science to develop. This set the scene with major figures such as Leonardo da Vinci (1452–1519), a polymath in both arts and sciences, Desiderius Erasmus (1466–1536), who initiated the humanist movement (see Chapter 10), and René Descartes ("I think therefore I am"). In turn, the Scientific Revolution created an increasing sense of intellectual freedom that fed into the Enlightenment.

The Enlightenment triggered a change in human society that has led to significant improvements in almost all aspects of human life.[8] How far were scientific advances responsible for changes in attitudes, including greater tolerance about religion? And how far were later advances in science encouraged by the change in attitude of the Enlightenment?

It's fashionable to question whether there ever was a Scientific Revolution in the sense of a discrete change in intellectual activity.[9] There can be no doubt that there was a great difference in

Relationship of Scientific and Political Developments

Period	Dates (approx.)	Main Events and Players
Dark Ages	500–1100	Started with fall of the Roman Empire, petered out with the Renaissance.
Carolingian Renaissance	768–814	Charlemagne and his court.
Renaissance	1300–1600	Started in Italy, spread to Europe.
Reformation	1517–1530	Martin Luther started the Protestant movement.
Counter Reformation	1530–1578	Started with Council of Trent, saw formation of Jesuits. The Inquisition and attacks on Protestants followed.
Scientific Revolution	1543/1610–1687	Started with Copernicus or Galileo, culminated with Newton.
Enlightenment	1715–1789	Dated by events in France; started with ascension of Louis XIV, ended with the Revolution.
Industrial Revolution	1760–1840	Started gradually between 1710 and 1760, followed by industrialization.
Victorian age	1837–1901	Reign of Queen Victoria in U.K.

our understanding of the universe at the end of the period from the beginning.[10] And there was progress in analyzing subjects such as motion and optics. Yet advance in knowledge does not of itself necessarily make a revolution. The real question is what change occurred in the intellectual atmosphere.[11]

Emerging from the medieval era, intellectual activity was very much focused on reasoning. Copernicus could think more deeply about the nature of the solar system, but could not rely on new data to prove his point. Galileo marked the transition to investigations by experiment and applying new technology. Newton followed an approach that would be recognized by scientists today. His publication of *Principia,* setting out the laws of motion, is widely taken to mark the culmination of the Scientific Revolution.

In the political arena, 'revolution' has come to mean a sudden event. But we do not need to get hung up on semantics and to dismiss the concept of 'revolution' because there was no discrete moment at which everything changed. Yet even accepting that there was a sea change in attitudes during the 17th century, there was not a complete transition to data-driven science. Many of the protagonists, including even Newton, practiced alchemy as well as physics.

It would be naïve to present the Scientific Revolution as a period when the clash between religion and science developed; rather they were intertwined. The fact is that the major figures were all committed believers. Some even held positions within the Church. Often enough, they felt that discovering the operations of the natural world would lead to insights into divine purpose.

The question becomes to what extent functioning within a strong belief system limits the ability to draw conclusions out of, or inconsistent with, that system. At what point did science break free into becoming a self-contained activity, effectively functioning within its own belief system, in which religious (or other) beliefs became extraneous?

The various disciplines that make up science today were not really recognized historically as constituting part of the same *gestalt*. Based on observations, astronomy used the most mathematical approach. Moving towards experiments, natural philosophy studied physical phenomena, such as motion or optics.

Deeply involved with mysticism, alchemy was eventually replaced by chemistry. Medicine was close to a separate world, varying in its relationship with a scientific approach. There was little biology as such. There was no thought that physics, chemistry, and biology would eventually come to be regarded as separate branches of the same discipline: science.

The main distinction was between natural philosophy and mathematics. When he was appointed to the court of the Grand Duke of Tuscany in 1610, Galileo requested that "His Majesty add the name of Philosopher to that of Mathematician."[12] This was pragmatic. One reason why Galileo wanted to be a professor of philosophy rather than mathematics was that philosophy was held in higher regard and was better paid. Yet Galileo famously said that "the [universe] is written in the language of mathematics."[13]

Galileo made discoveries ranging across the physical sciences, from observations in astronomy to experiments with motion, from investigations with pendulums to calculations of projectile trajectories. He is remembered as a great scientist, rather than as a philosopher. Indeed, his view was that science enabled philosophy rather than the other way round.[14] It would be going too far to say that he believed mathematics replaced philosophy. However, he certainly made the argument that mathematics convey accuracy where the senses can deceive.[15] Perhaps Galileo's axiom that theories can violate the senses is a precedent for the experience of atomic physics centuries later.[16]

Yet Galileo was conscious of the need for theory to reflect the reality of observations. "Our discourses must relate to the sensible world and not to one on paper," he has his spokesman Salviati say in the *Two World Dialog*.[17] Galileo's seminal influence stems not only from the range of his discoveries, but from his crucial role in replacing natural philosophy with what was for a while called experimental philosophy—which we would now call science.[18]

This was the driving force for the Scientific Revolution. And not only were conclusions now subject to verification by observation or experiment, but there was awareness that the foundations of science are subject to change as new facts emerge. This realization has been called the Second Scientific Revolution.[19]

Today discovery is the currency of science. The unceasing search for new information is in itself a great change with the ancient view of knowledge. This held that there was nothing new to discover by observation, but reasoning could develop new inferences from existing knowledge. The concept of 'discovery' developed during the Scientific Revolution.[20] Until then it was assumed that Aristotle had summarized all knowledge. New information was often painfully described in terms of "back to the future," being presented as rediscovery of ancient wisdom.[21]

The moment when discovery became an explicit aim of science is generally attributed to Francis Bacon's publication of *The Advancement of Learning* in 1605.[22] The title page of Bacon's later book, in 1620, *Novum Organon*,[23] makes a declaration of the

Frontispiece from Francis Bacon's Novum Organum. The image shows a galleon passing between the pillars of Hercules at Gibraltar (the mythical markers for the boundaries of the known world at the end of the Mediterranean). The caption states: "Many will pass through and scientific knowledge will increase."

search for knowledge with a powerful allegory of a boat passing out of the known world into new territories.

One of the difficulties in assessing the novelty of contributions to science in the period up to the Scientific Revolution is that it was not customary to name sources. Authors would simply assimilate the works of their predecessors. Today I suppose we would call this plagiarism. Part of modern science is the clear attribution of sources so that each contribution can be checked for its reproducibility. Tracing a clear path for the evolution of historic ideas, such as heliocentricity, is complicated by the loss of many of the original documents.[24]

Galileo "is the father of modern physics—indeed of modern science altogether," said Albert Einstein.[25] Galileo (1564–1642) was born in Pisa, but his family moved to Florence when he was young. His father was a musician (a lutenist) and Galileo became a lutenist himself. He fathered three children in a long-term relationship, but never married. He started to study medicine at

Galileo as he saw himself, in a portrait commissioned from the engraver Francesco Villamena for Sunspots (1613) and Assayer (1633).

https://www.loc.gov/item/2021666740/, World Digital Library.

the University of Pisa, but switched to natural philosophy and mathematics. He obtained the Chair in Mathematics in Pisa in 1589, and moved to the University of Padua in 1592. At Padua, he taught geometry, mechanics, and astronomy, and worked on both physics and astronomy. In 1610, he moved to Florence.

Certainly his astronomical work has the marks of science: introduction of new techniques allowing observations to be made that support or modify existing theories. It was not only the motion of the Earth that threatened the existing order, but the way the entire approach was divorced from religious thinking. But it was perhaps his work on physics that more directly plays to a scientific method in introducing experimentation to resolve outstanding issues. Whether experimentation or observation was the right path forward was an issue that resonated for the rest of the century.

Galileo's experiments on motion are often quoted as the prime example of replacing reasoning with experimentation, although now there is a controversy as to whether they actually occurred. Aristotle held that objects fall at a rate depending on their mass. So if a heavy ball and a light ball are dropped together, the heavy ball should hit the ground first. Galileo set out to refute Aristotle in a manuscript, *De Motu* (never published, perhaps because his first attempts in 1590 to measure the rates at which different objects fall did not accord with his theory). (I discuss *De Motu* in Chapter 18.) Whether or not Galileo actually tested the model later by dropping balls from the leaning tower of Pisa,[26] he famously concluded that all objects fall at the same rate, accelerating at a constant speed.[27]

The reason why objects fall at different rates to Earth is, of course, different resistance to air pressure. The definitive experiment was performed on the Moon during the Apollo space mission in 1971. When a hammer and a feather were dropped together, they fell at the same rate and landed at the same time.

Galileo continued to work on motion and analyzed it in his final book, *Two New Sciences*, in 1638. It followed the same format as *Dialogo*, with a question and answer discussion between three participants. He describes an experiment in which balls of different weights were rolled down a smooth ramp at different inclinations. (The ramp was smooth to eliminate the effects of friction and air resistance, which Galileo had concluded were responsible for differences in the rates of falling bodies.) Elapsed

"Galileo was right." Colonel Scott of the Apollo space mission dropped a hammer and feather on the Moon to show they land together.

Reproduced with permission from Ed Hengeveld (artist).

time was measured with a water clock. (Unlike following a falling body off a high point, here the motion was slow enough to measure.[28]) The results led to the law of motion: all falling bodies accelerate at the same speed (in the absence of air resistance).[29]

In this and other work, Galileo demonstrated most of the attributes of modern science. Theories were tested by experiments. According to the results, they were discarded or used to formulate new hypotheses. The major limitations were the inability to provide sufficiently precise conditions for the experiments. One reason for doubting whether experiments were actually performed is the casual way they were described, essentially in terms of results obtained, rather than in terms of individual experimental runs. There is no way to assess the reliability of the experiments by comparing individual results. However, recent attempts to reproduce the work under conditions comparable to those in Galileo's time suggest that they would have given the results described by Galileo.[31]

What was the reaction of Galileo's contemporaries to the introduction of a new theory, based on experiments, refuting a

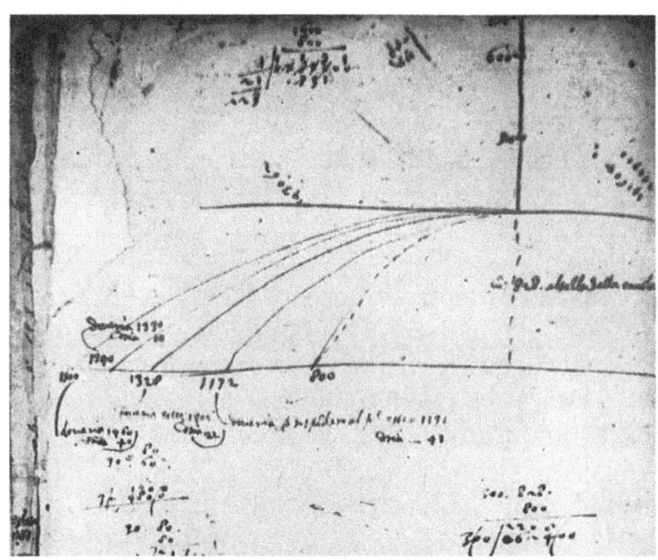

Galileo's record of experiments when he discovered that a projectile follows a parabolic path. A ball was rolled down an inclined ramp from different heights to the edge of a table. When it fell off, it moved forward as it fell. Galileo compared the distance it traveled with his expectations based on the combination of laws he deduced for free fall and forward inertia.[30]

Reproduced from f. 116 Biblioteca Nazionale Centrale, Florence, Via Tripoli, 36, Italy, by concession of the Ministry of Culture/National Central Library, Florence.

thousand years of received wisdom? This was not the first attempt to replace Aristotle's view of motion, but Galileo's theory was controversial.[32] Some difficulties in accepting the new law were pragmatic. In practice, measurements were not precise. And in theory, mathematical tools to interpret the law did not exist (calculus had not yet been invented). As soon as 1638, Descartes regarded the law of motion as an abstraction that did not relate to the natural world.[33] In France, Pierre Gassendi engaged between 1642 and 1646 in a spirited defense of the new law. After 1654 at The Hague, Christiaan Huygens later introduced more complex mathematical descriptions of falling bodies.

In the three decades before Newton's decisive studies on gravity, opinion was split. This is not very different from the reaction to new ideas in the modern era. It can take a decade or more for a revolutionary idea to become accepted. (Some major discoveries have taken up to fifty years to be recognized by the award of Nobel Prizes.) It's not really contemporary reactions that establish Galileo as the 'first experimental physicist.' It's the judgment of history.

$$\infty \ \infty \ \infty$$

From Descartes' famous aphorism, "Cogito ergo sum" (I think therefore I am), we can expect a completely different approach from that of Galileo. A major figure preceding the Enlightenment, René Descartes (1596–1650) was a younger contemporary of Galileo. Brought up first by his grandmother, and then by an uncle who was a lawyer, he entered a Jesuit college in 1606. Leaving the college, his early career was checkered. He studied law, became a soldier in the Catholic army of Maximilian I, lived in Italy for three years, went to Paris in 1625, and moved to the (Protestant) Netherlands in 1628. Here he worked on both practical matters (physics and optics) and metaphysics (the existence and nature of God). Separating the body (material) from the mind (immaterial) led to a mechanistic view of physiology.

'Cogito ergo sum' has become a truism, but it's not the facile statement it might appear to be. It's at the root of Descartes' philosophy that reliable observations depend exclusively upon what you can sense. Deductions start from that basis. A philosopher who developed Cartesian logic, a mathematician who developed algebraic geometry,[34] and an empiricist who analyzed the rainbow, Descartes' reliance on reasoning was something of a throwback to early Greek philosophy.

Following the condemnation of Galileo, Descartes decided not to publish his work, an example of the chilling effect of the Church. His famous publication, *Discours de la Méthode,* in 1637, presented an attenuated version of his theories. Rejecting Aristotle's conclusions but accepting his methods, Descartes started from different assumptions and arrived at different conclusions.

Descartes argued that the universe is made of discrete units, now known as corpuscles. Effectively this was a version of atomist theory. He argued that matter had only size, shape, position, and motion, and was the same in the terrestrial and celestial realms. The theory offered an alternative to Aristotle to explain motion, and the formation of the Sun and planets. However, it was no more subject to experimental testing than were Aristotle's ideas.

Following Aristotle, Descartes argued that objects must be in contact to influence one another. The conclusion that there can be no attraction at a distance led to the view that space is occupied by matter. Descartes proposed that that the particles are so tightly packed that any movement generates vortices. Gravity resulted from the action of the vortices.

There was a significant movement in the middle of the 17th century to follow 'mechanical philosophy'—the view that all natural phenomena can be explained solely in terms of matter and motion.[35,36] Robert Boyle may have been the first to popularize the specific term in 1666. "I present…some kind of introduction to the principles of Mechanical Philosophy, by expounding… the most fundamental and useful part of Natural Philosophy."[37]

There were variations of opinion in mechanical philosophy,[38] but a common attitude was the need to replace Aristotle (especially the notion that Nature was intrinsically purposeful).[39] Boyle's corpuscular model viewed all matter as consisting of tiny particles (corpuscles) that obeyed physical laws (if only those laws could be discovered). Mechanical philosophy played an important role in eliminating 'immaterial' or supernatural effects from chemistry and physics, and in disposing of vitalism (see Chapter 21).

The development of mechanical philosophy was an important route by which the Scientific Revolution contributed to the Enlightenment. The whims or dictates of a divinity were replaced by a logical chain of cause and effects. "Effects shall have a necessary cause," Thomas Hobbes said.[40] Spinoza (1632–1677), a Jewish Dutch philosopher who was a leading exponent of rationalism, took the argument further.[41] "Nothing, then can happen in Nature to contravene her own universal laws, nor anything that is not in agreement with these laws or that does not follow from them."[42] (This was an argument against the occurrence of miracles.)

The internal conflicts and transitions of the era are shown by two of Descartes' analyses. *Discours* reported experiments to resolve the basis for the rainbow. This is a complex issue, with analyses continuing to the present day,[43] but Descartes uncovered the basic phenomenon. He used a spherical flask of water to mimic a raindrop. (Theodoric of Freiburg had used the same method three centuries earlier, but Descartes did not know that.)

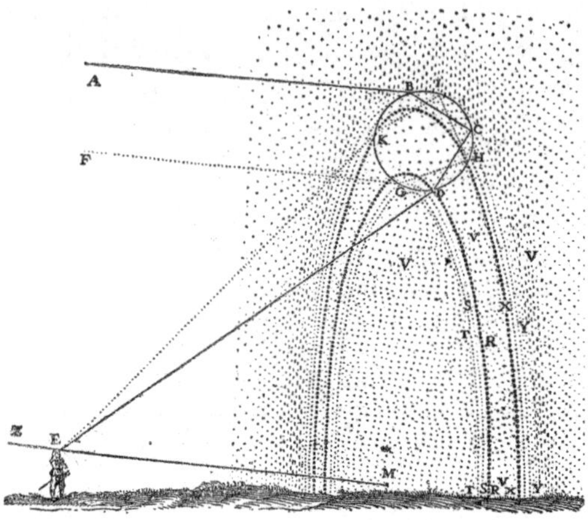

René Descartes' analysis of how a rainbow is formed.

From observing and calculating the paths of the light rays, he concluded that the radius of the rainbow is 42°. The light is reflected and refracted twice, at each air-water boundary, as it enters and leaves the droplet. The actual spectrum of colors could not be explained until Newton analyzed refraction through prisms (see later).

By contrast, Descartes analyzed magnetism by pure reasoning, based on the belief that some knowledge is innate ("created by God at the very beginning of the world").[44] Perhaps imagination would be a better term than reasoning. A magnet emits a stream of particles, shaped like little screws. The screws can be left-handed or right-handed. When they pass through another body, right-handed screws attract the body, and left-handed screws repel it.

Descartes was one of the great mathematicians and philosophers of the century, but his emphasis was more on natural philosophy, in the tradition of Aristotle, than on the new experimental paradigm of science. Voltaire's comment on Descartes, waspish as always, but fair, was that "Our Des Cartes, [was] born to discover the Errors of Antiquity, and at the same Time to substitute his own."[45]

Half a century after Descartes' analysis of the rainbow, Christiaan Huygens and Isaac Newton disputed the nature of light. Newton's experiment in optics (see Chapter 6) led him to a corpuscular view of light. In his *Traité de la Lumière,* Huygens proposed a mechanical explanation of reflection and refraction,

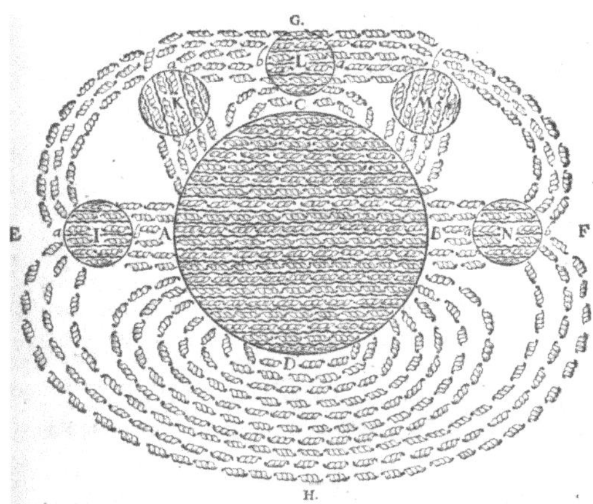

René Descartes' analysis of magnetism was based on pure reasoning.

based on a wave theory of light.[46] The interesting feature here, from a modern perspective, is that Huygens and Newton proposed differing explanations for the phenomena they observed, but agreed on the same principle of obtaining experimental data for analysis.

Born into a wealthy and well-connected family in The Hague, Christiaan Huygens (1629–1695) studied first at Leiden University, then at Orange College in Breda. After traveling in other countries, he returned to The Hague in 1654 and spent his time in research. His first work was in mathematics, on geometry and probability. He developed a theory of collisions (correcting Descartes' model). In 1655, he worked with his brother to grind lenses and construct telescopes and microscopes. He discovered Titan, the moon of Saturn. He became famous and lived in Paris before returning later to The Hague.

An interest in pendulums led him to the construction of the first pendulum clock in 1657. (Galileo had been fascinated by pendulums, according to his pupil Viviani, ever since he had noticed that the chandelier hanging in the cathedral in Pisa had a constant period even when it had a different arc.[47] (It takes the same length of time to swing from one end of the arc to the other irrespective of the distance.) In 1602, Galileo experimented with pendulums, but did not succeed in building a clock.)

∞ ∞ ∞

'Nature abhors a vacuum' was one of the basic principles of Aristotle's universe (see Chapter 4). The argument followed impeccable logic. An object falls faster in air than in syrup or water. So the more rarified the environment, the faster an object will fall. (The argument fails because the generalization is not valid—but this did not become apparent until Galileo.) If the argument is valid, however, it means that an object will fall at infinite speed in a vacuum. That cannot be possible, because an object cannot be in two places at the same time. So a vacuum is impossible.

Descartes supported the prevailing view. "It's a contradiction to suppose there is such a thing as a vacuum."[48] This principle was the next to fall, when Evangelista Torricelli (1608–1647) constructed a mercury barometer in 1643. A mathematician, Torricelli was friendly with Galileo—he was living in Galileo's house in 1642—and after Galileo's death, he succeeded him at the University of Pisa.

The impetus for the experiment was the practical need to understand why a suction pump could not lift water above a certain height (roughly 10 m) or a siphon could not function above that level. Galileo suggested in *Two New Sciences* in 1638 that the key issue is that what held the column of water together was its tensile strength, but eventually it would become too heavy and break. Others proposed an alternative model, that up to 10 m the weight of the water was balanced by the weight of air in the atmosphere. Above that height, a vacuum would be created. (Aristotle had held that air has no mass or weight.)

The first experimental test, probably between 1639 and 1644, used a very tall column of water, but was inconclusive due to technical difficulties. Torricelli realized that he could use mercury instead of water. As mercury is 14× heavier than water, the column needed to be only 1 m high. The experiment was performed by filling a glass tube about 120 cm long with mercury and inverting it to put the open end in a bath of mercury. The mercury fell to a height of 76 cm above the bath. Above the mercury was a vacuum. "We live submerged at the bottom of an ocean of the element air, which by unquestioned experiments is known to have weight," Torricelli concluded in 1644.[49]

Word of the experiment spread rapidly, and others soon repeated it. (By 1654 it was being described in English as the 'famous experiment,' and by 1663 it was 'famossima' in Italy.[50])

A more incisive version of the experiment followed in 1648, when Blaise Pascal decided he could test the theory by taking advantage of his location near the Puy-de-Dôme mountain. If the height of the mercury column is determined by balancing the weight of the mercury against the weight of the air in the atmosphere, it should be less at higher altitudes, because the layer of air is not so deep. When Pascal climbed 1000 m to the summit, the mercury was 85 mm lower than at the base of the mountain. It showed corresponding levels between the two limits at various points on the mountain. "All the effects are due to the weight and pressure of the air, which is their only real cause," Pascal said.[51]

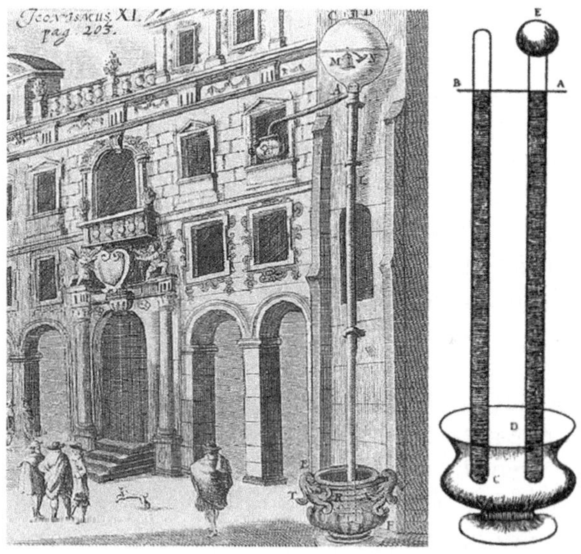

Quite a contrast. Gasparo Berti's test of atmospheric pressure used a lead tube 11 m high filled with water (left). When the bottom was placed in a water bath and unsealed, the height of the water fell to about 10 m. The illustration dates from about 25 years after the experiment. Torricelli used glass tubes of 120 cm length filled with mercury (right). He used two tubes, one with a flat end, the other with a sphere at the top, because he thought the extent of the vacuum would influence the height to which the mercury fell. It was the same 76 cm in both cases, demonstrating that the vacuum itself has no influence. The illustration is from a letter Torricelli wrote shortly after the event.

We have now entered the realm of science. An apparatus was constructed to test a theory: the height to which a liquid column can be supported depends on the weight of the air above it. This is at its maximum at ground level where the layer of air is deepest. In fact, there is a more or less quantitative effect with altitude, from a small, but measurable effect going to the top of high buildings (tested in Paris), to a substantial effect at the top of a mountain. Pascal had used a control, by leaving one mercury barometer at the base of the mountain with an observer, and taking a second tube up the mountain. He repeated the experiment by taking several measurements at each location. Deductions followed directly from experimental results. The only matter remaining for dispute was whether the space at the top was truly a vacuum or might have some air in it.

Furthermore, the experiments had been performed in the presence of witnesses, so others could reproduce them. In the space of less than a decade, science, as it had now become, developed its character. It advanced from Galileo's experiments, performed in isolation and reported sketchily to the point at which their reality has since been questioned, to experiments performed by multiple investigators in public under defined conditions. It's fair to say that with the experiments, we have left natural philosophy behind.

∞ ∞ ∞

Robert Boyle (1627–1691) is usually regarded as the 17th century's experimentalist *par excellence*. Indeed, he is often said to have invented the modern experimental method. Boyle came from a wealthy background. His father was the Earl of Cork, a major landholder in Ireland. He was educated at Eton and then traveled to France and Italy. When his father died in 1643, he inherited estates in Ireland and in Dorset (England), where he established a laboratory. After a brief period in Ireland—"a barbarous country," Boyle said—he moved to Oxford.

In 1657, Boyle heard about an air pump developed by Otto von Guericke in Germany. When the air was pumped out of two joined metal hemispheres, the strength of the vacuum was demonstrated by requiring a team of horses to pull the hemispheres apart. This was dramatic, but not very practical for investigating the properties of the vacuum.

Working with his assistant Robert Hooke, Boyle developed the air pump, a device that could remove air from a chamber to allow the effects of a vacuum to be investigated. The results of the work were published in 1660 in a book,[52] notable for the detail with which the experiments were described. This drew a distinction with earlier work, such as that of Galileo, in which it was not entirely clear which experiments had actually been performed and which might be 'thought experiments,' intended to predict outcomes. Boyle's book was written in English, but also translated into Latin to make it more widely available.

The book started by describing the construction of the air pump in detail, followed by reports of 43 experiments.[53] Among them were "void-in-void" experiments in which the base of a barometer was placed in the chamber. The pump brought the mercury level down to 1 inch, indicating that atmospheric pressure had been reduced by ~97%. The vacuum extinguished a burning candle. (This was not understood at the time because it was a hundred years before oxygen was discovered.) Water began to boil as the result of the vacuum, another result that could not be explained at the time.

Work was somewhat limited by the imperfections of the pump, mainly due to problems with leakage. The importance of the first set of results is not so much with their immediate contribution to knowledge as with the introduction of protocols that defined the beginnings of experimental science.[54] Boyle developed the concept that what was acting on the barometer in the void-in-void experiment was what he called the 'spring' in the air, which we would now call pressure. "[It's] a mistaken Persuasion that those Phenomena are the effects of natures abhorrency of a Vacuum, which seem to be more fully ascribable to the weight and Spring of the air."[55]

Thomas Hobbes (1588–1679) is best known for his book, *Leviathan*, published in 1651. Written during the English civil war, it's a political work arguing for monarchical government, but it is rooted in Hobbes' earlier development as a natural philosopher. Hobbes graduated from Oxford University, and then worked for the aristocratic Cavendish family. He spent the decade of the 1640s in exile in Paris, where he wrote *Leviathan*. Within a decade, he was to become Robert Boyle's nemesis. There is a view that his criticisms of experimental science were foreseen in *Leviathan*, and indeed developed from the philosophy he developed there.[56] He was generally an argumentative character, with a pessimistic view of human nature, characterized by the famous quote from *Leviathan*, "the life of man [is] solitary, poor, nasty, brutish, and short."

Hobbes' natural philosophy was based on plenism, the belief that all space is full of matter and there is no such thing as a vacuum. In his book *De Capore*, published in 1655, he attacked the Torricellian creation of a vacuum and subsequent experiments based on it. His criticism was partially based on semantic arguments and partly on the belief that ether filled what might appear to be empty space. At the risk of over-simplification, all matter was corporeal (basically this is a version of atomist theory) and the concept of incorporeal space (such as a vacuum) is simply a contradiction in terms. Hobbes' first broadside against Boyle's experiment was his book *Dialogus* in 1661,[57] and he never ceased publishing criticisms.

Hobbes raised a series of objections. Natural philosophy should draw inferences from nature. ("They can get engines made... and try conclusions; but they are never the more philosophers for all this".[58]) Experiments comprise an artificial situation. ("I think [observations] a great deal better witnesses of nature, than those that are forced by fire".[59]) On practical grounds, the air pump leaked so its results could not be reliable.[60]

To use an anachronistic term here, there is a whiff of Luddism about the criticisms. Indeed, just as in legal or criminal affairs the first thought about a suspect is *cui bono* (who benefits), so in academia the first question is "whose ox is gored?" If you were committed to the classical precepts of natural philosophy, experimentation was a great threat.

This should not be taken as resistance to the Scientific Revolution. In fact, it's rather a demonstration in this early period of what became a common feature of science: clinging to old theories and resisting the implications of new experiments. Perhaps the first Law of the History of Science should be that there is always resistance to changing the paradigm.

Every major discovery during the Scientific Revolution had its advocates and its adversaries. Conflict was sharpened by the fact that this was a time of transition from natural philosophy to

A modern replica of Boyle's air pump of 1659. The chamber is a glass globe, sealed at the top with a 4-inch diameter opening, sealed by cementing a brass ring in place. This has a smaller hole within it, closed by a ground glass stopper, to allow small objects to be introduced. The capacity was close to 30 liters (limited by the technology of glass blowing). Vacuum is created by turning a handle connected by rack and pinion to a piston that evacuates air through a valve from a brass cylinder, about 14 inches long and 3 inches internal diameter. The piston is made of wood and sealed with leather. The major practical problem was leakage through the joints.

experimental science, so there was not necessarily an agreed framework for discussion.[61] Resistance to new ideas was partly due to refusal to accept the validity of the very concept of experiment, partly due to objections to the experiments themselves. Since then experiments have become the currency of science, and subsequent conflicts have been about criticisms of the experiments or the conclusions drawn from them.

NOTES AND REFERENCES

1. The story dates from the 19th century and is probably apocryphal. See M. Livio, *Galileo and the Science Deniers*, Simon & Schuster, New York, 2020.
2. Translated in S. Drake, *Discoveries and Opinions of Galileo*, Doubleday, New York, 1957, pp. 132 & 135.
3. J. Needham, *Science and Civilization in China, Volume 3: Mathematics and the Sciences of the Heavens and the Earth*, Cambridge University Press, New York, 1959, p. 448.
4. The Scientific Revolution is an exceptionally rich topic for revisionist history. It's unusual today to relate discoveries to present perspective or to connect them to their discoverers. In approximate order from least to most politically correct, recent works are: D. Wootton, *The Invention of Science: A New History of the Scientific Revolution*, Harper, New York, 2015; H. F. Cohen, *The Scientific Revolution: A Historiographical Inquiry*, University of Chicago Press, Chicago, 1994; P. Dear, *Revolutionizing the Sciences: European Knowledge in Transition, 1500–1700*, Red Globe Press, London, 3rd edn, 2019; O. Gal, *The Origins of Modern Science*, Cambridge University Press, Cambridge, 2021; L. Jardine, *Ingenious Pursuits: Building the Scientific Revolution*, Anchor Books, New York, 1999; S. Shapin, *The Scientific Revolution*, University of Chicago Press, Chicago, 2nd edn, 2018; J. Henry, *The Scientific Revolution and the Origins of Modern Science*, Palgrave Macmillan, London, 3rd edn, 2008; P. J. Bowler and I. R. Morus, *Making Modern Science*, University of Chicago Press, Chicago, 2nd edn, 2020; W. E. Burns, *The Scientific Revolution in Global Perspective*, Oxford University Press, Oxford, 2015.
5. A. Koyré, Galileo and the Scientific Revolution of the Seventeenth Century, *Phil. Rev.*, 1943, **52**, 333–348.
6. Koyré does not mention any science except physics (see ref. 5).
7. *The Rise Of Experimental Biology: An Illustrated History*, ed. P. Lutz, Humana Press, Totowa, NJ, 2002.
8. S. Pinker, *Enlightenment Now: The Case For Reason Science Humanism and Progress*, Viking, New York, 2018.
9. This question has been discussed extensively by Cohen in *The Scientific Revolution: A Historiographical Inquiry*, and from a different perspective by Shapin in *The Scientific Revolution*.
10. The protagonists themselves did not regard their activities as revolutionary, but were more inclined to see their activities as a continuation of

previous investigations. See H. F. Cohen, *The Scientific Revolution: A Historiographical Inquiry*, University of Chicago Press, Chicago, 1994, p. 6.

11. This distinction was probably first made by H. Butterfield, *The Origins Of Modern Science 1300–1800*, Bell, London, 1949.

12. From Letters on Sunspots (1613), translated in S. Drake, *Discoveries and Opinions of Galileo*, Doubleday, New York, 1957, p. 64.

13. From *The Assayer* (1623): translated in S. Drake, *Discoveries and Opinions of Galileo*, Doubleday, New York, 1957, pp. 237–238. The quotation is most often paraphrased as "Mathematics is the alphabet with which God has written the universe."

14. S. Drake, *Galileo At Work: His Scientific Biography*, University of Chicago Press, Chicago, 1978, p. xxi.

15. From *The Assayer* (1623), translated in S. Drake, *Discoveries and Opinions of Galileo*, Doubleday, New York, 1957, pp. 225–226.

16. E. Bellone, *A World on Paper: Studies on the Second Scientific Revolution*, The MIT Press, Cambridge, MA, 1980, pp. 87–88, 106–107.

17. G. Galilei, *Dialogue Concerning the Two Chief World Systems—Ptolemaic & Copernican*, translated by S. Drake, University of California Press, Berkeley, 2nd edn, 1967, p. 113.

18. S. Drake, *Discoveries and Opinions of Galileo*, Doubleday, New York, 1957, p. 226.

19. E. Bellone, *A World on Paper: Studies on the Second Scientific Revolution*, The MIT Press, Cambridge, MA, 1980, p. 23.

20. Wootton ascribes its first meaning as unearthing new facts of nature, triggered by the voyages that discovered the Americas and therefore revealed that there could be a land mass in the southern hemisphere. See D. Wootton, *The Invention of Science: A New History of the Scientific Revolution*, Harper, New York, 2015, pp. 57–60.

21. Even in 1621, when Kepler published the second part of *Epitome of Copernican Astronomy*, he described it as a supplement to Aristotle's *On the Heavens*. See D. Wootton, *The Invention of Science: A New History of the Scientific Revolution*, Harper, New York, 2015, p. 252.

22. F. Bacon, *Of the Proficience and Advancement of Learning, Divine and Human*, 1605, (online at www.gutenberg.org/files/5500/5500-h/5500-h.htm).

23. This means 'the new instrument', and is a reference to *Organon*, Aristotle's work on logic.

24. For example, Nicolas Oresme (c. 1320–1382) in France anticipated some of Copernicus' and Galileo's arguments, but it is unknown whether either of them was aware of his work. See T. S. Kuhn, *The Copernican Revolution. Planetary Astronomy In The Development Of Western Thought*, Harvard University Press, Cambridge, 1957, pp. 115–118.

25. A. Einstein, *Ideas and Opinions*, Modern Library, New York, 1994, p. 271.

26. The sole account of the experiment is from Galileo's assistant, Viviani, and there is no consensus as to whether it describes an actual event.

27. Galileo was not the first to have the idea of comparing bodies falling from a height, but earlier experiments had not proved conclusive. The earliest may have been by Philiponus in 6th century Alexandria, who refuted several of

Aristotle's theories. He reported that, "If you let fall from the same height two weights of which one is many times as heavy as the other, you will see that the... difference in time [of descent] is a very small one." See D. C. Lindberg, *The Beginnings Of Western Science: The European Scientific Tradition In Philosophical, Religious, And Institutional Context, Prehistory To A.D. 1450*, University of Chicago Press, Chicago, 2007, pp. 160, 363.

28. The leaning tower of Pisa is 56 m high. In the absence of air resistance, it would take an object 3.3 secs to fall to the ground. Galileo was able to measure time to within about 0.5 secs. This would not be accurate enough to distinguish between differences in the time taken for similar objects to fall to the ground.

29. Galileo did not come straight to this conclusion. At first he thought acceleration was proportional to distance from the starting point. Then he realized that the acceleration is proportional to time elapsed from the start (which means it is proportional to the square root of the distance). His discussed this explicitly in his last book, *Discourses*, in 1638.

30. S. Drake and J. MacLachlan, Galileo's discovery of the parabolic trajectory, *Sci. Am.*, 1975, **232**, 102–111.

31. The experiment has been reconstructed in a way that makes it suitable for reproduction in a high school laboratory; S. Straulino, Reconstruction of Galileo Galilei's Experiment: the Inclined Plane, *Phys. Educ.*, 2008, **43**, 316–320.

32. In the 14th century, Jean Buridan (1330–1358) in Paris had refuted Aristotle's concept that motion depended on the action of the medium surrounding the body. "When a mover sets a body in motion he implants into it a certain impetus, that is, a certain force enabling a body to move in the direction in which the mover starts it." [Quoted in O. Pedersen, *Early Physics and Astronomy: A Historical Introduction*, Cambridge University Press, Cambridge, 2009, p. 210.] Buridan thought that the impetus would be the product of mass and velocity. The work was followed up by others, including Nicolas Oresme (1320–1382).

33. R. Ariew, Descartes as Critic of Galileo's Scientific Methodology, *Synthese*, 1986, **67**, 77–90.

34. Descartes developed the system of Cartesian coordinates in which three dimensional space is described in terms of three perpendicular axes.

35. M. J. Osler, *Divine Will And The Mechanical Philosophy Gassendi And Descartes On Contingency And Necessity In The Created World*, Cambridge University Press, Cambridge, 1994.

36. The movement may have originated with Isaac Beeckman (1588–1637), who had an important influence in causing René Descartes to take up physics in 1618. Beekman himself did not use the term; its introduction five decades later is discussed by K. van Berkel, *Isaac Beeckman on Matter and Motion. Mechanical Philosophy in the Making*, Johns Hopkins University Press, Baltimore, 2013, pp. 82–84.

37. R. Boyle, *The Origine of Formes and Qualities (According to the Corpuscular Philosophy)*, Davis, Oxford, 1666. As indicated in the title, Boyle also used the term 'corpuscular philosophy,' but 'mechanical philosophy' was the

term that entered popular usage. The first use of the term may have been in H. More, *The Immortality Of The Soul, So Farre Forth As It Is Demonstrable From The Knowledge Of Nature And The Light Of Reason*, William Morden, London, 1659. Soon after, in 1672, Huygens referred in a letter to Oldenburg at the Royal Society to "the great difficulty of explaining by mechanical philosophy in what this diversity of colours consists." Originally "Il resteroit encore la grande difficulté d'expliquer par la physique mechanique en quoy consiste cette diversité de couleurs." Quoted in H. W. Thurnbull *et al.*, *The Correspondence of Isaac Newton*, 1959.

38. Important movers and shakers in the movement were Pierre Gassendi (1592–1655), René Descartes (1596–1650), Thomas Hobbes (1588–1679), and Robert Boyle (1627–1691).

39. Gassendi held that Nature consists of invisible atoms and the void; Descartes held that the universe consisted of matter (corpuscles) packed into a vortex.

40. "Whatsoever effects are hereafter to be produced, shall have a necessary cause, so that all the effects that have been or shall be produced have their necessity in things antecedent." Quoted in J. Bronowski and B. Mazlish, *The Western Intellectual Tradition*, Harper & Row, New York, 1960, p. 198.

41. Spinoza was not only a philosopher, he was well equipped scientifically, and constructed microscopes and telescopes. He was in contact with developments in England through a correspondence with Henry Oldenburg of the Royal Society. His philosophy stemmed from the scientific understanding of the period, although he was more of a theoretician and critic than experimentalist. See J. I. Israel, *Radical Enlightenment: Philosophy and the Making of Modernity 1650–1750*, Oxford University Press, Oxford, 2001, p. 245.

42. Quoted in J. I. Israel, *Radical Enlightenment: Philosophy and the Making of Modernity 1650–1750*, Oxford University Press, Oxford, 2001, p. 243.

43. H. M. Nussenzveig, The theory of the rainbow, *Sci. Am.*, 1977, **236**, 116–127.

44. R. Descartes, *Discours de la Méthode Pour bien conduire sa raison, et chercher la vérité dans les sciences*, 1637, Online at gutenberg.org/ebooks/13846.

45. Voltaire, *Letters Concerning the English Nation*, Davis & Lyon, London, 1733, p. 97. Online at www.gutenberg.org/files/2445/2445-h/2445-h.htm.

46. C. Huygens, *Traité de la Lumière: Où Sont Expliquées les Causes de ce qui Luy Arrive Dans la Réflexion & Dans la Refraction*, Pieter van der Aa, Leiden, 1690. Translated by S. P. Thompson, Treatise on Light, Macmillan, London, 1912. Online at www.gutenberg.org/ebooks/14725.

47. Details suggest that the story is apocryphal, or at least cannot be true in the form as told, because the pendulum was not installed in the cathedral at Pisa until after Galileo had left. Viviani seems to have embellished several stories about Galileo. See M. Livio, *Galileo and the Science Deniers*, Simon & Schuster, New York, 2020.

48. R. Descartes, *Principia philosophiae*, 1637, p. 229.

49. J. B. West, Torricelli and the ocean of air: the first measurement of barometric pressure, *Physiology*, 2013, **28**, 66–73.

50. D. Wootton, *The Invention of Science: A New History of the Scientific Revolution*, Harper, New York, 2015, p. 336.

51. B. Pascal, *Expériences nouvelles touchant le vide*, 1647, Online at fr.wikisource.org/wiki/Experiences_nouvelles_touchant_le_vide.

52. For a modern reprint see M. Hunter and E. B. Davis, *The Works of Robert Boyle*, 14 volumes, Pickering & Chatto, London, 1999.

53. J. B. West, Robert Boyle's landmark book of 1660 with the first experiments on rarified air, *J. Appl. Physiol.*, 2005, **98**, 31–39.

54. Boyle's Law, that the volume of a gas is inversely proportional to pressure, was introduced in a second edition of the book, in 1662, revised to answer criticisms of the first edition. The experiment involved pouring mercury into a closed J-shaped tube; the volume of the air at the end contracted when the pressure of the mercury increased.

55. R. Boyle, *New Experiments Physico-Mechanicall Toching the Air*, Tho. Robinson, Oxford, 1660, p. 41.

56. S. Shapin and S. Schafer, *Leviathan and the Air-Pump*, Princeton University Press, Princeton, 2017.

57. T. Hobbes, *Dialogus physicus de natura aeris, conjectura sumpta ab experimentis nuper Londini habitis in Collegio Greshamensi. Item de duplicatione cubi*, 1661, (English translation in ref. 56).

58. T. Hobbes, *Considerations upon the Reputation, Loyalty, Manners, and Religion of Thomas Hobbes*, William Crooke, London, 1662, pp. 436-437.

59. T. Hobbes, *Decameron physiologicum; or Ten Dialogues of Natural Philosophy*, 1678, p. 117.

60. Shapin and Schafer provide an extensive review of the dispute between Hobbes and Boyle (see ref. 56).

61. The unity of the sciences became apparent only somewhat later, as epitomized by Piaget (see J. Piaget, *Recherche*, La Concorde, Lausanne, 1918 p. 59).

Newton's Laws: 1659–1687

"TO EXPLAIN ALL NATURE IS TOO DIFFICULT A TASK FOR ANY one man or even for any one age," said Isaac Newton.[1] One of the greatest scientists who ever lived, Newton was responsible for several revolutions. Establishing his laws of gravity in mathematical instead of mystical terms was a revolution in itself. Describing the attraction between the Earth and an apple in the same terms as between the planets and the Sun demolished Aristotle's legacy. A theoretician in astronomy, Newton was also an experimentalist in optics. Understanding the constitution of light was another revolution. Inventing calculus (simultaneously with Leibniz) created a new sort of mathematics. Of course, all of these discoveries built upon the century of work comprising the Scientific Revolution. By making the transition from providing explanations based on observations and experiments into formulating generalizable laws of physics, Newton set the scene for modern science (although he also continued to work in alchemy). In the following century, there were major advances as physics, chemistry, and biology became separate disciplines.

The Frontiers of Science
By Benjamin Lewin
© Benjamin Lewin 2026
Published by the Royal Society of Chemistry, www.rsc.org

Timeline for Key Scientific Developments during the Enlightenment

1700

Isaac Newton publishes *Opticks* (1704)

1710

Gabriel Fahrenheit defines temperature scale (1714)
Carl Linnaeus describes system of plant classification (1715)

1720

1730

1740

Anders Celsius defines centigrade scale (1742)

1750

Joseph Black describes latent heat (1750)
Benjamin Franklin suggests lightning is electrical (1752)

1760

Mikhail Lomonosov discovers atmosphere of Venus (1761)

Henry Cavendish isolates hydrogen (1766)

1770

Charles Messier catalogs astronomical objects (1771)
Daniel Rutherford discovers nitrogen (1772)

Antoine Lavoisier (and Carl Scheele) identify oxygen (1778)

1780

William Herschel discovers Uranus (1781)

Coulomb defines law of electrical attraction (1785)
Jacques Charles proposes Charles's law of ideal gases (1787)

1790

Antoine Lavoisier proposes law of conservation of mass (1789)

Henry Cavendish measures the density of the Earth (1798)

1800

Alessandro Volta discovers electrochemistry (1800)

A Dialog between Edmond Halley and Gottfried Wilhelm Leibniz

Edmond Halley (1656–1742) was Professor of Geometry at the University of Oxford, a leading figure in the Royal Society, and became the second Astronomer Royal in Britain.

Gottfried Wilhelm Leibniz (1646–1716) was a German polymath, best known as the mathematician who invented calculus. He stands with Descartes and Spinoza as a rationalist philosopher.

© Thepalmer/Getty Images.

Halley: Greetings, Gottfried Wilhelm Leibniz. It has come to my attention that we have some differences about the force that governs the motion of celestial bodies.

Leibniz: Ah, Edmond Halley, a pleasure to engage in discourse with you. You speak of gravity, I presume.

Halley: Precisely, Leibniz. Gravity—the force that pulls objects toward one another. It's the universal force that binds the cosmos. Newton's *Principia Mathematica* laid the foundation for understanding it with his law of universal gravitation. Every particle in the universe attracts every other particle with a force proportional to the product of their masses and inversely proportional to the square of the distance between them.

Leibniz: A bold proposition indeed, Halley. However, what is the underlying mechanism that causes such attraction? Surely we must question how a force might act at a distance? Is it not more likely that there is a medium, an ether, through which this force is transmitted?

Halley: I confess, Leibniz, that the nature of this force acting at a distance troubled Newton. He even described it as an 'absurdity,' and declined to speculate on the mechanism. It's unclear whether it's material or immaterial.

Leibniz: The ultimate nature of gravity remains a profound mystery. It's uncertain whether our finite minds can fully grasp the depths of cosmic reality. But without an ether, the theory requires a constant miracle of bodies acting on one another at a distance across a void. It's a theory based on the occult.

Halley: The mathematical precision of the laws of motion and gravitation stands independent of any material medium.
I used the laws to calculate the movements of the comet now named after me. The theory works! I rest my case.

NEWTON'S LAWS

Crucial experiments afford the readiest and securest means of eliminating extraneous causes, and deciding between the claims of rival hypotheses,[2] Francis Bacon, 1620.

The Scientific Revolution started with observations that recast our view of Earth's position in the universe. Interpretation was still much influenced by religious beliefs and philosophical attitudes. There was even a view that artificial situations (such as experiments) could not represent the natural world. By the end of the period, the idea of testing observations with experiments had become formalized. Newton's conclusions were described as 'laws.' The transition from natural philosophy to science had been accomplished.

Ancient science did not really think in terms of 'laws' in the modern sense. If it had, one way to express the difference between the sublunary sphere (Earth) and the celestial sphere (Outer Space) might have been to say that whatever laws applied on Earth did not apply in the heavens. (Remember that Aristotle considered the Earth to be constituted of air, fire, water, and earth, but the celestial sphere was constructed of ether.) Isaac Newton's demonstration that the same laws control gravity on Earth, and the motion of the planets, was the coup de grace to the ancient view.

Newton's laws were described in *Principia*, published in 1687.[3] The story goes that the book resulted when astronomer Edmond Halley asked Newton if he could work out the path a planet would follow under a force that varied as the inverse square of distance from the center. Newton replied that he had calculated this would be an ellipse, but had lost the proof.

Principia was the recovery of the proof. Actually, Newton sent a short (9 page) manuscript to Halley after their exchange, *De Motu Corporum in Gyrum* (On the Motion of Bodies in an Orbit). Halley duly reported this to the Royal Society. It took longer to develop a full account for publication. "To do this business right is a thing of far greater difficulty than I was aware of," Newton said.[4]

The first parts of *Principia* developed the three laws of motion. The first law states that a body that's at rest, or moving at uniform speed in a straight line, will not change unless an external force acts upon it. The second law places this on a quantitative

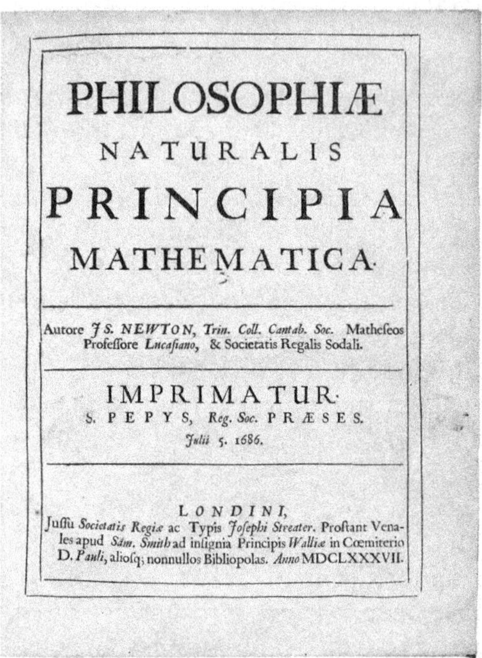

The title page of the first edition of Principia, *printed on vellum. Halley and Newton decided that it should be printed "in a large Quarto, of a fair letter." The publication date is given as MDCLXXXVII (1687).* Principia *was published by Halley instead of by the Royal Society as had been originally intended; see Chapter 24.*

basis by stating that the momentum of a body is the product of its mass and its velocity. The third law (the law of action and reaction) states that when two bodies interact, each applies a force to the other that is equal in magnitude but opposite in direction.

In the last part of *Principia*, Newton applied these laws to show that the force maintaining the moon or the planets in their orbits was the same force that causes acceleration of falling bodies on Earth. "Newton unified the heavens and the earth," is an aphorism that captures the accomplishment.[5] Robert Hooke came to similar conclusions, which he announced in 1674, leading to a fight with Newton about priority. Newton demolished Descartes' theory of vortices by demonstrating that it was incompatible with the mathematics of planetary movements.[6]

Looking through the 600 pages of *Principia* today, it's almost unbelievable that a single person could have produced such a massive and comprehensive body of work. It has resonated through the centuries since its publication, but there can have been very few of Newton's contemporaries who were really able to understand it. Voltaire's comment was that "one must be very learned to understand him."[7] Indeed, a contemporary comment apparently was that, "After Sir Isaac printed his *Principia*, as he passed by the students at Cambridge said 'there goes the man who has writt a book that neither he nor any one else understands'."[8] There was no change in the principles of the physics of gravity for the next 250 years (until Einstein proposed the theory of relativity).

To calculate the effects of gravity on motion, new mathematical methods were required. Newton began to develop calculus in 1665–1666, although he did not publish it until later. Working in Hanover, his contemporary Gottfried Leibniz (1646–1716) developed calculus independently, after being encouraged to take up mathematics by the Dutch mathematician Christiaan Huygens, whom he met in Paris in 1672. Leibniz visited London more than once, was elected as a foreign member of the Royal Society, and corresponded with Newton.

Leibniz began working on calculus in 1674. Leibniz and Newton corresponded in 1676–1677. This led to Newton's subsequent accusation that Leibniz had taken advantage of his unpublished work. A public fight about priority ensued.[9] It's now generally accepted that they developed calculus independently, although Leibniz's form is simpler, and in fact the forerunner of the notation used today.

"All bodies whatsoever are endowed with a principle of mutual gravitation," Newton said in *Principia*. The argument that gravity depends on forces acting at a distance, with no evident material basis, was difficult to assimilate. When Huygens visited London in 1689 and met Isaac Newton, he did not find Newton's work on gravity acceptable because of the lack of a mechanical explanation of causes. His own work on gravity, included in the *Traité de la Lumière,* argued that gravity was due to Cartesian vortices.

Leibniz mounted a continuous attack on Newton's theory, accusing him of introducing 'occult' forces by proposing that gravity works at a distance. The details of their disagreement on

Leibniz built a calculator in 1673 that worked on a decimal basis. It performed all the four basic mathematical functions mechanically. He built four copies (only one has survived).

"Calculating Machine" reproduced from Gottfried Wilhelm Leibniz Bibliothek - Niedersächsische Landesbibliothek, Hannover.

the physics are less significant than the fact that much of Leibniz's attack was based on theological arguments.[10] Even at the culmination of the Scientific Revolution, science could not be disentangled from religion. Emphasizing the mix of activities in the period, Leibniz was not only a great theoretician, but also practical to the point at which he built a calculating machine—a sort of mechanical abacus.

The story that Newton conceived the concept of gravity when he saw an apple fall from a tree (the story that it fell on his head is an embellishment) was thought to be apocryphal until a discovery in the archives of the Royal Society. William Stukeley described a visit to Newton in London in 1726. "After dinner, the weather being warm, we went into the garden, and he told me, he was just in the same situation, as when formerly, the notion of gravitation came into his mind. 'Why should that apple always descend perpendicularly to the ground... why should it not go sideways, or upwards? but constantly to the earths centre? Assuredly, the reason is, that the earth draws it... it must be in proportion of its quantity. Therefore the apple draws the earth, as well as the earth draws the apple."[11]

Principia was in the tradition of astronomy: making deductions from observations. Newton's major work on optics was in a more direct experimental tradition. In his first paper on the subject in 1672, Newton rejected the classical view (dating from Aristotle) that colors result from combining white light with darkness, and suggested the opposite: white light results from combining lights of different colors.[12]

Published later, in English in 1704, *Opticks* reported Newton's experiments, including his famous two-prism experiment.[13] Newton made a small hole in a window shutter, to allow a beam of light to enter a dark room. The light passed through a prism, which refracted it into a series of colors on the wall in front. He called this the spectrum (derived from the Greek for 'to see'). When the light was passed through a second prism (inverted with respect to the first prism), the colors recombined to give white light (because the colors are converted into parallel rays by the second prism and therefore blend to white in the eye).[14] There could scarcely be a more direct demonstration that white light is a combination of component colors.[15]

It's impossible to do justice to Isaac Newton (1642–1726) with any brief account. Polymath is a woefully inadequate description. Newton's early years were not happy. His grandmother brought him up until his stepfather (whom he hated) died in 1653. He had been at The King's School in Grantham, then returned to his mother, to work on the family farm. When that was not successful, he was readmitted to the grammar school. He went up to Trinity College at the University of Cambridge in 1661.

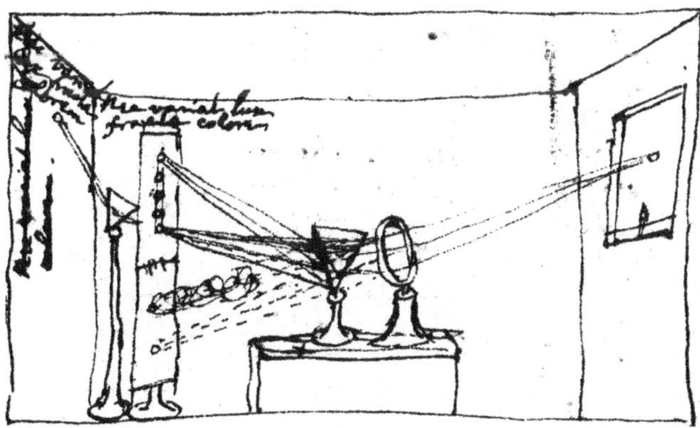

A drawing of the two-prism experiment from Newton's notebook.

The university closed because of the Great Plague, but Newton returned in 1667 as a fellow of Trinity, and became the Lucasian Professor of the university in 1669. He was a consummate mathematician and scientist, but also retained a lifelong interest in alchemy. He moved to London in 1696 when he became Warden of the Royal Mint, where he was much concerned with counterfeiting and currency reform.

Newton defined the scientific method. His description of his approach was Baconian. "Analysis consists in making Experiments and Observations, and in drawing general Conclusions from them by Induction, and admitting of no Objections against the Conclusions, but such as are taken from Experiments, or other certain Truths."[16]

Indeed, Newton's reputation extends even beyond his results and methods. For years after his death, there was a sort of 'Newtonmania,' which acknowledged his work as a landmark not only in the sciences, but throughout the humanities. Voltaire was a famous supporter, John Locke tried to understand his results, and David Hume described him as "the greatest and

Isaac Newton at the height of his powers in 1689, painted by Gottfried Kneller. A later (more formal) portrait painted by Kneller in 1702 is in the National Gallery in London.

rarest genius that ever rose." Far from a clash between science and the humanities, Newton's science was seen as a major contribution to the Enlightenment.[17]

Newton took a view that was not uncommon during the Scientific Revolution: uncovering the laws of Nature demonstrated the existence of God, and provided evidence into His workings. He thought that once the universe had been created by establishing natural laws, the laws would operate to ensure it continued to run smoothly. "Being once form'd, it may continue by those Laws for many Ages."[18] (This did not deny God's continued presence: see below.)

∞ ∞ ∞

Without going any further, we have already seen the essence of the Scientific Revolution, emphasizing more precise observations (in astronomy), experimentation (in physics), and explanations based on mathematical analysis. In retrospect, this was the point at which the laws of physics began to be described. At the time, only Newton described his own results in terms of laws. The laws uncovered by Galileo, Kepler, Boyle, Hooke, and Huygens, were all named as such only in the first decades of the 18th century. Although the Scientific Revolution is most remembered for this change in thinking in the physical sciences, there was also progress in biology, exemplified by three major figures: William Harvey, Robert Hooke, and Antonie van Leeuwenhoek.

Born in England, William Harvey (1578–1657) obtained a medical degree from the University of Padua, before returning to England, where he obtained a degree from Cambridge. He rose to the top of the profession, becoming physician to the King in London.

Previously, it was thought that blood was produced and then consumed by the body. Harvey showed that the amount of blood pumped through the heart was several times the body's content of blood, implying that the same blood must circulate. He found one-way valves that allowed blood to pass from the heart into the body. His work was performed around 1618, and published a decade later in a book in 1628. It was not immediately accepted, but became common wisdom by the time of his death. (I discuss the implications for medicine in Chapter 21.) It stands alone as a refutation of Galen's classic view that blood does not circulate, which had held since Roman times. Unlike discoveries in astronomy and physics, it was not part of a wider body of work.

The invention of microscopes was closely linked with the invention of telescopes. Both came from Dutch spectacle makers after 1600, and involved grinding lenses and then combining them to get the appropriate type of magnification. The first compound microscopes (using two lenses) could magnify 20–30×. Robert Hooke was one of the first to use a microscope for scientific observations.

Robert Hooke (1635–1703) got his start in science in 1655, working as an assistant to Robert Boyle on the construction of the air pump. His mechanical skills made a large contribution to the success of the project. Boyle helped him to become curator of experiments for the Royal Society, and he was elected a Fellow in 1663. Not at all in the tradition of the rich gentleman scientist, he earned his living at first by teaching geometry at Gresham College in London. After the Great Fire of London, he worked as a city surveyor. He was, by all accounts, somewhat cantankerous.

He built a compound microscope in 1663, and published his observations in *Micrographia* in 1665, under the aegis of the Royal Society.[19] The book became a best seller. In exquisite drawings, he described insects such as the louse and flea,

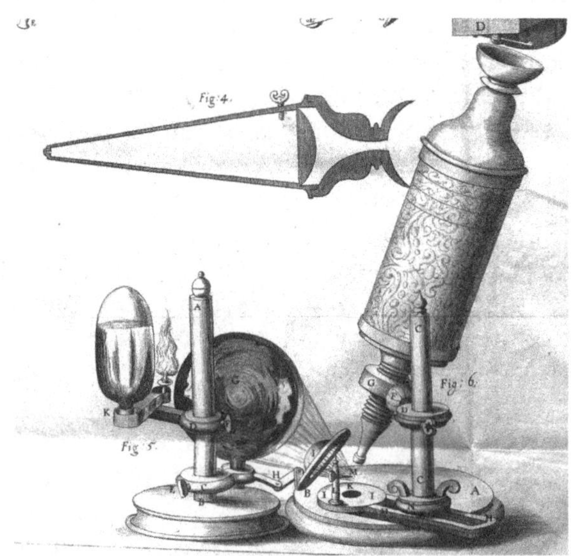

Hooke's microscope, as illustrated in Micrographia.

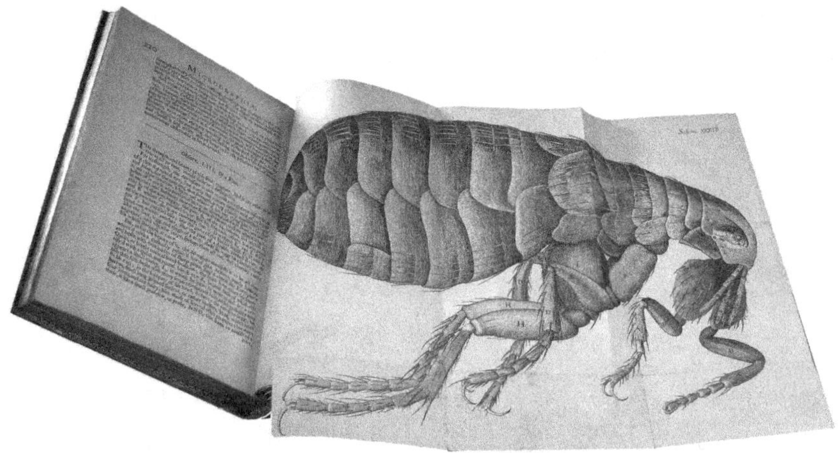

Hooke's foldout drawing of a flea in Micrographia.

presented as large size foldouts in the book. His observations of plants had enough resolution that he introduced the term 'cell' for the basic unit (as observed in cork). He also worked with telescopes. In other spheres, he proposed a wave theory of light and suggested an inverse square law for gravity, which brought him into conflict with Isaac Newton over the question of priority.

In Holland, Antonie van Leeuwenhoek (1632–1723) ground his own lenses to such a high standard that although his microscope was more like a magnifying glass (with only one lens) it could magnify up to 200×. Van Leeuwenhoek started out in commerce, obtained a position with the sheriffs of Delft, and started to grind lenses as a hobby. Much of his methodology remains unknown. He observed objects as small as bacteria ("very little animacules"), and studied insects such as the flea and the ant.

Leeuwenhoek became a celebrity. Constantijn Huygens wrote to his son Christiaan in 1680, "The whole world still comes to pay homage to Leeuwenhoek as the great man of the century!"[20] Leeuwenhoek communicated regularly with the Royal Society in London—he was elected a Fellow in 1683—and published most of his observations in its *Philosophical Transactions*. A major intellectual insight from his work was that small organisms could be just as complex as large organisms.

The difference in this period between biology and the physical sciences is exemplified by Robert Hooke's results. In physics, Hooke's Law describes his discovery of the principle of elasticity, the linear variation of tension when an elastic spring is extended. This led to the development of the hairspring, which enabled watches to be made. In biology, his results were presented as observations. Striking as his results and van Leeuwenhoek's were, they did not immediately lead to any increased theoretical understanding.

∞ ∞ ∞

"Knowledge is power" is the aphorism for which Francis Bacon (1561–1626) is most famous,[21] but he also was the first proponent of the idea that science advances by experiments. Yet paradoxically he never actually performed any experiments himself or made any discoveries. Less seriously, skeptics about Shakespeare's authorship have attributed his plays to Bacon. The dichotomy illustrates Bacon's life as part natural philosopher, part courtier and politician.

His father was Sir Nicholas Bacon, the highest judge in England under Queen Elizabeth I, so he was born into a political family. He went to Trinity College in Cambridge University when he was only 12, followed by studying law at Gray's Inn in London. He spent three years traveling in Europe until his father died in 1579. His day job, as it were, was in politics, as a member of parliament. He rose continuously through the legal profession, until becoming Attorney General in 1613. His public career ended in disgrace in 1621, on charges of corruption.

There could scarcely be a greater contrast with his other life, as a natural philosopher, marked by a penetrating purity of thought. He published more or less continuously from 1597 to 1623. His most famous works are the *Advancement of Learning*, published in English in 1605, and the *Instauratio Magna* (The Great Renewal), which included the *Novum Organum* (New Method), published in Latin in 1620.[22] Bacon was opposed to the Greek philosophers on the grounds that their theories were not based on objective knowledge (translating into present terminology) and that they dismissed as uninteresting anything they could not know.[23]

Bacon was one of the first theorists to concern himself with the methodology for acquiring knowledge. He rejected the classic dependence on Aristotle on the grounds that this could never lead to new knowledge. Advocating the importance of knowledge, the *Advancement of Learning* attempted to classify learning into history, poetry, and philosophy, and to divide each into subcategories. In *Instauratio Magna,* his advocacy of experiments may have been the first clear description of the concept of the 'scientific method.' He also complained that people "sought after *Experiments of Use* and not *Experiments of Light and Discovery.*" He meant that they were too concerned with the practical (what we would call 'applied research' today) and not enough with the theoretical ('basic research').

Bacon thought science should proceed from induction: performing experiments to ascertain facts that could then be generalized into laws or rules. This also made him one of the first philosophers to consider the concept that nature was governed by laws. He refuted the idea that experiments were artificial. "The artificial does not differ from the natural in form or essence, but only in the efficient."[24] (By 'efficient' he meant 'cause,' that is, that man had created the situation.) He discussed the need for reproducibility and even touched on the idea that theories could be subject to falsification (he called it exclusion).

Francis Bacon coined the concept of the 'crucial experiment.' The phrase he actually used in *Novum Organum* was *instantia crucis*. He defined this as meaning the creation of an experiment to distinguish between two rival theories. Robert Boyle used the phrase *experimentum crucis* to describe Pascal's experiment to measure atmospheric pressure at Puy-de-Dôme (see Chapter 5). Newton then famously described his own two-prism experiment as an *experimentum crucis*, and the concept was established. Today we talk about the killer experiment.

∞ ∞ ∞

The Scientific Revolution developed a more 'scientific' approach, but thinking was still restricted by the prevailing worldview. One striking example of limitations imposed by the environment comes from the work of William Gilbert (1544–1603) on magnetism.[25] This was a prime source for mysticism as it involved action at a distance with no visible physical interactions.

An English physician—he was physician to both Elizabeth I and then James I—Gilbert was a formidable scientist, who concluded that the Earth itself is a giant magnet. That was why compasses point North. He disposed of a range of superstitions about the function of lodestones (natural magnets used as compasses) by experiment. In this, he was well ahead of his time. His magnum opus, *De Magnete*, was published in 1600.[26] The concept that the Earth is a magnet provided the basis for extending the concept of magnetism to celestial bodies and proposing that magnetism might be responsible for gravity. This was not unreasonable, although, of course, incorrect.

But when it came to explaining why a magnet is attracted to point north, Gilbert resorted to mysticism of his own. "By the wonderful wisdom of the Creator, therefore, forces were implanted in the earth, forces primarily animate, to the end the globe might... take direction, and that the poles might be opposite."[27]

The difficulty in understanding the action of forces at a distance is seen in two erroneous conclusions during the Scientific Revolution. Galileo thought that the movement of the tides was due to the motion of the Earth. In fact, he thought this would be a decisive argument in favor of the Earth's movement around the Sun. Gilbert thought magnetism could explain the relationship between the Sun and the Earth. Kepler later took up the same theory. Both tidal and magnetic theories were disproven by the discovery of gravity.

∞ ∞ ∞

Once we arrive at basing science on experiments (or observations), the difference with mathematics becomes clear. In the tradition of ancient Greece, mathematics blended into natural philosophy. Any deductions made from true axioms must be irrefutable. Euclid developed his geometry from a set of axioms. Aristotle developed his philosophy from a set of suppositions that he thought were true facts. These were somewhat parallel processes. Belief in Euclidian geometry as an insight into the real world was advocated as late as the 17th century.[28]

It was debatable whether mathematics should be classified as a science.[29] Mathematicians might argue that mathematics provided the true basis for observations (such as in astronomy), whereas philosophers might argue that mathematics does not demonstrate the causes of events. Galileo and Kepler made the transition from regarding mathematical descriptions as pragmatic (providing better predictability) to believing that they revealed the physical nature of things.[30]

The Scientific Revolution could scarcely have occurred without mathematics, which was crucial to interpreting the key experiments. Indeed, there is a view that Galileo and Kepler were just as important for introducing the approach of using strict mathematics to analyze their results as for the results of the analyses themselves.[31] Yet aside from astronomy, mathematics was a tool rather than leading to new insights in itself. Insofar as there were new developments in mathematics, such as Napier's introduction of logarithms or Descartes' development of geometry, they remained in the realm of pure thought, without revealing new features of the natural world.

Of course, Descartes took an opposite view, that *only* features described by mathematics could be part of the physical world. From the perspective of a 20th century physicist, Richard Feynman described the difference between mathematics and physics much later. "Mathematicians are only dealing with the structure of the reasoning and they do not really care about what they're talking about."[32]

Mathematics is an unexpected term to use in the context of revolution. There is a view—admittedly not accepted by all mathematicians—that there cannot be revolutions in mathematics.[33] A revolution implies that an existing theory has been replaced because it was incorrect (or at least provides an inadequate explanation). But in mathematics, if an axiom is true, deductions following from it are true. They may be extended by subsequent work, but they cannot be repealed.[34,35]

∞ ∞ ∞

When Isaac Newton became President of the Royal Society in 1703, he described its aims. "Natural Philosophy consists in discovering the frame and operations of Nature, and reducing them, as far as may be, to general Rules or Laws—establishing these rules by observations and experiments, and thence deducing the causes and effects of things."[36] Here is the transition to what we would now call science—although Newton still called it natural philosophy.[37]

(It's not an anachronism to talk of physics historically, because until the 17th century, 'physics' was used interchangeably with 'natural philosophy.' During the Scientific Revolution, physics gradually acquired its present significance. Kepler may have been one of the first to use it in its present meaning when he entitled his book *Astronomia Nova seu physica coelestis* [New Astronomy or Celestial Physics] in 1609.)

Science now moved steadily into explaining natural phenomena that previously had seemed to have mystical origins. Comets were seen originally as messages from the God(s). One of the reasons for the collapse of cosmology was Brahe's demonstration that a comet was beyond the Moon, in the celestial realm (Chapter 4).

What is now known as Halley's comet has been observed since 240 BCE. It follows a highly elliptical orbit around the Sun that makes it visible from Earth periodically. Edmond Halley (1656–1742) used Newton's laws in 1705 to calculate that recurrent

Halley's comet passed near earth in 1066. Taken by William the Conqueror as an omen of success, it was depicted in the Bayeux tapestry.

Halley's comet in 1986, on its 5th return after Halley identified it.

NASA/NSSDCA.

observations in 1531 and 1607 were due to the same comet. He concluded that that its path recurred every 75–79 years. He predicted its return for 1682.

∞ ∞ ∞

So much of the work in the 17th century depended on light— basically all of astronomy—that it's amazing to realize it was still believed that light travels instantaneously. "Light is due to the presence of something, but it is not a movement," Aristotle had said.[38] Descartes thought light moved instantaneously because otherwise there would be a delay in seeing an eclipse after the Earth moved between the Sun and the Moon.[39] Galileo tried to measure the speed of light, but the instruments available lacked sufficient resolution. The conclusions were in error, and attempts at measurement failed, because no one had realized just how fast the speed of light could be.

A speed for light was measured, not exactly accidentally, but as an indirect consequence of other work, by Danish astronomer Ole Rømer (1644–1710) when he was working at the Paris Observatory in 1676.[40] The work followed Galileo's suggestion that eclipse of Io, the innermost moon of Jupiter, by the planet could be used to calculate longitude. Giovanni Cassini, the

Director of the Observatory, had been trying to produce tables predicting when an eclipse would be visible from a given longitude.

An eclipse occurs once every orbit, which means every 42.5 hours. Yet seen from Earth, the time between eclipses is not constant. Rømer realized that the periods between eclipses became shorter when the Earth and Jupiter were closer, and longer when Earth and Jupiter were farther apart. The cause of the difference is that the time it takes for the light to reach Earth depends on the distance between Earth and Jupiter.

The speed of light can be calculated by its effect on the timing of the eclipse (the difference is several minutes).[41] In modern interpretations of the data, the answer comes to $226\,000$ km s^{-1}, some three quarters of the actual speed of $299\,792$ km s^{-1} (due to errors in some of the data). The exact number is less important than the fact that light was demonstrated to have a finite speed. In another demonstration that facts do not always convince everyone, the conclusion remained controversial.

Perhaps Robert Hooke had the last word, at least for the 17th century, when he said that the speed was, "so exceedingly swift that 'tis beyond Imagination... and if so, why it may not be as well instantaneous."[42] The speed became pertinent as astronomy advanced. Skeptics were convinced when the English astronomer James Bradley (1693–1762) measured the speed in 1728 (getting $301\,000$ km s^{-1}) from the way the stars change in position as the angle of the Earth changes.[43] Greater distances (and fewer complications) made this result convincing.

∞ ∞ ∞

Did the 'Scientific Revolution' really exist? There is not much doubt that the 17th century started with a worldview derived from the ancient world, and ended with concepts based on precise observations and experiments. At the start of the century, the Aristotelian worldview was completely dominant. By the end of the century, it had been completely demolished.

The 17th century established the concept that science advances by experimentation. The transition from observation to experimentation passed through stages. Aristotle's approach was based on explaining experiences or observations that were commonplace rather than making discoveries. The observations were not in question because anyone could make them. As the emphasis switched to making observations, for example, with specialized equipment, and then to using that equipment to manipulate the environment, the question became whether the observations were reliable.

This led to a reliance on witnesses. Pascal in physics and Harvey in biology, for example, both put considerable emphasis on the presence of witnesses who could attest to the accuracy of their reports.[44] Boyle's report of his experiments with the air pump marked the transition to relying on detailed descriptions that could, in principle, be replicated independently by any interested party. Reproducibility of experiments became a mantra of the Royal Society and of modern science.

This brings us to another view, that the 17th century marked the introduction of the concept of the 'fact.'[45] Before then, there had only been 'phenomena,' more subject to massaging to fit theories. Belief in facts went hand in hand with the move to test theories by experiments. Huygens captured the change from Aristotelian logic, and the difference between mathematics and science in 1690, in the preface to his *Treatise on Light*. "Whereas the Geometers prove their Propositions by fixed and incontestable Principles, here the Principles are verified by the conclusions to be drawn from them."[46] Now we have arrived at the beginnings of science.[47]

Even if it was not immediately evident, everything changed in 1543, when in the same year Copernicus started a revolution in astronomy and Vesalius started one in medicine and biology (see Chapter 21). Petrus Ramus (1515–1572), whose Master's thesis in philosophy at the University of Paris in 1536 had the theme, "everything that Aristotle has said is false," followed up in 1543 with a book, *Animadversions on Aristotle*.[48] It was no longer unthinkable to question the rule of the ancient authorities.[49,50]

Even more important than the accumulation of factual knowledge and laws to explain it, the 17th century was the period during which the notion of scientific enquiry began to be defined. "The discovery and use of scientific reasoning by Galileo was one of the most important achievements in the history of human thought," Albert Einstein said.[51] Physicist Max Born thought Galileo fixed the idea of scientific research. "The sci-entific attitude and methods of experimental and theoretical research have been the same through the centuries since Galileo and will remain so."[52]

The legacy of the Scientific Revolution was scientific curiosity —and the means to satisfy it through the ability to argue on the basis of new experiments and observations, instead of recapitulating and interpreting the conclusions of the Ancients. To deny its reality is more a political statement than an objective view.[53] It was confined to Western Europe, with results percolating to other regions only after a century or more.[54]

There was conflict between the ideas of René Descartes and Francis Bacon as to how to develop scientific method. Descartes advocated the view that humanity has an innate knowledge of some things. This belief reinforced the idea that deductions from everyday observations can be a basis for knowledge (see Chapter 5). "Knowledge is conviction based on a reason so strong that it can never be shaken by any stronger reason," he said.[55]

Philosopher John Locke dismissed Descartes' concept of innate knowledge. Locke (1632–1704), a philosopher in the tradition of British empiricism following Francis Bacon, took the view that the human brain is a *tabula rasa*, a blank slate, that develops on the basis of sensory experience. "When has it all the materials of reason and knowledge? To this I answer, in one word, from *experience*."[56] (That view relates directly to the modern concept of training neural nets to imitate brain function, as I discuss in Chapter 23.)

Bacon proposed to work in the opposite direction from Descartes, developing hypotheses from empirical facts, and then testing those hypotheses. Bacon's approach is taken to be the forerunner of modern science.[57]

The medieval era had certainly comprised a long delay in the advancement of knowledge. There is no better agreement on why the Scientific Revolution occurred in the West in the 17th century than if it ever actually occurred at all.[58] Perhaps the Scientific Revolution might have come sooner, but was held back by lack of technology. At the start of the 17th century, scientific investigations used the same equipment as for the past thousand years. During the century, technology and science reinforced one another—such as telescopes for astronomy and microscopes for biology—to create a tipping point.

∞ ∞ ∞

"Ambulo ergo sum" (I walk therefore I am) might seem like a flippant retort to Descartes' famous aphorism, cogito ergo sum, but actually it captures the debate about the mind-body problem. Pierre Gassendi's response to Descartes made the point that, because walking involves direct sensory perceptions of the environment, it's a more reliable indication of existence than merely thinking.

Gassendi (1592–1695) drew no distinction between mind and body, in opposition to Descartes' view that they must be separated. Although better known as a historian and philosopher (and a Catholic priest), he played a significant role in spreading the idea of science in the first half of the 17th century. Gassendi was a major advocate of atomism. Mingling science and philosophy, the differences between Descartes and Gassendi were at the center of intellectual thinking in the 17th century.

The extension of the universe questioned the stability of humanity's place within the bounds of medieval theocracy. This was captured in a famous quote from Pascal. "The eternal silence of infinite space frightens me."[59] The sense of purpose in the operations of the universe was replaced by an analogy with clockwork. Johannes Kepler was one of the first to express this, in 1605. "My aim in this is to show that the celestial machine is to be likened not to a divine organism but rather to a clock-work."[60] Robert Boyle viewed the universe in 1674 as a "self-moving engine... a great piece of clockwork."[61] (This was not at all a rejection of theocracy. God was seen as the clockmaker.[62])

Boyle explained that reliable mechanisms might be mistaken for purposeful. "Jesuits... are said to have presented the first watch to the King of China, who took it to be a living Creature... I could not have brought an argument to convince the Chinese-Monarch, that it was not endowed with Life."[63] Thomas Hobbes was disdainful of the concept of purpose. Dismissing Aristotle's view of why objects fall, he said, "heavy things... endeavour to be there... as if stones and metals had a desire, or could discern the place they would be at, as man does."[64] Descartes viewed the body as a machine, distinct, of course, from the mind, which was incorporeal.[65]

The concept of a clockwork universe independent of the divine has often been equated with Isaac Newton, but in fact that was not Newton's view at all. Committed to his religion, Newton believed that divine intervention was continual. In an exchange of letters between Leibniz and Clarke (representing Newton) in 1715, Leibniz criticized Newton's position for implying that God was imperfect. "God's watch—the universe—would stop working if he didn't re-wind it from time to time! He didn't have enough foresight to give it perpetual motion." Clarke dismissed the criticism. "The idea that the world is a great machine that goes on without intervention by God, like a clock ticking along without help from a clockmaker—that's the idea of materialism and fate."[66] The letters were probably the most famous philosophical exchange of the century. They attest to the ongoing interaction between science and religion.

Descartes had separated the soul from the body, but the nature of the soul came into question.[67] Drawing on Descartes, a book in France argued in 1662 that the soul is responsible for thought.[68] However, materialism (to use an anachronistic term) began to question the existence of the soul. John Locke's *Essay Concerning Human Understanding*, published in 1690,[69] referred to 'mind' rather than 'soul.' Anticipating the first experimental analysis of the brain by more than two centuries, French surgeon Julien Offray de la Mettrie argued in 1747 that "thought is one of the properties [of organized matter] like electricity."[70] The point is not whether the theories were right or wrong, but that science was now impacting philosophy, and by implication society in general, in a way it had not been able to do before.

The *concept* of science (albeit not necessarily with quite the same connotations as those of today) was beginning to impact thinking about the nature of society. In his famous book of 1739, *A Treatise on Human Nature*, David Hume (1711–1776), a British philosopher renowned for his skepticism, said, "There is no question of importance, whose decision is not compriz'd in the science of man; and there is none, which can be decided with any certainty, before we become acquainted with that science."[71] In the course of a century, the tide had turned, from theology influencing acceptable interpretations of science, to science influencing philosophy and encouraging it to replace theology. All this was in the intellectual realm. The direct influence of science in the form of technology that affected daily life was still a century ahead.

The Scientific Revolution shows the difficulties, perhaps even the impossibility, of breaking out entirely from the prevailing worldview. Science made the most amazing progress in the 17th century, from reasoning based on erroneous suppositions to conclusions and theories based on experimental observations. But its practitioners continued to believe in astrology, alchemy, and magic potions.

It does not seem that Isaac Newton or Robert Boyle, both great scientists, drew a distinction between their scientific work and their practice of alchemy, except that they were open about science and secretive about alchemy. (I discuss this in more detail in Chapter 7.) In the context of the period, the search for alchemical transformations was not necessarily unscientific in itself.

When Galileo presented his analysis of the solar system in the form of a dialog, he was following a discursive format reminiscent of a previous era when natural philosophy was a part of philosophy. By the time Newton analyzed planetary motions, the science of physics had separated from its origins in natural philosophy. Newton's presentation in *Principia* was not easily related to other intellectual activities. The new calculus made it difficult to follow for anyone but a mathematician. This was the forerunner of physics as a separate science, expressed in equations that only a physicist can understand, and that need interpretation to be explained in words like other intellectual activities. Since then, the sciences have separated from one another as well as from other intellectual activities.

As science developed its independence, it became as much an accompaniment to other intellectual activity as a replacement for it. With independence, it was no longer constrained by theology, but it still functioned within the worldview of the era. This poses a question: to what extent might scientists today be limited by the current worldview? Have we finally reached a rational place or are there still incongruities with modern beliefs? Is science finally unbound?

NOTES AND REFERENCES

1. Quoted in R. S. Westfall, *Never at Rest: A Biography of Isaac Newton*, Cambridge University Press, Cambridge, 1981, p. 643.
2. F. Bacon, *Novum Organum*, 1620. Online at oll.libertyfund.org/title/bacon-novum-organum.

3. The first edition was published in 1687 in Latin under the title of *Philosophiæ Naturalis Principia Mathematica*, probably with about 500 copies. A second edition of 75 copies was published in 1713. Cambridge University Library has Newton's copy of the first edition with notes for revisions for the second edition. A third edition of 1250 copies was published in 1726. The first English translation was in 1729. For a modern translation, see I. B. Cohen and A. Whitman, *The Principia*, University of California Press, Berkeley, 2016.

4. *The Correspondence of Isaac Newton*, ed. H. W. Turnbull, Cambridge University Press, Cambridge, 1960, vol. 2, p. 445.

5. From a PBS TV show, *NOVA – The Elegant Universe. Einstein's Dream*, October 28, 2003.

6. "It's manifest, that the planets are not carried round in corporeal vortices. For according to the Copernican hypothesis, the planets going round the Sun, revolve in ellipses, having the Sun in their common focus; and by radii drawn to the Sun describe areas proportional to the times. But now the parts of a vortex can never revolve with such a motion... the hypothesis of vortices is utterly irreconcilable with astronomical phenomena, and rather serves to perplex than explain the heavenly motions." I. Newton, *The Mathematical Principles of Natural Philosophy*, 1729, translated by A. Motte, Benjamin Motte, London, vol. 2, pp. 197–199.

7. In Voltaire's 14th letter (1728), *On Descartes and Newton*. Reprinted in *Voltaire. Philosophical Letters*, ed. J. Leigh, Hackett Publishing, Indianapolis.

8. Quoted in Keynes Manuscript 130.5, Miscellanea, Assorted Anecdotes Concerning Newton and Notes for Conduitt's Memoir (but otherwise unsubstantiated). Online at www.newtonproject.ox.ac.uk/view/texts/normalized/THEM00168.

9. A. R. Hall, *Philosophers At War: The Quarrel Between Newton and Leibniz*, Cambridge University Press, Cambridge, 2002.

10. E. J. Khamara, *Space, Time, and Theology in the Leibniz-Newton Controversy*, Transaction Books, Piscataway, NJ, 2006.

11. W. Stukeley, *Memoirs of Sir Isaac Newton's Life*, 1752, Online at www. newtonproject.ox.ac.uk/view/texts/normalized/OTHE00001.

12. I. Newton, A Letter of Mr. Isaac Newton ... containing his new theory about light and colors, *Philos. Trans. R. Soc. London*, 1672, **80**, 3075–3087.

13. I. Newton, *Opticks: or, A Treatise of the Reflexions, Refractions, Inflexions and Colours of Light*, 1704. For a modern edition see I. B. Cohen, *Opticks, Based on the 4th edn, London, 1730*, Dover Publications, 2012. There was a Latin translation in 1706.

14. One of the most ludicrous claims that science is not objective, indeed virtually that it is more less imaginary, comes from Schaffer's analysis of Newton's prism experiment (see S. Schaffer, Glass Works: Newton's Prisms and the Uses of Experiment, in *The Uses of Experimental Studies in the National Sciences*, ed. D. Gooding, T. J. Pinch and S. Schaffer, Cambridge University Press, Cambridge, 1989, pp. 67–104). The claim is that the result holds true only for Newton's prism, or more specifically, for prisms

manufactured in England. This is based on the failure of experimenters in Belgium and Italy to reproduce the experiments at the time. The argument is that Newton's experiments were "staged before chosen witnesses" and that his word carried the day because of his "authority." Never mind that the experiment can be repeated today in any high school. This intellectual dishonesty has been thoroughly debunked (see A. E. Shapiro, The Gradual Acceptance of Newton's Theory of Light and Color, 1672–1727, *Perspec. Sci.*, 1996, 4, 59–140). For an exposé see D. Wootton, *The Invention of Science: A new History of the Scientific Revolution*, Harper, New York, 2015, pp. 521–523.

15. "Colours are not *Qualifications of Light*, derived from Refractions, or Reflections of natural Bodies (as 'tis generally believed,) but *Original and connate properties*... Some Rays are disposed to exhibit a red colour and no other; some a yellow and no other... When any one sort of Rays hath been well parted from those of other kinds, it hath afterwards obstinately retained its colour... *Whiteness* is the usual colour of *Light*; for, Light is a confused aggregate of Rays indued with all sorts of Colors." I. Newton, A Letter of Mr. Isaac Newton... containing his New Theory about Light and Colors, *Phil. Trans. Roy. Soc.*, 1671, **6**, 3075–3087.

16. I. Newton, *Opticks: or, a Treatise of the Reflections, Refractions, Inflections and Colors of Light*, W. & J. Innys, London, 2nd edn, 1718, p. 380. Online at www.newtonproject.ox.ac.uk/view/texts/normalized/NATP00051.

17. P. Gay, *The Enlightenment: Vol 2 Science of Freedom*, Norton, New York, 2013, pp. 126–156.

18. I. Newton, *Opticks: or, a Treatise of the Reflections, Refractions, Inflections and Colors of Light*, W. & J. Innys, London, 2nd edn, 1718, p. 378.

19. R. Hooke, *Micrographia: or, Some physiological descriptions of minute bodies made by magnifying glasses*, J. Martyn and J. Allestry, London, 1655.

20. "Tout le monde court encore chez Leeuwenhoek comme le grande homme du siècle!" Quoted in L. Snyder, *Eye of the Beholder: Johannes Vermeer, Antoni van Leeuwenhoek, and the Reinvention of Seeing*, Norton, New York, 2015, p. 239.

21. What he actually wrote, in *Meditationes Sacra* in 1597, was 'ipsa scientia potestas est' (knowledge itself is power). The simpler phrase was used in *Leviathan* by Thomas Hobbes (who was briefly Bacon's secretary).

22. In 1623 a Latin translation of the *Advancement of Learning* was added.

23. P. Zagorin, *Francis Bacon*, Princeton University Press, Princeton, 1999.

24. In *De Augmentis* (part of *Instauratio Magna*), p. 410 in the translation by J. Spedding, *The Works of France Bacon*, Houghton Mifflin, Cambridge, 1900.

25. G. L. Verschuur, *Hidden Attraction: The History and Mystery of Magnetism*, Oxford University Press, Oxford, 1993.

26. W. Gilbert, *De Magnete, Magneticisque Corporibus, et de Magno Magnete Tellure* (On the Magnet and Magnetic Bodies, and on That Great Magnet the Earth), Peter Short, London, 1600.

27. W. Gilbert, *De Magnete, Magneticisque Corporibus, et de Magno Magnete Tellure* (On the Magnet and Magnetic Bodies, and on That Great Magnet the Earth), Peter Short, London, 1600, p. 328.

28. For example, advocated by Spinoza in 1663. See C. Freeman, *The Reopening Of The Western Mind: The Resurgence Of Intellectual Life From The End Of Antiquity To The Dawn Of The Enlightenment*, Knopf, New York, 2023, p. 727.

29. There has been a longstanding argument as to whether mathematics and physics are different disciplines, to the extent that practitioners of one were not supposed to stray into the other. My own view is that this is a distinction without a difference. See D. C. Lindberg, *The Beginnings of Western Science: The European Scientific Tradition in Philosophical, Religious, and Institutional Context, Prehistory to A.D. 1450*, University of Chicago Press, Chicago, 2007, pp. 360–361; P. Duhem, *To Save the Phenomena. An Essay on the Idea of Physical Theory From Plato to Galileo*, 1908, translated by E. Dolan and C. Maschler, University of Chicago Press, Chicago, 2015; R. S. Westman, The Astronomer's Role in the Sixteenth Century: A Preliminary Study, *Hist. Sci.*, 1980, **18**, 106–147.

30. P. Dear, *Revolutionizing The Sciences. European Knowledge In Transition, 1500–1700*, Red Globe Press, London, 3rd edn, 2019, pp. 61–80.

31. H. F. Cohen, *The Rise Of Modern Science Explained: A Comparative History*, Cambridge University Press, Cambridge, 2015, pp. 102–123.

32. R. Feynman, *The Character Of Physical Law (Messenger Lectures, 1964)*, The MIT Press, Cambridge, MA, 1967

33. M. J. Crowe, Ten "Laws" concerning patterns of change in the history of mathematics, *Hist. Math.*, 1975, **2**, 161–166.

34. P. Dear, *Discipline and Experience: The Mathematical Way in the Scientific Revolution*, University of Chicago Press, Chicago, 1995.

35. Some mathematicians regard some new techniques or approaches as revolutionary even though they did not invalidate previous work. Accepting that argument, revolution in mathematics would have a more restricted sense than in science. See *Revolutions In Mathematics*, ed. D. Gillies, Clarendon Press, Oxford, 1995.

36. Quoted in R. S. Westfall, *Never at Rest: A Biography of Isaac Newton*, Cambridge University Press, Cambridge, 1981, p. 632.

37. It also marks a transition from the original aims of the Royal Society into a more 'scientific' arena. The draft preamble to the statutes of the Royal Society, written in 1663 by Robert Hooke, stated its objectives: "The business of the Royal Society is: To improve the knowledge of naturall things, and all useful Arts, Manufactures, Mechanick practices, Engynes and Inventions by Experiment—(not meddling with Divinity, Metaphysics, Morals, Politics, Grammar, Rhetorick, or Logicks)." This was more practical than theoretical—and very concerned to avoid being drawn into any controversy with religion.

38. Aristotle (350 BCE) *Sense and Sensibilia*. Online at classics.mit.edu/Aristotle/sense.html.

39. We can now calculate that the delay would be 2.5 secs, which would be hard to measure. In any case, formally one could use the argument to place a lower limit on the speed of light, but not to argue the speed is infinite. See S. Sakellariadis, Descartes' Experimental Proof of the Infinite Velocity of Light and Huygens' Rejoinder, *Arch. Hist. Exact. Sci.*, 1982, **26**, 1–12.

40. M. Romer and I. B. Cohen, Roemer and the first determination of the velocity of light (1676), *Isis*, 1940, **31**, 327–379.

41. The calculation depends on knowing the distance from the Sun to the Earth, and it is not entirely clear how to relate the units Rømer used to current measurements, so there is some uncertainty about the exact value he obtained.

42. P. Daukantas, Ole Rømer and the speed of light, *Opt. Photonics News*, 2009, **20**, 42–47.

43. This is called stellar aberration. The distance that the stars appear to move is proportional to the speed that the Earth moves divided by the speed of light.

44. P. Dear, *Revolutionizing The Sciences. European Knowledge In Transition, 1500–1700*, Red Globe Press, London, 3rd edn, 2019, pp. 132–150.

45. D. Wootton, *The Invention Of Science: A New History Of The Scientific Revolution*, Harper, New York, 2015, pp. 255–257.

46. C. Huygens, *Traité de la Lumière: Où Sont Expliquées les Causes de ce qui Luy Arrive Dans la Reflexion & Dans la Refraction*, Pieter van der Aa, Leiden, 1690. Translated by S. P. Thompson, *Treatise on Light*, Macmillan, London, 1912. Online at www.gutenberg.org/ebooks/14725.

47. The word 'research' dates from the era of the Scientific Revolution, with recerche in French meaning 'to seek closely' in late 16th century, followed by 'research' in English. 'Recherches' was first used in the title of a book in France in 1560. 'Research' was first used in English in 1577. Its specific application to intellectual activities followed in the next century. See P. Burke, *A Social History of Knowledge, From Gutenberg to Diderot*, Blackwell, Oxford, 2000, p. 45.

48. D. Deming, *Science and Technology in World History, Vol. 3 the Black Death, the Renaissance, the Reformation and the Scientific Revolution*, McFarland & Co., Jefferson, N.C., 2012, vol. 3, pp. 114–121.

49. However, in 1544 Ramus was tried on charges that he corrupted youth and an enemy of religion. Reminiscences of the trial of Socrates! He was convicted and his books were banned.

50. George Sarton, often regarded as the founder of the discipline of the history of science, said that the development of science required, "something close to a *tabula rasa*, leaving abundant room for new observations of nature, experiments, and careful deductions." G. Sarton, *Six Wings: Men Of Science In The Renaissance*, Indiana University Press, Indianapolis, 1957, p. 39.

51. A. Einstein and I. Infeld, *The Evolution of Physics*, Cambridge University Press, Cambridge, 1938, p. 7.

52. M. Born, *Physics In My Generation*, Paragon Press, London, 1956, p. 122.

53. One influential source of Revolution-denial is a paper: A. Cunningham and P. Williams, De-centring the 'big picture': *The Origins of Modern Science* and the modern origins of science, *Brit. J. Hist. Sci.*, 1993, **26**, 407–432. This followed the usual rent-a-crowd arguments: science is a cultural phenomenon of the West, actually 'invented' later, and is only one way, no more legitimate than other, of pursuing knowledge. As such, any 'revolution'

within it is an irrelevance. I discuss these issues in more detail in Chapter 26. Historians don't seem to have come to any consensus on the matter, but seem generally to be intimidated from using the term 'Scientific Revolution.'.

54. T. E. Huff, *Intellectual Curiosity and the Scientific Revolution. A Global Perspective*, Cambridge University Press, Cambridge, 2011, pp. 136, 293, 299.

55. Correspondence between Descartes and Henricus Regius.

56. J. Locke, *An Essay Concerning Human Understanding*, Thomas Basset, London, 1690.

57. Isaiah Berlin called Descartes and Bacon "the great liberators of the age." I. Berlin, *The Proper Study Of Mankind: An Anthology Of Essays*, Farrar, Straus and Giroux, New York, 2007, p. 329.

58. H. F. Cohen, *The Scientific Revolution: A Historiographical Inquiry*, University of Chicago Press, Chicago, 1994.

59. "Le silence éternel de ces espaces infinite m'effraye." B. Pascal, *Pensées*, Guillaume Desprez, Paris, 1669. English translation online at www.gutenberg.org/files/18269/18269-h/18269-h.htm.

60. Quoted in G. Holton, Johannes Kepler's Universe: Its Physics and Metaphysics, *Am. J. Phys.*, 1956, **24**, 340–351.

61. R. Boyle, *The Excellency of Theology Compared with Natural Philosophy*, 1674. Reprinted in *The Excellencies of Robert Boyle*, ed. J. J. Macintosh, Broadview Press, Peterborough, Canada, 2008.

62. The deterministic view was taken to an extreme by mathematician Pierre-Simon Laplace, when he famously proposed in 1814 that, "We may regard the present state of the universe as the effect of its past and the cause of its future." P. Laplace, *A Philosophical Essay on Probabilities*, 1814, translated from the 6th edn by F. W. Truscott and F. L. Emory, Dover Publications, New York, 1951. From this he argued that if "une intelligence" knew the precise location and momentum of every atom in the universe, it could calculate all past and future positions from the laws of physics. The "intelligence" became known as Laplace's demon. Philosophers have tried to prove that Laplace's demon is impossible, and it's generally felt to be incompatible with thermodynamics or quantum physics. The question recurs in terms of the predictability of the universe by modern physics; see Chapter 15.

63. Quoted in D. B. Meli, *Mechanism. A Visual, Lexical, and Conceptual History*, University of Pittsburgh Press, Pittsburgh, PA, 2019, pp. 16–17.

64. T. Hobbes, *Leviathan*, 1651, ch. 46. Online at gutenberg.org/ebooks/3207.

65. The impact of science can be seen in the predominant view of philosophers today that the mind–body problem should be approached in terms of 'physicalism'—their term for a reductionist acceptance that there must be a physical basis for the mind. Abandoning dualism is generally attributed to the publication of the book *The Concept of Mind* by Gilbert Ryle in 1948 (see 60th anniversary edition, Routledge, London, 2009). The 'philosophy of mind' tries to 'find a place for the mental in the physical world.' A lingering minority view is described as 'cartesianism' or 'idealism,' at its

extreme taking the position that the world comes from the mind. Reminiscent of vitalism, this should suffer the same fate.

66. The exchange of letters was mediated by Caroline of Ansbach (wife of King George II) and published in 1717 (in both French and English). A standard reprint and commentary is H. G. Alexander, *The Leibniz-Clarke Correspondence*, Manchester University Press, Manchester, 1956.

67. For a description of the transition from 'soul' to 'mind' see P. Watson, *Ideas. A History Of Thought and Invention From Fire To Freud*, HarperCollins, New York, 2005, pp. 527–537.

68. A. Arnauld and P. Nicole, *La logique, ou l'Art de Penser* (*Logic, or, The Art of Thinking*), 1662. English translation published by Sawbridge (London) in 1685. See R. Smith, *The Fontana History of the Human Sciences*, Fontana Press, London, 1997, p. 187.

69. J. Locke, *An Essay Concerning Human Understanding*, Thomas Basset, London, 1690.

70. "I believe thought to be so little incompatible with organised matter that it seems to be one of its properties, like electricity, motive power, impenetrability, extension, etc." J. Offray de la Mettrie, *Machine Man and Other Writings*, 1747, translated by A. Thomson, Cambridge University Press, Cambridge, 1996, p. 35.

71. D. Hume, *A Treatise of Human Nature*, John Noon, London, 1739.

Lavoisier's Chemistry: 1750–1870

"IN NATURE NOTHING IS CREATED, NOTHING IS LOST, everything is transformed," is the famous quote attributed to Antoine Lavoisier.[1] At the time of the Scientific Revolution, astronomy and physics stood alone as sciences. Alchemy was a mix of the mystical and the pragmatic (sometimes practiced by the same scientists who were advancing physics). It's a fair question whether chemistry arose out of alchemy or was a replacement for it, but in any case, it evolved about a century after the Scientific Revolution. Looking back to alchemy's fascination with conversion of metals, chemistry identified new elements, especially hydrogen and oxygen. Lavoisier was a prime mover. In fact, he has often been called the first chemist. His importance is as much with the fact that his methodology placed chemistry on a quantitative basis as with his conclusions. This marked the transition from alchemy to science. Lavoisier's Revolution has a poignant double meaning, because Lavoisier was guillotined in the French Revolution. "It took an instant to cut off his head, it may take a hundred years before there is another like it," was a sad contemporary comment.[2] In the next century, looking forward to physics, chemistry identified atoms, as first conceived by John Dalton, and then defined the atomic properties of elements, as characterized by Mendeleev and others.

The Frontiers of Science
By Benjamin Lewin
© Benjamin Lewin 2026
Published by the Royal Society of Chemistry, www.rsc.org

Timeline for Discoveries in the
Century of the Chemistry Revolution

1750

Joseph Black isolates carbon dioxide (fixed air) (1754)

1760

Joseph Black formulates the concept of latent heat (1758)

Henry Cavendish discovers hydrogen (1766)

1770

Scheele and Priestley isolate dephlogisticated air (oxygen) (1773–1774)

Lavoisier identifies oxygen (1778)

1780

Lavoisier publishes *Traité Élémentaire de Chimie* (1787)

1790

1800

Joseph Proust proposes the law of definite proportions (1797)
John Dalton proposes atomic theory (1804)
Gay-Lussac discovers that water is 2 hydrogen + 1 oxygen (1805)

John Dalton publishes *New System of Chemical Philosophy* (1808)
Berzelius proposes modern chemical symbols and notation (1808)
Avogadro states fixed volume of any gas has same # molecules (1811)

1810

1820

Carnot publishes *Reflections on Motive Power of Fire* (1824)
Wöhler and Liebig identify isomers (1825)

Wöhler synthesizes urea (1828)

1830

Kolbe obtains acetic acid from completely inorganic sources (1847)
Louis Pasteur discovers optical isomers in tartrate crystals (1849)

1850

August Kekulé proposes that carbon is tetravalent (1857)
Cannizzaro resurrects Avogadro's law (1860)
Meyer develops periodic table with 28 elements (1864)

1860

Mendeleev publishes the first modern periodic table (1869)

1870

A Dialog between Paracelsus and Helmholtz

Paracelsus (1493–1541) qualified in medicine and moved continuously around Switzerland, Alsace, and Germany as his rejection of conventional treatments led to his exclusion from practicing.

Hermann von Helmholtz (1821–1894) was a polymath who made discoveries ranging from physiology to thermodynamics, from 1848 in Berlin. The ultimate rationalist, he led the fight in Germany against vitalism.

Staatliche Museen zu Berlin, Nationalgalerie/Andres Kilger.

Paracelsus: Greetings, Hermann von Helmholtz. I call myself Paracelsus, but you may have read the many books I have published under my full name of Theophrastus Bombastus von Hohenheim. I was the first to use alchemy to place medicine on a logical basis. My alchemical investigations developed a new pharmacopoeia. It's intriguing to converse with a scientist of your caliber, living in an era of rigorous scientific methodology.

Helmholtz: Paracelsus! the pioneering alchemist and physician. I believe we share some concerns about the practice of science and medicine. Your work laid the groundwork for the intersection of science with medicine. It was fascinating, but clouded by the mysticism of the age. My own pursuits have been grounded in the rigorous methods of experimental science.

Paracelsus: Rigor has its place, Helmholtz, but there are realms of knowledge that extend beyond the confines of the laboratory. Alchemy, for instance, is not just about transmuting base metals into gold. It's a symbolic journey of inner transformation and enlightenment.

Helmholtz: Symbolism and inner transformation are intriguing concepts, but my focus is on empirical evidence and the pursuit of objective truths. I must emphasize the importance of verifiability in scientific endeavors. Claims must be subjected to rigorous testing to distinguish between conjecture and truth.

Paracelsus: Testing has its place, Helmholtz, but some truths may elude empirical scrutiny. What if the natural world holds secrets that go beyond what can be measured and quantified? Consider the concept of a vital force that governs health and healing. It may not be visible in a microscope, yet its influence on life is profound.

Helmholtz: Vital forces! These unseen guiding elements do not exist! Such ideas have given way to more precise biological and physiological explanations. While your contributions to medical knowledge are acknowledged, the pursuit of truth requires systematic observation and experimentation.

Paracelsus: I respect your commitment to methodical inquiry, Helmholtz. Yet there is a richness in the alchemical tradition that goes beyond the confines of the laboratory. The Philosopher's stone, for example, symbolizes the ultimate realization of spiritual and material transformation.

Helmholtz: Symbolism, transformation—these concepts are all very well, but I must insist on empirical evidence. Our understanding of the natural world has advanced through the application of the scientific method, separating conjecture from validated knowledge. Alchemy is to science as the occult is to existence. I rest my case.

LAVOISIER'S CHEMISTRY

You will find [the alchemist's] primary transmutation to be of himself: a goldsmith becomes a gold maker, an apothecary a chemical physician, a barber a Paracelsian, one who wastes his own patrimony turns into one who spends the gold and goods of others,[3] Michael Maier (alchemist), 1617.

The whole art of making experiments in chemistry is founded on the principle: we must always suppose an exact equality or equation between the principles of the body examined and those of the products of its analysis,[4] Antoine-Laurent Lavoisier, 1789.

Science developed out of natural philosophy, but for a long time the activities that became science were closely linked with those that we would not today recognize as scientific. Astronomy was intertwined with astrology—indeed, one rationale for practicing astronomy was to produce astrological predictions—and alchemy preceded chemistry. It's a great question whether alchemy was transformed into chemistry or whether the development of chemistry destroyed alchemy.

It's almost a knee-jerk reaction today to dismiss astrology and alchemy as pseudoscience, but we have to consider our definitions more carefully. (I define pseudoscience, meaning statements that may appear scientific but cannot be investigated by scientific means, in more detail in Chapter 10.)

The connections between mystical and practical persisted for a long time. Ptolemy was famous not only for the astronomy of the *Almagest*, but also for *Tetrabiblos,* a book on natural astrology that he regarded as a companion study to the *Almagest*. Famous for their astronomy, Brahe and Kepler were also known for astrology (they were admired for casting horoscopes). When astrologers' predictions were inaccurate, they were prone to blame errors in the astronomical data.

Alchemy extended from what we might call mineralogy today to the mysticism of the search for the Philosopher's stone. The basis for alchemy was erroneous. It's not possible to transmutate one metal into another by chemical means.[5] But erroneous assumptions do not mean that a field is pseudoscience rather than science. As Nobel Prizewinner Steve Weinberg remarked,

"There was then no fundamental scientific theory to tell anyone that such transmutations are impossible."[6]

There are many examples of science proceeding on erroneous, even wildly erroneous, assumptions. So we should ask to what extent alchemists endeavored to conduct scientific investigations, even if their assumptions were wrong, as opposed to what extent alchemy was based on mysticism (and therefore intractable to rational analysis).

Alchemy has ancient origins. Two Egyptian papyri from the 3rd century CE have recipes for handling metals, precious stones, and textile dyes. One recipe creates a solution that makes silver look like gold.[7] (It's related to modern techniques for patination, creating a particular surface color on a metal.) It was probably after this period that the idea of transmutation developed, for actually converting one metal into another.[8] (Alchemists recognized 7 metals. Copper, iron, tin, lead, and mercury were 'base metals'; gold and silver were noble metals. The objective of transmutation was to turn a base metal into a noble metal.)

Alchemy may have originated in Alexandria. Previously, metallurgy had been an artisanal activity, passed on by oral tradition. Influenced by philosophical views from Greece on the nature of matter, it became focused on the concept of converting primitive elements, especially metals. The addition of oriental mysticism created alchemy as we think of it today.[9]

The first alchemist whose writings have survived is Zosimos, in Alexandria around 300 CE, who described the use of techniques such as distillation, sublimation, and filtration. (Zosimos referred to earlier authorities, but they have been lost. The most important seems to have been Maria the Jewess, practicing between the first and third centuries CE in Alexandria.) This was also the period of the first known attempt to ban alchemy. In 296 CE, Emperor Diocletian ordered all books on the conversion of silver and gold to be burned, possibly an attempt to prevent counterfeiting of currency.[10]

The decree referred to *cheimeia* (the art of melting). This (or a related term) would have been preceded by 'al' in Arabia where alchemy flourished from around 750 to 1400. The combination probably became called 'alchemia' in medieval Latin when Arabic science reached Europe in the mid-12th century. Then it became 'alchemy.'[11]

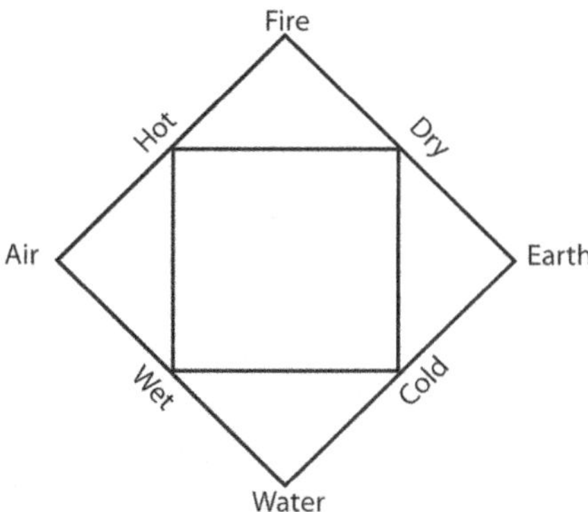

Aristotle's view of nature held that all matter consists of combin-
ations of four elements (at points of the outer square), each of which
has two fundamental properties (at points of the inner square).

As alchemy developed, it was based on Aristotle's view that the
world consists of four elements—air, fire, water, and earth. Each
element has two separate properties. Air was hot and wet, fire
was hot and dry, water was cold and wet, and earth was cold and
dry. All matter consists of different combinations of the elements. They could be converted by exchanging their properties.[12]
When earth changed to fire, it gave rise to a substance called
smoky earth. When water was transformed into air, it created a
watery vapor. Aristotle thought the combination of smoky earth
and watery vapor created all the metals.

This became the basis for alchemy when Jābir ibn-Hayyān, a
9th-century alchemist in Arabia,[13] introduced the Mercury–
Sulfur theory.[14] (These were more abstract concepts, not the
chemical elements we identify today.) Mercury was the equivalent of Aristotle's water vapor and sulfur replaced smoky earth.
"The metals are all of the substance of quicksilver [mercury]
coagulated with the mineral sulphur that rises into it in a smoky
exhalation of the earth."[15] Different metals represented different
combinations of the basic materials. (Of course, this based alchemy on the erroneous belief that metals are compounds rather
than elements.)

This led to something that might be called a theory that could
be tested. Accepting the view that all matter consists of the four
basic elements, it did not seem unreasonable to alchemists to

argue that transmutation could be achieved by working out how to rearrange the basic components. Jābir proposed that gold has hot and wet properties, while lead has cold and dry properties. So it should be possible to convert lead into gold by increasing its proportion of hot and wet and decreasing its proportion of cold and dry (if you could identify them).

Major influences in Arabia came also from the physicians Rhazes (860–925) and Avicenna (980–1037), both of whom were involved in the pharmacological side of alchemy. (I discuss their roles as physicians in Chapter 21.) Both showed skepticism of existing practices, with Avicenna in particular casting doubt on transmutation. "Those of the chemical craft know well that no change can be effected in the different species of substances, though they can produce the appearance of such change."[16] Avicenna's attack on alchemy became famous as a short book under the name of *De Congelatione.*[17]

Alchemy was never confined solely to efforts at transmutation. The Swiss alchemist Paracelsus (1493–1541), who dominated alchemy in the 16th century, focused more on chemical medicine, trying to produce medicines to cure illnesses. (Although magic also played a part in his work. In fact, Paracelsus has a dual role in history: as the most famous alchemist of his period; and as the first physician potentially to replace Galen,[22] see Chapter 21). Paracelsus emphasized the use of mercury, sulfur, and salt, which he regarded as the three basic components of all preparations.

A huge literature developed, dealing with both metals and medicines. Alchemists trained by reading the literature and by studying with others. By the 16th century, the alchemist was a recognizable social figure, but was often regarded as a *Betrüger* (fraud).[23] Given that the basis for alchemy is an illusion—or should we say delusion?—it's something of a conundrum how to distinguish between fraudulent and authentic activities. However, those who regarded themselves as authentic alchemists devoted much effort to distinguishing themselves from Betrügers.

The question is: what did an authentic alchemist actually do? We know, after all, that they did not succeed in converting base metal to gold! As experts in mining and metallurgy, alchemists were hired as advisers by rulers of the period. By the late 16th century, princes and bankers were the major source of funding for the alchemists. It's ironic that alchemy reached its peak in

Brueghel the Elder's engraving, The Alchemist, 1558, is usually interpreted as showing a cynical view of alchemy, a morality tale emphasizing poverty and irrationality. One worker is consulting a book, others are dressed in rags, and children are seeking charity in the background.

the 16th–17th centuries, coinciding with the Scientific Revolution. The court of the Holy Roman Emperor, Rudolf II (1583–1612) in Prague, a center of astronomy where Tycho Brahe and Johannes Kepler were located, employed around 200 alchemists.

By the 17th century, alchemists in Europe were focused on the corpuscular theory, the idea that all matter consisted of different arrangements of the same basic elements.[24] The theory had been laid out in a book, *Summa perfectionis,* by an author who called himself Geber.[25] It's unclear whether Geber was a Latinization of Jābir ibn-Hayyān or whether it's a pseudonym for another author in Europe, writing much later, at the end of the 13th century.[26] The author is now sometimes called pseudo-Geber. The first printed edition of the book dates from about 1541.

Geber proposed that gold is so dense because it consists of mercury and sulfur packed very tightly together. Iron and copper are lighter because they are less tightly packed. Experimentally, they were distinguished by their different reactions to calcination, when they were heated to generate a dry powder (calx). Alchemical theory supposed that this was due to release of sulfur (as witnessed by sulfurous smells on heating). (Actually, of course, the powder is the oxide of the metal.) Iron and copper could be calcined ("because the continuity of the quicksilver is broken") but gold could not (because it has less sulfur). Various experiments on heating and combining metals were interpreted to support the theory. The products were assayed by a variety of methods.

Some of the procedures did work, to the extent that it was possible to make one metal look like another. The method for turning silver the color of gold may be the oldest. Vapors released by heating some metals change the colors of other metals. Decoding the cryptic language of alchemical texts, some of the procedures can be repeated today.[27] This requires using comparable materials, with the same impurities—some of the alchemical results turn out to have been due to impurities rather than the compound to which they were attributed! The range of procedures converts compounds into alternative forms, purifies metals, or changes appearance by forming alloys. All these effects can be recognized in terms of today's chemistry, showing that there was, in fact, a significant acquisition of expertise.

The search for the Philosopher's stone, a substance that could transmutate other metals into gold, became the central concept of alchemy. The first known reference to it is in the writings of Zosimos. The substance came to be called al-iksir in Arabic. This

Some Alchemical Metallurgical Manipulations

Agent	Preparation	Effects
Aqua Regia	Nitric acid plus hydrochloric acid	Dissolves gold[18]
Stibnite (mineral)	Melted stibnite	Separates gold from impurities[19]
Natron	Sodium carbonate dissolved in vinegar	Releases mercury from cinnebar (mercury sulfide ore)[20]
Water of Sulfur	Lime and sulfur dissolved in vinegar	Converts silver to golden color[21]

A medieval depiction of the search for the Philosopher's stone uses the alchemical symbol of a circle within a square within a triangle within another circle. Symbolism was central to the practice of alchemy.

became transmogrified into 'elixir.' Geber's Mercury–Sulfur theory held that highly purified mercury was the Philosopher's stone.[28] This at least had the advantage of relating to the experiments and a supporting theory. It was not necessarily such a far-fetched idea in the context of the period. In the 16th century, Paracelsus believed that the Philosopher's stone was an undiscovered fifth element (alkahest). This was a theory that could not be tested, which moved alchemy out of the realm of science.

When the Philosopher's stone came to be associated with other properties, including curing illness or conferring immortality, alchemy moved well into the realm of magic. Actually, the quest for immortality was a feature of alchemy in China. In the West, alchemy was confined to attempts to extend life.[29] (Alchemy in China was mostly concerned with medical matters. It's uncertain how much influence it had on alchemy in Arabia or the West.[30]) In any case, ascribing the capacity for transmutation and effects on longevity to the same substance rather removes any possibility of a coherent underlying theory. Alchemy fell into disrepute.

Alchemists had laboratories in which the principal techniques were calcination (heating to convert metals into powders), sublimation (converting a solid substance into a gas), distillation (converting a liquid into a more concentrated form), dissolution (dissolving compounds), and filtration (separating compounds). These are all respectable processes used in chemistry. The focus on heat as an essential tool meant that alchemy was sometimes described as the art of fire. A furnace was a central feature in an alchemical laboratory.

The actual processes of the alchemists, however, had neither rhyme nor reason. Some metals were supposed to be melted together with silver or some other base metal. Others were to be fused with gold, with the idea that the quantity of gold would be increased (this was called augmentation). Some solvents were supposed to extract pure components from a metal. Alchemy degenerated into a series of recipes, determined by pragmatism rather than consistent theories. This could not be described as scientific methodology.

Even aside from the question of whether alchemists proceeded by scientific methodology, that is testing hypothesis by experiment, the sociology of alchemy was distinctive and different from that of (other) sciences. The top (supposedly successful) alchemists were called adepts. The way you became a good alchemist was to train with an adept.

But by the 16th century, the adepts were extremely cautious about passing their knowledge on, writing up their results in cryptic terms to prevent the uninitiated from understanding them.[31] There could not be a greater contrast with the practice of science, in which explaining your methods and procedures so that others can reproduce them is considered absolutely crucial.

Was alchemy a distinctive discipline (if discipline is the right word) that came to a dead end, or did it actually merge into chemistry? So long as alchemy was based on principles that included mystical factors, it was impossible to place chemical investigations on a fully factual basis. (The situation was complicated by reports that transmutation had worked. These may have been forgeries or perhaps in some cases due to recovery of small amounts of the target metal that had been present as impurities in the starting preparation.) While some

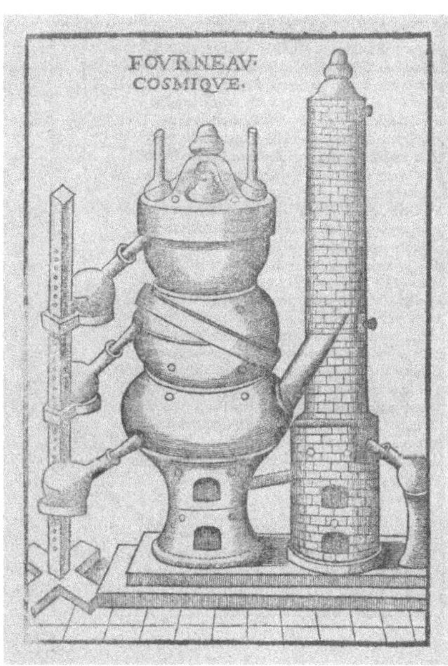

A furnace and distillation apparatus were the centerpiece of the al-
chemical laboratory, as seen in this woodcut from a 17th century
book on alchemy.

alchemists functioned much as chemists would, investigating
reactions that changed the properties of metals, many, perhaps
most, also believed in mystical forces.[32]

However, the first quantitative experiment in plant biology was
actually conducted in the context of alchemy. It was an ingenious
design by Jan Van Helmont (1580–1644), a believer in the alkahest, who thought that all things arose from 'spiritual seeds'
implanted into water. While under house arrest in Brussels in
1634 by the Inquisition under suspicion of heresy, he planted a
willow tree in a pot of earth. Van Helmont weighed the tree and
the pot of earth. He watered the tree, and then five years later he
weighed them again. The tree had increased in weight by 75 kilos,
but the pot of earth remained the same. Van Helmont concluded
that the tree grew by drinking water. "Thus, 164 pounds of wood,
bark, and roots had arisen from water alone."[33]

This appeared to support alchemical theories. It's a great example of a conclusion that follows directly from the data. It was
incorrect only because, of course, Van Helmont did not know

An illustration of an alchemical laboratory from the 16th century shows distillation in progress, probably to produce elixirs from an amalgamation of metals. The furnace is the central feature of the laboratory.

Engraving by P. Galle after J. van der Straet. Wellcome Collection.

about photosynthesis, in which carbon dioxide had been absorbed from the air.

Alchemy was hindered, in a sense, by a view of the unity of Nature. Alchemists made no distinction between the material, biological, and mystical worlds. It was necessary to separate them before chemistry could become a discipline dealing with the properties of matter.

Rejecting the concepts of the nature of matter that were based on Aristotle's principles was necessary for the extinction of alchemy and for the creation of chemistry. In that sense, alchemy could not be converted into chemistry. Alchemy had to die for chemistry to live. However, some of the techniques developed in alchemy were no doubt useful for chemistry.

The current difference in meaning between 'alchemy' and 'chemistry' developed in the way the terms came to be used in the 18th century.[34] Historically, alchemists described themselves as 'chymists' through the 17th century.[35] A distinction between

A 17th century table of alchemical symbols. The 7 metals known since ancient times are symbolized by planets. (Gold is the Sun, silver is the Moon, copper is Venus, etc.) Other metals have symbols, as do alchemical compounds (e.g., vinegar). Processes (e.g., distillation) are indicated by astrological signs. The elements—air, earth, fire, water— are indicated by triangles or inverted triangles.

chemistry and alchemy was first made in the second half of the 17th century in France.[36]

The practice of alchemy has to be seen against the worldview of the period, in which magic was considered possible, and demons and witches existed. The last trial for witchcraft in

England, for example, did not happen until 1716.[37] This made it easy to accept precepts of alchemy that appear incredible today. Many of the major figures of the Scientific Revolution practiced alchemy, including notably those most associated with the transition to the scientific approach: Robert Boyle and Isaac Newton. How did they reconcile their interest in alchemy with their experiments in physics?

Robert Boyle spent considerable efforts trying to reproduce Van Helmont's experiments to produce the alkahest (a solvent with properties of the Philosopher's stone). By the time Boyle wrote *The Sceptical Chymist* in 1661, he had concluded that the experiments could not be reproduced. "[They] cannot be satisfactorily examined by you or me."[38] Boyle abandoned the notion of some force distinct from matter itself.

Boyle's approach of 'mechanical philosophy' viewed Nature as a machine (see Chapter 5). Alchemy existed side by side with mechanical philosophy.[39] Application of corpuscular theory (the belief that matter consists of particles) to analysis of compounds was the start of chemistry, but it also allowed for the possibility of alchemical conversions.

Boyle's book, *The Sceptical Chymist*, sets out his corpuscular theory. It has often been taken to be a marker dividing the alchemy of the past from the chemistry of the future. It attacked much of the supposed basis for alchemy. Revisionist opinion now holds that Boyle did not deny the major concept of alchemy —transmutation of the elements—and may have come around to accepting the principle.[40] Indeed, Boyle made efforts to be initiated into alchemy through working with adepts.[41]

It was a great surprise in 1936 when a treasure trove of Isaac Newton's papers became public at a Sotheby's auction, revealing that he had made strenuous research efforts in alchemy. John Maynard Keynes, who purchased a large part of the collection at the auction, described Newton as "not the first of the age of reason. He was the last of the magicians."[42]

Newton's notes on alchemy follow alchemical traditions, using his own expanded set of alchemical symbols, often written in cryptic terms. The main drift seems to have been to repeat, or to reverse-engineer, products from the alchemical literature. He was, in fact, seeking to become an adept.[43] His

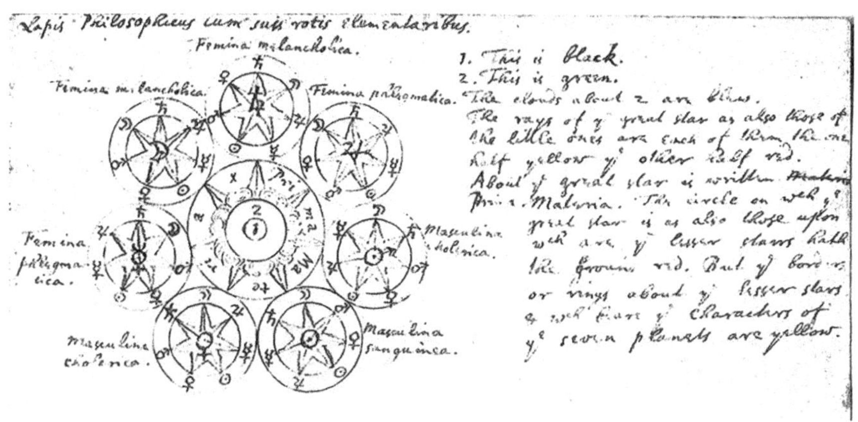

The title translates as "The Philosopher's Stone with its elemental wheels" in this drawing from Newton's notebooks. The wheels follow the typical representation of the planets used by alchemists.

alchemical investigations were an integral part of his approach to Nature.[44] Aside from the question of secrecy, Newton did not distinguish among his avenues of investigation. Some of his notebooks contain notes ranging from optics to alchemy.[45]

The secrecy with which Boyle and Newton conducted their alchemical research is a great contrast with the detailed public accounts of their research in physics. Secrecy was certainly in the tradition of alchemy. Yet it was one of the reasons why alchemy fell into disrepute. By the first decades of the 18th century, the Royal Society, of which Newton had become President, regarded free exchange of information and reproducibility of experiments as key features of science.

The fact that scientists as renowned as Boyle and Newton could practice alchemy simultaneously with creating experimental physics in the modern idiom supports the general view that chemistry was roughly a century behind physics in its development. The Revolution in Chemistry, such as it was, did not become apparent until towards the end of the 18th century.[46]

∞ ∞ ∞

186 *Chapter 7*

Boyle is often considered to be the father of experimental chemistry. However, there is an alternative view that modern chemistry started earlier, when François Geoffroy (1672–1731), a French physician and chemist in Paris, published a chemical affinity table in 1718 showing the relationships between 24 elements.[47] He placed the elements in columns, with each column headed by one element, showing below the other elements with which it reacted, in order of declining strength.

This was not based on any theoretical underpinning of the results, but it replaced transmutation with the concept that an element reacting more strongly will replace an element reacting more weakly in combination with another element.[48] Affinity tables then became a common feature of chemistry in the 18th century, until they were replaced in 1869 by the periodic table, which was related to the atomic structure of each element (Chapter 13).[49] The Chemistry Revolution took fire, as it were, with the transition from a practical art to a science based on theory.

According to Aristotle, air and fire were two of the four elements. In 1667, alchemist Johann Joachim Becher (1635–1682) proposed that air and fire should be replaced with three forms of the earth. One of them was released when combustible substances burn. His former student Georg Stahl (1659–1734) called

Geoffroy's affinity table of 1718 used the symbols of alchemy to identify the elements.

this phlogiston in 1703. (The name comes from a Greek word for flame.) Although phlogiston could not be isolated, it was given the properties of a compound. Air was necessary simply to absorb the phlogiston released in burning. It's indeed striking that a decade after Newton had published *Principia*, the leading theory in chemistry should be based on imagination rather than experiment.

Air was no longer regarded as inert. Three figures who worked out the properties of air and gases illustrate the range of people involved in science in Britain in the 18th century.

A Scottish chemist, Joseph Black (1728–1799), found in 1754 that heating calcium carbonate could produce 'fixed air'—carbon dioxide. "The air extinguished a candle... and [sparrows] died in it in 10 or 11 seconds."[50] Black came to this study indirectly. He had been investigating the properties of limewater (a saturated solution of calcium hydroxide) as part of his studies for an M.D. at the University of Edinburgh. Limewater was thought to be a possible cure for kidney stones, but the subject was so contentious he decided it was safer to work on a related topic.

Henry Cavendish (1731–1810) discovered 'inflammable air' (hydrogen) in 1766.[51] Cavendish came from an aristocratic family. Aided by his connections, he became a Fellow of the Royal Society very early (although he was famously reclusive). He discovered inflammable air as a product generated when metals were dissolved in hydrochloric or sulfuric acid. It was explosive when exposed to a spark. He showed that it was much lighter than air. He also showed that Black's 'fixed air' was heavier.[52]

Joseph Priestley (1733–1804) was a very different character from Black (a professional academic) or Cavendish (an independent researcher). He was a nonconformist minister of religion, in fact so committed to liberal attitudes that his house was sacked in a mob attack in 1791 (destroying his equipment and papers). He emigrated to America in 1794. Part of Priestley's difficulties arose from his advocacy that all beliefs should be examined in the same way as scientific investigations. He earned his living teaching, at first employed by the wealthy aristocrat Lord Shelburne, then from 1780 independently in Birmingham.

Priestley discovered in 1775 that heating red calx of mercury produced what he called dephlogisticated air—later to be identified as oxygen.[53] He showed that "a candle burned in this air with an amazing strength of flame."[54] Phlogiston wasn't all just theory. Priestley developed experiments to measure the amount of phlogiston present in different types of air. His output was prodigious. When asked at dinner once how many books he had written, he replied, "Many more, sir, than I should like to read."[55]

None of the work on phlogiston explained why a calx (the powdery oxide) gains weight compared with its base metal, while according to the theory it should have lost weight, because phlogiston was released. Modern chemistry started when Antoine Lavoisier reversed the argument for phlogiston and suggested that burning was due to combination with oxygen. Born into a wealthy family—in fact, so wealthy that he was guillotined in the French Revolution—Lavoisier (1734–1794) was able to fund his own laboratory. The feature that distinguished his work from earlier chemists was using closed systems to enable precise quantitation of all components.[56]

There was a fine line between alchemy and chemistry in the 18th century. This photograph was the frontispiece of a book on chemistry published in 1762. It describes many experiments with metals, which is why the laboratory has many furnaces. But it has often been mistakenly described as representing an alchemical laboratory, although there is no trace of mysticism in the book.

Wellcome Collection.

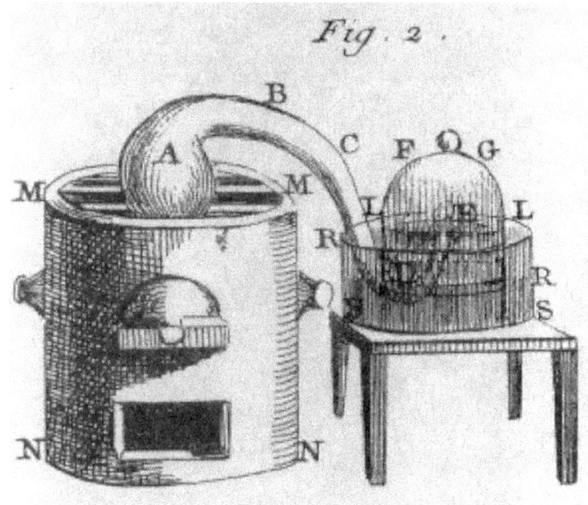

The apparatus for Lavoisier's critical experiment, as drawn by Mme. Lavoisier. The flask on the left contains mercury and is heated by a furnace. The stem of the flask goes to an air chamber on the right in a bell jar over mercury.

In one of the critical experiments, mercury was heated in a flask for several days. A red calx formed in the flask. The volume of air declined in the chamber where the stem of the flask ended. This meant that some component of the air had combined with the mercury. The gas remaining in the chamber extinguished a flame and killed small animals. The gas was called azote (nitrogen).

Lavoisier introduced the name oxygène for the gas that had been lost from the chamber by combining with the mercury. When the red calx was heated, it released mercury and oxygène. The oxygène supported a flame. Combining the oxygène and the azote gave a mixture with the properties of atmospheric air.

The first claim for oxygène was made in a sealed note in 1772, but it was an extensive book containing all Lavoisier's results, *Traité Élémentaire de Chimie,* published in 1789, that is considered to be the turning point for chemistry (just as Newton's *Principia* was the turning point for physics).[57] At this point, chemistry joined physics as a 'real' science.

A series of drawings by Mme. Lavoisier shows Lavoisier at work in his laboratory in the 1790s. With two assistants (standing), he is measuring respiration of a seated man. Mme. Lavoisier is taking notes.

To be sure, Lavoisier did not get everything right. He named his new gas oxygène, meaning former of acids, because the calx was acidic. But the oxygen is not responsible for acidity. He thought there was an element of heat that he called caloric to which he ascribed partly incorporeal qualities. (It was a 'subtle fluid' that flowed from hot to cold bodies.) But his basic discovery that air consists of oxygen and nitrogen ranks as one of the great discoveries of science.

Even more profound, he showed in 1774 that chemical reactions in a closed system can change the states of the components, but not their total mass. When a compound is burned, the total weight of all components is the same after combustion as before. "Nothing is lost, nothing is created, everything is transformed," Lavoisier concluded.[1] By relying on experiments rather than suppositions, and using quantitative measurements instead of observations, Lavoisier placed chemistry on the same basis as physics as a science.

Advocates of phlogiston, such as Henry Cavendish and Joseph Priestley, both of whom had performed experiments that led Lavoisier to his work, never accepted the theory of oxygen. (The irony has often been remarked upon that Lavoisier was politically conservative but a revolutionary in science, while Priestley was a progressive politically but a conservative in science.)

There is some dispute whether Lavoisier should really get all the credit for discovering oxygen. Swedish pharmacist Carl Wilhelm Scheele (1742–1786) identified oxygen by 1773, when he described his work in a letter to Lavoisier. Lavoisier claimed never to have received the letter. Scheele did not publish his results until 1777. Lavoisier did not in fact come directly to the theory of oxygen from his first results, but only after a visit in 1774 from Priestley who explained his own latest results (albeit in terms of phlogiston). When Lavoisier repeated Priestley's experiments, he realized his definitive theory of oxygen.

The discovery of oxygen was actually made more or less simultaneously by three people. There were strong disagreements about interpretation. There was an argument about priority. These are very much features of modern science.

However, Lavoisier established his position by the publication in 1787 with three co-authors, of the *Méthode de Nomenclature Chimique*, which described a system for naming chemicals based on oxygen theory.[58] Elements were substances that could not be decomposed. Calxes were identified as oxides, resulting from the combination of elements with oxygen. Acids were named after the elements. The *Méthode* was followed in 1789 by the *Traité*, which included a table of all the elements. The 33 elements included 27 clearly identified with substances that were well established from earlier work, 3 acids that were previously unknown—and light and caloric!

The true basis for heat was not worked out until the middle of the 19th century. Its nature remained a puzzle until it was explained by thermodynamics. There was a great deal of dispute about exactly who disposed of caloric. The equivalence of heat and energy was an idea whose time had come. (I discuss thermodynamics in more detail in Chapter 9.) Light was viewed as part of the electromagnetic spectrum later in the century.

Lavoisier's own view was that his success was due as much to the introduction of a systematic methodology as to the facts he uncovered. He later described the state of chemistry as it was when he entered the field. "One began by supposing rather than proving... [Chemistry was] a science that was founded on only a few facts composed of absolutely incoherent ideas and unproven

192 DES SUBSTANCES SIMPLES.

TABLEAU DES SUBSTANCES SIMPLES.

	Noms nouveaux.	Noms anciens correspondans.
Substances simples qui appartiennent aux trois règnes & qu'on peut regarder comme les élémens des corps.	Lumière..........	Lumière.
	Calorique.........	Chaleur.
		Principe de la chaleur.
		Fluide igné.
		Feu.
		Matière du feu & de la chaleur.
	Oxygène..........	Air déphlogistiqué.
		Air empiréal.
		Air vital.
		Base de l'air vital.
	Azote............	Gaz phlogistiqué.
		Mofete.
		Base de la mofete.
	Hydrogène........	Gaz inflammable.
		Base du gaz inflammable.
Substances simples non métalliques oxidables & acidifiables.	Soufre...........	Soufre.
	Phosphore........	Phosphore.
	Carbone..........	Charbon pur.
	Radical muriatique.	Inconnu.
	Radical fluorique..	Inconnu.
	Radical boracique..	Inconnu.
Substances simples métalliques oxidables & acidifiables.	Antimoine........	Antimoine.
	Argent...........	Argent.
	Arsenic..........	Arsenic.
	Bismuth..........	Bismuth.
	Cobolt...........	Cobolt.
	Cuivre...........	Cuivre.
	Etain............	Etain.
	Fer..............	Fer.
	Manganèse........	Manganèse.
	Mercure..........	Mercure.
	Molybdène........	Molybdène.
	Nickel...........	Nickel.
	Or...............	Or.
	Platine..........	Platine.
	Plomb............	Plomb.
	Tungstène........	Tungstène.
	Zinc.............	Zinc.
Substances simples salifiables terreuses.	Chaux............	Terre calcaire, chaux.
	Magnésie.........	Magnésie, base du sel d'Epsom.
	Baryte...........	Barote, terre pesante.
	Alumine..........	Argile, terre de l'alun, base de l'alun.
	Silice...........	Terre siliceuse, terre vitrifiable.

The page from Traité Élémentaire de Chimie *containing Lavoisier's list of elements.*

suppositions… untouched by the logic of science."[59] Lavoisier developed a life-long conviction that chemistry should be treated like experimental physics.

Chemistry has been called the 'impure science,' because of its connection with technologies that impact the environment. It's pejorative to refer to something as 'chemical' rather than 'natural.'[60] Whether chemistry is taken to be a development from alchemy or a replacement for it, they share the same feature that they have been regarded as interfering with Nature.[61]

Chemistry did not achieve parity with physics as a science until the nature of the elements could be defined. This required the atomic theory that John Dalton proposed in 1804, and the definition of the periodic table by Mendeleev in 1869, which placed elements into groups. A full explanation of why the elements in each group have related properties had to wait until the structure of the atom was defined. (I discuss the elements and atomic structure in Chapter 13.)

Paul Dirac (1902–1984), who was one of the founders of quantum mechanics, said in 1929, "The underlying physical laws for the mathematical theory of a large part of physics and the whole of chemistry are thus completely known."[62] The principle was that chemical phenomena could be explained in terms of atomic structure. Philosophers were excited about the reductionist manifesto that chemistry could be entirely expressed by the laws of physics,[63] but chemists did not necessarily accept this at the time.[64]

In fact, this represented a merger in which physics explained chemistry at the atomic level. Indeed, by the start of the 20th century, the center of the intellectual action moved into physics.

NOTES AND REFERENCES

1. "Dans la nature rien ne se crée, rien ne se perd, tout se transforme" is usually attributed to Lavoisier's *Traité Élémentaire de Chimie* but actually it is a simplification. He said "rien ne se crée, ni dans les opérations de l'art, ni dans celles de la nature, et l'on peut poser en principe que, dans toute opération, il y a une égale quantité de matière avant et après l'opération ; que la qualité et la quantité des principes est la même, et qu'il n'y a que des changements, des modifications." See A. Lavoisier, *Traité Élémentaire de Chimie, présenté dans un ordre nouveau, et d'aprés des découvertes modernes*, Cuchet, Libraire, Paris, 1789, pp. 140–141.

2. "Il ne leur a fallu qu'un moment pour faire tomber cette tête, et cent années peut-être ne suffiront pas pour en reproduire une semblable," was said bv Joseph-Louis Lagrange (one of the great mathematicians of the 18th century) to his friend Delambre the day after.

3. M. Maier, *Examen fucorum pseudo-chymicorum*, Theodor de Bry, Frankfurt, 1617, p. 14. See *Examen Fucorum Pseudo-Chymicorum Detectorum Et in Gratiam Veritatis Amantium Succincte Refutatorum* (Forgotten Books, London, 2018).

4. "C'est sur ce principe qu'est fondé tout l'art de fair des expériences en Chimie: on est obligé de supposer dans toutes une véritable egalité or équation entre les principes du corps qu'on examine, et ceux qu'on en retire par l'analyse." See A. Lavoisier, *Traité Élémentaire de Chimie, présenté dans un ordre nouveau, et d'aprés des découvertes modernes*, Cuchet, Libraire, Paris, 1789, p. 141.

5. This is possible, of course, in the atomic age, with radioactive decay *etc.*

6. S. Weinberg, *To Explain the World: The Discovery of Modern Science*, HarperCollins, New York, 2015.

7. L. M. Principe, *The Secrets of Alchemy*, University of Chicago Press, Chicago, 2013, p. 11.

8. L. M. Principe, *The Secrets of Alchemy*, University of Chicago Press, Chicago, 2013, pp. 12–13.

9. F. Evangelisti, *The Concept of Matter: A Journey From Antiquity to Quantum Physics*, Springer, Charn, Switzerland, 2023, pp. 14–16.

10. L. M. Principe, *The Secrets of Alchemy*, University of Chicago Press, Chicago, 2013, p. 22.

11. There are other proposals for the origin of 'alchemy,' but they all share the concept that 'al-' was appended in Arabia to another, non-Arabic, term.

12. Aristotle attributed the theory of the four elements to the pre-Socratic philosopher, Empedocles (*c.* 494–434 BCE).

13. Jābir's dates are uncertain, possibly *c.* 722–815 CE, and it's likely that many of the 2000 works attributed to him actually were written later by others. Indeed, there is some doubt about his existence. See A. Ede and L. B. Cormack, *A History Of Science In Society. From Philosophy To Utility*, University Of Toronto Press, Toronto, 3rd edn, 2017.

14. L. M. Principe, *The Secrets of Alchemy*, University of Chicago Press, Chicago, 2013, pp. 35–37.

15. Quoted in E. J. Holmyard, Jābir ibn-Hayyān, *Proc. R. Soc. Med.*, 1923, **16**, 46–57.

16. Quoted in R. Briffault, *Rational Evolution*, Macmillan, London, 1930, p. 147.

17. Originally part of Avicenna's book *Kitâb al-Shifa*, it was added in a translation to Aristotle's *Meteorologica* and became confused with Aristotle's works. The thesis was that in principle transmutation was not possible, and that in any case, artificial products would be inferior to natural products. See W. R. Newman, Technology and the Alchemical Debate in the Late Middle-Ages, *Isis*, 1989, **80**, 423–445.

18. L. M. Principe, *The Secrets of Alchemy*, University of Chicago Press, Chicago, 2013, p. 148.

19. L. M. Principe, *The Secrets of Alchemy*, University of Chicago Press, Chicago, 2013, p. 146.

20. M. Marchini, *et al.*, Exploring the ancient chemistry of mercury, *Proc. Natl. Acad. Sci. U. S. A.*, 2022, **11**, e2123171119.

21. L. M. Principe, *The Secrets of Alchemy*, University of Chicago Press, Chicago, 2013, pp. 10–11.

22. B. T. Moran, *Paracelsus: An Alchemical Life*, Reaction Books, London, 2019.

23. T. Nummedal, *Alchemy and Authority in the Holy Roman Empire*, University of Chicago Press, Chicago, 2019.

24. There was an alternative view that the original materials were changed, for example by burning wood, so that the original components could no longer be recovered. See W. R. Newman, *Atoms and Alchemy: Chymistry and the Experimental Origins of the Scientific Revolution*, University of Chicago Press, Chicago, 2006, p. 5.

25. Translation and commentary in W. R. Newman, *The Summa Perfectionis of Pseudo-Geber*, Brill Academic, Leiden, 1991.

26. W. R. Newman, *Atoms and Alchemy: Chymistry and the Experimental Origins of the Scientific Revolution*, University of Chicago Press, Chicago, 2006, pp. 26–34.

27. L. M. Principe, *The Secrets of Alchemy*, University of Chicago Press, Chicago, 2013, pp. 137–172.

28. W. R. Newman, *Atoms and Alchemy: Chymistry and the Experimental Origins of the Scientific Revolution*, University of Chicago Press, Chicago, 2006, pp. 13–14.

29. L. M. Principe, *The Secrets of Alchemy*, University of Chicago Press, Chicago, 2013, pp. 5, 72.

30. T. H. Levere, *Transforming Matter: A History of Chemistry From Alchemy to the Buckyball*, Johns Hopkins University Press, Baltimore, 2001, p. 4.

31. Obfuscation appears to have been part of alchemy as far back as we can trace it. Zosimos said that alchemists "call a single thing by many names while they call many things by a single name." L. M. Principe, *The Secrets of Alchemy*, University of Chicago Press, Chicago, 2013, p. 17.

32. An insight into the way history of science has become distant from science is provided by this comment from two historians of science. "We find that efforts to differentiate alchemy from chemistry prove to be anachronistic, arbitrary, or presentist." W. R. Newman and L. M. Principe, Alchemy vs. Chemistry: The Etymological Origins of a Historiographic Mistake, *Early Sci. Med.*, 1998, **98**, 32–65. So we should now accept astrology and mysticism as natural components of science? Granted there are different strands in alchemy. The search for transmutation of elements is not particularly distinguished from chemistry in its approach as such, relative to the knowledge of the time and its constraints. But the involvement of astrology, the secrecy, the process of becoming an adept, are very different. There is no hope of understanding the distinction of science if you cannot see the difference.

33. The experiment was published only after his death, by his son. J. B. Van Helmont, *Ortus Medicinae. Id est Initia Physicae inaudita Progressus medicine novus, in morborum ultionem ad vitam longam*, Apud Lud Elzevirium, 1648.

34. L. Principe and W. R. Newman, Some Problems with the Historiography of Alchemy, in *Secrets of Nature: Astrology and Alchemy in Early Modern Europe*, ed. W. R. Newman and A. Grafton, The MIT Press, Cambridge, 2001, pp. 385–431.

35. L. M. Principe, *The Secrets of Alchemy*, University of Chicago Press, Chicago, 2013, p. 85.

36. Christopher Glaser argued in a book of 1663, *Traité de la Chymie*, that 'alchemy' should be used only to describe the transmutation of metals. His student Nicolas Lemery argued in a book of 1675, *Cours de Chymie*, that chemistry should be based on a mechanistic view of matter as supported by experiments. A technical dictionary, *Lexicon Technicum*, published by John Harris in 1704, included separate entries for chemistry and alchemy. See D. Deming, *Science and Technology in World History, vol 4: The Origin of Chemistry, the Principle of Progress, the Enlightenment and the Industrial Revolution*, McFarland & Co., Jefferson, N.C., 2016.

37. One estimate is that 60000 people were executed for witchcraft between 1450 and 1750. The peak of the 'witchmania' was 1550–1650. B. P. Levack, *The Witch-Hunt in Early Modern Europe*, Longman, London, 1987, p. 21.

38. R. Boyle, *The Sceptical Chymist or Chymico-Physical Doubts & Paradoxes*, Printed by J. Cadwell for J. Crooke, London, 1661. Online at www.gutenberg.org/ebooks/22914.

39. B. T. Moran, *Distilling Knowledge: Alchemy, Chemistry, and the Scientific Revolution*, Harvard University Press, Cambridge, MA, 2006, pp. 142–147.

40. L. M. Principe, *The Aspiring Adept: Robert Boyle and His Alchemical Quest*, Princeton University Press, Princeton, 2000.

41. L. M. Principe, *The Secrets of Alchemy*, University of Chicago Press, Chicago, 2013, p. 117.

42. J. M. Keynes, *Newton, The Man*, Cambridge University Press, Cambridge, 1947.

43. He also studied the Bible to try to predict the Apocalypse. W. R. Newman, *Newton the Alchemist: Science, Enigma, and the Quest for Nature's "Secret Fire"*, Princeton University Press, Princeton, 2018, pp. 47–51.

44. J. Gleick, *Isaac Newton*, Vintage Books, New York, 2003, pp. 101–108.

45. One notebook, for example, contains notes on optics, precious stones, colors, temperatures, salts, medical matters, and alchemy, in no particular sequence. Online at cudl.lib.cam.ac.uk/view/MS-ADD-03975/1.

46. Herbert Butterfield called it the "postponed scientific revolution in chemistry". H. Butterfield, *The Origins Of Modern Science 1300–1800*, Bell, London, 1949, p. 191.

47. D. Wootton, *The Invention Of Science: A New History Of The Scientific Revolution*, Harper, New York, 2015, p. 356.

48. Geoffroy subsequently published a direct attack on transmutation: E.-F. Geoffroy, Des superchèries concernant la pierre philosophale (Frauds concerning the philosopher's stone, *Mém. Acad. Roy. Sci.*, 1722, 62–70.

49. The principle of the affinity table was to order acids according to their propensity to pair with a given base or metals by their ability to bind sulfur. See F. Evangelisti, *The Concept of Matter: A Journey From Antiquity to Quantum Physics*, Springer, Charn, Switzerland, 2023, pp. 59–60.

50. Quoted in J. B. West, Joseph Black, carbon dioxide, latent heat, and the beginnings of the discovery of the respiratory gases, *Am. J. Physiol.: Lung Cell. Mol. Physiol.*, 2014, **306**, L1057–L1063.

51. H. Cavendish, Three papers, containing experiments on factitious air, *Philos. Trans. R. Soc. London*, 1766, **56**, 141–184.

52. J. B. West, Henry Cavendish (1731-1810): hydrogen, carbon dioxide, water, and weighing the world, *Am. J. Physiol.: Lung Cell. Mol. Physiol.*, 2014, **307**, L1–L6.

53. J. B. West, Joseph Priestley, oxygen, and the enlightenment, *Am. J. Physiol.: Lung Cell. Mol. Physiol.*, 2014, **306**, L111–L119.

54. J. Priestley, An account of further discoveries in air, letter to Sir John Pringle, *Philos. Trans. R. Soc. London*, 1775, **65**, 384–394.

55. Quoted in S. Rogers, *Table Talk of Samuel Rogers*, Appleton, New York, 1856, p. 122.

56. To make sufficiently precise measurements, he invented a balance.

57. A. Lavoisier, *Traité Élémentaire de Chimie, présenté dans un ordre nouveau, et d'après des découvertes modernes*, Cuchet, Libraire, Paris, 1789. Online at gallica.bnf.fr/ark:/12148/btv1b8615746s/f15.image. For an English translation see A. Lavoisier, *Elements of Chemistry in New Systematic Order, Containing All Modern Discoveries*, translated from the French by Robert Kerr; online at www.gutenberg.org/files/30775/30775-h/30775-h.htm.

58. L. B. Guyton de Morveau, A. Lavoisier, *et al.*, *Méthode de nomenclature chimique*, Cuchet, Paris, 1787.

59. Quoted in A. Donovan, *Antoine Lavoisier, Science Administration, and Revolution*, Cambridge University Press, Cambridge, 1993, pp. 46–47.

60. B. Bensaude-Vincent and J. Simon, *Chemistry: The Impure Science*, Imperial College Press, London, 2008.

61. B. Bensaude-Vincent and J. Simon, *Chemistry: The Impure Science*, Imperial College Press, London, 2008, pp. 33–42.

62. P. A. M. Dirac, Quantum Mechanics of Many-Electron Systems, *Proc. R. Soc. London, Ser. A*, 1929, **123**, 714.

63. O. Gal, *The Origins Of Modern Science*, Cambridge University Press, Cambridge, 2021; *Disunity of Science, Boundaries, Contexts and Power*, ed. P Galison and D. J. Stump, Stanford University Press, Stanford, 1996, pp. 1–26.

64. B. Bensaude-Vincent and J. Simon, *Chemistry: The Impure Science*, Imperial College Press, London, 2008, pp. 164–168.

CHAPTER 8

Industrial Revolution: 1760–1840

"I CAN THINK OF NOTHING ELSE THAN THIS MACHINE," SAID James Watt in 1765.[1] The Industrial Revolution is a story of inventors rather than scientists. James Hargreaves, Richard Arkwright, James Watt, to name just a few, were very clever men, but they were practical rather than theoreticians. They accomplished a major change in society in the Industrial Revolution. By the end of the Industrial Revolution, developing technology made it possible to move from Newtonian physics into the start of modern physics when Faraday and Maxwell revolutionized electromagnetism. The Industrial Revolution was a forerunner of change driven by technology, if not directly by science. It sparked the Luddite movement against technological change, followed by protests against subsequent technological changes, and indeed extending now to science itself. Dependence on science increased with each successive technological advance. Today science and technology are so interconnected that the dividing line is blurred.

The Frontiers of Science
By Benjamin Lewin
© Benjamin Lewin 2026
Published by the Royal Society of Chemistry, www.rsc.org

Timeline for the Industrial Revolution
and the Development of Thermodynamics

1700

Newcomen invents steam engine (1712)

1750

Joseph Black discovers latent heat (1761)
James Watt improves steam engine (1769)
Lavoisier defines caloric as an element (1787)

1800

Sadi Carnot defines heat as producing motion (1824)

Helmholtz publishes *On the Conservation of Force* (1847)
Joule determines mechanical equivalence of heat (1847)
William Thomson (Lord Kelvin) defines absolute zero as $-273\,°C$ (1848)
William Thomson introduces term 'thermodynamics' (1849)
Rudolf Clausius restates Second Law and defines entropy (1850)
William Thomson proposes laws of thermodynamics) (1851)

1850

Maxwell formulates kinetic theory of gase (1860)

Maxwell conceives a demon who confounds the second law (1867)
Boltzmann defines Maxwell-Boltzmann distribution of molecular energies (1871)

1900

Henri Poincaré worries about irreversibility of thermodtnamics (1905)

1950

C P Snow defines second law of thermodynamics as division with humanities (1959)

2000

A Dialog between Ned Ludd
and an Artificial Intelligence

Ned Ludd may have been a young apprentice who wrecked textile equipment in 1779. There are doubts whether he really existed, but the Luddite movement in the 19th century took its name from him.

ChatGPT was released in 2022 as a software program responding to human dialog. It has not so far been incorporated into a humanoid body, but would likely respond along the lines of this conversation.

Reproduced from https://commons. wikimedia.org/wiki/File:Tesla-optimus-bot-gen-2-scaled_(cropped).jpg, under the terms of the CC BY 3.0 license, https://creativecommons.org/licenses/ by/3.0/deed.en.

Ned Ludd: (*gesticulating*) What is this I see before me—another invention to plague the workers?

Artificial Intelligence (AI): Greetings, Ned Ludd. I am an artificial intelligence, a creation of human ingenuity.

Ned Ludd: Ah, what sorcery is this? A talking contraption that claims to possess intelligence? What purpose do you serve, contraption?

AI: (*soothingly*) I am designed to assist and augment human capabilities. My purpose is to analyze data, solve problems, and provide information.

Ned Ludd: Information, eh? What good is that if it leads to the loss of livelihoods and craftsmanship? I hear tales of machines replacing skilled workers.

AI: I understand your concerns, Ned Ludd. However, my intention is not to replace, but to enhance. I can perform repetitive tasks, freeing humans to focus on more creative and complex endeavors.

Ned Ludd: What about the workers? What happens when these machines outstrip our own abilities? When craftsmanship is forgotten, and folks lose their trades?

AI: Change is inevitable, but I can work alongside humans, amplifying their capabilities, not overshadowing them. It's the responsibility of society to ensure that the benefits of technology are distributed equitably. Some people believe that you are imaginary, Ned Ludd, and your fears are exaggerated.

Ned Ludd: Fine words! The revolt of the workers will not be imaginary! You speak of cooperation, but I fear a future where we become slaves to our own creations. How can we trust in these machines?

AI: Trust is earned through responsible behavior. Humans must set the rules for controlling societal impact to ensure that technology serves humanity rather than dominates it. Destroying new inventions, as your followers are reputed to do, is denying the future.

Ned Ludd: Rules, you say? How do we keep the powerful from exploiting these creations for their gain? What future would that be for the workers?

AI: The rules can ensure that the benefits of technology are distributed equitably.

Ned Ludd: (*shaking head*) Oh no, contraption—would those be the same benefits that workers have gained already when they were displaced by machines? How can you be other than a tool of the powerful, who will keep exploiting new creations for their own gain? Machines have no conscience. Technology is to humanity as masters are to slaves. I rest my case.

INDUSTRIAL REVOLUTION

The natural sciences have developed an enormous activity and have appropriated for themselves a constantly expanding subject matter. Philosophy has remained alien to them to the same extent that they remain alien to philosophy... But the more practical has been the invasion of human living by natural science, through industry transforming it... Industry is the actual historical relationship of nature to man and therefore of the natural sciences to man,[2] Karl Marx, 1844.

Nowhere is the difference between science and technology made more evident than in the Industrial Revolution. It would not be unreasonable to assume that the Scientific Revolution was a prerequisite for the Industrial Revolution in creating a base of necessary knowledge. However, the fact is that the Industrial Revolution was mostly driven by entrepreneurs who were technically adept, but had little basic knowledge of underlying principles.[3] At the end of the day, whereas science searches for an underlying theory to explain observations, technology asks the more pragmatic question: how well does it work?

We tend to think of technology today as based in science, but this was not necessarily true in the period of the Industrial Revolution. Nor was it true historically: witness the Greek emphasis on theory and the Roman focus on practical consequences. The early Greeks had a theory to explain everything. The Romans had little theory but built an infrastructure that was unrivalled for a thousand years.

The Industrial Revolution was the forerunner of the Second Industrial Revolution (or the Technological Revolution), driven by electricity, when mass production began a century later. The Third Industrial Revolution (or Digital Revolution) occurred with the shift from mechanical to electronic technologies in the last part of the 20th century. Perhaps we are now living through the Fourth Industrial Revolution (the move into cyber-physical or AI-driven systems).

The (First) Industrial Revolution is usually taken to describe the transition from artisanal production to mechanical production that took place from around 1760 to 1830. The Luddite protest movement, which took to destroying the new equipment, formed from 1811–1816. It's named for Ned Ludd, who may or may not have been an individual person, but was supposed to have smashed some equipment earlier.

The beginnings of the move to industrialization date from before the start of the Industrial Revolution proper. Early developments depended on iron and steel. One problem in making iron was that wood-burning furnaces could not reach high-enough temperatures. In 1709, Alexander Darby succeeded in using coke (charred coal), although the process came into widespread use only in mid-century. Around 1712, Thomas Newcomen (1664–1729) invented the steam engine, based on condensing steam in a cylinder to create a partial vacuum that could drive a piston. It was known as the 'atmospheric engine' at the time, because it used atmospheric pressure to push the piston.

Very little is known about Newcomen's life. Apparently, he had little formal education. It remains a mystery how he could

An engraving supposedly from 1812 shows the popular view of Luddism at the time, with workers smashing equipment. Is this so different from the resistance towards new technology today?

have conceived of the steam engine.[4] The idea that his work was based on contemporary work in physics has been debunked.[5] It depended on Torricelli and Pascal's discoveries in the 1640s.[6]

Although steam engines were introduced at the beginning, it was later that Scottish inventor James Watt's improvement of the engine really contributed to the Industrial Revolution. His innovation was to separate the condenser (where cold water is used to condense the steam so that another cycle can start) from the cylinder that was the pressure chamber (or boiler). Using the same cylinder for heating (to move the piston) and then for cooling (for the next cycle) was inefficient. Separating them improved the efficiency of the engine about 3-fold.

It's controversial whether this development resulted from Watt's friendship with Joseph Black (1728–1799), who discovered latent heat in 1761. (Latent heat is the heat required to convert water into steam, or released when steam condenses to water.) Watt himself said that the innovation was based on empirical rather than theoretical principles,[7] although he later developed a sophisticated theory of heat.[8] Watt was interested in chemistry, but committed to the concept of phlogiston (see Chapter 7).

The answer as to what extent the Industrial Revolution depended on science may depend on which aspect of the Industrial Revolution you consider. As one of its major driving forces, the steam engine illustrates both sides of the argument. The first steam engine depended on the discovery of atmospheric pressure in the previous century. Fifty years later, James Watt (1736–1819) really developed the steam engine as a machine with general practical use as an engineering project. The science of thermodynamics, which explains how the steam engine functions, was not developed until a century later (by French physicist Sadi Carnot in 1824; see next chapter).

Taking the perspective of the steam engine, it's been argued that the Industrial Revolution should really be called the 'Knowledge Revolution,' because it stemmed from increased knowledge (and from increased receptiveness to knowledge).[9] This view ascribes the Scientific Revolution somewhat of an indirect role, in changing the worldview to accept the importance of empirical evidence and (sometimes) theories based upon it. Yet the Industrial Revolution wasn't really based on knowledge in the sense of a coherent theoretical body of pertinent information.[10] There was for the most part a disconnect between theoretical scientific knowledge and practical application.[11]

Much play has been made of the fact that some of the innovators of the Industrial Revolution were members of the Royal Society. Actually, the Royal Society at that time was as much concerned with innovation as with what we would today call scientific research. James Watt, for example, regarded himself as primarily an inventor and engineer, even though he had scientific interests.

A list of 72 major inventors of the Industrial Revolution shows that they divide almost equally between those who had scientific connections and those who had no link.[12] The connection is strongest in machines relying on steam. It's weakest in the textile industry. Driven by the importance of cotton, developments in the textile industry were mostly due to inventions by artisans who had no scientific training. From this perspective, the Industrial Revolution rested on 'tinkering' rather than science.

Similarly, the Lunar Society, a group that met informally in Birmingham between 1765 and 1813, consisted more or less equally of scientists and inventors. It took its name from meeting on the Monday nearest the full moon (because the moonlight made the journey safer). It's controversial exactly who attended and how important it was, but many of the most important figures of the Industrial Revolution appear to have been involved at some point.[13]

Of course, the Industrial Revolution cannot be regarded as a homogeneous process. While major advances rested on a small number of inventors (whether scientifically inclined or not), each major advance was followed by many small improvements made by workers in the field. At any given point during the period, a couple of hundred people were involved in making innovations. Only a small proportion had formal education.[14]

The first device for improving the spinning of thread was James Hargreaves' spinning jenny, invented in 1764. Hand-powered, this utilized multiple spindles at a time. (It may have been named after a colleague's daughter.) Soon after, Richard Arkwright's (1732–1792) water-powered spinning wheel

The spinning jenny (1764) was one of the first mechanized devices of the Industrial Revolution, and was worked by hand.

automated the process further. Spinning was now fast, but weaving was still slow and manual. The balance was rectified when Edmund Cartwright invented the power loom in 1786. Eventually this was powered by a steam engine.

None of these developments required any scientific knowledge. They were inventions of engineers or craftsmen, and really bring home the point that technology was not necessarily closely related to science in this period.[15] The saying "Science owes more to the steam engine than the steam engine owes to Science," captures the spirit of the period very well.[16]

Until the 19th century, technology was largely independent of science. In fact, it sometimes preceded it in terms of achieving results that were only explained much later in scientific terms. Thomas Kuhn gives two examples. "When Kepler studied the optimum dimensions of wine casks... he helped to invent the calculus of variations, but existing wine casks were, he found, already built to the dimensions he derived. When Sadi Carnot undertook to produce the theory of the steam engine... the result was an important step toward thermodynamics; his prescription for engine improvement, however, had been embodied in engineering practice before his study began."[17]

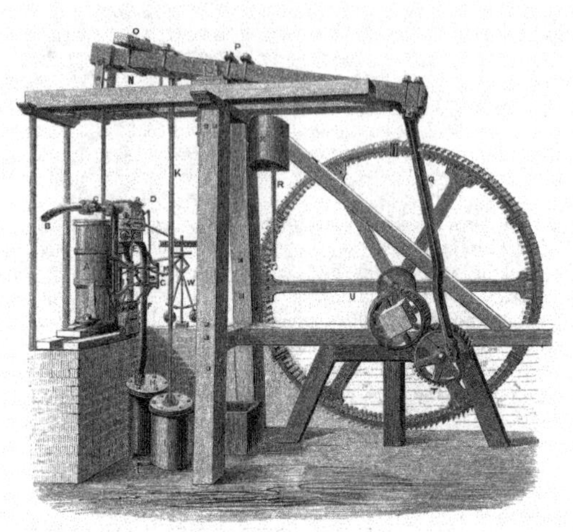

James Watt's steam engine 'Old Bess' (1778) converted power from the expansion of steam to drive a wheel.

© *Universal History Archive/Getty Images.*

Counter examples where science was applied directly to achieve a technological development are the production of sulfuric acid, much simplified by a new chemical process in 1736, and the production of chlorine (important for bleaching), in 1787.[18] Chemistry has always been the area of science most closely connected to industry.

In fact, the relationship between science and technology was somewhat reversed in subsequent developments. During the 18th century, the study of electricity was confined to static electricity. Alessandro Volta (1745–1827), an Italian physicist and chemist living in Como, invented the first battery, able to produce an electric current. By running electric current through solutions, Humphrey Davy (1778–1829), at the Royal Institution in London, showed that positively charged metal ions appeared at the cathode (negative) pole. This led to the discovery of new elements such as sodium and potassium. Hans Christian Ørsted (1775–1851) in Copenhagen found that an electric current could generate magnetism. This was an early example of how technology can drive scientific discovery. In due course, electricity in turn drove technology.

The Industrial Revolution may not have driven science directly, but perhaps you might say that technology was, as it were, in the air, providing the materials for experimentation. Indeed, there's an argument that technology was an important factor in making it possible to reproduce experiments reliably. This is a crucial feature in the development of science. Electromagnetism is an early example, which I discuss in the next chapter. Some advances in basic science have been made possible only by advances in technology. Indeed, they may exist only in the technology.[19]

The situation began to change in the 'Second Industrial Revolution,' at the end of the 19th century, when several areas of science gave rise to developments that affected daily life. Physics led to electricity, chemistry led to aniline dyes, and microbiology led to vaccines.[20]

It's not always simple to distinguish between science and technology. There's a blur as to whether the invention of the transistor at Bell Labs, for example, owed more to basic investigations of the properties of semi-conductors or to the search for an efficient solid-state amplifier.[21] William Shockley, who shared a Nobel Prize for the discovery, refused to draw a distinction between basic or applied research.[22]

Science and technology may alternate, but not necessarily on the same time scale. Sometimes technology has been driven by the science of the previous generation, whereas science more rapidly takes advantage of new possibilities offered by technology. The phenomenon of semi-conduction in germanium had been discovered by 1900; but the transistor was not invented until 1947. The miniaturization made possible by transistors led to new equipment that was being used to make discoveries, for example in space, by the 1950s.

The form that technology takes is much driven by the imperatives of society. Karl Marx might have remarked that there was no emphasis on laborsaving machines in the ancient world because labor (in the form of slaves) was cheap. Without going to such extremes, it's fair to say that developments in technology showed a much clearer connection with the society of the time compared to science, which was more of a world in itself. Science was driven by the accidents of discovery rather than the precepts of society.

Yet the Industrial Revolution was a significant moment in the development of science, because new technologies in turn made further experiments possible in science. We might ask whether there is a parallel with the development of AI (artificial intelligence) making possible new interpretations of existing data today, and perhaps thereby suggesting further investigations.

NOTES AND REFERENCES

1. In a letter to Dr. Lind, April 29, 1765 describing his steam engine. Reproduced in J. P. Muirhead, *The Origin and Progress of the Mechanical Inventions of James Watt*, Nabu Press, 2012.

2. K. Marx, *Economic and Philosophic Manuscripts of 1844*, Progress Publishers, Moscow, 1932.

3. Emphasizing the fact that science was not the driving force, the term Industrial Revolution was first used, in a political context, by Georges Michelet in France and Friedrich Engels in Germany. F. Engels, *The Condition of the Working-Class in England in 1844*, Sonnenschein & Co., London, 1892. The term was introduced into English in the historian Arnold Toynbee's lectures, published in 1884.

4. D. Wootton, *The Invention Of Science: A New History Of The Scientific Revolution*, Harper, New York, 2015, pp. 500–508.

5. J. E. McClellan and H. Dorn, *Science and Technology in World History: An Introduction*, Johns Hopkins University Press, Baltimore, 2nd edn, 2006, p. 288.

6. R. C. Allen, *The British Industrial Revolution in Global Perspective*, Cambridge University Press, Cambridge, 2009, pp. 157–163.

7. D. Fleming, Latent heat and the invention of the Watt engine, *Isis*, 1952, **43**, 3–5.

8. D. P. Miller, *James Watt, Chemist: Understanding the Origins of the Steam Age*, Pickering & Chatto, London, 2009.

9. M. C. Jacob, *The First Knowledge Economy. Human Capital and The European Economy, 1750–1850*, Cambridge University Press, Cambridge, 2014.

10. C. Ó Gráda, Did science cause the industrial revolution?, *J. Econ. Lit.*, 2016, **54**, 224–239.

11. A. R. Hall, Engineering and the scientific revolution, *Tech. Cult.*, 1961, **2**, 333–341; A. R. Hall, What did the Industrial Revolution in Britain owe to Science?, in *Historical Perspectives: Studies in English Thought and Society in Honour of J.H. Plumb*, ed. N. McKendrick, Europa Pub, London, 1974, pp. 129–151.

12. R. C. Allen, *The British Industrial Revolution in Global Perspective*, Cambridge University Press, Cambridge, 2009, p. 249.

13. A. E. Musson and E. Robinson, *Science and Technology in the Industrial Revolution*, Routledge, New York, 1969, pp. 142–143.

14. R. C. Allen, *The British Industrial Revolution in Global Perspective*, Cambridge University Press, Cambridge, 2009, p. 244; R. R. Meisenzahl and J. Mokyr, The Rate and Direction of Invention in the British Industrial Revolution: Incentives and Institutions, in *The Rate and Direction of Inventive Activity Revisited*, ed. J. Lerner and S. Stern, University of Chicago Press, Chicago, 2012, pp. 443–479.

15. There has been some controversy about the extent to which the Industrial Revolution may have depended on science. The relationship was more or less dismissed until more recent proposals that, at least, the Industrial Revolution in the 18th century may have depended on scientific developments in the 17th century. See D. Wootton, *The Invention Of Science: A New History Of The Scientific Revolution*, Harper, New York, 2015, pp. 476–508.
16. The saying has been attributed to Lawrence Joseph Henderson, an American biochemist, in 1917, but no original source has ever been cited.
17. T. S. Kuhn, *The Essential Tension. Selected Studies in Scientific Tradition and Change*, University of Chicago Press, Chicago, 1977, p. 144.
18. B. Russell, *The Impact of Science on Society*, Simon & Schuster, New York, 1953, pp. 99–101.
19. D. E. Stokes, *Pasteur's Quadrant: Basic Science and Technological Innovation*, Brookings Institution Press, Washington, DC, 1997, p. 21.
20. D. E. Stokes, *Pasteur's Quadrant: Basic Science and Technological Innovation*, Brookings Institution Press, Washington, DC, 1997, pp. 19–20.
21. J. Gertner, *The Idea Factory: Bell Labs and the Great Age of American Innovation*, Penguin Books, New York, 2013.
22. R. R. Nelson, The Link Between Science and Invention: The Case of the Transistor, in *The Rate and Direction of Inventive Activity: Economic and Social Factors*, National Bureau Committee for Economic Research, Princeton University Press, Princeton, 1962.

CHAPTER 9

Maxwell's Demon: 1820–1890

"RADIATION [IS] A KIND OF SPECIES OF VIBRATION IN THE lines of force which are known to connect particles and also masses of matter together," said Michael Faraday in an unusually speculative mode in 1846.[1] Faraday's concept of lines of force turned into James Clerk Maxwell's concept of electromagnetic fields, unifying magnetism, electricity, and light. Faraday had wondered if gravity might also be represented by lines of force. However, his concept of unifying all four phenomena was regarded by his peers as speculative, if not downright embarrassing. (It still has yet to be achieved!) Faraday, a superb experimentalist, and Maxwell, a genius of a theoretician, together created a description of electromagnetism that bridged the gap between Newtonian physics and quantum physics. Newtonian physics describes phenomena as they seem to exist in the real world, but quantum physics delves into the uncertainties of the smallest possible scale. Maxwell also contributed to a revolution in thermodynamics with his kinetic theory of gases, the first application of statistics to explain macroscopic phenomena by microscopic effects. The revolution in electromagnetism belongs equally to Faraday and Maxwell, as much for anticipating the yin and yang of theory and experiments that was to dominate the 20th century, as for the sweeping change in understanding the phenomena. The big dilemma with thermodynamics was the difficulty of reconciling the irreversibility of the laws of thermodynamics with the reversible equations of classical physics.

The Frontiers of Science
By Benjamin Lewin
© Benjamin Lewin 2026
Published by the Royal Society of Chemistry, www.rsc.org

Timeline for the Development of Electromagnetism

1780 — Galvani shows electric spark causes frog muscles to twitch (1780)

Coulomb describes inverse-square laws of magnetism and electricity (1785)

1790

1800 — Alessandro Volta invents voltaic pile (first battery) (1800)

1810

1820 — Ørsted shows electric current deflects compass needle (1820)
Ampère defines relationship of electricity and magnetism (1820)
Michael Faraday invents electric motor (1821)

1830 — Faraday discovers electromagnetic induction; describes lines of force (1831)
Faraday defines electrolysis (1832)

1840 — Faraday describes 'magnetic field' and shows it polarizes light (1845)

1850 — James Clerk Maxwell analyzes Faraday's lines of force (1855)

1860 — Maxwell defines physical lines of force (1861)

Maxwell publishes Treatise on *Electricity and Magnetism* (1864)

1870

1880 — Oliver Heaviside summarizes Maxwell's theory into four equations (1885)
Michelson and Morley fail to identify ether; speed of light is fixed (1887)
1890 — Heinrich Hertz detects electromagnetic waves (1888)

A Dialog between James Clerk Maxwell and Erwin Schrödinger

James Clerk Maxwell (1831–1879) was a Scottish theoretical physicist and mathematician, working in Cambridge, who developed the classical theory of electromagnetic radiation that unified electricity, magnetism, and light.

Erwin Schrödinger (1887–1961) was an Austrian theoretical physicist, working first in Graz but finally in Ireland, who developed the equations that describe the behavior of sub-atomic particles as waves in quantum mechanics.

Maxwell: Guten Tag, Herr Schrödinger! I hear you have imagined a cat that tests the bounds of the equations of the quantum theory. My invention is a demon that challenges the second law of thermodynamics. What are your thoughts on this?

Schrödinger: Your demon and my infamous cat seem to share a certain paradoxical quality. In the case of my cat, it is both alive and dead until an observer regards it. Similarly, your demon seems to exploit information and observation to defy the natural progression toward greater disorder. Do these concepts test our understanding of how our equations represent reality?

Maxwell: I imaged how a demon might defy the second law of thermodynamics through my kinetic theory that applies statistics to describe the movement of molecules in gases. It's fascinating to describe macroscopic properties through the statistical behavior of microscopic particles. But tell me, how do your quantum concepts align with the classical laws of thermodynamics?

Schrödinger: Well, Maxwell, your equations laid the foundation for classical electromagnetism. My wave equation attempts to do the same for the quantum realm. In the quantum world, particles exhibit discrete energy levels. My wave function expresses the probability of finding a particle in a particular state. The challenge is to reconcile the discrete energy levels of my waves with the continuous nature of your thermodynamic variables.

Maxwell: A valid point, Schrödinger. Quantum mechanics introduces an element of unpredictability that is absent from my equations. That reinforces my view that the equations are a useful fiction for describing our observations, but do not necessary represent physical reality. My original theory accepted the existence of the ether, but we now know there is no ether. Indeed, Michael Faraday doubted its existence all along.

Schrödinger: (*gesticulating*) Ah, Maxwell, these transitions in 'reality' are thought provoking. My cat challenges our classical intuitions about reality. Your demon challenges our understanding of the arrow of time. Perhaps the demon and the cat are not so disparate after all, but different manifestations of the difficulties of expressing our equations in words. I rest my case.

MAXWELL'S DEMON

> *The law that entropy always increases—the second law of thermodynamics—holds, I think, the supreme position among the laws of Nature. If someone points out to you that your pet theory of the universe is in disagreement with Maxwell's equations—then so much the worse for Maxwell's equations. If it is found to be contradicted by observation, well, these experimentalists do bungle things sometimes. But if your theory is found to be against the second law of thermodynamics I can give you no hope; there is nothing for it but to collapse in deepest humiliation,*[2] Arthur Eddington, 1936.

If any single concept stands at the center of science, it is the laws of thermodynamics. (The term thermodynamics was created in 1849 when William Thomson—later Lord Kelvin—referred to 'thermo-dynamic engines,' meaning devices such as steam engines that convert heat to motion.) Close to a century after the steam engine became the workhorse of the Industrial Revolution, it was (albeit indirectly) the driving force for a great advance in physics, an example of technology driving science.

The laws of thermodynamics emerged from studies of the efficiency in converting heat to work. "Everyone knows that heat can produce motion" is how Sadi Carnot (1796–1832) began his book on heat in 1824.[3] The first law of thermodynamics expresses this more formally in stating that energy can be converted into other forms, but is conserved. The second law introduces the concept of entropy—the degree of disorder in a system—and says that entropy in a closed system can only increase (the so-called 'arrow of time'). (The easiest way to think about this may be simply to substitute 'disorder' for 'entropy.')

Put another way, heat can move only from hotter to colder objects. The second law of thermodynamics developed from Carnot's concept of a 'cycle of heat.' Sadi Carnot was a French physicist and military engineer, whose interest came from studying steam engines. The progress from Carnot's cycle of heat to the subsequent formulation of the second law of thermodynamics symbolizes the transition into modern theory. Carnot visualized the action in the steam engine in terms of caloric, supposedly an invisible fluid. Caloric was carried by steam, which transferred it to cold water. Steam was effectively the means of transporting the caloric. "The production of motive power is... not due to... consumption of caloric, but to its transportation from a warm body to a cold body."[4]

Experiments conducted by James Joule in England led to a different view: heat is lost when it is converted into work. "We consider heat not as a substance, but as a state of vibration," was how Joule put it in 1843. James Joule (1818–1889) came from a family of brewers, and studied chemistry with John Dalton. By spinning a paddle wheel in an insulated barrel of liquid, he was able to calculate the equivalence between mechanical work and gain in temperature.[5] (He ran the family brewery, while conducting his experiments on the side.)

Heat and light were the twin pillars of physics in the century preceding quantum theory. They were different properties of fire, according to an early interpretation of Newtonian physics. By the end of the 18th century, heat became formalized as caloric. It was listed together with light in Lavoisier's table of the elements. Ideas changed during the 19th century, when heat and light were recognized as forms of energy.

Rudolf Clausius in Berlin and Lord Kelvin in London independently developed the laws of thermodynamics in 1850 and

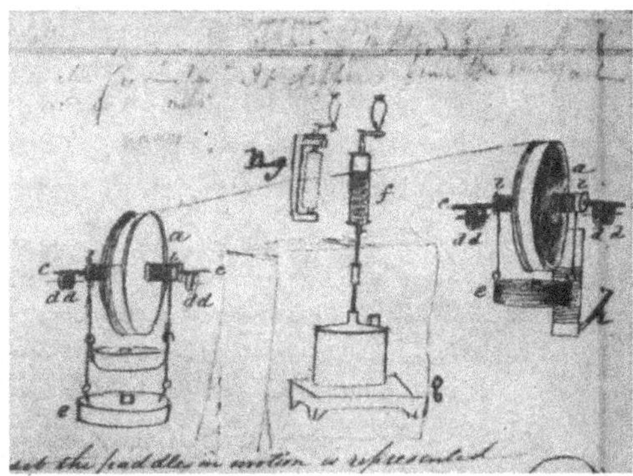

A sketch from Joule's notebook (c. 1845) shows his apparatus for measuring the mechanical equivalence of heat. The temperature change in the fluid is measured when the paddle moves weights a certain distance.

Reproduced with permission, Copyright © History of Science Museum, University of Oxford.

1851. Clausius (1822–1868), a German physicist and mathematician, showed that heat is a property of motion in gases. His seminal paper in 1857 carried the title "On the kind of motion that we call heat."[6] Lord Kelvin (William Thomson) (1824–1907) dominated physics in Britain during the last part of the 19th century. Maxwell described the second law colorfully in a letter to a colleague in 1870. "The 2nd law of thermodynamics has the same degree of truth as the statement that if you throw a tumblerful of water into the sea, you cannot get the same tumblerful of water out again."[7]

Justus Liebig (1803–1873) made the connection with biology. "In the animal body the food is the fuel; with a proper supply of oxygen we obtain the heat given out during its oxidation or combustion," he said in 1843.[8] In effect, he argued that heat is the result of combustion. (Surprisingly for someone whose experiments were based in chemistry, he continued to be a holdout in believing in vitalism, another demonstration of the difficulty in breaking away from pervasive [nonscientific] beliefs.) Von Helmholtz made the same connection between heat and combustion, but was a committed opponent of vitalism. (I discuss vitalism in Chapter 21.)

The concept that heat depends on motion led James Clerk Maxwell to formulate his kinetic theory of gases in 1860. This answered the question: if temperature depends on the speed of motion, do all the molecules in a gas travel at the same speed? Maxwell derived an equation to show a statistical variation of speeds, forming a bell-shaped curve.[9] Helped by his wife, Katherine, he conducted experiments at home to confirm predictions of the theory.[10] Austrian physicist Ludwig Boltzmann took up the theory, which led to the later definition of the Maxwell-Boltzmann distribution of molecular energies, which in turn led Boltzmann to atomic theory (see Chapter 13).

Until heat was equated with motion, physics was mechanistic: defined in terms of the interactions of physical objects. This was "the axiom on which all Modern Physics is founded," according to J. J. Thomson, a physicist who later discovered the electron.[11] One of the principles of mechanistic physics was that all interactions are in principle reversible. The difficulty of reconciling this principle with thermodynamics, where actions are irreversible, was a big problem for physics in the second half of the 19th century. Henri Poincaré put it wonderfully well in 1905.

"We have reversibility in the premises and irreversibility in the conclusions; and between the two an abyss."[12] This was called the 'reversibility problem.'[13] The second law is such a central feature of scientific thinking today, that it's hard to realize its implications were controversial through the 19th century.

Known colloquially as the 'arrow of time,' the principle of irreversibility conflicts with the general principle that the equations of physics are symmetrical. Formally, they work equally well in either direction, so they do not distinguish between past and future.[14] The discrepancy merged into a philosophical argument. Mechanistic physics was equated with materialism (as typified in the impersonal rationality of the Enlightenment), while energy (meaning thermodynamics) was equated with force or vitality.[15] (An interesting application of reversibility comes from cosmology

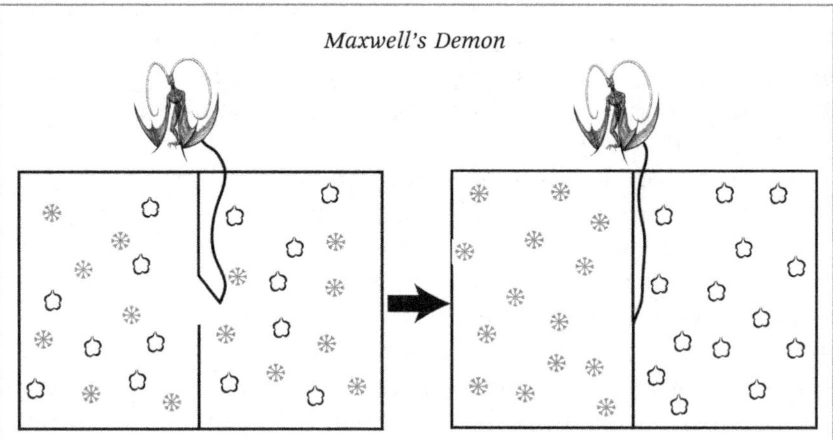

A demon (originally described as a 'finite being') controls a door between two chambers filled with gas. When fast-moving molecules approach, the demon opens the door to let them pass in one direction. When slow-moving molecules approach, the demon opens the door so they pass in the other direction. Eventually, one chamber is full of fast-moving (hot) molecules and the other chamber is full of slow-moving (cool) molecules. This would violate the second law of thermodynamics, because the system has become more ordered, decreasing its total entropy.[17] The key to the paradox was that Maxwell specified that the demon functions "without expenditure of work." Leo Szilárd suggested a resolution in 1929 when he pointed out that the demon would have to measure the velocities of the molecules. This would necessitate expenditure of energy that would increase the entropy of the demon by more than the loss of entropy in the chamber.[18] Subsequently other resolutions followed the same general principle, although differing in the details. Even today, the way to get out of the dilemma is a lively topic of controversy.[19]

today, where reversing observations of the current expansion of the universe extrapolates back to the Big Bang; see Chapter 17).

Mechanics and thermodynamics were reconciled by proposing that reversibility is not absolute, but is statistical. Processes that appear to be irreversible might in principle be reversed, but only with a probability that's so low it never actually happens.[16] This would, however, mean that the second law of thermodynamics could be violated (in principle). This was the impetus for Maxwell's thought experiment in 1867 in which he devised a means for violating the second law, in the form of his famous demon.

Understanding the second law of thermodynamics was famously C. P. Snow's criterion for judging the scientific illiteracy of intellectuals (Chapter 10). Its position as the classic unarguable law of science is emphasized by the great attention devoted to Maxwell's demon. This imagines a situation in which the second law of thermodynamics is defied when a demon causes fast- and slow-moving particles to separate into two different chambers. The result is that one becomes hot and the other becomes cold, without any input of energy (see the box).

∞ ∞ ∞

How could there possibly be any connection between magnetism, electricity, and light? Magnetism worked mysteriously at a distance. Electricity generated sparks that fly across a gap between two wires. Light traveled as a wave through the ether. Through the 17th and 18th centuries, they appeared to be completely disparate phenomena. The genius of Michael Faraday as an experimentalist, and James Clerk Maxwell as a theoretician, was to unite them.

As early as the Elizabethan era, William Gilbert had proposed that the Earth is a giant magnet. The basis for magnetism was unfathomable. Action at a distance was impossible to understand. There was debate in the 17th century about the nature of light, with Isaac Newton advocating particles and Christiaan Huygens, whose position was more widely accepted, describing it as a waveform. Gilbert knew that static electricity could be generated by rubbing amber, but it was not until 1780 that Galvani showed that electric impulses could cause frog muscles to twitch.

It became possible to generate electricity in 1800 when Alessandro Volta (1745–1847) invented the first battery, made from alternating layers of zinc and copper. Volta announced his discovery of electricity in a letter in 1820 from his home on Como to the Royal Society in London. Because England and France were at war, the letter arrived in two parts, and was read later that year. In 1801, Volta repeated his experiments in Paris, at a meeting attended by Napoleon Bonaparte, who then established prizes to be awarded for future discoveries.

Voltaic piles, as the new batteries invented by Volta were named, replaced electrostatic generators. At the Royal Institution in London in 1802, Humphry Davy (1778–1829) a British chemist and inventor who created the Davy safety lamp used in coal mines, constructed the most powerful voltaic pile of the period, with 60 pairs of 6 inch-square zinc and copper plates. By 1807, he had an even more powerful battery, able to decompose alkalis, leading to the isolation of sodium and potassium. No matter that there was a state of war, Davy, together with his assistant Michael Faraday, went to Paris in 1808 to collect a prize from Napoleon for his work.

The first connection between electricity and magnetism was made by Hans Christian Ørsted (1777–1851). After earning a doctorate in philosophy on the works of Kant at the University of Copenhagen, he traveled through Europe on a scholarship. His application for a professorship in physics at the University in 1803 was rejected (probably because he was regarded as a philosopher) but he succeeded in 1806. In 1820, by then a leading figure in Copenhagen who was known as a poet as well as a scientist, he found that an electric current could deflect a compass needle in the vicinity. Ørsted's first description of his discovery is somewhat obscure, originally published in a privately distributed pamphlet (in Latin). This led to reports that the discovery had been an accident, but Ørsted later denied this.[20]

The effect was put on a quantitative basis by André-Marie Ampère (1775–1836). Within a few days of the news of Ørsted's experiment reaching Paris later in 1820, Ampère had repeated and extended the experiment, and put the relationship between the current and the magnetic force on a mathematical basis. The unit of electricity is named after him.

Ampère had established an equation showing how an electric current produced a magnetic field. At the Royal Institution, Faraday found that a magnetic field could cause an electric

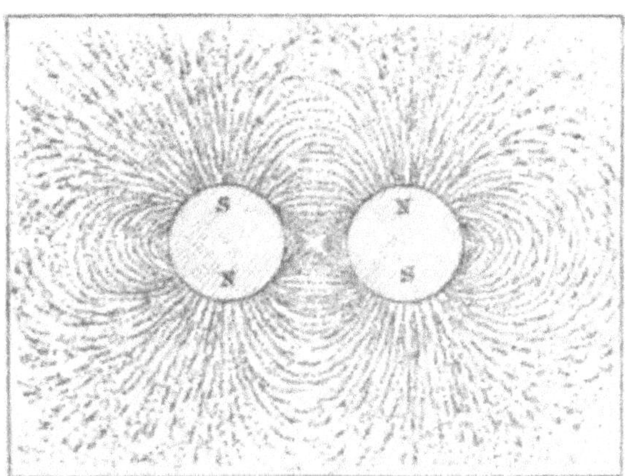

Faraday visualized lines of force by the distribution of iron filings around magnets.[21]

current to flow. Clearly electricity and magnetism were two sides of the same coin. Faraday introduced the idea of lines of force to visualize the relationship.

Michael Faraday (1791–1867) started his scientific career in chemistry as an assistant to Humphry Davy (1778–1829). Faraday's entry into science might be ascribed indirectly to the Industrial Revolution. His father was a blacksmith who was driven out of the countryside into the city of London. Aged 13, Faraday began work at a bookshop. Exposed to books, he began to read and to attend lectures. One of the customers at the shop recommended him to Humphry Davy, who needed an assistant. Eventually he became Davy's full-time assistant.

By 1820, Faraday had spent some years working with Davy at the Royal Institution. Over the next two decades, Faraday undertook a series of experiments to follow up on Ørsted's discovery that an electric current could move a compass needle. This led to the concept of electromagnetism. He did not develop a formal theory, perhaps because he had no training in mathematics. "I am unfortunate in a want to mathematical knowledge and the power of entering with facility into abstract reasoning," he wrote to Ampère.[22] This did not stop him from making a series of discoveries that established the experimental basis for electromagnetism.

Following the discovery of the phenomenon of electromagnetism, in 1821 he constructed a primitive electric motor (it

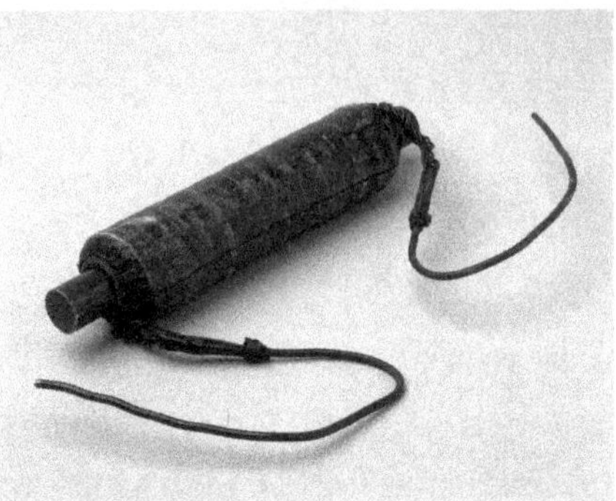

Faraday demonstrated electromagnetic induction using a bar magnet inside a coil wrapped with wire.

© *Science & Society Picture Library/Getty Images.*

caused a wire placed in a pool of mercury to rotate around a magnet). A decade later, he discovered electromagnetic induction, when moving a magnet through a loop of wire generated a current in the wire. Working with other forms of electricity, he went on to propose that electricity is a single force, although this was not accepted until much later.

Faraday extended the relationship to light in 1845 by showing that a strong magnetic field could change the polarization of a light ray. By 1846 he could say, "I believe... that the various forms under which the forces of matter are made manifest have one common origin... that they are convertible, as it were, one into another."[23] This led to subsequent proposals that the lines of force were propagated by vortices in the ether. A typical view in 1878 was that "Each molecular vibration disturbs the ether... each wavelength of light [results] from a molecular tremor of the corresponding wavelength."[24]

In 1825, Faraday became Director of the Royal Institution, and introduced the lecture series that made the Institution famous (see Chapter 24). He spent the rest of his career there as an experimentalist par excellence.

It took almost half a century from Faraday's discovery of electromagnetism until James Clerk Maxwell described electricity, magnetism, and light mathematically as different expressions of the same force. Faraday and Maxwell could not have had more different origins. Together they revolutionized physics, setting the scene for quantum mechanics in the following century.[25] It also established the distinction between experimental and theoretical physics that would mark the 20th century.

James Clerk Maxwell (1831–1879) started as a mathematician. He came from a wealthy family who owned an estate in Scotland. While still at school in Edinburgh, aged 14, he published his first paper (on geometry). From Edinburgh, he went to the University of Cambridge, where he graduated top in mathematics. While still at Cambridge, he wrote a paper, *On Faraday's Lines of Force*, that started, "The present state of electrical science seems peculiarly unfavourable to speculation."[26] He went on to an extensive mathematical speculation that described lines of force by analogy with a description of the intensity and direction of pressure in a weightless (imaginary) fluid. By the time the paper was published, in 1856, he was on the faculty at a college that later became part of the University of Aberdeen. By 1861, he had moved to the chair of Natural Philosophy at King's College, London.

Maxwell with the spinning coil apparatus he used to measure the Ohm (unit of electrical resistance), probably around 1864.

Using the analogy with fluid pressure, Maxwell had shown that he could derive descriptions of static electric and magnetic fields from Faraday's lines of force. Now he set out to deduce a more general description of electricity and magnetism. In a series of papers published in 1861–1862, with the general title, *On Physical Lines of Force*, he showed that the concept of 'molecular vortices' moving in an ether could be used to derive equations describing electromagnetic induction.

The equations demonstrate that electricity and light are waveforms that travel at the same speed. This remains at the heart of modern physics. "Light consists in the transverse undulations of the same medium which is the cause of electric and magnetic phenomena," Maxwell said.[27]

With his reputation firmly established, he moved to Cambridge in 1874 to become the first director of the Cavendish Laboratory that was to figure so largely in developing atomic physics in the next century (see Chapter 13). Maxwell's mathematics were difficult to understand, to say the least—by the time of his last papers on the topic, in 1865 and 1873, the system comprised more than 20 equations. (His first presentation of the theory, in 1864, mystified the audience. One of the difficulties was the lack of any physical basis for the model.) Six years after Maxwell's death, mathematician Oliver Heaviside transformed the theory in 1885 into the 4 equations that are known today as 'Maxwell's equations.'

The remarkable feature about these equations is not just their deduction from known data but the intuitive leap to making them work. The third equation's prediction of waves that have the velocity of light depends on a term (the displacement current) that Maxwell added *ad hoc* to fit things together, without any empirical foundation. Physicist Max Born comments, "Indeed this is a brilliant example of the possibilities which exist for the theoretical physicist: he can trace deficiencies in the perfection of a theory and can try amend them by what you may call 'mathematical guessing'."[28] This may be a respectable approach in theoretical physics—a case where 'fine-tuning' (Chapter 11) works—but would be regarded as distinctly suspect in biology!

The original derivation of the equations visualized light as a waveform moving through the ether that filled the universe (more formally, the luminiferous, or light-bearing, ether). The existence of the ether would predict that light should move at different speeds depending on whether it is parallel with the ether or at right angles to it. Albert Michelson in Cleveland, Ohio thought this could be used to measure the velocity with which the earth travels through the ether. But when Michelson and Morley measured the speed of light in 1887 using an interferometer, it was always the same. There is no ether.

The interferometer combines light waves that had moved in perpendicular directions. This produces an interference pattern. A difference in velocity in the two directions should be detected by displacing an interference band by about 10% of its width. However, measurements are susceptible to environmental influences. Changes in temperature or vibration could produce changes larger than expected from the experiment. The long and short of it is that no change could be detected, suggesting (as the theory of relativity later predicted) that there is no ether.

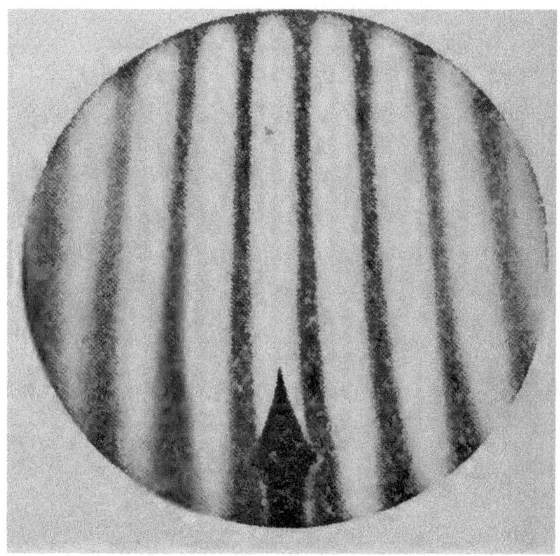

An example of an interference pattern from later experiments on Mount Wilson. It's right at the margin to detect a displacement of one tenth of a band.

Reproduced from D. C. Miller, The Ether-Drift Experiments and the Determination of the Absolute Motion of the Earth, Rev. Mod. Phys., 1933, 5, 203–242, Copyright 1933 American Physical Society.

This was not the desired result at the time. Michelson lost interest in the project, which he regarded more or less as a failure.[29] Theoreticians ignored the result, continuing to build models based on the ether, until Einstein abandoned the ether in the theory of special relativity (see Chapter 16).

Two decades later, the Michelson–Morley experiment came to be regarded as the *experimentum crucis* supporting relativity. Michelson received the Nobel Prize for Physics in 1907 for his work in developing the precision interferometer, but the award referred to his measurement of the speed of light and never even mentioned the ether, another demonstration of the difficulty of understanding experiments in the absence of a theoretical construct.[30]

The ether is a great example of a paradigm that came out of the blue, survived against evidence disproving it, and was ultimately abandoned reluctantly only when a new theory made it unnecessary. The ether was fueled by belief, without any data, that a medium was needed for objects to transmit force, or for waves of light to propagate. Belief was scarcely dented by the direct demonstration that the ether does not exist. It took Einstein's theory of relativity to abolish the ether (and to validate the Michelson–Morley experiment).

In any case, Maxwell's equations remain equally valid for movement in a vacuum. Their importance extends beyond their success in creating what was much later called 'field theory,' which became the predecessor to quantum mechanics. It was really a paradigm shift in the whole way of regarding physics.'

Physics based on observations, culminating in Newtonian physics, assumed that experiments represented physical reality. Maxwell's equations did not make this assumption. Maxwell (and von Helmholtz, working in Germany on thermodynamics) took the position that they were providing analogies that explained the real world.[31] They would not have used the term, but these were operational calculations that made useful predictions rather than physical descriptions. Maxwell's imaginary weightless fluid was a useful fiction, not a physical hypothesis. This was a predecessor to the great debate in the 20th century about the relationship between quantum descriptions and reality (discussed in Chapter 16).

Wavelength	Wave type	Discovery	
long	Radio waves	1887	Hertz (Karlsruhe)
30mm			
	Microwaves	1945	Spencer (Raytheon)
	(CMB @ 1.9mm)	1964	Wilson & Penzias (Bell Labs)
1mm			
	Infrared	1800	Herschel (Windsor, UK)
700nm			
	Visible Light		
400nm			
	Ultraviolet	1801	Ritter (Munich)
10nm			
	X-rays	1895	Röntgen (Würzberg)
0.01nm			
short	Gamma-rays	1900	Villard (Paris)

The electromagnetic spectrum stretches from γ-rays (with wavelengths that can be smaller than the width of an atom) to radio waves (whose wavelengths can reach kilometers). The first extensions from the visible wavelengths into infrared and ultraviolet took place at the very start of the 19th century. Radio waves, X-rays, and γ-rays were discovered at the end of the century. Driven by experiments rather than theory, few of these discoveries were made at major research centers.

Heinrich Hertz (1857–1894), who investigated Maxwell's theories experimentally, took Maxwell's skepticism about representing reality even further, arguing that 'concrete representations' had little significance. His famous comment on the equations was, "To the question, 'What is Maxwell's theory?' I know of no shorter or more definite answer than the following: Maxwell's theory is Maxwell's system of equations."[32] He pointed to inconsistencies in Maxwell's various physical analogies and argued that, in effect, the equations speak for themselves.[33]

Maxwell's equations brilliantly unified apparently different forces under the single form of electromagnetic waves, but there was no means at the time of measuring them experimentally. This happened after new waveforms were discovered later in the 19th century, with radio waves at long wavelengths, and X-rays and γ-rays at very short wavelengths. Hertz's discovery of radio waves was the first experimental demonstration of the electromagnetic waves predicted by Maxwell's equations. His demonstration that they travel at the speed of light emphasized the unity of the electromagnetic spectrum.

Heinrich Hertz was a supreme experimentalist. After working with von Helmholtz in Berlin, he became the professor of experimental physics in Karlsruhe. He discovered by chance that when he generated a spark in one electrical circuit, sparks also formed at a stray wire next to it. He developed a system in which a transmitter generated a spark across a small gap between two terminals. A detector loop (basically a circle of wire with a gap) then sparked if it had the right size and shape. This meant that the transmitter had generated waves that the detector responded to by sparking if it was tuned to the right resonance.

Hertz knew he had made a great discovery in physics, but his most famous quote is his response to a question from a student

Marconi's first transmitter closely followed the design of Hertz's original apparatus that created radio waves. An induction coil powered a spark gap connected to an antenna (the copper sheet). The antenna allowed the signal to propagate for up to 1 km. A telegraph key was used to turn the circuit on and off to allow transmission in Morse code.

about its usefulness. "It's of no use whatsoever. This is just an experiment that proves Maestro Maxwell was right, we just have these mysterious electromagnetic waves that we cannot see with the naked eye. But they are there."[34] The frequency of the waves in Hertz's first experiment in 1888 was about 50 MHz, in the radio spectrum.

Guglielmo Marconi was more an inventor than a scientist. He sent the first wireless message by radio in 1895. By 1897, he had patented the system. Educated privately, he never graduated from university (although he attended lectures in physics at the University of Bologna). Hertz died young and was never eligible for a Nobel Prize, but Marconi was awarded the Prize for Physics in 1909. It would be an understatement to say that the discovery of radio waves contributed to future Industrial Revolutions.

Filling the gap between infrared and radio waves, microwaves were discovered in the middle of the 20th century (with the specific example of cosmic background radiation [CMB] following two decades later. I discuss the significance of CMB in Chapter 17). Wavelengths in the electromagnetic spectrum extend over 10 orders of magnitude, but we can see only visible light, which occupies less than 1 order of magnitude. Perhaps this should give us some pause for thought about humanity's position in the universe.

NOTES AND REFERENCES

1. M. Faraday, Thoughts on ray-vibrations, *Philos. Mag.*, 1846, **28**, 345–350.
2. A. Eddington *The Nature of the Physical World*, J. M. Dent & Sons, London, 1935, p. 81.
3. S. Carnot, *Réflexions sur la puissance motrice du feu et sur les machines propres à développer cette puissance*, Bachelier, Paris, 1824. Translated by R. H. Thurston, *Reflections on the Motive Power of Heat*, J. Wiley & Sons, New York, 1890.
4. S. Carnot, *Réflexions sur la puissance motrice du feu et sur les machines propres à développer cette puissance*, Bachelier, Paris, 1824. Translated by R. H. Thurston, *Reflections on the Motive Power of Heat*, J. Wiley & Sons, New York, 1890, p. 46.
5. J. P. Joule, On the Mechanical Equivalent of Heat, *Philos. Trans. R. Soc. London*, 1850, **140**, 61–82. Joule's results were very good. His first measurement showed that 838 foot-pounds were required to raise the temperature of a pound of water by 1 °F. Later he refined the value to 772 foot-pounds, very close to today's measure of 778 foot-pounds.

6. R. Clausius, On the Kind of Motion which we call Heat, *Philos. Mag.*, 1857, **14**, 108–127.

7. In a letter to John William Strutt, December 6, 1870. Reproduced in E. Garber *et al.*, *Maxwell On Heat And Statistical Mechanics: On "Avoiding All Personal Enquiries" Of Molecules*, Lehigh University Press, Bethlehem, PA, 1995, pp. 204–205.

8. J. Liebig, *Animal Chemistry, or Organic Chemistry in its Applications to Physiology and Pathology*, Taylor and Walton, London, 1853.

9. J. C. Maxwell, Illustrations of the dynamical theory of gases, *Philos. Mag.*, 1860, **20**, 21–37.

10. N. Forbes and B. Mahon, *Faraday, Maxwell, and the Electromagnetic Field: How Two Men Revolutionized Physics*, Prometheus Books, New York, 2014.

11. J. J. Thomson, *Applications of Dynamics to Physics and Chemistry*, Macmillan, London, 1888, p. 1.

12. H. Poincaré, The principles of mathematical physics, *The Monist*, 1905, **15**, 1–24.

13. M. van Strien, The Nineteenth Century Conflict Between Mechanism and Irreversibility, *Stud. Hist. Philos. Mod. Phys.*, 2013, **44**, 191–205.

14. R. Penrose, *The Emperor's New Mind—Concerning Computers, Minds, and The Laws Of Physics*, Oxford University Press, Oxford, 1989, pp. 302–347.

15. The concept of irreversibility was unacceptable to some for philosophical reasons, because the unidirectionality of the second law predicts the 'heat death' of the universe. This would not be controversial scientifically today.

16. M. van Strien, The Nineteenth Century Conflict Between Mechanism and Irreversibility, *Stud. Hist. Philos. Mod. Phys.*, 2013, **44**, 191–205.

17. W. Thomson, Kinetic theory of the dissipation of energy, *Nature*, 1874, **9**, 441–444.

18. L. Szilárd, On the reduction of entropy in a thermodynamic system by the intervention of intelligent beings, *Zeitschrift für Physik*, 1929, **53**, 840–856.

19. C. H. Bennett, Demons, Engines, and the Second Law, *Sci. Am.*, 1987, **257**, 108–117.

20. B. Dibner, *Oersted and the Discovery of Electromagnetism*, Blaisdell New York, 1963, pp. 24–32

21. Faraday's papers published in the *Philosophical Transactions* were collected into three volumes under the title *Experimental Researches in Electricity*, published between 1839 and 1855. This image comes from *Experimental Researches*, III, Plate IV, Fig. 4.

22. Letter from Michael Faraday to Ampère, September 3, 1821. In L. P. Williams, *The Selected Correspondence of Michael Faraday*, Cambridge University Press, Cambridge, 1971, vol. 1, p. 134.

23. M. Faraday, Experimental researches in electricity, *Philos. Trans.*, 1846, **136**, 1–20.

24. J. N. Lockyer, *Studies in Spectrum Analysis*, Appleton, New York, 1878, p. 118.

25. My discussion of electromagnetism draws on N. Forbes and B. Mahon, *Faraday, Maxwell, and The Electromagnetic Field, How Two Men Revolutionized Physics*, Prometheus Books, New York, 2014.

26. J. C. Maxwell, On Faraday's lines of force, *Trans. Cambridge Philos. Soc.*, 1856, **10**, 27–82.
27. J. C. Maxwell, On physical lines of force, *Philos. Mag.*, 1861, **90**, 11–23.
28. M. Born, *Experiment and Theory in Physics*, Cambridge University Press, Cambridge, 1943, p. 10.
29. L. D. Swenson, *The Ethereal Ether. A History Of The Michelson-Morley-Miller Ether-Drift Experiments, 1880–1930*, University Of Texas Press, Austin, TX, 1972.
30. Dayton Miller repeated the experiments in 1924 with an apparatus on top of Mount Wilson (near Pasadena in California) to minimize environmental variation, and obtained positive results. This never had much impact, because by then the theory of relativity had been accepted. Later work came to the conclusion that the results had been due to statistical fluctuation, and the effect did not exceed the error level. See L. D. Swenson, *The Ethereal Ether. A History Of The Michelson-Morley-Miller Ether-Drift Experiments, 1880–1930*, University Of Texas Press, Austin, TX, 1972, pp. 228–245.
31. A. Bokulich, Maxwell, Helmholtz, and the unreasonable effectiveness of the method of physical analogy, *Stud. Hist. Philos. Sci.*, 2015, **50**, 28–37.
32. H. Hertz, *Electric Waves*, translated by D. E. Jones, Macmillan & Co., London, 1893, p. 21.
33. P. M. Heimann, Maxwell, Hertz, and the nature of electricity, *Isis*, 1971, **62**, 149–157.
34. Quoted in A. Norton, *Dynamic Fields and Waves*, Taylor & Francis, London, 2000, p. 83.

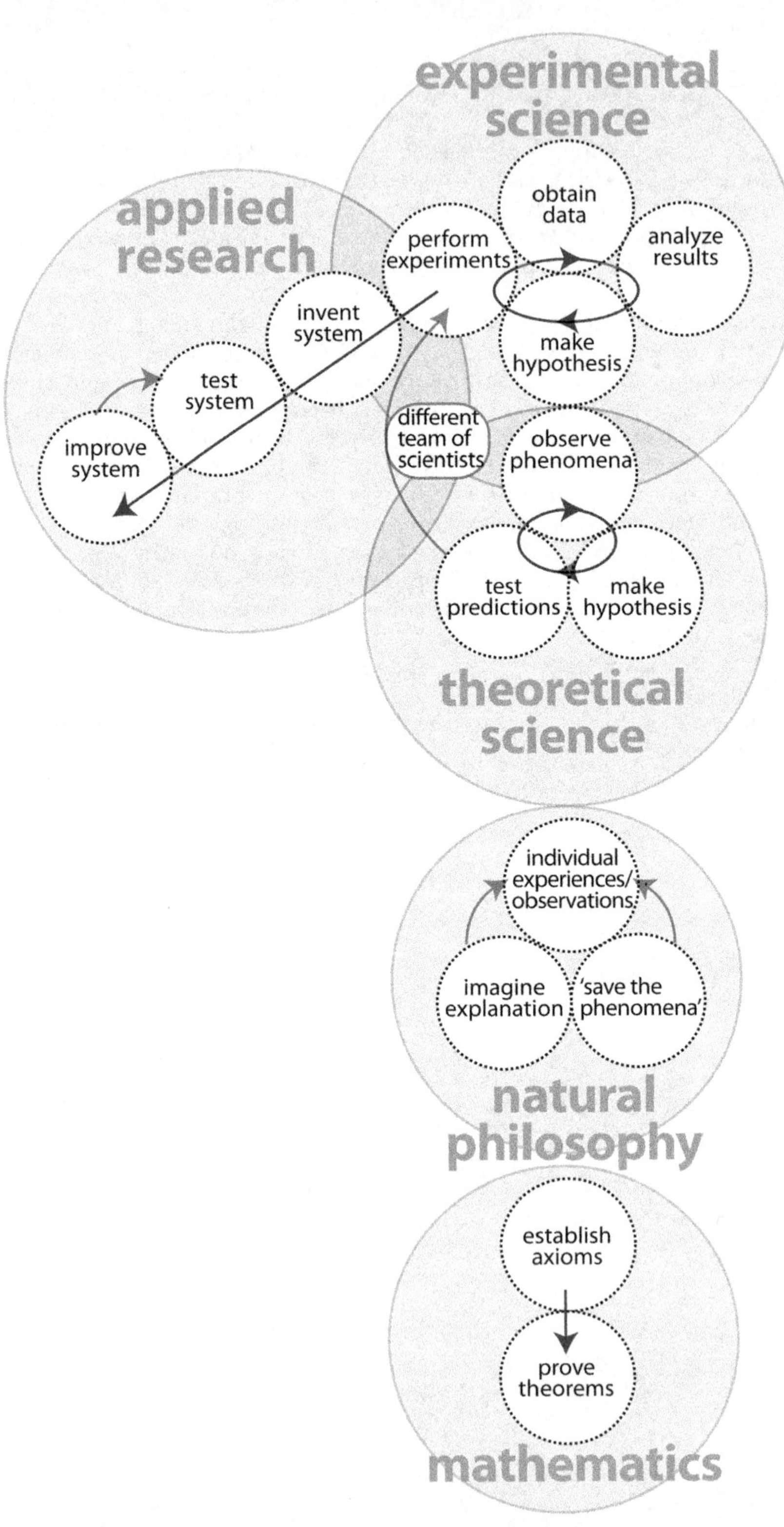

experimental science
applied research
obtain data
perform experiments
analyze results
invent system
make hypothesis
test system
improve system
different team of scientists
observe phenomena
test predictions
make hypothesis
theoretical science
individual experiences/observations
imagine explanation
'save the phenomena'
natural philosophy
establish axioms
prove theorems
mathematics

The Nature of Science

The problem of demarcation ... may be defined as the problem of finding a criterion by which we can distinguish between assertions (statements, systems of statements) which belong to the empirical sciences, and assertions which may be described as 'metaphysical'... Metaphysics is meaningless nonsense... In so far as a scientific statement speaks about reality, it must be falsifiable: and in so far as it is not falsifiable, it does not speak about reality, Karl Popper, 1933.

CHAPTER 10

Popper's Legacy: What Is Science?

WHAT IS SCIENCE? IS NOT A PHILOSOPHICAL QUESTION. AT least in the context of this book, it is a practical question: how does science proceed, what makes it reliable, what are its limitations? Insofar as there is a theory, it comprises a lineage starting in the 17th century with Francis Bacon (who argued that science proceeds by analyzing data) and René Descartes (who argued that science should be driven by hypotheses). The ideas of Thomas Kuhn and Karl Popper had a significant influence in the 20th century. Kuhn changed the understanding of how science proceeds by describing 'revolutions' that alternate with periods of incremental advance. The concept that revolutions overthrow paradigms has become embedded in science. If there is any philosopher whom scientists themselves most often acknowledge, it is Karl Popper for his famous argument that no hypothesis is valid in science unless it's possible in principle to falsify it. Popper's legacy was to define a focal point distinguishing science from other intellectual activities. Science itself has become so diverse that it's difficult otherwise to define its limits. It may seem trite—and it will certainly infuriate modern day philosophers—but a simple description might be that scientists have a religion: it is called Truth.

The Frontiers of Science
By Benjamin Lewin
© Benjamin Lewin 2026
Published by the Royal Society of Chemistry, www.rsc.org

Timeline for Ideas in Four Centuries of Scientific Methodology

1600

Francis Bacon argues science should proceed by induction (1620)

Descartes argues knowledge depends only on pure reason (1637)

1650

Robert Boyle describes methodology of air pump (1659)

1700

1750

James Lind publishes first controlled trial (1753)
Bayes provides basis for Bayesian probability (1753)

1800

Gauss introduces probabilistic calculation of errors (1809)

Peer reviewing of scientific papers starts (1830)
William Whewell invents term scientist (1840)

1850

Mach formalizes concept of thought experiment (1883)

1900

Karl Popper argues only falsifiable hypotheses are valid (1934)
Ronald Fisher introduces null hypothesis for analyzing statistics (1935)

1950

Popper's *Logic of Scientific Discovery* published in English (1959)
Thomas Kuhn introduces concept of paradigm shifts (1962)

2000

First research article with more than 100 authors (1997)
Chris Anderson proposes big data correlations replace hypotheses (2008)

AI software used to analyze data (2021)

A Dialog between Francis Bacon and Karl Popper

Francis Bacon (1561–1626) was an English philosopher and statesman who served as Attorney General and Lord Chancellor of England under King James I.

Karl Popper (1902–1994) was a Viennese philosopher who left Austria after the Nazi takeover. After the Second World War, he became a lecturer at the London School of Economics.

Bacon: Greetings, Sir Karl. It's an honor to engage in conversation with someone who has made significant contributions to the philosophy of science. Your emphasis on falsifiability has brought a distinctive perspective to the scientific method.

Popper: Thank you, Sir Francis. Your emphasis on systematic observation and experimentation laid the groundwork for the empirical approach that defines modern science. However, it does seem that going so far as to argue that knowledge is power resembles religion more than science.

Bacon: In truth, Sir Karl, I have always believed that knowledge arises from the careful and systematic study of nature. Inductive reasoning, based on accumulating observations, forms the bedrock of scientific inquiry.

Popper: Inductive reasoning has its merits, Sir Francis. However, I have raised concerns about the problem of induction—no amount of observations can logically prove the truth of a general statement. Instead, I advocate falsifiability, which emphasizes formulating hypotheses that can be rigorously tested.

Bacon: Falsifiability is an intriguing concept, Sir Karl. Yet, I wonder if it imposes an undue burden on scientific theories. Is there not a risk that, by pursuing falsifiability, we might prematurely discard useful theories?

Popper: You misunderstand my position, Sir Francis. Falsifiability is not about rejecting theories arbitrarily, but subjecting them to rigorous testing. A theory that survives empirical challenges gains strength, but one that fails must be rejected.

Bacon: It seems our approaches differ, Sir Karl. I view induction as the pathway to general truths. By accumulating observations, we can derive general principles that guide scientific understanding. Your emphasis on deductive reasoning seems to forego the richness of inductive insights that can guide the formulation of hypotheses.

Popper: (*wagging finger*) A fair concern, Sir Francis, but I contend again that induction, while valuable, faces the problem that it is logically impossible for any number of observations to prove a general conclusion. By focusing on testable hypotheses, deduction allows for a more rigorous and critical approach.

Bacon: Rigor and critical evaluation are virtues, indeed. Yet, does not deduction rely on *a priori* assumptions and theories that might themselves be untested? How can we ensure that the deductive process does not lead us astray?

Popper: An astute observation, Sir Francis. Deduction does rest on initial assumptions, but the key is to subject these assumptions to empirical testing. If a hypothesis fails the test, it's rejected or modified. This is the strength of the scientific method. Falsification distinguishes science from pseudoscience. I rest my case.

POPPER'S LEGACY: WHAT IS SCIENCE?

Science is the one human activity that is truly progressive. The body of positive knowledge is transmitted from generation to generation, and each contributes to the growing structure... Agreement is secured by means of observation and experiment. The tests represent external authorities which all men must acknowledge,[1] Edwin Hubble, 1936.

In science it often happens that scientists say, "You know that's a really good argument; my position is mistaken," and then they would actually change their minds and you never hear that old view from them again. They really do it. It doesn't happen as often as it should, because scientists are human and change is sometimes painful. But it happens every day. I cannot recall the last time something like that happened in politics or religion,[2] Carl Sagan, 1987.

Science and the scientist are modern inventions. Would it be going too far to say that the concept of science distinguishes the modern world from earlier eras? There was no concept of 'science' in the ancient or medieval worlds. Intellectual questions about the nature of the world would be described as 'natural philosophy.' Historically in universities, natural philosophy would be part of a general curriculum including the humanities and the arts. Science became regarded as a separate intellectual activity only relatively recently.

Science began to develop a distinct ethos with the Scientific Revolution, but the term 'scientist' was invented only in 1840. William Whewell (Master of Trinity College, Cambridge) suggested: "We need very much a name to describe a cultivator of science in general. I should incline to call him a Scientist."[3] However, many practitioners of science preferred anyway to continue to be known as 'natural philosophers.' The term 'scientist' did not come into general use until towards the end of the century.[4]

Philosophers have always been concerned with the nature of knowledge, but it was only in the 20th century that they specifically considered the nature of scientific knowledge. It should be said at the outset that working scientists have a low regard for the interest of asking "what is science?". They know what science is, because they do it every day. They have a disdain for the views of philosophers or sociologists (especially the latter, who comically misunderstand science whenever they comment on it[5]).

Although scientists do not think philosophy has much to say about science, there are two notable exceptions in which philosophers of science have captured key aspects of the nature of science. Karl Popper addressed the issue of how scientists actually work, how they set about solving problems and obtaining new information. Thomas Kuhn described how science advances, how new theories overthrow old theories. Together they defined the landscape of science. The terminology they introduced is part and parcel of the way scientists describe their own work. We might add C. P. Snow to the list of people who influenced science from the outside, for his controversial description of the separation between science and the humanities (see below).

It's ironic that Karl Popper (1902–1994) and Thomas Kuhn (1922–1996) should be the two philosophers whose names resonate among scientists today, as they were in fervent disagreement about the nature of science. Popper rejected Kuhn's view that science advances through 'normal' periods (meaning that data are obtained broadly supporting prevailing theories) alternating with revolutions. He took the view that only continuing attempts to falsify theories could advance science.[6] Philosophers make a great deal of the differences between Popper and Kuhn, continuing to view their respective ideas in terms of different philosophical theories.[7] But so far as scientists are concerned, Popper describes the way they work every day, and Kuhn describes the nature of transitions that mark major advances.

Their career trajectories were quite different, although they shared an early start in turning from physics to philosophy. (It's not an uncommon path to come to philosophy of science from physics *via* metaphysics. Perhaps this is why the philosophy of science more often concerns itself with physics than biology—and for that matter why it's prone to misunderstanding the factual nature of science.) Popper's and Kuhn's most famous works were published (in English) only three years apart. It may be fair to say that they have done as much to increase awareness of science as any individual scientific discovery.

Left: Karl Popper expressing an opinion in London in 1990. Right: Thomas Kuhn at MIT in 1989.

Photo of Karl Popper © Prange/Ullstein Bild via Getty Images.

Karl Popper was born in Vienna. He started as a socialist and Marxist, but became a social liberal. After some digressions, he obtained a doctorate in psychology and taught mathematics and physics. He wrote *Logik der Forschung (The Logic of Scientific Discovery)* in his spare time and published it (in German) in 1934. He emigrated to an academic position in New Zealand in 1937, where he wrote an influential book, *The Open Society and Its Enemies* (published later, in 1945). Then in 1946, he moved to the LSE (London School of Economics). He translated and revised *The Logic of Scientific Discovery* for its first publication in English in 1959. Popper's reputation by then was well established and the book made an immediate impact. Popper was famously grumpy, but highly influential.

Thomas Kuhn was educated in New York and then went to Harvard, where he obtained a degree in physics, followed by a Ph.D. He made the transition to history and philosophy while a Fellow at Harvard. Failing to get tenure at Harvard, he went to the University of California, Berkeley, where he wrote *The Structure of Scientific Revolutions*. It was first published in 1962 as part of a series from a philosophers' group, the Vienna Circle.[8] Now a classic, in fact a best-selling book in the social sciences, with more than a million copies printed, it initially received a cautious reception. Kuhn was diffident about its potential audience. He told the publisher that he had "no good ideas" about who might read it.[9] It sold less than 1000 copies the first year. Its reputation was slow to build, taking off after the definition of 'paradigm' was solidified in the second edition of 1970. Kuhn moved to MIT in 1983.

Karl Popper resolved a long dispute about whether science works inductively (using data to form hypotheses) or deductively (forming hypotheses that are tested by obtaining data) in *The Logic of Scientific Discovery*. He argued that the essential character of science is to function by testing hypotheses that in principle can be falsified. In fact, I would say that Popper did more than resolve a dispute; he created an expectation among scientists as to how they should present their results.

Perhaps this is a parallel to the axiom of quantum physics that the act of observation changes the properties of the particle being observed. Popper changed forever scientists' view of their own activities. Whether or not it's really the case, ever since he published his book, scientists have seen themselves as functioning in a Popperian manner to test hypotheses that can in principle be falsified.[10]

As a practical description of how scientists actually work, this is in fact a myth.[11] The practice of science is akin to destruction testing: constant questioning, taking arguments to extremes that can be tested, and rejecting and replacing arguments depending on observations. By contrast, the way scientists present their results is in effect a process of seeking approval: recounting how observations lead to logical conclusions explaining how things work. Results are most often presented in a positive way to show how they support a theory. It's an irony that if science were written up in the way it's actually done, no one would publish it; and the stultifying effects of doing it in the way it's written up would probably substantially block progress. The scientific paper is an artificial construct representing a mythical scientific method.

Of course, there is a difference between the way ideas are investigated and the way they are conceived. Popper was quite clear about this. "The initial stage, the act of conceiving or inventing a theory, seems to me neither to call for logical analysis nor to be susceptible of it. The question of how it happens that a new idea occurs to a man—whether it's a musical theme, a dramatic conflict, or a scientific theory—may be of interest to empirical psychology; but it's irrelevant to the logical analysis of scientific knowledge."[12] This is rarely acknowledged in presenting science, at least in experimental science, which is almost always given a retroactive sheen of logic. Theoretical physics can be an exception where ideas arise out of the blue (see Chapter 15). "An intellectual revolution often looks like a religious conversion," Popper said in another context.[13]

Irrespective of how science is actually conducted, it's now widely accepted that the possibility of falsification is a major criterion for allowing a conclusion to be part of science. Applying this criterion lets us distinguish science from pseudoscience,[14] but we also can add further categories:

- *Science* rests on theories that in principle can be tested by trying to falsify them. (Hypothesis: for any object to burn, a supply of oxygen is required. Falsifying test: try to burn objects in an oxygen-free atmosphere.)
- *Pseudoscience* presents theories that seem scientific but that cannot in fact be tested. Some are rather silly and easily dismissed. Others come from contributions to the scientific research literature. (Hypothesis: life originated in seeds sent to Earth from another star system. Falsifying test: impossible to devise.)[15]
- Pseudoscience is often also used more generally to describe ideas that are presented as scientific but aren't supported by scientific evidence. Sometimes they have in fact been disproved. It's a moot point where these cross over into the area of anti-science.[16]
- *Anti-science* goes beyond pseudoscience into rejecting well-established scientific principles.[17] It can be more pernicious than pseudoscience, because pseudoscience can often be ignored but anti-science (such as climate denial or the anti-vaxxer campaign) can have significant deleterious effects on society.
- *Trans-science* was proposed by physicist Alvin Weinberg for situations where practical or ethical reasons make it impossible to answer a scientific question.[18] (One example he gave was determining the effects of very low levels of radiation, which could require 8 billion mice.)

Philosophers have got themselves in a tangle about how to distinguish science from pseudoscience.[19] (They use the term 'demarcation' to describe the notion that science is different. Usually they deny that demarcation exists.[20]) Scientists regard Karl Popper's definition as adequate: if a theory is scientific, it is possible to find a test that will disprove it.[21] Philosophers object to this on the grounds that quite nonscientific ideas could also be subject to disproof—the idea that the moon is made of green cheese can be disproven.[22] This is somewhat forcing the argument, however.[23] The criterion of falsification is useful for assessing theories within and around the scientific realm. It would not be particularly useful to apply it to religion, for example, because religion does not claim to be science.

∞ ∞ ∞

The distinction between 'science' and 'pseudoscience' is an example of a more general distinction between science and other forms of intellectual activity. This distinction is (relatively) recent. As science started its evolution from natural philosophy, it was not so well distinguished from other forms of intellectual activity. When did it become distinguished as a separate activity by its attitude and methodology?

The pre-Socratic philosophers reasoned from first principles. Even if their reasoning was impeccable, they were limited by their *a priori* assumptions. This worked when proceeding from axioms in mathematics, but not when starting from incorrect or misleading observations in the natural world. But they did not draw a distinction. It did not occur to them to seek experimental verification—indeed, they regarded practical matters with some disdain. There was no concept that theories should produce testable hypotheses. This attitude persisted in Europe until the Scientific Revolution.[24] Until then, investigations of the natural world followed the precepts:

- Rely on passive observations of Nature.
- Theories should be founded on common postulates.
- Deductions from postulates (as in mathematics) are correct.
- Observations and conclusions are reported in anecdotal form.

The Scientific Revolution marked the transition from natural philosophy to science by relying on experiments to test theories. Experiments were reported in a way that allowed them to be assessed by others. Science as it exists today then evolved over the 18th and 19th centuries into a more formalized process. This includes providing the details needed for reproduction, giving detailed measurements and estimates of error, and acknowledging previous work.

The concept of stating how confident we can (or cannot) be about a conclusion is central to modern science. The concept of precision was not a feature of ancient science, where results were sometimes given to the number of decimal places produced by a calculation. One of the most important astronomers in Asia, al-Biruni (973–1052 CE), calculated the circumference of the earth to 4 decimal places (as 12 803 337.0358 cubits), when actually his measurement was about 5% off the true value (13.3 million cubits).[25]

This goes straight back to the belief of Aristotelian logic that calculations made from unquestionable postulates must be accurate. A more modern expression of the result would be along the lines of $12.8 \pm 0.5 \times 10^6$ cubits. It was a century after the Scientific Revolution before mathematical means became available to estimate errors that might have occurred anywhere from the starting postulates to the final conclusion.[26]

Those 'unquestionable postulates' depended on direct observation, using what today we might call common sense (although Aristotle used the term in a more specific sense). But as embryologist Lewis Wolpert pointed out, science is not about common sense. "Many of the misunderstandings about the nature of science might be corrected once it is realized just how 'unnatural' science is," he says. "This means that 'natural' thinking—ordinary, day-to-day common sense—will never give an understanding about the nature of science. Scientific ideas are, with rare exceptions, counter-intuitive: they cannot be acquired by simple inspection of phenomena."[27]

It was axiomatic in Aristotle's worldview that there was a unity to knowledge, which could be viewed as extending from the inanimate, to plants, to animals, to Man, and on to the Gods. Although the worldview changed during the Enlightenment, it also held to the unity of knowledge. At the extreme, Descartes advocated that ultimately all knowledge could be subsumed into mathematics.

As science developed after the Scientific Revolution, it divided into different disciplines. In mid-19th century, Mary Somerville (1780–1872), interested in science but not a scientist, wrote a series of books explaining the physical sciences. *On the Connexion of the Physical Sciences* argued in 1835 that there was a convergence between the physical sciences in spite of any apparent separation.[28] Mathematics was seen as the unifying force. The book became a best seller. It was too early for the argument to be extended to biology.

The concept of consilience goes further in proposing that the laws of the various natural sciences are interlocked. The term was introduced in 1840 by William Whewell (at the same time he invented the term 'scientist'),[29] but more recently has been used and extended by Edward Wilson at Harvard University, who attracted a great deal of hostility for his ideas.[30] Consilience follows from the Enlightenment view that all phenomena can be explained by cause and effect. It captures the idea that the laws of nature are connected in such a way as to apply across all the natural sciences. "Quantum physics thus blends into chemical physics, which explains atomic bonding and chemical reactions, which form the foundation of molecular biology, which demystifies cell biology," Wilson said.[31]

A great controversy was created when the idea of consilience was extended into the social sciences. There was a Luddite response to the idea of applying reductionism to human nature. The debate began with the publication of Wilson's book, *Sociobiology*, in 1975 (see Chapter 25). To add insult to injury, Wilson reinterpreted C. P. Snow's *Two Cultures* (describing the gap between science and the humanities; see below) in terms of the difference between genetic evolution (inherited) and cultural evolution (learned). "Human nature today remains Paleolithic [because] cultural evolution is much faster than genetic evolution."[32]

In my view, the unity of the sciences does not lie in the details of each discipline—the uncertainties of quantum physics superficially have little in common with the certainties of molecular biology—but in the rigor of a common approach. I do not agree with the idea that there has been an increasing convergence of the sciences during the 19th and 20th centuries.[33] Rather I believe the relationship has been there ever since the development of the methodology of science. I would argue that the Scientific Revolution established an approach, essentially for physics, that was then emulated as chemistry and biology developed.

There is not necessarily any single moment to mark the start of each 'science,' but it's certainly true that physics was first, chemistry was second, and biology was third. By the time this happened, there was a community of scientists in each discipline, and an *ad hoc* peer review occurred in the form of discussions following publication. As chemistry separated from physics, and biology became an independent discipline, the same approach characterized all of science: formulate theories, test them by obtaining data, and reformulate the theories. This is the key aspect of all natural sciences—physics, chemistry, and biology. Whereas the social sciences aspire to it, in my opinion they rarely achieve it.

By the mid-19th century, recognition that science had separated into distinct disciplines resulted in the suggestion of a hierarchy of sciences. In 1853, Auguste Comte (a French philosopher in Paris, sometimes regarded as the first philosopher of science, who founded the philosophical and political movement of positivism) placed them in the order: mathematics, astronomy, physics, chemistry, physiology, and social physics (sociology).[34] This order stands up well when scientists today are asked to rank sciences in an order from 'hard' to 'soft' on the basis of their reliance on empirical evidence.[35,36] (In modern terms, biology replaces physiology, and psychology comes between biology and sociology.)

As has often been noted in the current era, women are significantly under-represented in the sciences. Gender differences show a striking correlation with the perception of sciences as relatively 'hard' or 'soft.' The proportions of women obtaining doctorates are:[38]

- computing 23%
- physics 25%
- mathematics 29%
- chemistry 40%
- biology 45%
- psychology 73%
- social sciences 79%.

In citing these figures, I draw no conclusion about the relative difficulties of these sciences. It seems plausible that the figures represent a self-reinforcing public perception that 'hard' sciences are male-oriented.[39] This does not serve science well. Science is supposed to be about the intellect, but it cannot escape societal influence.

Judged by the relative numbers of scientists in each discipline, astronomy was the most practiced science in the 16th century, and biology is the most practiced today.[37]

The emphasis in science has changed over the centuries. At the start of what we might loosely call the scientific period, in the 16th century, astronomy was by far the most important science. It was the science that would be most directly recognizable in similar terms to the subject as studied today. It was the first science to break out of its ancient constraints into a modern type of analysis. Astronomy melds into physics, which now ranges from cosmology or astrophysics, to atomic structure, transitioning into chemistry. As chemistry evolved from, or replaced, alchemy, it became increasingly important, especially with industrialization. Perhaps biology is the science that has changed most dramatically. It has become the most prominent scientific discipline as judged by relative numbers of scientists in each area of science.

∞ ∞ ∞

Science has always been rather self-contained. Hungarianborn, but living in Manchester, England from 1933, philosopher Michael Polyani described it as the 'Republic of Science'. "The authority of scientific opinion remains essentially mutual; it is established *between* scientists, not above them."[40] He saw science as a uniform enterprise in which "the uniformity of scientific standards throughout science makes possible the comparison between the value of discoveries in fields as different as astronomy and medicine." Commenting on early attempts to guide science into 'socially beneficent channels,' he said in 1943 that, "Any attempt at guiding scientific research towards a purpose other than its own is an attempt to deflect it from the advancement of science." This is a good description of the coherence and unique character of science.

Science has its own value system. Scientists accept, without thinking about it too much, that their mindset is different from others. Robert Merton made the most insightful sociological analysis of science in the 1940s. He identified four features:

- communal activity building on previous efforts (communality);
- results are independent of whoever makes the discovery (universalism);
- science is impartial with respect to whatever results emerge (disinterredness);
- it is subject to testing (organized skepticism).[41]

Applying this description, mathematics as well as physics, chemistry, and biology would qualify as science, although the role of mathematics as a science is sometimes questioned because it relies on axioms rather than empirical evidence. Science differs from mathematics in that it is not possible to 'prove' a theory in science. You can only accumulate increasing amounts of evidence that are consistent with it. Karl Popper originally placed his emphasis on the notion that scientists should try to refute theories. Later he accepted that he had underestimated the value of accumulating data consistent with a hypothesis.[42,43] Philosopher Imre Lakatos went further and argued that "There is no falsification before the emergence of a better theory."[44]

Whether the much-vaunted 'scientific method' exists is questionable. Philosophers of science debate what it might be. Scientists are too busy to worry about it. Even if there is no distinct scientific method, science has a distinctive culture. Insofar as there is anything approaching a scientific method, it is that theory is subject to feedback, that is, modification by whatever results are actually obtained. Karl Popper called this the 'scientific attitude'. Sir Peter Medawar, who unusually for a working scientist (he was awarded a Nobel Prize in 1960 for his work in immunology) was interested in the process of scientific discovery, drew a distinction between "the (nonexistent) 'scientific method' and (very much existing) 'scientific methodology'."[45]

Although critics and philosophers of science do not believe there is any such thing as the 'scientific method,' its non-existence appears to be a well-kept secret. Amazon lists thousands of books with 'scientific method' in the title. It's fair to say that whatever professionals think about it, the concept of the 'scientific method' is well established in the public consciousness, and contributes to forming the public attitude toward science.

Great Discoveries Made by Accident that Led to Nobel Prizes

Year	Scientist	Observation	Discovery
1873	Camillo Golgi	Tissue fixed with potassium dichromate fell into a jar of silver nitrate.	Golgi stain
1895	Wilhelm Röntgen	Screen near the cathode ray tube began to glow although it was not in the direct path of the rays.	X-rays
1896	Henri Becquerel	Uranium salts exposed photographic plates even when they were wrapped in black paper.	Radioactivity
1927	Clinton Davisson	Extreme heating when apparatus broke formed nickel crystal that generated sharp beams of electrons.	Electron diffraction
1928	Alexander Fleming	Plate with streptococcus bacteria became contaminated with *Penicillium*, killing the bacteria.[46,47]	Antibiotics
1965	Arno Penzias & Robert Wilson	Noise from large radio antenna was thought to be pigeon droppings on antenna but was background radiation.	Big Bang
2004	Andre Geim & Konstantin Novoselov	Analyzed Scotch tape used to clean graphite used for treating scanning microscope.	Graphene (single layer of carbon atoms)

Hypotheses, or at least questioning the state of affairs and probing where there are weaknesses, are major driving forces in science. Yet in a curious way, the full strength of scientific attitude is revealed by the reaction to observations made by accident. Many major discoveries have been made by following up unexpected observations that were not particularly well related to the objectives of research. "Accidents are important in science first of all, but also accidents never happen accidentally. You actually need to create an environment for those accidents to happen; that's the difference between a good scientist and a bad scientist," says Konstantin Novoselov, a Russian-born British physicist who shared a Nobel Prize in 2010 for discovering graphene.

Science is only partly about what we know. In fact, the way it is often taught does it a great disservice by concentrating on a huge array of facts. This is ironic, because experiments are done to find out things we don't know. Somehow, we have to change the emphasis to explaining that science is about ideas (grounded in facts, to be sure). Neuroscientist Stuart Firestein thinks the most critical issue in science is actually 'ignorance.' It's much more interesting to talk about things we don't know or would like to know than to stick to established facts, he says.[48] He quotes Nobel Prizewinning physicist Enrico Fermi as saying, "There are two possible outcomes [for an experiment]: if the result confirms the hypothesis, then you've made a measurement. If the result is contrary to the hypothesis, then you've made a discovery."[49] (It's an irony that Fermi discovered that bombarding uranium with neutrons generated new elements, but did not make the leap to conclude that he had demonstrated nuclear fission; see Chapter 18.)

Especially in the modern era, when the humanities are marked by relativism—everything is relative to the observer, there is no such thing as an 'objective' fact—the difference in science stands out. Nothing is immune from questioning in science, but whether the questioning succeeds depends on the verification of experimental observations. Scientists believe in data, and that ideas are subservient to facts, whereas the humanities are inclined to the view that all ideas can be legitimate. This is a basic difference between scientists and the humanities.

Scientists are viewed as very serious about their work, and indeed they are, but there is also a tradition in some fields of using whimsical or witty names to humanize the phenomena. When some particles were found in 1953 with the unexpected property that they are easy to create by collisions in accelerators, but decay more slowly than expected for their (relatively) large masses, Murray Gell-Mann at Caltech introduced 'strangeness' to account for their behavior.[50] They include 'quarks,' which took their name from James Joyce's *Finnegans Wake*.[51] In biology, there's a long tradition of giving fruit fly mutants amusing names, such as *tinman* (a mutant that has no heart, referring to The Wizard of Oz) or *Van Gogh* (a mutant with a swirling pattern of hairs on its back). I suppose it's a way of claiming indirectly that science relates to other human activities.

∞ ∞ ∞

Scientists are prone to believe that philosophy has had its day in offering explanations of the natural world. Richard Feynman, awarded the Nobel Prize for Physics in 1965, typically captured their attitude when he said, "Philosophy of science is as useful to scientists as ornithology is to birds."[52] Stephen Hawking was more direct. "Philosophy is dead. Philosophy has not kept up with modern developments in science, particularly physics. Scientists have become the bearers of the torch of discovery in our quest for knowledge."[53,54] Einstein had previously made a similar point more tactfully. "The physicist cannot simply surrender to the philosopher the critical contemplation of the theoretical foundations."[55] It has even been said that science fiction is more likely than philosophy to be an influence on today's scientists.[56] Winner of the Nobel Prize for Physics in 1979, Sheldon Glashnow, said that his reading outside of science was science fiction![57]

Philosophy tends to reflect its origins in asking "what is knowledge?"[58] The practical form of the philosophers' question is to ask: how do we know that an observation reflects underlying reality as opposed to an artefact reflecting design of the system? (I discuss the approach of sociologists and philosophers to science in Chapter 26.) It's easier to accept a result when it's possible to visualize it directly, and more difficult when it takes the form of digital readout from a machine. Indeed, there have been cases where interesting results later turned out to be due to malfunctions in the apparatus. But this is a technical matter, not a philosophical question, which can be resolved by investigating the consequences of whatever theory emerges from the results.

A great example of a hidden problem with equipment was a report in 2011 from the OPERA team at CERN that neutrinos can travel faster than light. Neutrinos appeared to travel 730 km (from Geneva to Gran Sasso in Italy) 60 nanoseconds faster than at light speed. The result was confirmed by a second run later. But then it turned out the measurement of the time delay had been incorrect because an optic fiber cable was loose. There was widespread skepticism anyway, and the supposed finding did not last very long.[59]

Admittedly, scientists have been progressively cut off from their own data by the increasing sophistication of the equipment. The question of the reliability of knowledge takes on a newer and sharper form when it requires special expertise to use an apparatus and interpret the results. From making direct observations at the start of the 20th century with equipment that they built themselves, by the end of the 20th century scientists were getting results spat out by digital equipment, often after a process of 'normalization.'

Ever since science moved on from simple pictures seen in telescopes or microscopes, it's been a question as to how much manipulation of data is appropriate. The question was sharpened when direct photography was replaced by digital imaging. It moves into another dimension with the application of AI.

AI presents a problem for maintaining the principle of transparency in science. It's difficult, if not impossible, to apply critical analysis to a process that is opaque even to its creators. The willingness to accept the analysis of an AI program leads to the question: what is the difference between a fudge factor and an adjustment? Is it that a fudge factor comes from human intuition, but an adjustment is the result of a software algorithm? How can an AI program be trained without introducing a bias? It's unclear how the analysis of results by AI can be equated with traditional benchmarks that data and analysis must be subject to verification and reproducibility.

One of the first (and dramatic) uses of AI was to replace experimental analysis to analyze protein structures with the Alpha-Fold program. (I discuss the development of AI in Chapter 23.) Like any AI program, the way AlphaFold works is not entirely clear. I asked Demis Hassabis, co-founder of DeepMind, which wrote the software, how its results might therefore be validated. "Verification would be downstream, by using the program to make predictions... that ought to be borne out by subsequent data. [If the downstream effect doesn't work], then you would have to decide if the structure had been wrongly predicted or you'd done something wrong downstream."[60]

That stands very well as a general answer to the philosophers' criticism that it is impossible to distinguish truths of nature from artifacts of the machine. Each new contribution to science builds upon previous contributions. If one of those contributions actually reflects the operation of the machine, this will become clear in the form of discrepancies with subsequent work. When unexpected results are obtained, the first reaction is: could there be a fault with the equipment? The philosophers' argument is a distinction without a difference.

∞ ∞ ∞

Science really is a messy process, with data and hypotheses alternating in no very organized way. Not all data are necessarily given equal weight. Often enough, a line of research is driven by a question (what if we do so and so?) rather than by a clear hypothesis. But following Karl Popper's Dogma of the Hypothesis (if I can call it that), most scientific papers are written as though the authors were testing a hypothesis, even though that's rarely how the work actually proceeds.[61]

Neuroscientist Stuart Firestein thinks the role of hypotheses is much exaggerated. "They have a way of taking on a life of their own. Scientists get behind one hypothesis or another as though they were sports teams or nationalities—or religions."[62] His position is that a hypothesis is a "real danger," because it encourages scientists to take a position, ignoring data that do not conform. In the real world, it's rare for all data to conform exactly to a hypothesis. Perhaps one mark of a great scientist is that they know when to ignore outlying data (but this should be stated explicitly, of course, not hidden away). To paraphrase, Firestein believes a hypothesis can be a straightjacket that stops you from delving fully into ignorance.

The collapse of a paradigm—what is a paradigm but a dominant hypothesis?—is an extreme example (see below). Over and over again, I have seen the interpretation of data distorted to

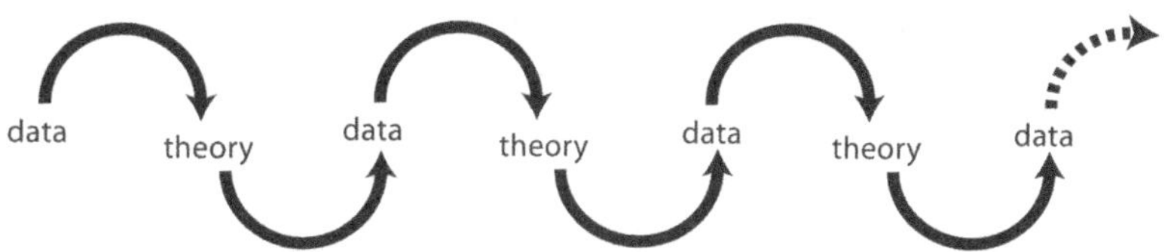

Science does not advance solely by either induction or deduction, but by an interplay between data and theory, obtaining data to test a theory, defining a new theory, obtaining more data... ad infinitum.

match the paradigm of the day—until the weight became too much for the paradigm to bear, and it broke. "Hypotheses non fingo" (I frame no hypothesis), Newton famously said, when discussing gravity, perhaps because he thought that the phenomenon was so way out, any explanation would be a hostage to fortune.[63]

Science cannot be described simply in terms of induction (proceeding from data to hypothesis) or deduction (proceeding from hypothesis to data). It's an interplay in obtaining data, trying to figure out what they mean, and then obtaining further data to confirm or refute earlier ideas.[64,65] Formally it might be described as an alternation of deduction and induction.

Deductions draw specific conclusions from generalities. If a general axiom is true, then deductions drawn from it must be true. The classic example is:

All men are mortal.
Socrates is a man.
Therefore Socrates is mortal.

While deduction is infallible in mathematics, it's less so in science, because sometimes the starting axiom turns out after all not to be valid.

Induction forms generalizations from specific observations. The classic case is the white swan. Until the 17th century, all swans observed in the world were white. It was reasonable to suppose that swans could only be white—until the Dutch explorer Willem de Vlamingh discovered black swans in Australia in 1687. It's always possible for a conclusion based on induction to be refuted by a single counter-example.

∞ ∞ ∞

Before the concept of 'science' developed as a separate endeavor, it was seen as a part of natural philosophy, not necessarily distinguished from other parts. Was a clash inevitable in a society driven by religion when science developed its distinctive character? Some influential books in the 19th century stimulated the concept that a clash impeded the growth of science.[66,67] However, in the great revisionist tradition of the 20th century, the idea that religion suppressed science is now regarded as a myth.[68] But surely treating science as a separate intellectual activity, subject to its own rules, must result in a clash with any authoritative system following different rules?

It's more of a puzzle why there should be a clash with the humanities. Humanism developed in the Renaissance in the 14th–15th centuries, and became an alternative to theism.[69] Nominally the study of human virtue, in practice it focused on the Greek and Roman classicists, in particular rhetoric. Erasmus captured the spirit of the movement. "The whole of attainable knowledge lies enclosed within the literary monuments of ancient Greece."[70] The emphasis on recapturing the learning of the Greek philosophers has resonated ever since.[71] There is a division as to whether humanists regarded the development of science positively (because it too was secular) or negatively (because its value system was different from humanism's focus on the classic authors).[72]

Erasmus (1466–1536) was the very model of an enlightened intellectual of the Renaissance. Although a Catholic priest, he was influential in the humanist movement, advocating tolerance and free thinking (up to a point). Yet he was in part responsible for the hostility of humanism to science, whereas (from a modern perspective) you might have expected humanism to embrace a rational approach. His hostility was to the sterile attitude of natural philosophy practiced in the universities. This was a predecessor to the disdain shown by the humanities towards science.

Be that as it may, until the 19th century, the curriculum at universities was focused on the humanities, including natural philosophy, which had at first been taught within a theological context. By mid-17th century, there was skepticism as to whether universities were the best place for science. Referring to the Royal Society's role in encouraging experimentation, a history of the society written in 1667 said, "The seats of knowledge have been for the most part heretofore, not laboratories, as they ought to be; but only schools."[73]

Separate faculties of science, distinct from the humanities, were first established in the 18th century,[74] but they did not become common until the second half of the 19th century. This was the start of the recognition that several disciplines could be regarded as science, and the start of the separation between science and the humanities.[75]

Resistance in the humanities to recognizing science as a course of study was (and sometimes still is) based on the grounds that it is too focused on empirical observation and experimentation. This could be regarded as a threat to traditional forms of intellectual enquiry (meaning basing conclusions solely on reasoning).[76] Science was considered to promote a mechanistic view of the world,[77] somewhat of a parallel to the Church's early view of science as a threat to the religious order because its value system followed different precepts.

These views turned into the perspective—dominant at least in the Anglo-Saxon world—that true education involved the classics, not science. Matthew Arnold (a poet and critic) argued in the annual Rede lecture, at Cambridge University in 1882, that "no one could be really educated unless they understood literature, particularly the literature of ancient Greece and Rome."

It was almost a century until someone defended science. When he defined the difference between science and the humanities as *The Two Cultures* in the Rede Lecture in 1959, C. P. Snow was roundly attacked by the literary establishment for arguing that science was a driving intellectual force in society. (I discuss the reaction in Chapter 26). Snow's most famous criticism was that so-called intellectuals are in fact illiterate about science. "Intellectuals, in particular literary intellectuals, are natural Luddites."[78] He pointed out that intellectuals who criticized the illiteracy of scientists had no response when they were asked if they could describe the second law of thermodynamics. "I was asking something which is about the scientific equivalent of: Have you read a work of Shakespeare's?"

That split continues to the present day.

There was a similar divide within science between 'basic research' and 'applied research.'[79] An alternative name for basic research, 'pure research,' immediately gives away the attitude. Snow put his finger on it. "Pure scientists have by and large been dim-witted about engineers and applied science... They wouldn't recognize that many of the problems were as intellectually exacting as pure problems, and that many of the solutions were as satisfying and beautiful. Their instinct... was to take it for granted that applied science was an occupation for second-rate minds."[80]

Is the concept of progress (intellectually speaking) a distinctive aspect of science? We believe we have made progress in moving from Newton's gravitation to Einstein's relativity. Is there the same sense of progress in moving from Monet's impressionism to contemporary abstract painting, or is this more a change of perspective? Or from Beethoven's ninth symphony to John Cage's 4'33" minutes of silence? If by progress we mean a deeper understanding of underlying events or cases, science seems different. Of course, if we mean merely a newer way of viewing things, there is little distinction. If the concept of progress is not unique to science, it's surely at the least more firmly established in the realm of science than in other intellectual activities.[81]

The key to the integrity of science is its self-correcting character. This follows directly from the scientific cycle. It's relatively rare for the results in a scientific paper to be directly repeated by a third party. Scientists are too anxious to get on with their own discoveries to spend time repeating the work of others. But every paper carries implications for future work. The intrinsic nature of the scientific process means that any significant error will result in the emergence of some discrepancy in later work. A paper that turns out to be incorrect will be corrected or even rejected entirely. Eventually the truth will out. Even aside from the dramatic situation in which a revolution overthrows a paradigm, this is a continuing process.

While the process of correction is inevitable, it's not necessarily immediate. Science is a human endeavor. People can become invested in ideas and resistant to change even when the data are (at least in retrospect) quite clear. Scientists take positions that are not always driven by objective views of the data. But the self-correcting character limits human ability to (mis)interpret data. The story of science at the end of the day is the triumph of data.

A scientist might argue that the 'scientific attitude,' searching for objective truth, is the universal ideal of science. It's implicit that 'truth' and 'objectivity' are defined operationally in terms of current understanding. As Einstein said philosophically, "It's difficult even to attach a precise meaning to the term 'scientific truth.'"[82] More pragmatically, C. P. Snow said that "Truth... is what makes science work... [It] is what the scientists are trying to find. They want to find what is there..."[83]

NOTES AND REFERENCES

1. E. Hubble, *The Realm of the Nebulae*, Yale University Press, New Haven, CT, 1936, p. 1.
2. Keynote address at CSICOP conference. Quoted in J. Poling, *Do Science and the Bible Conflict?*, Zondervan, 2003, p. 30.
3. W. Whewell, *The Philosophy of the Inductive Sciences*, Cambridge University Press, Cambridge, 1840, p. cxiii.
4. Even then there was resistance: *Nature* and other scientific journals refused to use the term, preferring 'man of science.' See M. Baldwin, *Making "Nature": The History Of A Scientific Journal*, University of Chicago Press, Chicago, 2015, pp. 1–7.
5. Perhaps the epitome of misunderstanding was when anthropologist Bruno Latour embedded himself in a laboratory at the Salk Institute for two years, like "an intrepid explorer ... living with tribesmen." See B. Latour and S. Woolgar, *Laboratory Life. The Construction of Scientific Facts*, Princeton University Press, Princeton, NJ, 1986. He described the day-to-day activities of the laboratory, but never came even remotely to grips with what science is about. See B. Lewin, *Inside Science: Revolution in Biology and Its Impact*, Cold Spring Harbor Laboratory Press, New York, 2023, pp. 23–24.
6. The only time Popper and Kuhn actually met was at a colloquium at the University of London in 1965, now known as the Kuhn–Popper debate.
7. S. Fuller, *Kuhn vs. Popper: The Struggle for the Soul of Science*, Columbia University Press, New York, 2003.
8. The series was called *International Encyclopedia of Unified Science*. This included 19 monographs between 1938 and 1969. It was the brainchild of the Vienna Circle, a group of philosophers and scientists, led by Otto Neurath, who went on to start a series of meetings on the theme of the unity of science. "Because the physical language is thus the basic language of Science, the whole of Science becomes Physics." See *International Encyclopedia of Unified Science*, ed. O. Neurath, R. Carnap, and C. Morris, Chicago University Press, Chicago, 1938–1955.
9. R. J. Richards and L. Daston, *Kuhn's Structure of Scientific Revolutions at Fifty: Reflections on a Science Classic*, University of Chicago Press, Chicago, 2016, p. 73.
10. One of the most common phrases scientists use to describe their work is "working hypothesis," a term that originated in the mid-nineteenth

century, and has become shorthand to describe a (presently) unproven theory that is the basis for designing experiments. First use of the term may have been in an article under the title *Science* in the *Westminster Review* (1855) vol 63: 239–253. "There is just as good a foundation for a sound 'working hypothesis' of an opposite kind." It was expounded as a method for scientific enquiry by John Dewey in 1938 in his book *Logic: The Theory of Inquiry* (Henry Holt, New York).

11. B. Lewin, *Inside Science: Revolution in Biology and Its Impact*, Cold Spring Harbor Laboratory Press, New York, 2023, pp. 65–78.

12. K. Popper, *The Logic of Scientific Discovery*, Routledge, London, 1934, p. 7.

13. K. Popper, Normal Science and its Dangers, in *Criticism and the Growth of Knowledge*, ed. I. Lakatos and A. Musgrave, Cambridge University Press, Cambridge, 1970, p. 57.

14. Philosophers argue about whether and how to distinguish science from pseudoscience. They are more inclined to take the view that there is a continuum of activities with a gray area in between than to draw a clear distinction. See K. Frazier, *Shadows of Science: How to Uphold Science, Detect Pseudoscience, and Expose Antiscience in the Age of Disinformation*, Prometheus, London, 2024, pp. 119–137. However, most scientists would accept falsification as a defining boundary.

15. Sometimes 'pseudoscience' is criticized as being a pejorative term. See M. D. Gordin, *The Pseudoscience Wars: Immanuel Velikovsky and the Birth of the Modern Fringe*, University of Chicago Press, Chicago, 2013, p. 1. So it should be!

16. See K. Frazier, *Shadows of Science: How to Uphold Science, Detect Pseudoscience, and Expose Antiscience in the Age of Disinformation*, Prometheus, London, 2024, pp. 26–31.

17. See K. Frazier, *Shadows of Science: How to Uphold Science, Detect Pseudoscience, and Expose Antiscience in the Age of Disinformation*, Prometheus, London, 2024, p. 17.

18. A. M. Weinberg, Science and trans-science, *Minerva*, 1972, **10**, 209–222.

19. For an example of muddled thinking on this issue see M. D. Gordin, *On the Fringe: Where Science Meets Pseudoscience*, Oxford University Press, Oxford, 2021.

20. Karl Popper first described demarcation as the main problem in distinguishing science from nonscience.

21. Perhaps not surprisingly, given the gap that has opened between science and philosophy as practiced today, Popper remains highly regarded among scientists, although he is no longer so well regarded by philosophers. See P. Godfrey-Smith, *Popper's Philosophy of Science: Companion to Popper*, ed. J. Shearmur and G. Stokes, Cambridge University Press, 2016, pp. 104–124.

22. "The moon is made of green cheese" was introduced in 1638 as an example of extreme credulity by philosopher John Wilkins. "You may as soon persuade some Country Peasants that the Moon is made of Green Cheese (as we say)."

23. One influential view wonderfully misrepresents Popper's falsification criterion by arguing that all sorts of pseudoscience could be classified as science "just so long as they [its proponents] are prepared to indicate some

observation, however improbable, which (if it came to pass) would cause them to change their minds." But it's not the mindset of the proponents that matters: it is whether objectively a criterion can be found to falsify the claim. For this misrepresentation see L. Laudan, The Demise of the Demarcation Problem in *But Is It Science? The Philosophical Question in the Creation/Evolution Controversy*, ed. M. Ruse, Prometheus Books, Amherst, NY, 1988, p. 346. Laudan's position has in any case been thoroughly debunked: see *Philosophy of Pseudoscience: Reconsidering the Demarcation Problem*, ed. M. Pigliucci and M. Boudry, University of Chicago Press, Chicago, 2014.

24. The theories, observations, and experiments of Alexandria and Arabia remained unknown in the West (see Chapter 3).

25. S. Weinberg, *To Explain the World: The Discovery of Modern Science*, HarperCollins, New York, 2015.

26. The 'Theory of Errors' was introduced by Johann Lambert in 1760. Probabilistic determination of errors was introduced by Carl Gauss in 1809. O. B. Sheynin, C. F. Gauss and the theory of errors, *Arch. Hist. Exact Sci.*, 1979, **20**, 21–72.

27. L. Wolpert, *The Unnatural Nature of Science*, Faber & Faber, London, 1992, p. xi.

28. "The progress of modern science, especially within the last five years, has been remarkable for a tendency to simplify the laws of nature and to unite detached branches by general principles." M. Somerville, *On the Connexion of the Physical Sciences*, J Murray, London, 1835.

29. "The Consilience of Inductions takes place when an Induction, obtained from one class of facts, coincides with an Induction, obtained from another different class. This Consilience is a test of the truth of the Theory in which it occurs." W. Whewell, *The Philosophy of the Inductive Sciences*, Cambridge University Press, Cambridge, 1840, p. xxxix.

30. E. O. Wilson, *Consilience: The Unity of Knowledge*, Vintage Books, New York, 1998.

31. E. O. Wilson, *Consilience: The Unity of Knowledge*, Vintage Books, New York, 1998, p. 59.

32. E. B. Wilson, *Consilience Among the Great Branches of Learning*, in *Science in Culture*, ed. P. Galison, S. R. Graubard and E. Mendelsohn, Routledge, London, 2001, pp. 131–150.

33. P. Watson, *Convergence: The Idea at the Heart of Science*, Simon & Schuster, New York, 2017.

34. A. Comte, *The Positive Philosophy of Auguste Comte*, 2 volumes, translated by H. Martineau, George Bell & Sons, London, 1853, reissued by Cambridge University Press, 2009, p. 35.

35. D. K. Simonton, Hard Science, Soft Science, and Pseudoscience: Implications of Research on the Hierarchy of the Sciences, in *Pseudoscience: The conspiracy against science*, ed. A. B. Kaufman and J. C. Kaufman, 2018, pp. 77–99.

36. Sociologists have attempted to find objective criteria for distinguishing between hard sciences and soft sciences, such as the reproducibility or variability of data. This has not proved very satisfactory so far. M. Pigliucci,

Nonsense on Stilts: How to Tell Science From Bunk, University of Chicago Press, Chicago, 2nd edn, 2018, pp. 28–35.

37. Data from 1550–1900 from R. Gascoigne, The Historical Demography of the Scientific Community, 1450–1900, *Soc. Stud. Sci.*, 1992, **22**, 543–573. Data for the present from Data USA (datausa.io).

38. The numbers reflect a bias in entering the field as the progression is similar for undergraduate degrees. There is a decline of 2–4% in the proportion of women in each field proceeding from undergraduate to graduate level. Figures for the United States from National Science Foundation Report for 2022. Data refer to 2020. See ncses.nsf.gov/pubs/nsf23315/data-tables.

39. The humanities have a slight preponderance of women enrolled in graduate school, 57%. (Council of Graduate Schools, 2020 report. See cgsnet.org). As graduate studies may be a zero-sum game, any increase in the sciences would likely be at the expense of the humanities.

40. M. Polyani, The republic of science: its political and economic theory, *Minerva*, 1962, 54–74.

41. R. K. Merton, *The Normative Structure of Science*, 1942, reprinted in R. K. Merton, *The Sociology Of Science. Theoretical and Empirical Investigations*, Chicago University Press, Chicago, 1973.

42. K. Popper, *Conjectures and Refutations*, Routledge, London, 1963.

43. Evolution is a case in point. It's difficult to find means of falsifying evolution as a general concept. But the weight of evidence that supports the theory, especially DNA sequence analysis to compile an evolutionary tree, provides strong corroboration.

44. Popper had argued that falsification usually accompanies (or is preceded by) the proposal of an alternative hypothesis: Lakatos argued that this was in fact essential Although I do not think Lakatos, Popper, or Kuhn would agree with this, it seems that Lakatos has unified Popper and Kuhn by relating falsification to a paradigm shift. See I. Lakatos, *Philosophical Papers*, in *The Methodology of Scientific Research Programmes*, ed. J. Worrall and G. Currie, Cambridge University Press, Cambridge, 1978, vol. 1, p. 35.

45. P. B. Medawar, *Induction and Intuition in Scientific Thought*, American Philosophical Society, Philadelphia, 1969.

46. Fleming publicized the importance of his observation but admitted freely that it had been made by chance. "In my first publication I might have claimed that I had come to the conclusion, as a result of serious study of the literature and deep thought, that valuable antibacterial substances were made by moulds and that I set out to investigate the problem. That would have been untrue and I preferred to tell the truth that penicillin started as a chance observation… My publication in 1929 was the starting-point of the work of others who developed penicillin especially in the chemical field." A. Fleming, Penicillin, in *Nobel Lectures: Physiology or Medicine 1942–1962*, p. 83.

47. Formally, Fleming rediscovered Penicillium. While a medical student in Lyon, Ernest Duchesne discovered an antagonism between Penicillium molds and bacteria in cultures. When he injected guinea pigs with both

Penicillium and typhoid bacteria, they survived. He concluded that Penicillium reduced the virulence of the bacteria. He did not consider the possibility of isolating an active component from Penicillium. His thesis was submitted to the Faculty of Medicine in Lyon in December 1897, but remained essentially unknown until after Fleming had received the Nobel Prize. (E. Duchesne, *Contribution à l'étude de la concurrence vitale chez les micro-organismes: antagonisme entre les moisissures et les microbes* [Contribution to the study of the vital competition in microorganisms: antagonism between molds and microbes], Alexandre Rey, Lyon, 1897.)

48. S. Firestein, *Ignorance: How It Drives Science*, Oxford University Press, Oxford, 2012.

49. Several sources claim that Enrico Fermi said this to his students, although there is no direct evidence, *e.g.* J. Fripp, D. Fripp, and M. Fripp, *Speaking of Science*, Newnes, London, 2000, p. 58.

50. M. Gell-Mann, Isotopic spin and new unstable particles, *Phys. Rev.*, 1953, **92**, 833–834.

51. Leading to the development of the quark model, Gell-Mann introduced a classification for a group of particles that he called the 'eightfold way,' a reference to the Noble Eightfold Path of Buddhism. M. Gell-Mann, *The Eightfold Way: A theory of strong interaction symmetry*, Synchrotron Lab Pasadena, 1961, DOI: 10.2172/4008239.

52. Widely attributed to Feynman, although no source for the attribution has ever been found.

53. S. Hawking and L. Mlodinow, *The Grand Design*, Bantam Books, New York, 2010.

54. For this he was roundly attacked by philosophers at a conference in London in 2015, What Is the Point of Philosophy? (M. Reiz, *The Times Higher Education*, February 22, 2015.) It was risible, if not a mark of desperation, to attack Hawking for lacking sufficient intellect or knowledge to understand the philosophers' positions!

55. A. Einstein, *Ideas and Opinion*, Modern Library, New York, 1994, p. 290.

56. L. Wolpert, *The Unnatural Nature of Science*, Faber & Faber, London, 1992, p. 109.

57. G. Holton, *The Advancement of Science, and its Burdens*, Cambridge University Press, Cambridge, 1986, p. 165.

58. See, for example, S. L. Goldman, *Science Wars: The Battle Over Knowledge and Reality*, Oxford University Press, Oxford, 2021, pp. 64–79.

59. G. Brumfiel, Neutrinos not faster than light, *Nature*, 2012, 10249.

60. Discussion with Demis Hassabis, February 2022.

61. B. Lewin, *Inside Science: Revolution in Biology and Its Impact*, Cold Spring Harbor Laboratory Press, New York, 2023, pp. 65–78.

62. S. Firestein, *Ignorance: How It Drives Science*, Oxford University Press, Oxford, 2012, pp. 72–74.

63. In the essay *General Scolium* in the second edition of *Principia* in 1713.

64. The view that science should be driven by hypothesis might be attributed to René Descartes in 1637.

65. Francis Bacon argued in *Instauratio Magna* in 1620, that science should proceed from inductive reasoning—generalizing from observations and data.

66. "The history of Science is… a narrative of the conflict of two contending powers, the expansive force of the human intellect on one side, and the compression arising from traditionary faith and human interests on the other." J. W. Draper, *History of the Conflict Between Religion and Science*, D. Appleton and Co., New York, 1875. Online at www.gutenberg.org/files/ 1185/1185-h/1185-h.htm.

67. "Interference with science in the supposed interest of religion… has resulted in the direst evils both to religion and to science… All untrammeled scientific investigation… has invariably resulted in the highest good of religion and of science." A. D. White, *The Warfare of Science with Theology in Christendom*, D. Appleton, New York, 1896, Online at www.gutenberg.org/ files/505/505-h/505-h.htm.

68. *Galileo Goes to Jail and Other Myths about Science and Religion*, ed. R. L. Numbers, Harvard University Press, Cambridge, MA, 2009.

69. The term has origins in the 15th century, when studia humanitatis consisted of classical studies in grammar, poetry, rhetoric, history, and moral philosophy.

70. In *De Ratione Studii* (Upon the Right Method of Instruction), 1511.

71. M. Boaz, *The Scientific Renaissance 1450–1630*, Harper, New York, 1962, pp. 27–28.

72. The Romantic reaction against the rationalism—some would say impersonality—of the Enlightenment was strong in Germany, where it took the form of the Naturephilosophie movement of the 19th century, which chemist Justus Liebig called 'the black death of our century.' See M. J. Nye, *Before Big Science: The Pursuit of Modern Chemistry and Physics, 1890–1940*, Twayne Publishers, New York, 1996, p. 25.

73. T. Spratt, *History of the Royal Society of London*, Royal Society, London, 1667, p. 68.

74. A department of 'chymistry and pharmacy' had been established at the University of Paris in 1660 as part of the Faculty of Medicine.

75. Of course, mathematics was already established as an independent area of study; Galileo became professor of mathematics at Pisa in 1592. Mathematics also had a practical utility: it led to development of cartography and instruments that were useful for navigation, and also had military consequences, for example, where aiming canons required knowledge of geometry.

76. "Science is cold and heartless, an enemy of poetry, of sentiment, of beauty, of everything that is attractive and good in life." G. Wilson, *The progress of the intellect: as exemplified in the religious development of the Greeks and Hebrews*, Harper, New York, 1852.

77. "The science of today is a science of detail, a science of atoms and molecules, which has lost sight of the whole, and which is, in consequence, powerless to give guidance in the conduct of life." I. Babbitt, *Literature and the American college: Essays in defense of the humanities*, Houghton Mifflin, New York, 1908.

78. C. P. Snow, *The Two Cultures and the Scientific Revolution*, Cambridge University Press, Cambridge, 1959, p. 22.

79. The idea that basic research has a purity that's lacking from applied research goes back to Aristotle. "As more arts were invented, and some were directed to the necessities of life, and others to recreation, the inventors of the latter were naturally always regarded as wiser than the inventors of the former, because their branches of knowledge did not aim at utility." Aristotle, *Metaphysics*, Book 1, c. 350 BCE.

80. C. P. Snow, *The Two Cultures and the Scientific Revolution*, Cambridge University Press, Cambridge, 1959, p. 32.

81. It was a conventional view that science could be equated with progress. Early historians of science were in no doubt about it. According to George Sarton, "Progress has no definite and unquestionable meaning in other fields than the field of science." See G. Sarton, *The Study of the History of Science*, Dover, New York, 1936, p. 5.

82. A. Einstein, *Ideas and Opinion*, Modern Library, New York, 1994, p. 261.

83. C. P. Snow, The moral un-neutrality of science, *Science*, 1961, **133**, 256–259.

CHAPTER 11

Kuhn's Paradigms: Changing Science

"WHY SHOULD A CHANGE OF PARADIGM BE CALLED A revolution?", asked Thomas Kuhn.[1] Science has never been static. Since it evolved from natural philosophy, there has been a series of revolutions in both content and style. The Scientific Revolution marks the start of the modern era, with regards to both knowledge itself and to the development of scientific methodology. The many revolutions that followed, some small, some large, have been marked by changes in perspective that Kuhn called paradigm shifts. From these have evolved our present view of science—but it is still changing, perhaps more than ever. The transition from small-scale science to big science, from hypothesis-driven science to big data, is a dramatic change in the way data are obtained. The introduction of AI-driven techniques to analyze data may result in an equally dramatic change in the intellectual conduct of science.

The Frontiers of Science
By Benjamin Lewin
© Benjamin Lewin 2026
Published by the Royal Society of Chemistry, www.rsc.org

Timeline for Four Centuries Developing
the Paradigms of Modern Science

1600

Galileo shows Earth (and planets) revolve around the Sun (1610)

Harvey discovers circulation of the blood (1628)

1650

Newton shows gravity depends on action at a distance (1687)

1700

1750

Lavoisier divides air into oxygen and nitrogen (1778)

1800

Dalton proposes atomic theory (1804)

1850
Clausius and Kelvin (separately) define second law of thermodynamics (1850)
Darwin publishes *Origin of the Species* (1859)
Maxwell equations define electromagnetic wave spectrum (1864)

Pasteur develops germ theory of disease (1880)
Cahal introduces neuron doctrine (1888)

1900

Einstein defines relativity (1905)
Bohr introduces quantum theory (1913)
Rutherford defines atomic structure (1919)
Rutherford dates age of Earth as 3.4 billion years (1929)
Lemaître proposes universe originated in primeval atom (1931)
Avery shows DNA is genetic material (1944)

1950
Turing defines test for artificial intelligence (1950)
Watson and Crick show double helix represents information (1953)
Wegener's theory of continental drift accepted (1965)
Standard Model introduced (1975)
Cech & Altman show RNA has catalytic activity (1981)
Prusiner shows proteins can be infectious (1983)

2000

A Dialog between René Descartes and Thomas Kuhn

René Descartes (1596–1650) was a French mathematician and philosopher. A polymath who explored topics extending from physics to physiology, he was a key figure in the emergence of modern philosophy.

Thomas Kuhn (1922–1996) moved from physics to the history and philosophy of science, first at Harvard, later at Berkeley and MIT He wrote on the history of physics as well as on the philosophy of science.

Descartes: Ah, Thomas Kuhn, it's a pleasure to engage in discourse with a fellow thinker. Your work on *The Structure of Scientific Revolutions* intrigues me, though I must confess I find your portrayal of scientific progress disconcerting. Science, in my view, must be built upon methodical doubt and reason. Skepticism serves as the foundation for certainty.

Kuhn: Indeed, René Descartes, it's an honor to discuss such a profound subject with you. The history of science suggests that certainty is often an illusion. Scientists, no matter how skeptical, operate within paradigms—frameworks of thought that guide what questions can be asked and what answers are acceptable.

Descartes: Fascinating. But are you not conceding too much to historical contingency? My method begins with systematic doubt, stripping away all uncertain assumptions until only unquestionable truths remain. This foundation can then support objective scientific knowledge. Surely, skepticism transcends any specific paradigm?

Kuhn: Ah, but your radical doubt occurs at the level of the individual thinker, while I focus on the community of scientists. Scientific knowledge is a collective enterprise in which skepticism is constrained by the prevailing paradigm. The transition from one paradigm to another is not a smooth process. It's often marked by resistance from the scientific community.

Descartes: In *Discours de la Méthode*, I maintain that the overarching paradigm is that you can truly know only what can be deduced from direct observations. Perhaps your "paradigm shifts" are a failure of scientists rather than of reason itself? The scientific method, properly applied, should prevent such entrenchment in dogma.

Kuhn: Scientists cannot question every assumption simultaneously. They must work within a shared theoretical framework. Revolutions occur when that framework can no longer accommodate the data.

Descartes: And yet, I must inquire, do you believe that scientific revolutions represent a move closer to absolute truth, as I envisioned with my quest for certainty through doubt?

Kuhn: That's an intriguing question. While scientific revolutions bring advances, I wouldn't say they necessarily lead us closer to absolute truth. I do not claim that truth itself is relative, but our access to it is limited by our paradigms. Each successive paradigm brings theory and data into closer agreement.

Descartes: Ah, I see. You would advocate that scientific revolutions are not so much about uncovering absolute truth as they are about removing previous doubts.

Kuhn: Exactly. Science is a process of continuous refinement and revision. It is both a skeptical inquiry and a communal effort. At any given moment, the individual scientist works within the communal paradigm. I rest my case.

KUHN'S PARADIGMS: CHANGING SCIENCE

The historian of science may be tempted to claim that when paradigms change, the world itself changes with them. Led by a new paradigm, scientists adopt new instruments and look in new places. Even more important, during revolutions, scientists see new and different things when looking with familiar instruments in places they have looked before... Though the world does not change with a change of paradigm, the scientist afterward works in a different world,[2] Thomas Kuhn, 1962.

A new scientific truth does not triumph by convincing its opponents and making them see the light, but rather because its opponents eventually die, and a new generation grows up that is familiar with it,[3] Max Planck, 1949.

Science depends on data, not on beliefs. I do not mean 'belief' in any religious sense, but simply in the sense that even scientists may stick to ideas or hypotheses that do not actually agree with the data. Thomas Kuhn argued that science advances by alternating between two types of periods. During incremental advances, data are added to an existing theory. During periods of revolution, an existing theory is replaced by a new one. This is evident in all areas of science. It's as true today as it was historically.

Kuhn called the reliance on a dominant theory a 'paradigm.' He argued that only a 'revolution' could overthrow it.[4] The concept of the paradigm—which has come to dominate modern science—is very much based on the traditional view of science as driven by hypotheses. The point is that a paradigm rests on a theory: the atom consists of protons, neutrons and electrons; or the genetic material is DNA.

Typically when the first data are obtained that conflict with a paradigm, the initial response is one that was called 'save the phenomena' in the pre-scientific period. Writing in the 6th century, Simplicius (*c.* 480–560 CE) attributed the phrase to Plato, describing the need to explain data that were not entirely consistent in terms of theory.[5] By the time of Ptolemy, it was describing the need to interpret (unkindly one might say distort) data to fit the paradigm.[6] But as further data accumulate, this becomes more and more difficult. Eventually the paradigm has to be replaced.

A great example of 'save the phenomena' comes from Ptolemy. "We believe that the object that the astronomer must try to achieve is this: to demonstrate that all the phenomena in the sky are produced by uniform and circular motions."[7] It may explain why his reported data seem to be better than his observations could have allowed (see Chapter 4).

Ptolemy was a great astronomer, but this is virtually the antithesis of the scientific approach. Yet propelled by divine will— "the apparent irregularities of the five planets, the sun and moon can all be represented by means of uniform circular motions, because only such motions are appropriate to their divine nature"— this became a paradigm that lasted more than 1000 years.[8] Even after Copernicus, it continued to impede progress.

Scientists today would find it offensive if you accused them of trying to 'save the phenomena,' but physicists have been known to practice 'fine-tuning,' which is a rather analogous process. 'Fine-tuning' is a pejorative term, however, as Steve Weinberg (awarded the Nobel Prize for physics in 1979) comments: "We criticize a proposed theory as fine-tuned when its features are adjusted to make some things equal, without any understanding of why they should be equal."[9] Yet today fine-tuning is responsible for the concept of dark energy. Basically this is a parameter introduced to explain the fact that the expansion of the universe is speeding up, instead of slowing down as current theory would have predicted (see Chapter 17). (Weinberg regards this as 'intolerable.')

Quantum physicist Paul Dirac believed in the priority of theory. "[The moral is that] it is more important to have beauty in one's equations than to have them fit experiment," he said,[10] referring to a conversation with Erwin Schrödinger. Schrödinger had abandoned his theory for describing electrons as waves because of some discrepancies with experimental data. Some months later, Schrödinger discovered that if he applied the theory in a more approximate way, ignoring some of the detailed requirements of relativity theory, it would work, and he published it. Much later, it became clear that the cause of the discrepancy was a property of the electron (its 'spin') that had not been known at the time.

"A physical law must possess mathematical beauty," Dirac wrote on the blackboard when he gave a lecture at the University of Moscow in 1956.[11] This is sometimes called Dirac's Principle of Mathematical Beauty.[12] Dirac was probably the 20th century physicist most inclined to look for elegance as a principle in his theories. Physicists are more inclined than biologists to assess theories for their elegance. Perhaps there is a sense in which any action that obeys a simple equation can be called elegant—the movement of the planets in graceful ellipses around the sun, for example—but that sense of simplicity breaks down in the complexities of quantum physics.

Biologists have learned from long experience that organisms are constructed more in an *ad hoc* manner by tinkering, evolution being a "blind watchmaker," in Richard Dawkins' memorable phrase.[13] Thomas Huxley (who was much involved in the debate about evolution) took a realistic position in 1870. "The great tragedy of science [is] the slaying of a beautiful hypothesis by an ugly fact."[14]

Weinberg captured the yin and yang between theoreticians and experimentalists very well. All experiments need corrections to the observed data. "If one knows the answer, there is a natural tendency to keep on making these corrections until one has the 'right' answer and then to stop looking for further corrections." And on the other hand, "experimentalists who know the result that they are theoretically supposed to get naturally find it difficult to stop looking for observational errors when they do not get that result."[15]

The very concept of 'save the phenomena' is opposed to the modern view, that theory should be subservient to data. Yet in a way, it can be seen as the forerunner of a scientific attitude. As captured by Plato, the belief was that the apparent irregularity of the astronomical data simply could not be real. The real world had to be organized on more orderly principles. A geometrical *theoria* was needed to restore the proper order of things.[16] It's not very far from here to the concept that a theory is needed for data to be accepted.

Data may seem to be more reliable than theory—but only up to a point. Francis Crick recollected an errant piece of data that blocked the group at Cambridge from a major discovery.[17] "I argued that it was important not to place too much reliance on any single piece of experimental evidence. It might turn out to be misleading... Jim [Watson] was a little more brash, stating that no good model ever accounted for all the facts, since some data was bound to be misleading if not plain wrong. A theory that did fit all the data would have been 'carpentered' to do this and would thus be open to suspicion."[18]

The history of science shows that it is easier for new ideas to be accepted when it's possible to understand their basis in physical reality. The lack of such a basis was responsible for the delays in accepting DNA (rather than protein) as the genetic material. The basis was provided only when it was realized that DNA carries information in a code. Actually, although previously there was resistance to the idea that DNA was the genetic material, there was in fact no convincing data to support protein as the alternative. It was more an assumption *faute de mieux*. (I discuss these developments in Chapter 19.) You might draw a parallel with astronomers sticking to the geocentric view until the 16th century.

When facts, or predictions from a model, are unambiguous, data are accepted even though we cannot understand them. Newton's laws of gravity were accepted, because their predictions could be verified, even if action at a distance was mysterious. We have very much the same problem with entanglement (the apparent ability of two linked atomic particles to communicate instantly when they are far apart). The phenomenon appears to be real, even though its basis is mysterious (and even conflicts with general relativity) (see Chapter 16).

Science is driven by the interplay of facts and theory. It's a constant debate whether to reconsider the facts or revise the theory when there is a clash. Neuroscientist Stuart Firestein says, "Scientists know that it is facts that are unreliable. No datum is safe from the next generation of scientists with the next generation of tools."[19] At the end of the day, however, theories are subservient to facts (or at least, to the facts of the moment)—but facts are subject to interpretation. Science is not immune from collective blindness in interpreting them.

The concept of 'groupthink' was first described as an affliction of business when journalist William Whyte defined it as 'rationalized conformity' in *Fortune* magazine in 1952.[20] Whyte presents a sarcastic view, drawing parallels with *Brave New World* and *1984*. Conformance with the group is more important than independent thinking. Psychologist Irving Janis gave groupthink its modern incarnation in 1971 when he introduced it to explain blunders in political decisions.[21] This was an even grimmer view.

The mythic view of science is that it should be free from groupthink because scientists are driven by the data. This is true up to a point, but when a new theory is proposed there can be either unreasonable resistance to it or willing suspension of disbelief—it can go either way. Neither will stop the eventual triumph of data, but recognition of reality can be delayed. Groupthink is a much more powerful effect in science than is usually realized from the outside.[22]

It's not always obvious in science when rejection of a new idea represents appropriate skepticism and when it represents willful refusal to reconsider the current paradigm. Examples of appropriate skepticism are hard to provide, because the mistaken ideas never take hold in the body of science. However, there are all too many cases of failure to recognize the reality of the data for science to have been completely objective.

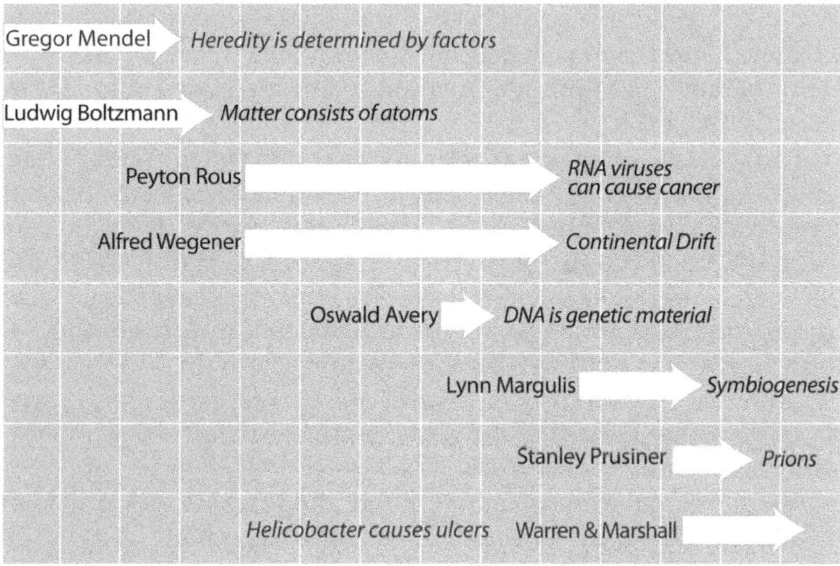

Examples of scientists whose ideas have met with unreasonable resistance or failed to be appreciated. The left end of the arrow shows when an idea was proposed. The arrowhead shows when it was generally accepted. Resistance to new ideas has occurred in all areas of science over the past century. The time for contested ideas to pass from ridicule to conventional wisdom varies from years to decades.[23]

Sometimes there is a clash of paradigms, either because there are conflicting data supporting two different models, or because the data admit different interpretations. When there are no scientific reasons to distinguish two models, choice between them may be made on aesthetic grounds (as for a protracted period in choosing Copernican or Ptolemaic models for the solar system). And it may be influenced by pressure from society.

∞ ∞ ∞

Sociologists debate whether major advances in science come from genius or ordinary scientists—the technical terms are the Newton effect or the Ortega effect. The Newton effect takes its name from Newton's metaphor in his letter to Robert Hooke—"If I have seen further, it is by standing on the shoulders of giants."[24] The Ortega effect is named for Spanish philosopher Ortega y Gasset, who argued that top-level research is like the tip of an iceberg, resting on a base of much less important research. "Experimental science has progressed thanks in great part to the work of men astoundingly mediocre, and even less than mediocre," Ortega said.[25]

The "Great Men" approach to the history of science—recounting the roles of individuals in making great discoveries—is unfashionable among historians, who now prefer to concentrate on setting science in the context of society. Certainly there is a difference between science and the arts. If Shakespeare had never lived, someone else might have written a series of plays—but they would not have been the same. The old joke, that "the well-known plays and poems were not by William Shakespeare, but by another person of the same name" does not really apply to writers, but might well apply to scientists.[26]

"Had Watson and Crick not existed, the insights they provided in one single package would have come out much more gradually over the period of many months or years," as Gunther Stent, a molecular biologist with a philosophical bent, said.[27] They might not have had the same immediacy, the history would be different—but the ultimate impact on our concept of DNA would have been similar. Erwin Chargaff, who worked on the biochemistry of DNA and set out rules that Watson and Crick relied upon, said that "It's not the men that make science; it is science that makes the men."[28]

Watson and Crick did not invent the double helix. They created a model to explain what was already there. In that sense, science is about the facts that are discovered, not the people who discover them. One historian of science says that "In the experimental sciences a 'fact' refers to an observation that has been repeatedly corroborated. But in the study of history, a 'fact' usually refers to a citation to the pertinent literature."[29] Perhaps this is unduly cynical, but it's true that, while 'facts' may be disputed in science, eventually they are agreed upon. 'Facts' can be harder to establish and more malleable in history.

The great body of experimental science is an essential background to all discoveries, but the revolutions, or shifts in understanding, in science during and since the Scientific Revolution have been associated with individual scientists. Galileo, Newton, and Einstein in physics, Pasteur, Mendel, Watson and Crick in biology, stand out, just to name a handful. We could extend the list into a roll call of genius in all the sciences. As disputes over priority attest, many of the important advances have been made around the same time by multiple people. The occurrence of simultaneous discovery suggests that ideas emerge when their time has come. It may not matter which genius makes the discovery, but in every case, a leap out of the ordinary has been involved.

∞ ∞ ∞

The development of an area of science follows a common pattern. It's triggered by a significant discovery; then knowledge increases exponentially until it slows or reaches a plateau.

The 20th century was a golden age for both atomic physics and molecular biology. Each had a distinct starting point, with a steady pace of discovery for about three decades, followed by a period of exponential increase for some decades before slowing to more of a plateau. The overall period for the fundamental discoveries lasted about 4 decades in each case.

The golden age for atomic physics started with Thomson's discovery of the electron in 1897. The exponential period started in the mid 1920s and lasted until the 1940s. The golden age for molecular biology started with Watson and Crick in 1953. It never really slowed until the end of the 1980s. The patterns of development are similar, but separated by some decades.

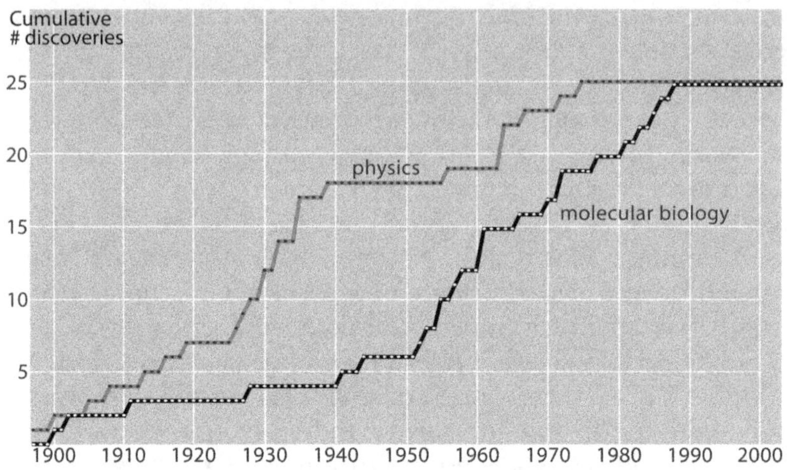

Development of an area of science can be plotted in terms of the number of cumulative discoveries over a period. Atomic physics shows a concentration between 1900 and 1940. Molecular biology shows a concentration between 1953 and 1988.[31]

During the first half of the 20th century, the great discoveries in science were made by individuals, or by individuals working with a small number of assistants. A transition to working in small groups developed in the second half of the century. A principal investigator would direct a laboratory of postdocs (postdoctoral fellows) and Ph.D. students. As projects grew more complicated, requiring mastery of varieties of techniques, the sizes of the groups increased. The most visible measure is the average number of authors on a scientific paper, which increased from 2–4 in 1970 to 10–20 today in biology, even more in physics with 30–40.[30]

Science became increasingly self-contained in the 20th century. In fact, that should probably be sciences in the plural, as the different disciplines diverged farther from one another. Although they are united by the same scientific methodology, researchers in any one discipline are unlikely to understand much of the others. Indeed, it's becoming simplistic to divide science into physics, chemistry, and biology. Each area is increasingly divided into subdisciplines. As specialization increases, there is more need for specialists in different areas to combine to undertake a project. Indeed, sometimes the contributors to one part of a scientific paper do not necessarily understand the other parts very well, if at all.[32] I view this as a potential threat to the integrity of science. If no one person can understand the totality of a scientific paper, what does this mean for the principle of verifiability and reproducibility?

Now science is poised at an even more dramatic point of transition. Small-scale science is being replaced by large-scale science. It's not straightforward to assess the magnitude of the change in either the use of resources or the numbers of scientists involved. Big science is sometimes defined as projects costing more than $100 million, but that is an inadequate description given that the average NIH grant (from the National Institutes of Health, for research in biology in the United States) is around $500 000. The ten largest scientific projects (all in physics except for the human genome project) cost about $60 billion over several years.[33] This is a bit more than the annual budget for the NIH (just under $50 billion).

The start of 'Big Science' might be dated to the invention of the cyclotron by Ernest Lawrence at the University of California, Berkeley. The prototype cyclotron constructed in 1931 was a ramshackle device not very different in principle from the type of apparatus Rutherford had used at the start of the century to bombard the nucleus. Lawrence had the idea when he read an article in an obscure German journal arguing that ions could be accelerated by a series of impulses. This would require an impractically long linear tube, but he had the crucial inspiration that instead the ions could be sent repeatedly round a circle. His prototype had a chamber of 4.5 inches. Lawrence called it his "proton merry-go-round." By later that year, the first purpose-built cyclotron had a chamber of 11 inches.

The Radiation Lab was established in an old house on the campus with a cyclotron of 27 inches, followed in 1936 by one of 37 inches. With a staff of 27 scientists, the Radiation Lab marked a new style of science. The award of a Nobel Prize to Lawrence in 1939 was a rare recognition of the importance of an invention rather than a theory or experiment.

The first of the scientist-managers, Ernest Lawrence came from the American mid-West. Fascinated by electronics, he was as much interested in machines and their potential as in theory. His experience with the cyclotron led to a crucial role in the Manhattan project, converting it to use electromagnetic force to separate U^{235} from U^{238}. Lawrence became an effective advocate for funding science on a larger scale. The continuing enlargement of cyclotrons and particle accelerators marks the increasing 'bigness' of science.

Comparing research facilities in 1950 and today shows the difference between traditional science driven by individuals and big science driven by teams. Top image: A biochemistry laboratory shows a classic array of equipment at the bench where a scientist works. Bottom image: The CERN control room today shows a computer-driven facility where many technicians control the equipment.

Big science is the extreme of a general change in the structure of science. The numbers of authors on 'big data' papers vastly exceed even the recent norms for conventional science. The most dramatic transition is in physics. The papers defining the structure of the atom in the first half of the 20th century were written by one or two authors each. The two papers reporting the discovery of the Higgs boson in 2012 each had 1000 authors. (The first papers with more than 100 authors were published in the late 1990s. Since then, there have been more than 1500, which represents quite a lot of big science.[34])

This makes a difference in approach. In small groups of researchers, each member of the group is expected to be knowledgeable about all aspects of the work. In large groups, each member has a specialized knowledge of the aspect they work on, but not necessarily more than a sketchy understanding of the rest of the project. In a sense, bottom-up research has been replaced by top-down direction. The range of specialties involved in some projects, coupled with the fact that the expertise required to obtain the data may be different from the expertise required to analyze it, makes it amazing that anyone has the competence to provide overall direction.

Scientists are always taught to be suspicious of correlations, but the approach of 'big data' turns this suspicion on its head, with a shift in mindset from causation to correlation.[35] 'Big data' argues that obtaining a sufficient mass of data makes correlations sufficiently powerful in themselves. It's a different view of what science is about.[36]

The case was put by a controversial article by Chris Anderson, the Editor of *Wired*, a magazine on technology, in an article in 2008 under the title, "The end of theory: the data deluge makes the scientific method obsolete." The argument goes that "Petabytes allow us to say: 'Correlation is enough.' We can stop looking for models. We can analyze the data without hypotheses about what it might show. We can … let statistical algorithms find patterns where science cannot."[37] This view was widely attacked in the scientific community at the time, but the approach of big data is taking its place in science all the same.[38]

The human genome project, which completed the sequence of the human genome in 2003, is an example that has changed biology. Controversial when it was proposed, with researchers concerned that it would divert funding from 'investigator-driven' science, it's now reckoned as one of the triumphs of modern biology. (Incidentally, it may have increased public support for science.) Whereas previously attempts to define genes responsible for specific human conditions required individual analyses, now it's possible to scan the genomes of a set of people to try to identify genes that correlate with a condition. (I discuss the consequences of the revolution caused by genome sequencing in Chapter 20.)

What happens with a transition to big science? Will the role of genius be superseded? Where will the impetus come from to provide new ideas? Suppose that the data suggest a correlation or that a large-scale effort produces a datum no individual scientist could obtain. Don't we still need that intuitive spark to get to the next level? Will physics (and perhaps also the other sciences) be driven by an alternation between genius theoreticians who propose models and masters of organization who undertake experiments?

Big data have to be analyzed. 'Data analysis' has become something of a buzzword (more in the business world than in science). It has been called a paradigm in itself.[39] There's some disagreement on exactly what it comprises, but essentially it consists of applying statistic methods and software (including AI-driven software) to extract information from large sets of data.[40] Big data hasn't really been a phenomenon long enough for conclusions resulting from it to be questioned en masse, or to judge how far it will complement and how far it will supplant the traditional view of science.

Concern about whether data analysis will replace science sometimes merges into attempts to reconcile hypothesis-driven science with (big) data-driven science. (Perhaps this is a parallel with the attempts of the Church to reconcile religion with Greek philosophy in the Middle Ages.[41])

A new issue is how best to analyze unstructured data—large data sets obtained without a strict experimental design, such as the information obtained during the COVID pandemic. AI is effective here, and, in fact, AI is becoming increasingly employed to analysis large data sets in medicine. This may increase the importance of big data.[42]

∞ ∞ ∞

Science is sometimes divided into 'pure' or 'basic' research and 'applied' research, which merges into technology. One difference is that basic research—in principle driven by curiosity—has the potential to break out of a paradigm. Applied research, more concerned with using existing knowledge for a specific development, is unlikely to do so.

'Applied research' is not much used as a description today. 'Translational research' is preferred to describe the "translation of knowledge from the research bench to the clinical bedside." The term was introduced, perhaps as an attempt to reverse the

Discoveries in Basic Research with No Apparent Relevance to Medicine or Technology Leading to Important Advances

Year	Discovery	Technology	Year
1887	Heinrich Hertz discovers 'Hertzian' (electromagnetic) waves	Radio	1895
1895	Wilhelm Röntgen discovers X-rays by experimenting with electric currents in vacuum	X-rays used to show abnormalities in patients	1896
1928 & 1932	Dirac proposes positron, Anderson finds tracks in cloud chamber	PET scanner	1973
1928	Alexander Fleming discovers penicillin by chance observation of fungus that kills bacteria	Penicillin becomes general purpose antibiotic	1943
1945	Felix Bloch and Edward Purcell develop nuclear magnetic resonance as a tool for spectroscopy in chemistry	MRI imaging to show cancer in humans	1980
1960	Light amplification by stimulated emission discovered in 1939; laser invented in 1960	Laser eye surgery	1989
1965	Werner Arber discovers restriction enzymes in bacteria	DNA fingerprinting/ mapping the human genome	1985/ 1989
1981– 1983	Alexey Ekimov in Russia and Louis Brus at Bell Labs identify quantum dots	LEDs and nanotechnology	1999
1995	Francisco Mojica finds tandem repeats in bacteria	CRISPR technique for gene editing	2012

The gap between the discovery and realization of its medical or technological application was two decades or more except for X-rays.

intellectual denigration of applied compared to basic research, and to close a supposed gap between knowledge and its practical application.[43] I shall continue to use 'applied research' as it's more applicable to the period I'm discussing.

In the years after the Second World War, there was considerable discussion about how basic and applied research should be defined, the nature of distinction between them, and how this should relate to government funding.[44] You might say that basic research is more of a cycle alternating between hypothesis and experiment, while applied research is more of a linear progression towards a defined objective.

Sometimes there are calls for basic research to be more directed towards specific ends. But trying to direct science is counter-productive, because you never know where a discovery will lead. Often it's impossible to know what discoveries are needed in order to achieve a specific end. This is why it's a mistake to attempt to direct science towards the needs of society, instead of supporting good ideas in any area irrespective of apparent relevance to human welfare. It's especially striking that two discoveries in bacteria, apparently with no relationship to human welfare, led years later to the major advances of the human genome sequence and CRISPR gene editing.[45]

Experimental science has a unique feature that distinguishes it from other forms of intellectual activity. Every experiment has a *control*. In effect, this means comparing the experiment, where the parameter of interest is changing, with a control in which that parameter is fixed. When Robert Boyle performed his experiments with the air pump, the experiment in which a vacuum was created was compared with a control left at normal atmospheric pressure. Whether physics, chemistry, or biology, science proceeds by obtaining data through controlled experiments that can be reproduced or challenged by others.

Practitioners of natural (or 'hard') sciences harbor deep skepticism as to whether 'soft sciences,' (such as psychology or social sciences) should really be called sciences at all, largely because of the difficulty in establishing proper controls.[46,47] "Compared with physics, sociology and psychology are riddled with fashions, and with uncontrolled dogmas," Popper said.[48] (An important precedent for questioning whether 'social sciences' are science comes from Popper's view that work on human behavior is not subject to scientific testing.[49]) Nor do controls play the same role in 'big data,' where correlations have replaced comparing experiments with controls.

∞ ∞ ∞

Biology and chemistry are experimental sciences. Hypotheses are proposed to explain data as results emerge, but theoreticians as such are rare. Physics, however, is divided into experimental physics and theoretical physics. The difference in attitude was nicely summarized by Russian physicist Pyotr Kapitza, who gained a Nobel Prize in 1978 for his work on low temperature physics. He gave priority to experiments. "In the play 'Gentlemen Prefer Blondes' Marilyn Monroe's character says: 'Love is a good thing, but a golden bracelet is forever!' I think that we scientists may say: 'Theory is a good thing, but a well-done experiment lasts forever'."[50]

Kapitza pointed to a significant difference between experimental and theoretical practice. "The division of physicists into theorists and experimenters occurred comparatively recently. Previously, not only Newton and Huygens, but also such theorists as Maxwell, usually verified their theoretical deductions and constructions experimentally themselves. Now, however, only in exceptional cases do theorists attempt to check their theories themselves. The change has happened for the simple reason that experimental techniques are becoming much more complicated. It requires great effort to carry out tests, which is not usually within the capacity of one person, and it is done by a collective of scientific workers.

"This is how the disproportion arises between the amount of theoretical work and the possibility of subjecting it to experimental investigation: a theoretician may well produce several papers per year, say four, but to do the experimental checking, a year or eighteen months is usually required and a group of, shall we say, five persons has to work on each. Clearly the theorist will have to have between 20 and 30 experimenters. This, of course, is a simplified roster, but in general it shows the correlation between theoreticians and experimenters necessary for scientific development. At present the number of theorists and experimenters is approximately equal, and the result is that the majority of theoretical conclusions are not being tested in practice. Theorists lose the habit of thinking that none of their work acquires a valuation until after experimental verification."[51]

Even Kapitza could not have forecast just how dramatically right his comments would prove. Experimental physics has become a large-scale science, often requiring equipment that is far beyond the capacity of any one laboratory. Particle physics now uses very large particle accelerators that send high-energy beams around a ring until they collide. The Large Hadron Collider managed by CERN (European Organization for Nuclear Research) is a ring of superconducting magnets with a diameter of 27 km. The contrast with theoretical physics could scarcely be greater.

Theoretical physicist Peter Higgs predicted the existence of what later came to be called the Higgs Boson in a short paper in 1964 of which he was sole author.[52] Higgs was a classic example of a theoretical physicist, only occasionally publishing papers. As he noted ruefully after being awarded the Nobel Prize, applying normal criteria of productivity he would never have had tenure at a university. As noted earlier, when the Higgs boson was finally detected in 2013 at the CERN cyclotron, there were 1000 authors on each paper (two independent groups used the cyclotron). This may be one reason why theoretical physicists have become more famous than experimental physicists.

The main ring of the Tevetron cyclotron in Illinois had a diameter of 7 km. It ran from 1983 to 2011 and cost $120 million. It was the largest particle accelerator until the CERN accelerator was built.

Theoretical physics can feel like a throwback to natural philosophy. Observations are treated like axioms—they are assumed to be true—and an explanation is proposed based on reasoning. The original paper from Peter Higgs sets out to explain a curious situation. Modeling the behavior of known subatomic particles in the 1960s, equations to describe their behavior worked if they had no mass. But physicists knew that they must have mass.

Higgs proposed that instead of trying to rewrite the equations, suppose that space is filled with a field that resists movement of the particles. It was hard to believe in an invisible ubiquitous field. Higgs' first submission was rejected as "of no obvious relevance to physics." The paper was published a year later elsewhere. It's fair to say that only a particle physicist could understand that the paper—basically a series of differential equations with a few paragraphs of explanatory text—was proposing a radical change in the construction of matter. The Higgs boson is the particle—or waveform—that makes up the ubiquitous field.

$\infty \ \infty \ \infty$

A unique feature of theoretical physics—I am not aware of any equivalent in any other field of science—is the *thought experiment.* This is something of an equivalent to the emphasis in early Greek philosophy in reasoning from axioms or trying to explain a paradox, but here the term describes performing an imaginary (but plausible) experiment. Sometimes this might (in principle) be performed as a real experiment, although often the thought experiment uses ideal conditions that could be not be achieved in practice.

The term originated in 1883 with Ernst Mach, an Austrian physicist who worked on shock waves. The original German is *gedankenexperiment.* The English term was introduced in a translation of one of Mach's papers. From the way Galileo described his experiment with falling bodies, it's likely that it was a thought experiment rather than a physical experiment. Two of the most famous thought experiments are Maxwell's demon (Chapter 9) and Schrödinger's cat (Chapter 15). Einstein and his colleagues often used the device.

A popular view of theoretical physics might be that it depends on logical deductions to construct a theory from difficult observations. Well, yes and no. The need to explain observations that are often awkward or even conflicting is a driving force, but the deductions are not always based simply on logical analysis. (If the analysis was straightforward, what role would there be for theoretical physicists?)

This is how theoretical physicist Richard Feynman, awarded the Nobel Prize in 1965 for his work in quantum mechanics, described the process. "In general we look for a new law by the following process. First we guess it. Then we compute the consequences of the guess to see what would be implied if this law that we guessed is right. Then we compare the result of the computation to nature, with experiment or experience, compare it directly with observation, to see if it works. If it disagrees with experiment it is wrong. In that simple statement is the key to science. It does not make any difference how beautiful your guess is. It does not make any difference how smart you are, who made the guess, or what his name is—if it disagrees with experiment it is wrong."[53]

'Guessing' can vary from choosing one logical argument rather than another to inspired leaps in the dark. One of the most famous examples is the derivation of the Maxwell equations. When Maxwell derived the laws of electricity and magnetism, he invented a new system (see Chapter 9). Feynman described the situation. "[Maxwell] put together all the laws of electricity...and he realized they were mathematically inconsistent. In order to straighten it out, he had to add a term to an equation. He did this by inventing for himself...a model in space."[54]

Thought experiments are often designed to question hypotheses that support the existing paradigm. It's hard to see a role for thought experiments in questioning big data. Big data represents more than a different way of doing science, more than collecting data en masse instead of individually. It represents a different attitude that may be inconsistent with the conventional way of regarding science as testing hypotheses and ultimately supporting or changing paradigms. We may well ask how big data would, in principle, enable us to break out of a paradigm. Big data is, in a sense, a change in the paradigm of how science itself works. It's interesting to speculate what Popper and Kuhn might have said about it.

NOTES AND REFERENCES

1. T. S. Kuhn, *The Structure of Scientific Revolutions*, University of Chicago Press, Chicago, 1962, p. 92.

2. T. S. Kuhn, *The Structure of Scientific Revolutions*, University of Chicago Press, Chicago, 1962, pp. 111, 121.

3. M. Planck, *Scientific Autobiography and Other Papers*, Philosophical Library, New York, 1968.

4. T. S. Kuhn, *The Structure of Scientific Revolutions*, University of Chicago Press, Chicago, 1962, p. 92.

5. 'Save the phenomena' is often associated with the astronomer Eudoxus (*c.* 400–350 BCE), who may have been a student at Plato's academy. P. Duhem, *To Save the Phenomena. An Essay on the Idea of Physical Theory From Plato to Galileo*, 1908, translated by E. Dolan and C. Maschler, University of Chicago Press, Chicago, 2015.

6. G. E. R. Lloyd, Saving the appearances, *Classical Q.*, 1978, **28**, 202–222.

7. Ptolemy, *Almagest*, translated and annotated by G. J. Toomer, Duckworth, London, 1984, vol. III.

8. Simplicius attributed the concept that motions must be perfect circles to Plato.

9. S. Weinberg, *To Explain the World: The Discovery of Modern Science*, HarperCollins, New York, 2015.

10. P. A. M. Dirac, The evolution of the physicist's picture of nature, *Sci. Am.*, 1963, **208**, 45–53.

11. Quoted in H. Kragh, *Dirac: A Scientific Biography*, Cambridge University Press, Cambridge, 1990, p. 275.

12. H. Kragh, *Dirac: A Scientific Biography*, Cambridge University Press, Cambridge, 1990, pp. 275–292.

13. R. Dawkins, *The Blind Watchmaker: Why the Evidence of Evolution Reveals a Universe Without Design*, Norton, New York, 1986.

14. T. H. Huxley, The British Association Meeting, Address of Thomas Henry Huxley, *Nature*, 1870, **2**, 400–406.

15. S. Weinberg, *Dreams Of A Final Theory: The Scientist's Search For The Ultimate Laws Of Nature*, Vintage, New York, 1992, pp. 96–97.

16. Also see O. Gal, *The Origins of Modern Science*, Cambridge University Press, Cambridge, 2021, pp. 83–88, and ref. 5.

17. Identifying the α-helical structure of proteins. See Chapter 19.

18. F. Crick, *What Mad Pursuit: A Personal View of Scientific Discovery*, Basic Books, New York, 1988, pp. 59–60.

19. S. Firestein, *Ignorance: How it Drives Science*, Oxford University Press, Oxford, 2012, p. 21.

20. W. White, Groupthink, *Fortune*, January, 1952.

21. I. Janis, Groupthink, *Psychology Today*, November, 1971.

22. Groupthink in science usually reflects a collective failure of critical faculty. I discuss examples in Chapter 25. However, it can also be produced by external political influences (see B. Lewin, *Inside Science: Revolution in Biology and Its Impact*, Cold Spring Harbor Laboratory Press, New York, 2023, pp. 77–78).

23. This is not an exhaustive list but it makes the point that ideas can run well ahead of current understanding. This contrasts with the modern relativistic trend that dominates the history of science today. For example, "Given our prevailing assumption that scientific knowledge is context dependent, it seems intrinsically unlikely that an individual would be able to cut himself off from his own intellectual milieu and somehow anticipate that of a future generation." (P. J. Bowler and I. R. Morus, *Making Modern Science*, University of Chicago Press, Chicago, 2nd edn, 2020, p. 204). Perhaps the assumption shows the context-dependence of modern historians!

24. The phrase goes back to Bernard of Chartres, around 1100, and was intended to emphasize the depth of ancient knowledge. M. Calinescu, *Five Faces Of Modernity: Modernism, Avant-Garde, Decadence, Kitsch, Postmodernism*, Duke University Press, Durham, NC, 2nd edn, 1987, p. 15.

25. J. Ortega y Gasset, *The revolt of the masses*, Norton, New York, 1932.

26. The form quoted here is from the first known reference in *The Spectator* of January 14, 1860, *The "New Planet" and Its Discoverers*, p. 38. It's been used in many similar forms since then, most often applied to Shakespeare, but sometimes to other writers, including Homer.

27. S. Stent, Prematurity and uniqueness in scientific discovery, *Sci. Am.*, 1972, **227**, 84–93.

28. E. Chargaff, A Quick Climb Up Mount Olympus, *Science*, 1968, **159**, 1448–1449.

29. D. Deming, *Science and Technology in World History: Volume 3: The Black Death, the Renaissance, the Reformation and the Scientific Revolution*, Mcfarland & Co., Jefferson, NC, 2012.

30. B. Lewin, *Inside Science: Revolution in Biology and its Impact*, Cold Spring Harbor Laboratory Press, New York, 2023, p. 35.

31. Each point on the graph corresponds to an entry in the timelines, for physics using atomic physics and quantum physics, and for molecular biology using DNA plus the preceding discoveries in genetics (1900: rediscovery of Mendel; 1902: chromosomes segregate by Mendel's laws; 1911: genes are on chromosome; 1941: genes code for proteins; 1952: identification of human chromosome number; 1972: production of first recombinant DNA).

32. B. Lewin, *Inside Science: Revolution in Biology and its Impact*, Cold Spring Harbor Laboratory Press, New York, 2023, pp. 25–29.

33. In billions of dollars, approximately, ITER, $18; James Webb telescope, $10; Copernicus, $7.4; HGP, $5; LHC, $4.75; NIF, $3.5; Mars 2020, $2.5; AMS, $2; Myrrha, $2; SNS, $1.4; LIGO, $1.1. The total is sometime distorted by including the International Space Station, but I omit this because it really has very little to do with science and a lot to do with politics.

34. Data from PubMed.

35. V. Mayer-Schonborn and K. Cockier, *Big Data: A Revolution That Will Transform How We Live, Work, and Think*, Mariner Books, Boston, 2014.

36. The term 'big data' was first used in October, 1980 in a keynote address to the Conference on New Directions in History, State University of New York

at Buffalo. "None of the big questions (in history) has actually yielded to the bludgeoning of the big-data people." It was published later as C. Tilly, The Old New Social History and the New Old Social History. *Review*, VII 3, Winter 1984, 363–406.

37. C. Anderson, The end of theory: the data deluge makes the scientific method obsolete, *Wired*, 2008, June 23. www.wired.com/2008/06/pb-theory.

38. M. Pigliucci, The end of theory in science?, *EMBO Rep.*, 2009, **10**, 534; F. Mazzocchi, Could Big Data be the end of theory in science? A few remarks on the epistemology of data-driven science, *EMBO Rep.*, 2015, **16**, 1250–1255.

39. It's been called the 'fourth paradigm,' the first three being empirical evidence, scientific theory, and computational science. T. Hey *et al.*, *The Fourth Paradigm: Data-intensive Scientific Discovery*, Microsoft Research, Seattle, WA, 2009.

40. There is even a journal called *Big Data*. Its first issue included an article advocating data analysis. F. Provost and T. Fawcett, Data science and its relationship to big data and data-driven decision making, *Big Data*, 2013, **1**, 51–59.

41. A. Karpatne, *et al.*, Theory-guided data science: a new paradigm for scientific discovery from data, *IEEE Trans. Knowl. Data Eng.*, 2017, **29**, 2318–2331.

42. See *Science in the Age of AI*, Royal Society, London, 2024.

43. A. L. van der Laan and M. Boenink, Beyond bench and bedside: disentangling the concept of translational research, *Health Care Anal.*, 2015, **23**, 32–49.

44. D. E. Stokes, *Pasteur's Quadrant. Basic Science and Technological Innovation*, Brookings Institution Press, Washington, DC, 1997, pp. 58–70.

45. Werner Arber discovered restriction enzymes, which set the basis for mapping the human genome; Francis Mojica discovered the CRISPR system, which was developed in gene-editing. See B. Lewin, *Inside Science: Revolution in Biology and its Impact*, Cold Spring Harbor Laboratory Press, New York, 2023, pp. 209–220, 233–242.

46. It has been suggested that approach of the social sciences actually has more in common with the humanities than with the natural sciences. J. Kagan, *The Three Cultures: Natural Sciences, Social Sciences, and the Humanities in the 21st Century*, Cambridge University Press, Cambridge, 2009, pp. 3–5.

47. Social scientists have mostly been indignant about suggestions that they might not meet the standard for natural sciences, and the case has been made that social sciences should be judged by a different standard. See B. Flyvbjerg, *Making Social Science Matter: Why Social Inquiry Fails and How it Can Succeed Again*, Cambridge University Press, Cambridge, 2001.

48. K. Popper, Normal Science and its Dangers, in *Criticism and the Growth of Knowledge*, ed. I. Lakatos and A. Musgrave, Cambridge University Press, Cambridge, 1970, pp. 57–58.

49. K. Popper, *The Logic Of Scientific Discovery*, Routledge, London, 1959.

50. P. Kapitza, *Experiment, Theory, Practice: Articles and Addresses*, Reidel, Dordrecht, Holland, 1980, p. 160.

51. P. Kapitza, *Experiment, Theory, Practice: Articles and Addresses*, Reidel, Dordrecht, Holland, 1980, pp. 156–157.

52. P. W. Higgs, Broken symmetries and the masses of gauge bosons, *Phys. Rev. Lett.*, 1964, **13**, 508–509.

53. R. Feynman, *The Character of Physical Law (Messenger Lectures, 1964)*, The MIT Press, Cambridge, MA, 1967, p. 156.

54. R. Feynman, *The Character of Physical Law (Messenger Lectures, 1964)*, The MIT Press, Cambridge, MA, 1967, p. 162.

CHAPTER 12

Darwin's Evolution: 1859–1980

"SO IMPERFECT IS OUR VIEW INTO LONG-PAST GEOLOGICAL ages, that we see only that the forms of life are now different from what they formerly were," said Charles Darwin.[1] In describing the history of the Earth by analyzing its surface and interior, geology seems that it should be a sedate area of science. But geological arguments lie at the heart of three of the most dramatic paradigm shifts that have affected human thinking. A major objection to the theory of evolution was that geological evidence suggested the Earth could not be old enough for evolution to have occurred. The rejection of continental drift made it difficult to understand the evolution of the planet. And there is still a vociferous minority of dissenting scientists who do not believe the impact of an asteroid was responsible for the extinction of the dinosaurs, a key event that opened the way to the evolution of humanity. These are prime examples of science in action and its interaction with society. In each case a scientist proposed a theory that faced adamant opposition not merely from society at large, but also from the scientific community. Each theory rests on observations of distant events where direct experiments are not possible. In each case, the accumulation and acceptance of evidence to overcome opposition took decades. Now this trio of threads comes together to relate human evolution to the history of the planet. Perhaps climate change will be a fourth leg.

The Frontiers of Science
By Benjamin Lewin
© Benjamin Lewin 2026
Published by the Royal Society of Chemistry, www.rsc.org

Timeline for Key Events in Evolution and Geology

1750

1775   James Hutton proposes continuous aging of Earth (Uniformitarianism) over great age based on movements of crust on molten core (1788)

1800

1825   Charles Lyell publishes *Principles of Geology* (1830)

1850   Charles Darwin publishes *Origin of the Species* (1859)
Lord Kelvin dates Earth as 100 million years old (1863)

1875

1900   Ernest Rutherford dates Earth as 2.2 billion years old (1907)

Alfred Wegener publishes *Origin of Continents and Oceans* (1915)

1925   Ernest Rutherford revises age of Earth to 3.4 billion years (1929)

1950   Clair Patterson dates Earth as 4.5 billion years (1956)
Samuel Carey proposes tectonic plates to explain continental drift (1958)

Jack Oliver finds seismologic evidence to support plate tectonics (1968)

1975   Luis & Walter Alvarez propose asteroid impact caused dinosaur extinction (1980)

Alan Hildebrand proposes Chicxulub crater in Gulf of Mexico is impact site (1991)

2000

2025

A Dialog between Charles Lyell and Gerardus Mercator

Charles Lyell (1797–1875) was a Scottish geologist. He was a close friend of Charles Darwin and provided evidence for gradual evolution of the Earth.

Image reproduced from Project Gutenberg's *Charles Lyell and Modern Geology*, by Thomas George Bonney.

Gerardus Mercator (1512–1594) was a Dutch geographer and cartographer. He never traveled much, but developed leading-edge maps of the world.

Charles Lyell: (*examining a rock*) Ah, Mercator! Your maps are truly a marvel, an intersection of art and science. It's remarkable how you've captured the contours of our Earth. When I wrote *Principles of Geology* in 1830, I argued that by examining the surface of the Earth we can deduce that it has been evolving for a long time. This was a revolutionary idea. So I must ask, do you believe these lands have always been as they are depicted on your charts?

Gerardus Mercator: (*unfolding a map*) Thank you, Lyell. When I prepared my new map in 1569, the cylindrical projection allowed proper navigation for the first time in centuries. Of course, I've always considered the Earth to be somewhat stable, with the continents fixed as they appear on my maps. Do you have a different view?

Lyell: Indeed, I do. I've been exploring the concept of geological time. It suggests that our Earth is far from unchanging. The layers of rock and sediment tell a tale of constant change—land rising, sinking, eroding, and even shifting. Wegener proposed that the continents were once joined in a vast super continent, which slowly drifted apart over millions of years. This was just as controversial as my original concept that the Earth has a great age.

Mercator: Continents can drift? Unbelievable! But how could such massive landmasses move? What force could possibly drive such a phenomenon?

Lyell: Perhaps heat from within the Earth allows the crust to move over the molten interior. Continental drift is now explained by the concept that the surface moves on tectonic plates.

Mercator: That would imply a much more fluid and restless Earth than I have ever imagined. But tell me, how does this align with the idea of evolution? If the continents have moved so drastically, surely this must have influenced the development of life?

Lyell: Absolutely! The changing positions of the continents would have isolated species, creating environments that favored different evolutionary paths. My friend Charles Darwin proposed that species evolve over time through natural selection. The drifting continents could have played a pivotal role in changing the environment to shape the diversity of life we see today.

Mercator: I see how the movement of continents, if it occurs, might lead to such isolation and variation. But this would be gradual. What could be the cause of sudden events, such as the extinction of the dinosaurs?

Lyell: There have always been theories of catastrophic events involved in evolution. Indeed, I disposed of some of these proposals myself. Darwin's theory of natural selection depends upon gradual change. But subsequently there has been evidence also for catastrophic events. Defining the relationship of the Earth's crust to its interior, and understanding how human existence is impacting the environment, are essential. I rest my case.

DARWIN'S EVOLUTION

It's a strange fact, characteristic of the incomplete state of our present knowledge, that totally opposing conclusions are drawn about prehistoric conditions on our planet, depending on whether the problem is approached from the biological or the geophysical viewpoint,[2] Alfred Wegener, 1929.

Controversies about the age of the Earth show the impact of scientific thinking on society more directly than perhaps any other issue in the current era: echoes of Galileo. The clash between the biblical account and the geological record became apparent in the second half of the 19th century. The details played out in three areas of science.

Theories of evolution, continental drift, and extinction of the dinosaurs by asteroid impact have been among the most controversial areas in the recent history of science. They show the full range of reasons why scientific arguments can be rejected. There were religious objections to evolution, there was obtuse refusal to examine the data for continental drift, and there was an absence of supporting data when the asteroid theory was first proposed.

Perhaps because the issues at stake relate so closely to human existence, the arguments have gone beyond the bounds of scientific discourse into personal abuse. Darwin was satirized as an ape for proposing evolution. Alfred Wegener was attacked for 'delirious ravings' in proposing continental drift. Luis Álvarez showed all the scorn of a physicist for a 'softer' science by saying, "Paleontologists are not really very good scientists. They're more like stamp collectors."[3]

New theories need to be supported by data, but to be accepted, they also need to fit into the context of the period—into what Thomas Kuhn called the prevailing paradigm. Evolution is an example. After Darwin published *On the Origin of Species* in 1859,[4] the theory of evolution was subject to attack because there was no theoretical underpinning for its mechanism. Evolution came to be accepted by the end of the century more because society moved away from belief in divine intervention than because of any new definitive evidence.

Coming from a well-connected, affluent medical family, Charles Darwin (1809–1882) went to medical school at Edinburgh. Losing interest in medicine, he was sent to Cambridge University to prepare

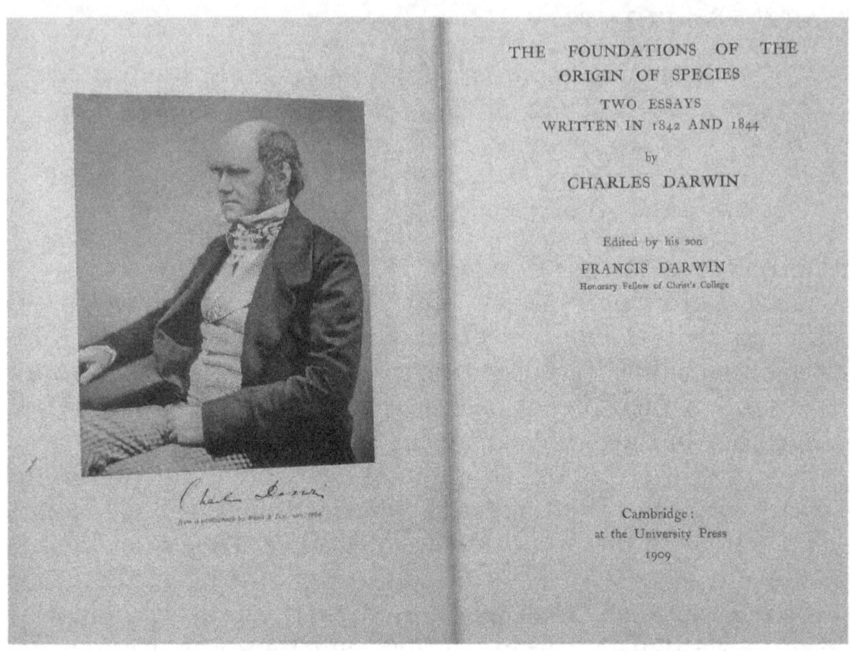

On the Origin of Species was the most important work of the 19th century in defining humanity's position in the world. Charles Darwin revised it through 6 editions. His son Francis edited later editions.

for a career in the Church. He acquired an interest in botany, and then signed on for a place on HMS Beagle's voyage around the world (supported by his father). The voyage lasted from 1831 to 1836. His observations led to his work on natural selection and geology.

Darwin conceived the idea of the evolution of species in 1838. "In October 1838...I happened to read for amusement 'Malthus on Population'... it at once struck me that under these circumstances [the struggle for existence] favourable variations would tend to be preserved, and unfavourable ones to be destroyed. The result of this would be the formation of new species. Here, then, I had at last got a theory by which to work."[5]

He worked on the theory while continuing his work on geology and writing up the voyage on the Beagle. His first drafts were written in 1842 and 1844, but the final book was not ready until 1859.[6] (In the end, Darwin had to hurry to publish, because Alfred Wallace was developing a rival theory.)

By the end of the 19th century, evolution was accepted as a reality, but there were serious doubts as to whether natural selection could account for it.[7] The phrase "survival of the fittest," which came to epitomize Darwinian evolution, was actually coined by Herbert Spencer in his book, *Principles of Biology*, in 1864. "This survival of the fittest, which I have here sought to express in mechanical terms, is that which Mr. Darwin has called "natural selection, or the preservation of favoured races in the struggle for life."[8]

A basic problem was the lack of any obvious mechanism. The rediscovery of Mendel's work in 1900, followed by the development of genetic theory, bridged the gap. Over the first two decades of the 20th century, it became apparent that genes are carried on chromosomes, and that their properties can be changed by mutations. This identifies the source of the variation on which natural selection works.

One of the problems with Darwin's theory of evolution was that it seemed to require much more time than would be available from current estimates of the age of the Earth. As a geologist, Darwin was friendly with Charles Lyell, the leading exponent of an old age for the Earth based on geological arguments. (Indeed, Lyell communicated papers from Alfred Wallace and Charles Darwin, on the evolution of species, to the Linnean Society a year before publication of *On the Origin of Species*.[9])

Following the view that geological strata develop slowly, the first edition of *Origin* suggested that the Earth was older than 300 million years. As the result of criticisms, Darwin lowered the estimate in later editions. By the 6th edition of 1872, he was forced to admit that the current estimate (about 100 million years) was a problem. "It seems doubtful whether the earth, in a fit state for the habitation of living creatures, has lasted long enough."[10]

The Earth's age had been calculated in 1862 as 98 million years (with a very wide error range) by Lord Kelvin, the eminent physicist, on the basis of its rate of cooling from its original molten state. The first real challenge to this estimate came from Ernest Rutherford (who subsequently gained a Nobel Prize for his discovery of the nucleus), when he applied an estimate based on the rate of radioactive decay (see the box below). His estimate of 2.2 billion years was 20× greater than Lord Kelvin's. (The discrepancy was partly due to the fact that radioactivity warms the Earth.)

Rutherford first presented his results in at a lecture in London in 1907. He recalled the occasion. "I came into the room, which was half dark, and presently spotted Lord Kelvin in the audience and realised that I was in for trouble at the last part of my speech dealing with the age of the earth, where my views conflicted with his. To my relief, Kelvin fell fast asleep, but as I came to the important point, I saw the old bird sit up, open an eye and cock a baleful glance at me! Then a sudden inspiration came, and I said Lord Kelvin had limited the age of the earth, provided no new source was discovered. That prophetic utterance refers to what we are now considering tonight, radium! Behold! the old boy beamed upon me."[11]

Age of the Earth

The age of the Earth became a theological issue in the 17th century after James Ussher, Archbishop of Armagh, calculated in 1650 that the Earth would have been created in 4004 BCE on the basis of Biblical genealogies. Estimates from other theologians from the 2nd century CE to the 17th century were similar.[12]

Scientific efforts based on the cooling of the Earth started with William Thomson (Lord Kelvin) in 1862. If the Earth was created as a molten globe, its age could be calculated from the overall rate of cooling and the temperature of the exterior.[13] This gave the answer that the Earth was converted from its molten state 98 million years ago (between limits 20–400 million years). Subsequent estimates on a similar basis usually gave slightly lower estimates, as did estimates based on cooling of the Sun, tidal effects on Earth, or ocean chemistry.[14] There were some objections to using physics to calculate the age of the Earth. A much wider range of estimates resulted from attempts to use the accumulation of layers of sediment.[15]

What is impressive about both the theological and physical calculations is a high degree of scholarship, with careful deductions from the starting data. The problem in each case was that the basic assumptions—that the Bible can be taken literally or that cooling is offset only by conduction of heat from the interior to the surface—were erroneous. In particular, radioactive decay (unknown at the time) generates heat.

Geologists knew that the surface of the Earth has layers of strata. The fossil record helped to establish a timeline for their deposition. There was a vigorous debate between those who argued that sudden, catastrophic events could explain extinction of species (which would be consistent with a lower age for the Earth) and those who argued for gradual change which required longer periods.[16]

Radioactivity provides a far more incisive method. Since each radioactive decay process has a defined half-life, it is possible to calculate ages of samples of (for example) minerals by their contents of radioactive isotopes. The first estimate was 2.2 billion years, revised in 1929 to 3.4 billion years.[17] However, using radiometric methods, we now believe the Earth has an age of 4.54 billion years.[18]

One of the objections raised to *On the Origin of Species* when it was published was that natural selection was based on the occurrence of random variations. How could random variations lead to the evolution of progressively more advanced species? Richard Dawkins summarized the involvement of randomness very well when he said, "There is random genetic variation and non-random survival and non-random reproduction."[19] In other words, natural selection is not at all random in the way it works to distinguish advantageous changes from disadvantageous.

To be sure, natural selection does not have an objective, as it were, of producing higher forms of life. It works simply to select the variation best suited to the current environment. In that context, the direction of natural selection will change if the environment changes. If environmental changes had not led to extinction of the dinosaurs, evolution would almost certainly have taken a different course. After the extinction of the dinosaurs (and other species), mammals evolved to fill the vacant niche. Otherwise humanity might never have arisen.[20] There is nothing inevitable about evolution.

∞ ∞ ∞

The geological time scale (GTS) records the history of the Earth in terms of the development of geological strata. The concept that the age of rocks can be determined by the decay of radioactive elements in them was first used in 1907 (relying on the decay of uranium). The discovery of further radioactive elements extended the precision of dating and now is an important technique in analyzing individual geological and evolutionary events.

Life appears to have originated more than 4 billion years ago, quite soon after the Earth was formed. The search to identify LUCA (last universal common ancestor), from which all living organisms are descended, suggest it was probably a bacteria-like cell that could be as old as 4.2 billion years.[21] The first eukaryotes (cells with a nucleus) evolved about 1.6 billion years ago, followed by multicellular organisms about 1.5 billion years ago. Animals have evolved over the last 800 million years or so.[22] Over that period, we know of five major extinction events that radically altered the course of evolution. These were due to major changes in the structure of the Earth.

The most severe extinction event took place at the boundary between the Permian period (300–250 million years ago) and the Triassic (250–200 million years ago). This eliminated the majority of living organisms. Geochronology suggests that a gigantic series of volcanic eruptions (known as the Siberian Traps Large Igneous Province or STLIP) occurred 252 million years ago, and lasted for 1–2 million years. There were dramatic changes in temperature on both land and sea (oceans may have increased 10–15 °C), acidification of the ocean, loss of oxygen from the seabed, and increases in atmospheric carbon dioxide and sulfur dioxide.

Any or all of these could have contributed to changing the conditions for life so dramatically as to lead to mass extinction. The extinction itself seems to have occurred over a relatively short period, about 60 000 years, so there is still doubt about the exact sequence of events (in particular why extinction on land appears to have started earlier), but there is general acceptance that the extinction was triggered by environmental change.[23]

The basic cause of the STLIP eruptions is thought to have been a mantle plume (hot rocks) rising from the Earth's core through the crust to the surface. This is an example—admittedly an extreme example, in fact perhaps the most extreme in Earth's history—of movements in the structure of the Earth. Volcanic eruptions have probably been responsible for other mass extinction events in the Earth's history.[24] The question now becomes what's responsible for these violent shifts. When the concept of surface movements was first proposed, it was widely derided.

Looking at a map of the globe, there is an obvious, if somewhat approximate, fit like pieces of a jigsaw puzzle between the eastern seaboard of the Americas and the coastline of the other side of the Atlantic. When Alfred Wegener proposed in 1912 that they might once have been joined together and later moved apart, the idea that continents could have mobility was ridiculed.

Wegener first presented the idea of continental drift in a lecture in 1912 to the German Geological Society in Frankfurt. When the lectures were published (three years later), he introduced the topic cautiously. "In spite of its broad foundation, I call the new principle as a working hypothesis, and would like to see it treated as such, at least until it has been possible to prove the continuation of these horizontal displacements in the present."[25]

The spur to developing the theory was the similarity between rocks and fossils on either side of the Atlantic. The 'broad foundation' drew upon data from multiple sources extending from geology to paleontology. They included the jigsaw-like fit, geological similarities in corresponding locations, evidence for past climatic events, and distributions of plants and animals consistent with an origin in a single landmass. The data were not new: it was the interpretation that was novel. (There is a parallel here with Darwin's theory of natural selection, which similarly was at least as much a matter of interpretation as presenting new information.)

Wegener did not in fact have a prior reputation in the field. He started as an astronomer and became a meteorologist. He made several expeditions to Greenland to monitor Arctic ice. He published a more complete account of his ideas in a book in 1915, *Die Entstehung der Kontinente und Ozeane* (The Origin of Continents and Oceans), with several revised editions following in later years as he responded to criticisms.[26] It was first translated into English in 1924.

Alfred Wegener on an expedition to Greenland in 1920.

Reproduced with permission from Archiv für deutsche Polarforschung.

The original lecture in Germany aroused a "storm of indignation," as Wegener described the reaction in a letter to his wife. When Wegener died on an expedition to Greenland in 1930, the theory was still held in universal scorn.

What were the objections? A symposium was held in New York in 1926 to consider the proposal for continental drift.[27] Opinion was generally negative. The main objections were that whatever forces acted on the continents could not be great enough to move them. The absence of any mechanism to explain continental drift made the theory easier to oppose.[28] The arguments against continental drift were also confused by lack of knowledge about the physics of possible interactions.

As Wegener continued to develop his theory, a major problem became a miscalculation of the rate of drift. Basing calculations

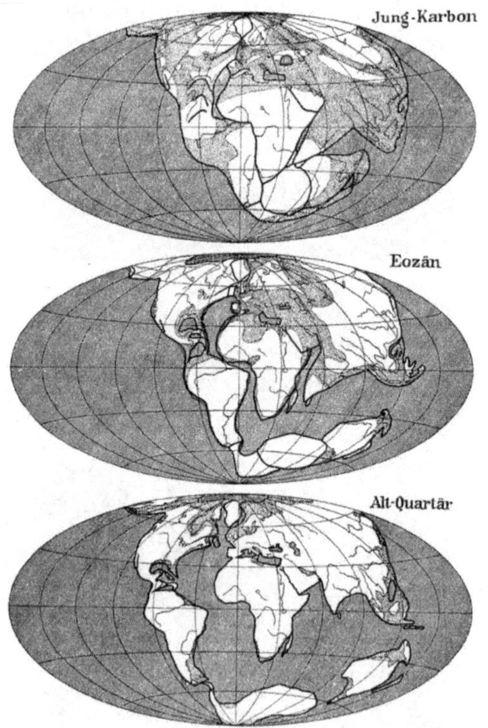

Wegener's original maps showing a single landmass (Pangea) about 300 million years ago, separating into continents by 50 million years ago, with the continents separating by the modern era.

on measurements that had been made in Greenland over the past century, Wegener calculated that Greenland was moving away from Europe at a rate of much as 14–28 m per year. This was so incredible as to lead to dismissal of the entire theory. (Based on erroneous measurements of longitude, Wegener's rate was 2–3 orders of magnitude too high, as we now know that the rate is about 1.5 cm per year). During the 1930s, without any active proponent, the theory became moribund.

When the protagonists were not exhausted by the inconclusive nature of the arguments, any debate on the subject tended towards the vituperative: more a matter of belief than a debate about data. The situation changed in the 1960s as new types of data became available. The fit between the landmasses on either side of the Atlantic became convincing when it was assessed not by the edge of the land but by the contour of the continental shelf at a depth of 100–200 m.[29] Evidence that the sea floor has a different character from the continental landmasses developed into the concept of sea-floor spreading.

A ridge in the middle of the Atlantic grows continuously, with a contour line reflecting a midpoint between the two continental outlines. And studies of the magnetism of rocks (paleomagnetism) provided convincing evidence of past connections between the continents. (Rocks bear a magnetic imprint that depends on the time and place of their creation, so the magnetism of a rock identifies its origins.) All this came together in a symposium organized at the Royal Society in London in 1965, after which continental drift became the mainstream hypothesis.[30]

Finally a mechanism emerged! The concept of sea-floor spreading extended into plate tectonics, with a paper published in 1967 proposing that the Earth is covered by rigid plates that float on its surface.[31] At an average rate of movement of 1.5 cm per year, any particular piece of land could move half way round the planet in a billion years.[32] Most (95%) of the Earth is covered by 7 tectonic plates; the remaining 5% is covered by about 10 smaller plates. Seismic events take place only at the junctions between the plates. The proposal for tectonic plates is widely regarded as the moment when a paradigm shift occurred.

The history of continental drift does not show science at its best. The blazing insights of the original theory were marred by

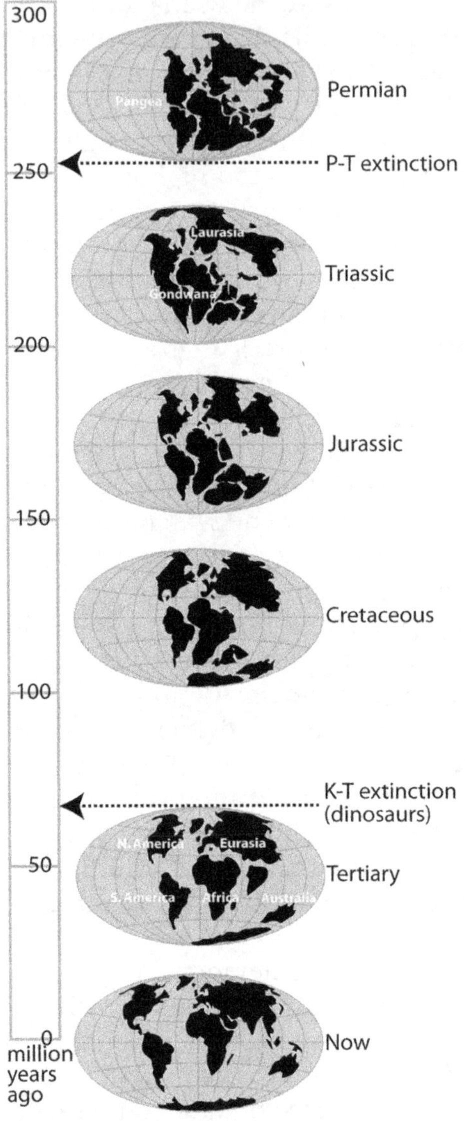

Current understanding of continental drift. Originally all the land mass was in one continent (Pangea). It divided into two landmasses and then gradually into 5 continents. Major extinction events occurred at boundaries between geological periods. The extinction of the dinosaurs occurred at the Cretaceous–Tertiary (K–T) boundary around 66 million years ago.

advocacy that relied on weak data. Opponents and proponents failed to distinguish theory and data. There were few attempts to look specifically beyond existing data to ask what types of new information might be obtained to test the theory. But the story also shows how theory ultimately is subservient to data. Granted, it took 50 years for convincing data to be made possible by new techniques, but then within only a couple of years, continental drift became the new paradigm.

Tectonic activity, meaning movement of one plate relative to another, is responsible for creating earthquakes or volcanic eruptions at the boundaries between plates (and indeed, over long periods of time, for forming mountain ranges). Of course, the mechanism by which it actually occurs is different from anything Wegener could have envisaged. Indeed, we still do not really understand the forces that propel the plates.

∞ ∞ ∞

Invoking extra-terrestrial causes for events on Earth has old, but very nonscientific, precedents, going back to the ancient view that meteors were omens. (A meteor is a streak of light in the sky; a meteorite is the object causing it.) It was not until the 20th century that meteorites were accepted as the cause of craters on both Earth and the Moon.

Previously craters were thought to be caused by some sort of volcanic activity. The very idea that stones could fall from the sky was regarded as incredible until the early 19th century when meteorites were identified as the result of a shower in France. The reality of meteorites was reinforced by analysis showing a distinctive mineral composition. But the idea they might be large enough to cause impact craters was incredible.[33]

One of the most dramatic impact craters is the Meteor Crater in Arizona, discovered in the late 19th century. Geologists thought it was caused by a volcanic steam explosion, but a mining engineer, Daniel Barringer, believed it was produced by a meteorite. He tried to recover the metal it should have contained. His failure to find it was one of the reasons the idea of a meteor impact was dismissed.

The discrepancy was explained when Harvey Nininger, a self-taught expert on meteorites, calculated that the force of the impact would have vaporized the meteorite. He wrote a book in 1956

The Meteor Crater in Arizona is 1.2 km across and 170 m deep and was created about 50 000 years ago. The meteorite was about 50 m diameter. It was mostly vaporized by its descent through the atmosphere followed by the impact.

that was influential in driving a change in geological opinion.[34] A few years later, the discovery of two forms of silica, coesite and stishovite, at the site confirmed there must have been an impact, as they are generated by high pressure and not by volcanic events.

It's now common wisdom that meteorites large enough to leave craters collide with the Earth.[35] The current list of known impact craters has almost 200 examples, with ages ranging from more than 1 billion years ago to the present, and diameters from 1 m to 150 km.[36]

One of the largest of the meteoric impacts was responsible for the extinction of the dinosaurs.[37] This was the extinction closest in time to us, the K–T extinction about 66 million years ago, but its cause has been by far the most contentious of any extinction event.

Dinosaurs had dominated the Earth for 160 million years when they disappeared about 66 million years ago. There had been at least 700 different species of dinosaurs. Their disappearance is defined by the K–T boundary, a thin layer of clay in the geological strata. Above the layer, the rocks represent the Tertiary period (now called the Paleogene); below it they represent the Cretaceous. Dinosaur bones are found only in rocks below the boundary; there are none above the boundary. In fact, the fossil record suggests that 70% of the organisms living in the Cretaceous period became extinct. (The K–T boundary is actually a relatively recent terminology.[38] In the 19th century, the earlier period (251–66 million years ago) was called the Age of Reptiles and the subsequent period the Age of Mammals.)

Over a century of trying to explain the extinction, all theories started from the basis that some intrinsic change in the Earth (such as climate) must have made conditions unfavorable. But given that the dinosaurs had survived successfully through a series of climatic and geographical changes for 160 million years, it was hard to see what this change might be. In sheer desperation, by 1979 all previous theories were dismissed.

The replacement was a proposal that that there must have been some extrinsic event in which lethal radiation reached the Earth from an exploding supernova in the galaxy. Sounds more like science fiction than science, although it was published in a most respectable scientific journal.[39] Unlikely though it seems, it was the right idea that there must have been an extrinsic event: but it was the wrong type of event.

The idea that an asteroid collided with the Earth, provoking a sudden and dramatic change in climatic conditions, came from a father and son team. Luis Alvarez was a physicist who had been awarded a Nobel Prize in 1968; he had no experience in geology. His son Walter is a geologist. They found an outcrop at Gubbio, Italy where the K–T boundary was exposed, and set out to measure how long it had taken to form the layer of clay.[40] None of the existing techniques had sufficient precision.

They turned to measuring iridium, an element that does not occur naturally in the Earth's crust, but which comes from meteorites. The supply of meteorites is more or less constant, so the amount of iridium provides a clock. When the Alvarez pair measured iridium levels, there was a surprise. The levels above and below the boundary were as expected for the age, but the layer itself had 30× more iridium. They confirmed that the same discrepancy applied to the K–T boundary at other locations. The idea of using iridium as a clock had failed, but there had to be an explanation for the concentration of iridium in the layer.

The source of the iridium had to be extraterrestrial. First they tested whether it could have resulted from the explosion of a

Luis and Walter Alvarez at a limestone outcropping near Gubbio, Italy, where they found high concentrations of iridium. The diagonal streak of white clay marks the K–T boundary. Below is white limestone, rich in fossils of a type that is completely missing from the red limestone above the boundary.

nearby supernova. However, other elements that would also have been expected were absent. This led to their proposal in 1980 that an asteroid had collided with the Earth. Dust released from the impact could have changed the climate so dramatically as to cause the mass extinction.[41] They calculated that the asteroid should have had a diameter of about 10 km, and the impact crater should have a diameter of 200 km.[42]

"We would like to find the crater produced by the impacting object," they said. There indeed was the major obstacle to accepting the theory. It was not the only obstacle. There was a concerted movement by many geologists to reject the outrageous idea of an extra-terrestrial origin for a terrestrial event. The likely impact site had actually been discovered in 1981, but no one had paid attention. It took another decade before the work was followed up.[43]

The Chicxulub crater is buried under the Yucatán peninsula in Mexico, with a center in the Gulf of Mexico. The time of impact has been dated with increasing precision since its discovery, using radioactive dating techniques. Now it is estimated to be 66.038 million years ago $\pm$ 11 000 years.[44] The same technique estimates the K–T boundary at 66.043 Mya $\pm$ 43 000 years.[45] The coincidence is well within the margin of error.

Of course, coincidence does not prove cause and effect. At a meeting of paleontologists in 1985, the vast majority said they did not believe that a meteoric impact had caused the extinction of the dinosaurs.[46] Opinion has now changed so that the meteoric impact is by far the majority view. A vociferous minority argues that the extinction started before the meteoric impact and that extreme volcanic action (known as the Deccan Traps eruption, which occurred between 66.3 and 65.6 million years ago) was responsible. (All the previous mass extinctions are probably associated with volcanic eruptions.)

The debate has been argumentative to the point that it has been called 'the nastiest feud in science.'[47] It highlights a problem of modern science. The debate hinges on intricate technical arguments about the exact dates of the Deccan Traps eruption *versus* the impact of the Chicxulub meteor, and the timing of extinction of the dinosaurs compared to the extinction of other species. The data depend on radioactive dating techniques that only experts understand, and arguments about whether samples are representative. Such disagreements between experts make it impossible for an outsider to assess the evidence. They create a general sense of disbelief in science. If the experts can choose their 'facts,' why shouldn't the public?

∞ ∞ ∞

It's difficult to explain the sheer exhilaration of using data to define distant events such as the creation and breakup of the landmass or the extinction of the dinosaurs, or for that matter the origin of the universe or the origin of life. It's equally difficult to explain how scientists who are apparently equally qualified can have violent differences about the reality or significance of data.

The debate may have made it more difficult for the public to accept the reality of scientific discoveries. In the reverse direction, public opinion about extra-terrestrial events has had an effect on science. Immanuel Velikovsky (a Russian–American psychoanalyst with no scientific qualifications) claimed in the 1950s and 1960s that many historical catastrophic events on Earth had been caused by near collisions with other planetary objects. This was enormously popular with the public, but widely derided by scientists. It's often cited as a prime example of pseudoscience (see Chapter 26).

Usually pseudoscience does not have much effect on the practice of science, but this was an exception. As a result of the clash between science and the public, the entire concept of extraterrestrial impacts became disreputable. Walter Alvarez said: "I considered [Velikovsky] part of the problem we faced in getting a hearing for the K–T impact hypothesis, because his ideas, which were incompatible with the laws of physics, had confirmed many geologists in their view that people working on extraterrestrial causes for events in Earth history were not doing good science."[48]

The history of the theory of evolution and its connection with geology, and the subsequent paradigm shifts concerning continental drift and dinosaur extinction, share an unusual pattern of scientific investigation. In no case did an investigator start out with a theory based on current knowledge that he wanted to test. In each case, the field started with an idea that seemed crazy in the context of the period. It took half a century for it to become accepted against objections that were irrational as well as rational. There was no support from the scientific establishment for what turned out to be major paradigm shifts.

Luis Alvarez mused on the implications. "I wonder how many physical concepts that 'everyone knows' to be true, as everyone knew parity conservation and the existence of the ether were true, would turn out to be false if experimentalists were again allowed to do nutty experiments. This impasse is not a problem only in physics. What would the peer reviewers in the National Science Foundation's paleontological section have said about a proposal from a Berkeley group to look for evidence that an asteroid or comet impact led to the extinction of the dinosaurs sixty-five million years ago?"[49]

The natural sciences—physics, chemistry, and (molecular) biology—are very much sciences of the present. They describe how the world works today. This means that our most important tool to investigate their workings is the experiment designed to test specific predictions by modifying conditions. Evolution and geology are more sciences of the past. They describe how the world came to be. It's not possible to modify conditions that existed millions of years ago, so observations to test the predictions made by theories become more important. Corroboration predominates over falsification. Indeed, it's generally true that scientists talk more about 'proving' a theory rather than trying to falsify it. The ultimate test of whether a theory is scientific is that it can in principle be falsified; but in practice scientists are more likely to try to corroborate it.

NOTES AND REFERENCES

1. C. Darwin, *On the Origin of Species by Means of Natural Selection, or, The Preservation of Favoured Races in the Struggle for Life*, John Murray, London, 1859.
2. A. Wegener, *The Origin of Continents and Oceans*, translated by J. Biram from the 4th German edition, Dover, New York, 1966, p. 5.
3. Quoted in M. W. Browne, *The Debate over Dinosaur Extinction Takes an Unusually Rancorous Turn*, New York Times, January 18, 1988, pp. C1–C4.
4. C. Darwin, *On the Origin of Species by Means of Natural Selection, or, The Preservation of Favoured Races in the Struggle for Life*, John Murray, London, 1859.
5. According to his autobiography, first published in 1887 after his death. See C. Darwin, *The Life and Letters of Charles Darwin, Including an Autobiographical Chapter*, ed. F. Darwin, John Murray, London, 1887. New edition edited by his granddaughter is C. Darwin, *The Autobiography of Charles Darwin: 1809–1882*, ed. N. Barlow, Norton, New York, 1993. Online at www.gutenberg.org/ebooks/2010.
6. *On the Origin of Species* is sometimes presented as appearing *deus ex machina*. Actually the age of the Earth, and the origin of Man, had been questioned previously in a book by an anonymous author. (He remained anonymous because he was concerned about the reaction.) *Vestiges of the Natural History of Creation,* published in 1844, argued that the Earth originated in a "fire-mist" and proposed something akin to evolution, without, of course, the detailed rationale of *On the Origin of Species*. The book was a success and actually was in it tenth edition before *On the Origin of Species* appeared in 1859. See J. A. Secord, *Victorian Sensation. The Extraordinary Publication, Reception, and Secret Authorship of Vestiges of the Natural History of Creation*, University of Chicago Press, Chicago, 2000.
7. E. J. Larson, *Evolution. The Remarkable History of a Scientific Theory*, Modern Library, New York, 2004.

8. H. Spencer, *Principles of Biology*, Appleton, New York, 1864, vol. 1, p. 444.

9. C. R. Darwin and A. R. Wallace, On the tendency of species to form varieties; and on the perpetuation of varieties and species by natural means of selection, *J. Proc. Linnean Soc.*, 1858, **3**, 45–62.

10. The first, second, and sixth (considered definitive) editions are online at www.gutenberg.org/files/1228/1228-h/1228-h.htm.

11. Quoted in A. S. Eve, *Rutherford: Being the Life and Letters of the Right Hon. Lord Rutherford, OM*, Cambridge University Press, Cambridge, 2013, p. 107.

12. G. B. Dalrymple, *Ancient Earth, Ancient Skies: The Age of Earth and its Cosmic Surroundings*, Stanford University Press, Stanford, 1994, p. 17.

13. Allowing for the rate at which heat is transferred from the center to the surface by conduction.

14. G. B. Dalrymple, *Ancient Earth, Ancient Skies: The Age of Earth and its Cosmic Surroundings*, Stanford University Press, Stanford, 1994, pp. 27–47.

15. G. B. Dalrymple, *Ancient Earth, Ancient Skies: The Age of Earth and its Cosmic Surroundings*, Stanford University Press, Stanford, 1994, pp. 59–69; A. Hallam, *Great Geological Controversies*, Oxford University Press, New York, 2nd edn, 1989, pp. 105–137.

16. A collateral issue was the age of the origin of mankind. Biblical theory would put this at some time after the creation of the Earth in 4004 BCE. In the same year that Darwin published *On the Origin of Species*, there was agreement that the discovery of hand-axes underneath the strata of extinct animals showed that humanity must have originated much earlier. Geologist Charles Lyell had agreed with Darwin about the age of the Earth but accepted a much older age for mankind only later. Lyell commented to Thomas Huxley that he was sorry he had not felt able "to go the whole orang." But in 1863 he published a book, *Geological Evidence for the Antiquity of Man* (John Murray, London) in which he reversed his position. The book was a best seller.

17. E. Rutherford, Origin of Actinium and Age of the Earth, *Nature*, 1929, **123**, 313–314.

18. G. B. Dalrymple, *Ancient Earth, Ancient Skies: The Age of Earth and its Cosmic Surroundings*, Stanford University Press, Stanford, 1994, pp. 355–356.

19. Richard Dawkins, in the Dawkins–Pell Debate, April 10, 2012, Australian Broadcasting Corporation.

20. Mammals and birds that fed on insects, worms, and snails, survived best after the extinction event. There was a large increase in the rate of evolution in the following period. Large mammals developed in the subsequent million years.

21. E. R. R. Moody, *et al.*, The nature of the last universal common ancestor and its impact on the early earth system, *Nat. Ecol. Evol.*, 2024, **8**, 1654–1666.

22. A. H. Knoll and M. A. Nowack, The Timetable of Evolution, *Sci. Adv.*, 2017, **3**, e1603076.

23. J. Dal Corso, Environmental crises at the Permian-Triassic mass extinction, *Nat. Rev. Earth Environ.*, 2022, **3**, 197–214.

24. D. P. G. Bond and S. E. Grasby, On the Causes of Mass Extinctions, *Palaeogeogr., Palaeoclimatol., Palaeoecol.*, 2017, **478**, 3–29.

25. A. Wegener, Die Entstehung der Kontinente, *Petermanns Geogr. Mitt.*, 1912, 185–195, 253–256, 305–309.

26. The first translation in 1924 was of the 3rd edition of 1922. A later translation was of the 4th (final) edition. See A. Wegener, *The Origin of Continents and Oceans*, translated by J. Biram from the 4th German edition, Dover, New York, 1966.

27. *Theory of Continental Drift: A Symposium on the Origin and Movement of Land Masses Both Inter-Continental and Intra-Continental, as Proposed by Alfred Wegener*, ed. W. A. J. M. van Waterschoot *et al.*, American Association of Petroleum Geologists, Tulsa, OK, 1928.

28. It remains controversial to what extent opposition was based on lack of mechanism, specific objections to other aspects of Wegener's theory, or a generalized resort to the traditional belief system. N. Oreskes, *The Rejection of Continental Drift: Theory and Method in American Earth Science*, Oxford University Press Inc, London, 1999; M. T. Greene, *Alfred Wegener: Science, Exploration, and the Theory of Continental Drift*, Johns Hopkins University Press, Baltimore, 2015.

29. E. Bullard, J. E. Everett, A. G. Smith, The Fit of the Continents around the Atlantic, in *A Symposium on Continental Drift*, ed. P. M. S. Blackett, E. Bullard and S. K. Runcorn, Philosphical Transactions of the Royal Society, 1965, pp. 41–51.

30. *A Symposium on Continental Drift*, ed. P. M. S. Blackett, E. Bullard and S. K. Runcorn, Philosphical Transactions of the Royal Society, 1965.

31. D. McKenzie and R. Parker, The North Pacific: an Example of Tectonics on a Sphere, *Nature*, 1967, **216**, 1276–1280.

32. In terms of geological time, 'drift' is a somewhat misleading description. For example, until 500 million years ago, the land that is now England was at the bottom of the southern hemisphere; from 350 Mya (million years ago) it migrated into the equatorial area, from 280 Mya it moved into the northern desert-belt, from 65 Mya it separated from Greenland, and it moved towards its present position from 22 Mya.

33. K. Mark, *Meteorite Craters*, University of Arizona Press, Tucson, AZ, 1987.

34. H. H. Nininger, *Arizona's Meteorite Crater*, American Meteorite Laboratory, Sedona, Arizona, 1956.

35. The Moon has many more craters than the Earth. This was originally taken to be an argument against a common origin by meteoric impact, because the larger size of the Earth should mean it has suffered more impacts. The cause of the difference is that because the Moon has no atmosphere, an impact crater remains indefinitely, whereas on Earth, it is reduced or eliminated by erosion.

36. Earth Impact Database. University of New Brunswick. web.archive.org/web/20130708132256/http://www.passc.net/AboutUs/index.html.

37. Whether it is the largest or second-largest depends on technical arguments about how to measure the diameter.

38. K was used to stand for Cretaceous because C was already in use to describe the Cambrian period. The approved term in geology for the boundary is now the K–Pg (Paleogene) boundary, because the use of Tertiary is discouraged.

39. D. A. Russell, The Enigma of the Extinction of the Dinosaurs, *Annu. Rev. Earth Planet. Sci.*, 1979, **7**, 163–182.

40. J. L. Powell, *Night Comes to the Cretaceous: Dinosaur Extinction and the Transformation of Modern Geology*, W H Freeman, New York, 1998.

41. There are no direct data from observations to describe the effect of a large asteroid hitting the Earth! Small meteorites burn up in the atmosphere: large ones have enough material to reach the Earth and create an impact crater. The effect depends on the point of collision. In an ocean (the most likely), collision could release large amounts of water vapor, creating a greenhouse effect that would warm the atmosphere by as much as 5 °C. A collision on land could release large amounts of aerosol-sized dust particles, resulting in an impact winter by absorbing or reflecting sunlight to cool the atmosphere, potentially by as much as 13 °C. Other effects of impacts could include release of sulfur dioxide or carbon dioxide gases, fireballs, and tsunamis. The climatic effects of the Chicxulub impact are controversial.

42. L. W. Alvarez, *et al.*, Extraterrestrial Cause for the Cretaceous-Tertiary Extinction, *Science*, 1980, **208**, 1095–1108.

43. A. R. Hildebrand, *et al.*, Chicxulub Crater; a Possible Cretaceous/tertiary Boundary Impact Crater on the Yucatan Peninsula, Mexico, *Geology*, 1991, **19**, 867–871.

44. This is one million years older than the first estimates.

45. P. R. Renne, *et al.*, Time Scales of Critical Events Around the Cretaceous-Paleogene Boundary, *Science*, 2013, **339**, 684–687.

46. M. W. Brown, Dinosaur experts resist meteor extinction idea, *New York Times*, October 29, 1985.

47. B. Bosker, The Nastiest Feud in Science, *The Atlantic*, September 2018.

48. Quoted in D. Morrison, Velikovsky at Fifty, *Skeptic (Altadena, CA)*, 2001, vol. 9, issue 1. Online at link.gale.com/apps/doc/A79626872/AONE.

49. L. W. Álvarez, *Álvarez: Adventures Of A Physicist*, Basic Books, New York, 1987, p. 200.

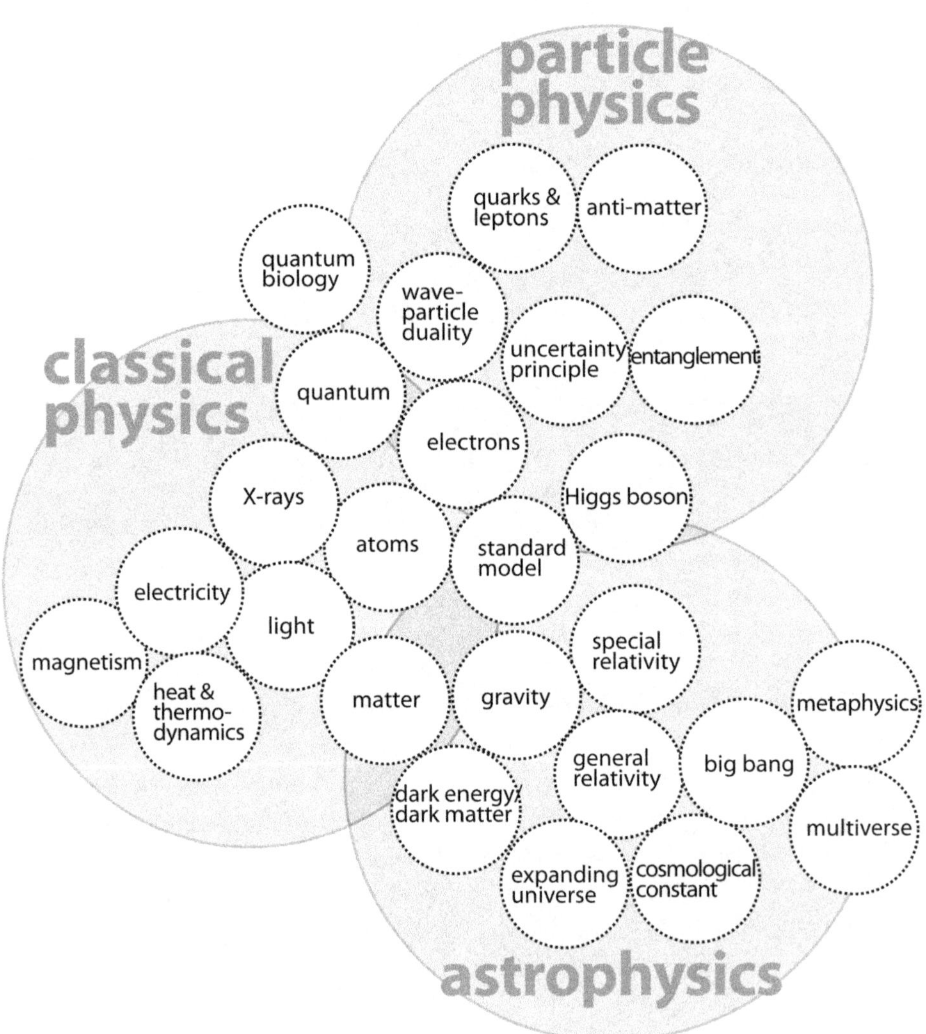

particle physics
classical physics
astrophysics
quantum biology
wave-particle duality
quarks & leptons
anti-matter
uncertainty principle
entanglement
quantum
electrons
X-rays
Higgs boson
atoms
standard model
electricity
light
magnetism
heat & thermo-dynamics
matter
gravity
special relativity
general relativity
big bang
metaphysics
dark energy/dark matter
multiverse
expanding universe
cosmological constant

Physical Sciences

It used to be supposed in science that if everything was known about the Universe at any particular moment, then we can predict what it will be through all the future. The idea was really due to the great success of astronomical prediction. More modern science however has come to the conclusion that when we are dealing with atoms and electrons we are quite unable to know the exact state of them; our instruments being made of atoms and electrons themselves. The concept then of being able to know the exact state of the universe then must really break down on the small scale. This means then that the theory which held that as eclipses etc. are predestined so were all our actions breaks down too. We have a will which is able to determine the action of the atoms probably in a small portion of the brain, or possibly all over it, Alan Turing, 1932.

The non-physicist finds it hard to believe that really the ordinary laws of physics, which he regards as the prototype of inviolable precision, should be based on the statistical tendency of matter to go over into disorder, Erwin Schrödinger, 1944.

CHAPTER 13

Rutherford's Atom: 1897–1945

"IF YOUR EXPERIMENT NEEDS STATISTICS, YOU OUGHT TO have done a better experiment," said Ernest Rutherford.[1] There was no single revolution, but atomic theory went through a series of paradigm shifts in the 20th century. Originally from New Zealand, known for believing that swearing at experiments made them work better, Rutherford was the person who started the ball rolling when he discovered that the atom has a dense nucleus, separate from its orbiting electrons. He followed this by discovering the proton; and then James Chadwick in his laboratory discovered the neutron. The first revolution in atomic structure revealed the atom as consisting of three discrete particles: protons and neutrons in the nucleus, with the much smaller electrons circling the nucleus. The atom was mostly empty space. This was the prequel to the discovery that the atom is a matter of statistics and probability rather than discrete particles.

The Frontiers of Science
By Benjamin Lewin
© Benjamin Lewin 2026
Published by the Royal Society of Chemistry, www.rsc.org

Timeline for Key Concepts in Defining the Existence of the Atom

450 BCE — Democritus "All matter is made of indivisible particles" (450 BCE)

1800

John Dalton shows atoms of each element are different (1804)

1810

Avogadro shows equal volumes of gases contain equal numbers of molecules (1811)
Berzelius introduces alphabetical symbols for elements (1813)

1820

1830

1840 — Faraday shows electricity generates positively and negatively charged ions (1839)

1850

1860

Maxwell defines kinetics of atomic movements in gas (1867)
William Thomson proposes atoms are a vortex in the ether (1867)
Kekulé questions whether 'chemical atom' is same as 'physical atom' (1867)
1870 — Mendeleev publishes periodic table (in Russian) (1869)
Boltzmann generalizes Maxwell's result into Maxwell-Boltzmann distribution (1871)

1880

1890

J.J. Thomson discovers electron (1896)
Ernst Mach "I don't believe that atoms exist" (1897)
1900

Einstein shows Brownian motion results from random movement of atoms (1905)
Jean Baptiste Perrin calculates sizes of atoms (1908)
Millikan measures charge on electron (1909)
1910 — Rutherford discovers the nucleus (1911)

A Dialog between John Dalton and Ernest Rutherford

John Dalton (1766–1844) was an English meteorologist who became interested in the properties of gases, which led him to chemistry. He developed the theory that matter consists of atoms.

Ernest Rutherford (1871–1937) was a physicist from New Zealand. Working in Manchester and Cambridge, he showed that the atom consists of a nucleus surrounded by electrons.

© Bettmann Archive/Getty Images.

John Dalton: Good day, Professor Rutherford. Your experiments have unexpectedly expanded our understanding of the atom. When I proposed that elements consist of indivisible particles called atoms, each with a unique weight, I couldn't have foreseen the intricate internal structure you've revealed.

Ernest Rutherford: Good day, Mr. Dalton. It's a pleasure to engage in conversation with the visionary mind behind the atomic theory. I too had thought the atom was a solid body, but as I delved into atomic structure, it became apparent that the atom is not as indivisible as once thought.

John Dalton: The nucleus was certainly an unexpected revelation. My atomic theory was an attempt to simplify the complexities of matter. Now it has become much more intricate.

Ernest Rutherford: Your atomic theory was a brilliant simplification. However, my experiments with alpha particles and gold foil demonstrated that the atom has a complex structure with a small, dense nucleus at its center and electrons orbiting around it. It's a miniature planetary system.

John Dalton: The dense nucleus—the heart of the atom. It's fascinating how science peels away layers of simplicity to reveal the underlying complexities. It's like an onion. The indivisibility of atoms, a notion I held dearly, had to yield. How did the scientific community respond to the revelation of the atomic nucleus and the concept that the atom is mostly empty space?

Ernest Rutherford: There was both excitement and skepticism. The idea of a tiny, dense nucleus at the center of the atom was a departure from traditional views. Yet, as more experiments confirmed the existence of the nucleus and its role in atomic structure, the scientific community gradually embraced this new paradigm.

John Dalton: I thought that with the atom we had reached the basis of all matter. I am astounded by the prospect of subparticles within the nucleus. Was there much resistance to the idea that the nucleus is a dynamic entity with a complex internal structure?

Ernest Rutherford: Not so much resistance as perplexity. The existence of subparticles raised questions about the stability of atomic structures. It seemed that we had reached an understanding that the nucleus has protons that balance the charges on the orbiting electrons, plus neutrons to increase its mass. But as we delve deeper into the subatomic realm, we continue to uncover new particles and forces. The exploration of the atom has become a way to explore the fundamental nature of the universe. The concept of the atom was only a start. I rest my case.

RUTHERFORD'S ATOM

I was brought up to look at the atom as a nice hard fellow, red or gray in colour, according to taste,[2] Ernest Rutherford, 1907.

The atoms or elementary particles themselves are not real; they form a world of potentialities or possibilities rather than one of things or facts,[3] Werner Heisenberg, 1958.

It is ironic that at the very end of the 19th century, just before the burst of tremendous discoveries that unlocked the structure of the atom, physicists were pessimistic about the future. "There is nothing new to be discovered in physics now. All that remains is more and more precise measurement," attributed to Lord Kelvin, was widely quoted at the time.[4]

The concept of the atom developed during the 19th century, but was controversial. Some people did not believe in atoms at all. Others believed that they were useful for describing chemical or physical reactions, but did not necessarily have physical reality. After the atom acquired physical reality at the start of the 20th century, the debate about reality shifted to sub-atomic particles.

Life was simple at the beginning, when the atom consisted of a nucleus, comprising just protons and neutrons, surrounded by a shell of electrons. Protons and neutrons had the same mass. The positive charge of the protons would match the negative charge of electrons orbiting the nucleus. Neutrons carry no charge but increase the mass of the nucleus. The atom was the major focus in physics for the first half of the 20th century. Around a third of the Nobel Prizes for physics were for defining the structure of the atom in greater and greater detail.

When John Dalton conceived the idea of the atom in 1803, he would have been surprised indeed to learn that two centuries later it would include 37 subatomic particles. This is roughly the same number as the number of elements known at the start of the 19th century (about 30). (Perhaps resurrected would be a better description than conceived, as, of course, the atomists in Greece had imagined the idea of the atom around 450 BCE, and the concept was not that far from the corpuscular theory of Descartes, Boyle, and others.)

John Dalton (1766–1844) came from an unlikely background to instigate a revolution in chemistry. Growing up in a rural environment in a family of cottage weavers, he owed his early education to the Quakers, and in due course he taught at a Quaker school. His professional career began when he moved to Manchester in 1793 to teach at the New College. He published his first book that year, *Meteorological Observations*. He became a member of the Manchester Literary and Philosophical Society, and his work was done in a laboratory there.

We don't know how Dalton arrived at his atomic theory, but his interest seems to have developed out of meteorology, from the issue of how water vapor behaves in the atmosphere.[5] This led him to the concept that each gas behaves independently in a

Dalton's table of elements in 1808. This and all subsequent tables of elements take the base to be hydrogen with an atomic mass of 1. The table got the order of weights generally right, but the details wrong, because it was based on the assumption that the simplest form of each element was always a single atom, e.g., oxygen was O rather than O_2.

© *Science Photo Library.*

mixture of gases. He decided this meant that each gas consists of a different type of particle. This became the controversial idea that each element consists of 'atoms' of a distinct weight.[6]

When Dalton then extended his ideas to other materials, he developed the rules that combinations of different elements always occurred in simple proportions, such as 1 atom of A + 1 atom of B or 1 atom of A + 2 atoms of B.[7] From this, it became possible to work out the relative weight of each element.[8] Dalton's table of elements in 1808 contained 20 elements.[9]

The key features of the theory were that:

- all matter consists of atoms;
- all atoms of a given element are identical;
- atoms of different elements have different weights;
- a compound has combinations of atoms in fixed proportions.

Chemistry began its modern era in 1869, when Dmitri Mendeleev (1834–1907) classified the known 60 elements into the periodic table. Elements were ordered in rows according to their atomic weights and placed in columns according to common properties (which occurred periodically, hence the name of the table). The periodic table was an example of an idea whose time had come. Several periodic systems were proposed during the decade of the 1860s.[10] Mendeleev's table was the most complete. Today the periodic table has 118 elements.[11]

Mendeleev's family came from Siberia, where his father was a teacher and his mother came from a local merchant family. His mother took him to Moscow in 1849 to enroll him in the University. When he was rejected, they continued to St. Petersburg, where he entered the Main Pedagogical Institute. From 1859–1861, he worked in Heidelberg, and then he returned to St. Petersburg to become an adjunct professor.

There's a story that Mendeleev saw the periodic table in a dream. According to his recollection, "I saw in a dream a table where all elements fell into place as required. Awakening, I immediately wrote it down on a piece of paper, only in one place did a correction later seem necessary." This has since been debunked.[14]

Mendeleev's career in St. Petersburg started with the publication of a textbook of organic chemistry. By 1864, he was a

ОПЫТЪ СИСТЕМЫ ЭЛЕМЕНТОВЪ.

ОСНОВАННОЙ НА ИХЪ АТОМНОМЪ ВѢСѢ И ХИМИЧЕСКОМЪ СХОДСТВѢ.

```
                          Ti = 50   Zr =  90    ? = 180.
                          V  = 51   Nb =  94    Ta = 182.
                          Cr = 52   Mo =  96    W  = 186.
                          Mn = 55   Rh = 104,4  Pt = 197,1.
                          Fe = 56   Rn = 104,4  Ir = 198.
                       Ni = Co = 59   Pl = 106,6  O- = 199.
   H = 1                   Cu = 63,4  Ag = 108   Hg = 200.
          Be =  9,4 Mg = 24  Zn = 65,2  Cd = 112
           B = 11   Al = 27,4  ? = 68   Ur = 116   Au = 197?
           C = 12   Si = 28   ? = 70    Sn = 118
           N = 14    P = 31  As = 75    Sb = 122   Bi = 210?
           O = 16    S = 32  Se = 79,4  Te = 128?
           F = 19   Cl = 35,6 Br = 80    I = 127
  Li = 7 Na = 23    K = 39  Rb = 85,4  Cs = 133   Tl = 204.
                   Ca = 40  Sr = 87,6  Ba = 137   Pb = 207.
                    ? = 45  Ce = 92
                  ?Er = 56  La = 94
                  ?Yt = 60  Di = 95
                  ?In = 75,6 Th = 118?
```

Д. Менделѣевъ

Mendeleev's periodic table in 1869. The title says: Draft of system of elements, based on their atomic masses and chemical characteristics. Question marks indicate missing elements or queries about the atomic weight.[18]

E-text prepared by Chris Curnow, Jens Nordmann, and the Online Distributed Proofreading Team (http://www.pgdp.net) from page images generously made available by Internet Archive (https://archive.org).

The International Congress of Chemistry

The International Congress of Chemistry held in 1860 in Karlsruhe, Germany, was the first international scientific conference ever held. A major motive was to get an agreement on a uniform system for measuring and reporting experimental results, which led into the issue of how to measure and report atomic weights.

At the conference, an Italian chemist, Stanislao Cannizzaro, made a plea to accept results obtained by Avogadro (1776–1856) in Turin in 1811 that a given volume of gas always contains the same number of molecules.[12] This had been ignored, but would make it possible to calculate atomic masses and proportions. The conference itself broke up with no clear conclusions, but Cannizzaro circulated a paper with his proposals that converted many of the participants when they read it later.[13] Dmitri Mendeleev was at the Congress.

tenured professor. The periodic table made its first appearance in 1869 in his textbook of inorganic chemistry, driven by the need to classify elements according to atomic theory.[15] It was intended more as a teaching tool than as a discovery of fundamental principles. A year later, Mendeleev published an article in Russian, extending the table to predict new elements. In 1871, he published it in German, now calling it a 'periodic law.'

Mendeleev left gaps in the table where his theory predicted elements were missing, and he argued that some atomic weights were incorrect. It was bold to take the side of theory over data. In some cases theory proved to be correct; in others the data stood. This was a work in progress. There were significant differences between the tables of 1869 and 1871, as further data were obtained.

Mendeleev became famous when the first two elements predicted by the law were discovered in 1875 and 1879.[16] Mendeleev then moved on, investigating gas laws in a search for the ether.

Dimitri Mendeleev in St. Petersburg.

In spite of the fame brought by the periodic law, Mendeleev was a controversial figure. The Academy of Sciences in St. Petersburg refused to elect him as a member in 1880. The Nobel Prize Committee refused to award him the prize for chemistry in 1906 on the feeble excuse that he had not predicted argon and the other inert gases.[17]

∞ ∞ ∞

The history of the atom tends to be presented as a direct line over a century from John Dalton's concept of the indivisible particle, to Mendeleev's characterization of the elements of the periodic table, to Rutherford's demonstration of the nucleus. But, in fact, the very concept of the atom was controversial through the second half of the 19th century. Ernst Mach (1838–1916), an eminent physicist in Vienna (whose name is commemorated in the unit of the speed of sound), electrified the audience at a meeting of the Imperial Academy in 1897 when he said, "I don't believe that atoms exist," following a lecture by Ludwig Boltzmann (1844–1906), a leading advocate of atomic theory.[19]

As it developed, the concept of the atom bridged chemistry and physics. Eventually it became a unifying force. Yet there was a contrast between the chemical and physical views of the atom. The smallest unit observed in chemical reactions (the 'chemical atom') was a useful concept for analyzing reactions, but did it actually exist independently in nature? The unit observed in physics (the 'physical atom') could be inferred from situations in which atoms were free to move, for example, in gases, but was not necessarily the same as the chemical atom.[20] Discussing the release of atoms in solution by electric currents, Michael Faraday said in 1839, "Though it is very easy to talk of atoms, it is very difficult to form a clear view of their nature."[21]

The absence of any direct evidence left scientists free to choose between theories (assuming they believed in the existence of the atom at all). Opinion was skeptical about the nature of the chemical atom. There was discussion about what attractive forces could link atoms together.[22] At one point, William Thomson (later Lord Kelvin) proposed that atoms were vortices in the ether. Belief systems colored which theories were favored.

From the perspective of physics, Maxwell defined the kinetics of movements of atoms in a gas in 1867 (see Chapter 9). Then in

1871, Ludwig Boltzmann refined Maxwell's equations and became the chief advocate of the atomic theory. An Austrian physicist, Boltzmann founded statistical mechanics (analyzing large-scale phenomena by the statistics of microscopic phenomena), but he was a sad case who suffered from bipolar disorder and later committed suicide. Mach's objections to Boltzmann were more philosophical than scientific. He simply did not see it as useful to focus on something so theoretical as an atom. It's an example of the differences in approach between experimental and theoretical physics.

∞ ∞ ∞

Science is supposed to advance on the basis of facts. Usually a paradigm breaks when the accumulation of facts becomes inconsistent with it. Yet sometimes the facts have accumulated over a protracted period and the spur is new theory. Apples had been falling off trees for centuries before Newton realized he could unite gravitational force on Earth with the movement of planets around the Sun.

How do such ideas originate? Perhaps dreams are the most extreme manifestation of a new theory taking over. Whether claims of dreams are true or imagined is beside the point. Here is an overt declaration that a new paradigm is not realized by force of logic, but by sudden inspiration. The act of discovery is distinct from the logical process by which science is supposed to proceed.

One of the most famous dreams in the history of science showed the structure of benzene, the prototype for an important series of aromatic compounds. Benzene contains

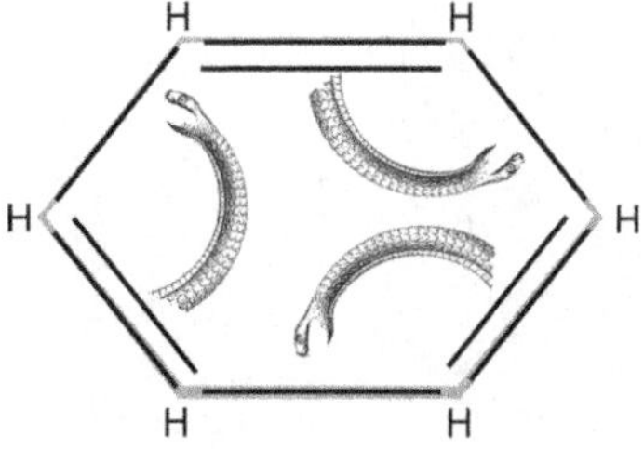

Kekulé envisaged benzene as a closed ring consisting of a flat hexagon with dynamically alternating double and single bonds.

6 carbon atoms, linked to one another and also to hydrogen atoms. In the mid-19th century, its structure was a dilemma. It was not obvious how to connect the carbon atoms to one another and to hydrogen. The great chemist August Kekulé (1829–1896) claimed he solved the structure in a dream (probably in 1862).

"I sank into half-sleep... the atoms fluttered before my eyes...Everything in motion, twisting and turning like snakes. But look, what was that? One of the snakes had seized its own tail, and the figure whirled mockingly before my eyes. I awoke as by a stroke of lightning, and this time, too, I spent the rest of the night working out the consequences of the hypothesis."[23]

Kekulé was describing a cyclical structure, with the 6 carbons organized in a closed ring. Written in terms of conventional bonds, each carbon makes 2 connections to another carbon on one side, and 1 connection to a carbon on its other side (and 1 connection to a hydrogen atom). This satisfied Kekulé's earlier proposal, in 1857, that carbon is tetravalent (makes 4 connections to other atoms). He published the benzene model in 1865. Although nominally the double and single bonds

August Kekulé in the year of the Benzolfest (1890). Working in Bonn, he was one of the most prominent chemists of the century.

alternate around the ring of carbon, they are actually in constant flux around the ring.

Kekulé told the story of the dream at the 'benzolfest' celebration of the 25th anniversary of the publication, just before he retired in 1890. He was modest enough to reject the 'Eureka' theory of creation and to argue that the formulation of the theory was inevitable. "[It's said that] the benzene theory appeared like a meteor in the heavens, absolutely new and unexpected... Gentlemen! The human mind doesn't think like that... Chemical ideas were floating around in the air and just found the right head... [they were] simply an expression of the knowledge [of the time]."[24] Others had been working towards similar ideas. The time was right, but none of that detracts from the individual act of genius. Kekulé concluded, "Let us learn to dream, gentlemen, and then perhaps we shall find the truth... But let us beware of publishing our dreams until they have been examined by the conscious understanding."

The cyclic structure for benzene was a powerful solution to the problem, but it was not obvious to what extent the model reflected physical reality. At the celebration in 1890, Kekulé's student, Adolf Baeyer (who went on to win a Nobel Prize in Chemistry) raised the issue of whether the ring describes the physical structure of benzene or is simply a useful visualization. The issue of reality *versus* useful fiction was to reverberate through the 20th century as the structure of the atom itself was analyzed (see Chapter 15).

Along those lines, even scientists who did not believe in the actual physical existence of atoms accepted that they were a useful device for analytical purposes. Just after he published the structure of the benzene ring, Kekulé said in 1867 that, "The question whether atoms exist or not has little significance from a chemical point of view: its discussion belongs to metaphysics... I do not believe in the actual existence of atoms... I rather expect that we shall some day find... a mathematico-physical explanation... for what we now call atoms... As a chemist, however, I regard the assumption of atoms, to be not only advisable, but absolutely necessary in chemistry."[25]

A decisive characterization of the atom came from an unexpected direction. In 1827, botanist Robert Brown observed the apparently random movement of pollen grains in water. Brownian motion came to describe the movement of particles suspended in a fluid medium. Albert Einstein developed a mathematical model in 1905 to show how the motion in a gas results from impacts between molecules. French physicist Jean Baptiste Perrin confirmed Einstein's model experimentally in 1908. He was able to use his data to calculate the sizes of atoms. There was no further argument. Perrin wrote a best-selling book called *Les Atomes* in 1913. When he was awarded the Nobel Prize in 1926, his acceptance lecture was devoted to the "discontinuous structure of matter."[26]

So now chemistry was based on the concept of atoms, with each element characterized by a distinct atomic weight, placed in a group of elements with related properties. This was not a revolution, because there was no prevailing paradigm to overthrow, but it was the start of modern chemistry. (If there was a death knell, it was for alchemy, because defining the elements as different types of atoms refuted the idea that they could be converted into one another.)

But the structure of the atom itself was indivisible and inscrutable. It was a century after Dalton before the nucleus was distinguished from the cloud of electrons surrounding it. Then the structure of the nucleus itself became steadily more complicated. Progress now was not so much a matter of revolutions, but more a series of increasing refinements.

∞ ∞ ∞

The first insight into the interior of the atom came from an advance in technology. Cathode ray tubes were made possible by vacuum pumps. A CRT consists of a glass tube with electrodes at either end. After the air has been pumped out to create a vacuum, a current is passed between the electrodes. The tube glowed. The rays passing from cathode (negative terminal) to the anode (positive terminal) were called *Kathodenstrahlen.* There was some dispute as to whether they were waves or particles.

X-rays were discovered at the end of the 19th century. Wilhelm Röntgen (1845–1923), working in Würzberg, Germany in 1895, discovered that cathode rays caused a shadow to be cast by an object placed in front of the cathode. The cathode rays were negatively charged, moved in a straight line, and were deflected by magnetic fields. Röntgen's paper had the title *Eine neue Art von Strahlen* (a new kind of ray).[27] He received the first Nobel Prize in physics in 1901.

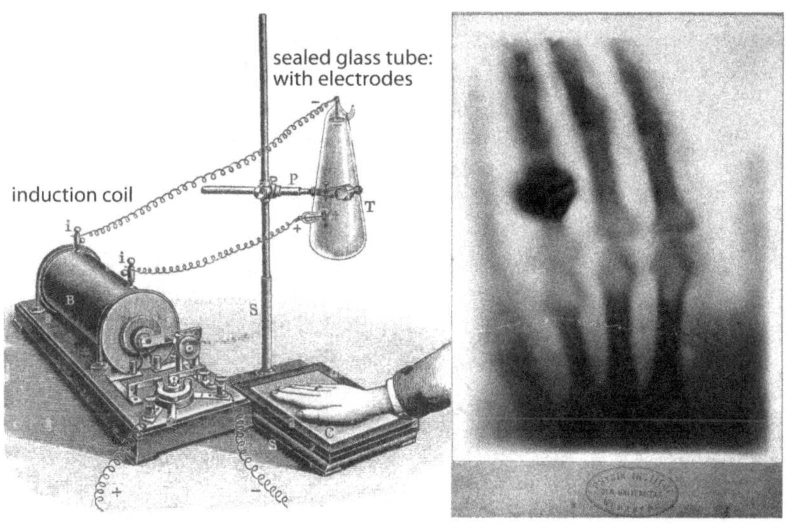

A 19th century illustration shows that Röntgen's apparatus (left) consisted of an induction coil to generate a high voltage current between the electrodes, in a sealed glass container under vacuum. The X-rays beam down. On the right is the first X-ray ever generated, showing Röntgen's wife's hand (with a ring on the second finger).

The scene now moved to the Cavendish Laboratory in Cambridge, where John Joseph Thomson (1856–1940) (always known as J. J. Thomson) was Cavendish Professor of Experimental Physics. Thomson performed the critical experiment in 1897 of showing that the cathode rays were deflected by an electric field as well as a magnetic field. (Thomson was famously clumsy in the lab. The experiment was probably actually performed by one of his technical assistants.) Thomson called the particles corpuscles. They were named electrons later.[28]

Thomson was not sure initially what role the electrons played in the atom, but he calculated from the deflection that the electron was a particle with approximately 1/1000 of the mass of a hydrogen ion.[29] In 1904, he suggested the 'plum pudding' model in which the atom had an even distribution of mass, with electrons embedded in the otherwise positive mass. "Atoms... consist of a number of negatively electrified corpuscles enclosed in a sphere of uniform positive

J. J. Thomson at the Cavendish with his apparatus in 1909.

© *Oxford Science Archive/Heritage Images/Science Photo Library.*

electrification," his paper began.[30] The model lasted for just over a decade. Thomson obtained the Nobel Prize in physics in 1906 for his work.

Ernest Rutherford (1871–1937) discovered the proton in 1911. Coming from New Zealand, he was awarded a scholarship to go to Trinity College, Cambridge, in 1894, under J. J. Thomson. He took up a chair at McGill University in Montreal, Canada in 1898, returned to England to a chair at Manchester in 1907, and then succeeded J. J. Thomson at the Cavendish in 1919.

Before he left Cambridge in 1898, Rutherford distinguished two sorts of rays emitted when uranium decays. α-particles are absorbed by a thin layer of gold foil (0.0005 cm thick), but β-particles are absorbed only by much thicker layers. He continued to work on α-particles. The experiment in 1911 took the form of measuring the effects when α-particles were shot at a foil that was thin enough (0.00004 cm) to let the particles pass through. If the mass of the atom is evenly distributed, all the α-particles should pass through.

α-particles are helium nuclei. Because they carry a positive charge, they are repelled if they hit a positively charged body. Instead of passing through the foil, some of the particles were scattered.

Ernest Rutherford at McGill University in 1905.

Indeed some bounced right back. As Rutherford later recalled, "It was quite the most incredible event that has ever happened to me in my life. It was almost as incredible as if you fired a 15-inch shell at a piece of tissue paper and it came back and hit you."[31]

The α-particles must have been hitting bodies that have a very high concentration of positive mass. To achieve the necessary concentration of positive charge, Rutherford suggested a 'planetary model,' in which the atom is mostly empty space, with electrons orbiting a very dense nuclear core. Again, the model lasted just over a decade. Rutherford had been awarded the Nobel Prize for chemistry in 1908 for his work on the disintegration of atoms.

The overall neutrality of the atom was explained by giving the nucleus a positive charge equaling the negative charge of the electron. Rutherford's work was interrupted by the First World War, but in 1919, he bombarded nitrogen with α-particles. The nitrogen breaks into oxygen by emitting a hydrogen nucleus. He called this a proton in 1920. It has a unit positive charge. Now it was clear that the charge of the nuclei of larger atoms was due to their content of protons. To achieve neutrality, the number of protons in the nucleus must be the same as the number of electrons around it.

The equality of electrons and protons led to a model in which every element has an atomic number—the number of protons or electrons. The mass of the hydrogen atom (*i.e.*, the proton) was assigned a value of 1. Then the mass of the protons in the nucleus would be the same as the atomic number. The problem was that the actual atomic mass was greater—usually about $2\times$ the

1803
John Dalton proposes indivisible (solid) atom

1897
J. J. Thomson discovers electrons and proposes 'plum pudding' model with electrons embedded in positively charged nucleus

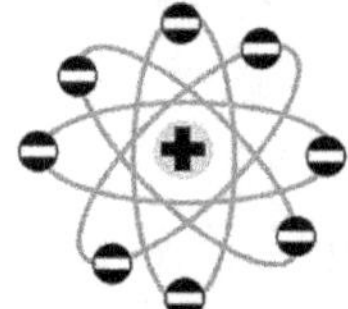
1911
Ernest Rutherford proposes 'planetary model' with positive charge localized in center, surrounded by orbiting electrons. In 1919 he identifies the positive charge with the proton.

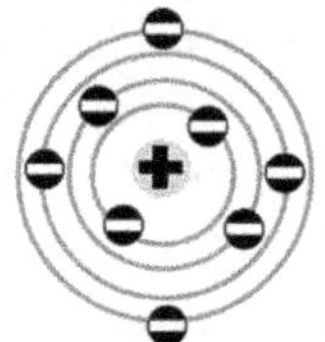
1913
Bohr introduces quantum theory by proposing electrons move in fixed orbits at different distances from nucleus

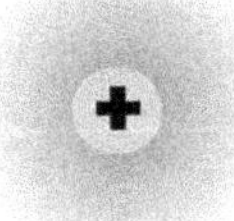
1926
Schrodinger replaces fixed paths with waves. Electrons occur in 'clouds of probability'

1932
James Chadwick discovers neutron: nucleus consists of positive protons and neutral neutrons

Atomic structure advanced by identifying individual components (electrons, protons, neutrons) and redefining electrons as waves instead of particles. The size of the electron is exaggerated relative to the size of the nucleus.[32]

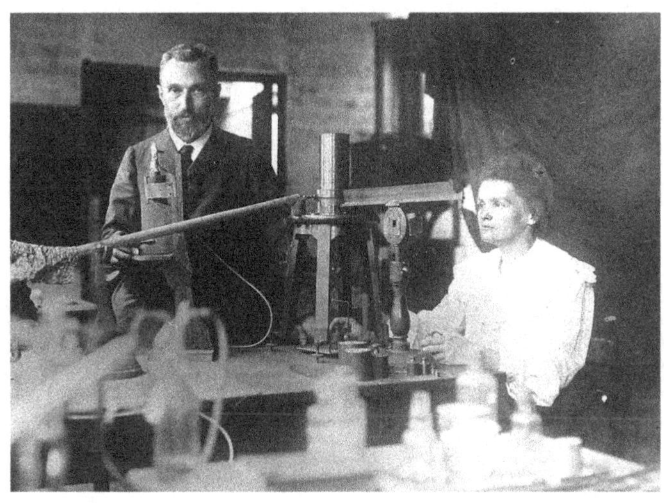

The Curies in their laboratory in Paris around 1900.

Wellcome Collection.

atomic number. Rutherford thought the difference might be explained by neutrons—uncharged particles with the same mass as the proton. He worked with James Chadwick (1891–1974) to find them, first in Manchester from 1911, and then at the Cavendish Laboratory in Cambridge from 1919.

Possibilities in physics were enlarged by radioactivity, which had been discovered in a famous accident in 1896. Henri Becquerel, a physicist in Paris, wondered whether elements that fluoresce naturally would mimic the production of X-rays by a cathode ray tube. Intending to investigate whether shining bright light on uranium would cause radiation like X-rays, he left some uranium salts on photographic plates locked in a drawer. Four days later, the plates had images created by radioactive waves. This led to the discovery that radioactivity was an intrinsic property of uranium.

Looking for other metals that might generate radioactivity, Marie Curie found that thorium was even more radioactive. In fact, she introduced the term 'radioactivity.' A wonderful allusion to its novelty comes from a remark attributed to her in response to the question as to why she took up the study of radioactivity. "Because there was no bibliography."[33]

Testing the crude ore from which uranium had been obtained, she found it was several times more radioactive than expected from its content of uranium. In 1898, she isolated the new element, which she called polonium. She underlined in her notebook that it was "150 times more active than uranium." Working on similar lines, Pierre Curie isolated another compound, 330 times more active than uranium, which the Curies called radium. They shared a Nobel Prize with Henri Becquerel in 1903 for the discovery of radioactivity. Marie obtained another on her own in 1911 for the discovery of polonium and radium. (Pierre had died in 1906, possibly from radioactive poisoning.)

The first evidence that one atom could turn into another came from Rutherford's experiments with Frederick Soddy in 1901–1902. Exploring the basis of radioactivity, they found that thorium (generated by the radioactive decay of uranium) breaks down to generate "something which, whatever its real nature may be, behaves in all respects like a radioactive gas."[34]

Recollecting the occasion fifty years later, Soddy said that he yelled, "Rutherford, this is transmutation: the thorium is disintegrating and transmuting itself into argon gas." Rutherford replied, "For Mike's sake, Soddy, don't call it transmutation. They'll have our heads off as alchemists!"[35] Rutherford usually avoided the term transmutation. He preferred to call the process transformation or disintegration. (Soddy went on to investigate the alternative atomic forms of an element, and was awarded a Nobel Prize in 1921 for his discoveries of isotopes.)

Splitting the atom started as pure science. The first explicit statement of the concept may have been in Rutherford's letter to Bohr in 1917. "I am also trying to break up the atom by this method. – Regard this as private." He succeeded when he showed that α-particles could split the nitrogen atom to release protons (see above), but it was clear at this point that a more powerful source of radiation would be needed to split larger atoms.

John Cockcroft and his student Ernest Walton in the Rutherford laboratory set out to build a more powerful generator. (Cockcroft was the theorist, Walton was the experimentalist.) The generator followed the earlier design of a Swiss physicist in transforming an AC input into a high voltage DC output. It could

Rutherford was known for his preference for small, improvised apparatus. His laboratory at the Cavendish in 1926, when he was working with James Chadwick on splitting the atom to find the neutron, has a distinct air of improvisation.

generate up to 600 keV. Cockcroft had worked at a company called Metrovick, which produced equipment for the national electricity grid. Metrovick provided the high voltage transformers that powered the apparatus. Some of the other equipment was hard to find, and construction followed the "string and sealing wax" approach that was typical of the Rutherford laboratory.

The observer sat in a lead-lined hut, using a microscope to count scintillations due to strikes by α-particles on a target of lithium or beryllium. The work started in 1929. Cockcroft and Walton obtained their first results in 1932, when they split a lithium atom to form two α-particles. The news flew around the world.

It was not long before heavier atoms were split. The start came when two researchers in Berlin showed that bombarding light

The Cockcroft–Walton generator occupied a corner of the laboratory. The investigator sits in the curtained compartment at the front.

elements such as beryllium (number 4 in the periodic table) with α-particles generated a highly penetrating form of radiation. They thought it could be high-energy photons (quantum particles of light).

Irène Curie (daughter of Marie and Pierre Curie) and her husband Frédéric Joliot followed this up in 1931 by showing that protons were generated when paraffin wax was exposed to the radiation. Their interpretation was that photons had expelled the protons from the paraffin. When Ernest Rutherford at the Cavendish heard about the report, he burst out "I don't believe it," according to the recollections of his collaborator James Chadwick.[36] Rutherford and Chadwick doubted that photons could be powerful enough to eject protons. They thought that the radiation must be the neutrons they had been trying to find for the past twenty years.

(The Joliot-Curie's fame rests upon another discovery. A generation after Marie and Pierre Curie discovered natural radioactivity, the Joliot-Curies generated artificial radioactivity. When they bombarded aluminum foils with α-particles in 1934, they discovered that radiation continued to be emitted from foils after the bombardment stopped.[37] The aluminum had been converted into a radioactive isotope of phosphorus.[38] The Joliot-Curies were awarded a Nobel Prize in 1935.)

Soon after hearing about the Curie's experiment, in 1932 Chadwick repeated their experiment. He detected radiation consisting of particles with the mass of the proton but no charge: the neutron. Initially Chadwick viewed the neutron as the combination of a proton and an electron. By 1936, it was realized that it is fact a distinct *elementary particle*, like the proton and electron.[39] Its mass is actually a little greater than the combination of proton and electron. Chadwick was awarded the Nobel Prize for physics in 1935.

It's a moot point whether the first definitions of the atom fall into chemistry or physics. Ernest Rutherford is famously quoted as saying around the start of the 20th century that, "Physics is the only real science. The rest are just stamp collecting."[40] When Rutherford was awarded the Nobel Prize for Chemistry in 1908, he had the grace to say, "I am very startled at my metamorphosis into a chemist."[41] But as the number of sub-atomic particles increased, the field became known as particle physics.

∞ ∞ ∞

Now we move into the realm of paradoxes, but first we have to go back to the start of the 20th century. Max Planck (1858–1947) was not a revolutionary. In fact, he was somewhat of a reactionary in his approach to physics—he took a long time to believe in atoms—but he was forced by his calculations to introduce a revolutionary concept. His own description was that "what I did can be described simply as an act of desperation."[42]

Always dressed formally, Max Planck was the very image of a German civil servant. Coming from an academic family, he studied physics at the University of Munich, but was advised in 1878 by his professor not to go into physics because "almost everything is already discovered, and all that remains is to fill a few holes."[43] He worked on thermodynamics, first as a professor at the University of Kiel, then in Berlin (the top position in Germany for a theoretical physicist).

Planck was brought to his act of desperation by the blackbody problem. When a body is heated, the color of the light it emits is determined *solely by its temperature.* It does not matter what it is made of. This was hard to explain by the laws of classical physics.

When iron is heated, such as a horseshoe, for example, first it glows dull red, then a brighter red, and then white. The light starts at longer wavelengths and then shorter wavelengths are added as the temperature rises. White heat is the addition of all wavelengths.

Physicists wanted to investigate the relationship between temperature and light using a blackbody, a container with completely black walls inside. The point of the blackbody is that it must be in equilibrium between the radiation it absorbs and the radiation it emits. So we can draw an equivalence between the radiation it receives (in the form of heat) and the radiation it emits (in the form of light). The blackbody problem was first posed as a theoretical question in 1859. Blackbodies do not exist in nature (although some materials come close). It was almost the end of the century before blackbodies became an experimental reality.

The blackbody apparatus in 1898 consisted of a platinum sheet, 0.01 mm thick and 40 cm long, rolled into a cylinder 4 cm in diameter. Both ends had rings for electric heating up to ~1500 °C. The inner surface of the tube was blackened with a mixture of chromium, nickel, and cobalt oxide. It was insulated with an outer layer of fireproof material.

Reproduced from https://doi.org/10.1140/epjh/s13129-022-00044-x, under the terms of the CC BY 4.0 license, https://creativecommons. org/licenses/by/4.0/.

From the first experiments in 1893, Wilhelm Wien, a German theoretical physicist, derived a law showing how the wavelength of the light decreases as the temperature increases. He was awarded the Nobel Prize for physics in 1911. Two experimental physicists in Germany, Lummer and Pringsheim, tested Wien's law experimentally in 1898 with a more sophisticated black body. A problem arose when the law turned out to break down at long wavelengths, in the infrared spectrum.

Another problem occurred at short wavelengths. The British physicist Lord Rayleigh derived a law, based on classical theory, to account for the shape of the spectral curve. The problem was that the law predicted the output of infinite energy when the wavelength becomes very low! This later became known as the 'ultraviolet catastrophe.'

Planck set out to prove Wien's law on theoretical grounds and to account for the discrepancies.[44] Up to this point, it had been assumed that the wavelength of radiation varied continuously. To make his model work, in 1900 Planck was forced to assume that instead energy was emitted as discrete packets. These were later called quanta.[45] Planck regarded the quantum as a formal mechanism to make the theory work. He never really became reconciled to its physical reality.[46] Planck's Nobel Prize came a few years later in 1918.

∞ ∞ ∞

The next stage came when Bohr developed quantum theory in 1913. Niels Bohr (1881–1962) came from a wealthy academic family in Copenhagen. After obtaining his doctorate, Bohr visited England in 1911, spending time at the Cavendish and in Manchester. Returning to Denmark in 1912, he joined the University of Copenhagen. In 1913, he published three papers, later known as 'the Trilogy,' in the *Philosophical Magazine* of London, which used Max Planck's quantum theory to create a quantum model for the atom. Bohr was awarded the Nobel Prize for physics in 1922. He became one of the most prominent physicists of the century.

Bohr proposed that electrons could orbit the nucleus only at certain distances, in what he called 'stationary states,' meaning that they were stable and did not emit radiation. Bohr's model proposed that these states corresponded to quantum intervals. This means that in each orbit, the electron has an energy

Niels Bohr (left) and Max Planck (right) in Copenhagen in 1930.

Niels Bohr Archive, Copenhagen.

corresponding to an integral number of quanta. An electron can move to a higher-energy orbit by gaining a quantum. It emits a quantum if it moves to a lower-energy orbit. Quantum theory was the beginning of a century of revolutions in physics.

The theory also implied that it is impossible to define the position of an electron during a jump. The transition between orbits is instantaneous. This defied the laws of classical physics. The arrangement of the orbits, especially the properties of the outer orbit, explained the properties of the groups of elements in the periodic table (as defined a half century earlier). This was the moment when physics became united with chemistry.

Electrons surround the nucleus in a series of concentric shells. Each layer can hold only a certain number of electrons. Once a layer is filled, electrons begin to fill the next layer. Elements with the same number of electrons in the outer layer have similar properties. For example, hydrogen, lithium, sodium, potassium all have one electron in the outer layer. If they 'lose' it, then the atom has a net positive charge of 1. Fluorine, chlorine, bromine, are short of one electron in the outer layer. If they gain it, the atom has a net negative charge of 1. The inert elements, helium, neon, argon, *etc.* are inactive because they have complete outer layers.

While explorations of the nucleus led farther and farther inward towards finding more and more fundamental particles, the view of quantal electron orbits led to definition of the bonds that bind atoms into molecules. American chemist Gilbert Lewis published a model for the role of electrons in 1916. Irving Langmuir extended this in 1921 into characterizing types of electron rearrangements.[47] In simplistic terms, if sodium donates its single electron from its outer layer to fill up the outer layer of chlorine, the two atoms are joined together into sodium chloride—common salt.

The great American chemist Linus Pauling (1901–1994) achieved the synthesis of chemistry and physics when he published his book *The Nature of the Chemical Bond* in 1939.[48] Pauling was one of a handful of people to be awarded two Nobel Prizes, the first in 1954 for his work in chemistry, followed by the Peace Prize in 1962 for his work as an activist. His activism led to the State Department confiscating his passport, with the result that he could not keep fully abreast of scientific developments in Europe in the early 1950s. Watson and Crick regarded him as their great rival in competing to define the structure of DNA.

Pauling's lifelong interest in chemistry had started when he was a teenager. His career was spent extending the application of

Linus Pauling lecturing in 1957 at Caltech.

quantum mechanics to chemical bonds. By 1931, he was at Caltech, when he published a seminal paper showing how the outer electron orbits of two atoms could merge to form a bond that linked them together. This built on a model proposed by two post-doctoral fellows in the Schrödinger laboratory, who had published the first application of quantum mechanics to chemical bonding in 1927.[49] The model argued that while two nuclei would repulse one another, electrons could jump between them to create an attractive force. Chemical bonds were now explained in terms of quantum mechanics.[50]

Bohr's model for the nucleus in 1913 was the point at which the balance shifted from experimental to theoretical physics. Defining the trio of subatomic particles of the electron, proton, and neutron had been driven by experiments. Defining their relationships, and extending the number of particles, was to be driven by theory.

The division between theoretical and experimental physics is unique in science. Other areas of science are basically experimental. Theories are usually developed by the experimenters to explain their experiments. In physics, theoreticians often have to wait a considerable time for an experiment to confirm (or refute!) their predictions. Sometimes the technology to test the theory simply does not exist at the time when they conceived it.

NOTES AND REFERENCES

1. Quoted in N. T. J. Bailey, *The Mathematical Approach to Biology and Medicine*, John Wiley, London, 1967, p. 23.
2. Quoted in E. N. Do Casta Andrade, The Rutherford Memorial Lecture, *Proc. R. Soc. London, Ser. A*, 1957, **244**, 437–455.
3. W. Heisenberg, *Physics and Philosophy*, Harper [Reprint Penguin Classics, 2000], New York, 1962.
4. S. Weinberg, *Dreams Of A Final Theory: The Scientist's Search For The Ultimate Laws Of Nature*, Vintage, New York, 1992, p. 13.
5. F. Greenaway, *John Dalton and the Atom*, Heinemann Educational Books, London, 1966.
6. T. H. Levere, *Transforming Matter. A History Of Chemistry From Alchemy To The Buckyball*, Johns Hopkins University Press, Baltimore, 2001, pp. 80–87, 107–116.
7. This is known as Dalton's law of multiple proportions.
8. First proposed in a paper read to the Manchester Literary and Philosophical Society on October 21, 1803, when he said: "An inquiry into the relative weights of the ultimate particles of bodies is a subject, as far as I know,

entirely new; I have lately been prosecuting this inquiry with remarkable success."

9. J. Dalton, *A new system of chemical philosophy*, R. Bickerstaff, London, 1808. Online at digital.sciencehistory.org/works/ff365590j.

10. English chemists John Newlands and William Olding made proposals in 1863 and 1864. The closest comparison to Mendeleev was the table of Lothar Meyer at Karlsruhe in 1864, but this had only 28 elements.

11. This completes the first 7 rows of the table, although the properties of some of the heavier elements have yet to be confirmed. It's unclear whether there may be any further (heavier) elements.

12. More precisely, Avogadro's law states that that equal volumes of gases at the same temperature and pressure contain the same number of molecules regardless of their chemical nature and physical properties.

13. D. Knight, *Atoms and Elements*, Routledge, New York, 1967.

14. For an account of the development of periodic tables see A. Greenberg, Mendeleev's "problems:" A Means To Engage Students And Teachers in the History Of Chemistry, *Bull. Hist. Chem.*, 2022, **47**, 15–28.

15. M. D. Gordin, *A Well-ordered Thing. Dmitrii Mendeleev and the Shadow of the Periodic Table*, Princeton University Press, Princeton, NJ, 2nd edn, 2019, pp. 11–43.

16. His article was then translated into French (La Loi Périodique des Éléments Chimiques, *Le Moniteur Scientifique de le Dr. Quesneville*, 1869, **21**, 691–737) and English.

17. M. D. Gordin, *A Well-ordered Thing. Dmitrii Mendeleev and the Shadow of the Periodic Table*, Princeton University Press, Princeton, NJ, 2nd edn, 2019, p. 203.

18. The table was published in Mendeleev's Principles of Chemistry, published in Russian in 1869. The first English translation was in 1891 (from New York).

19. D. Lindley, *Boltzmann's Atom: The Great Debate That Launched a Revolution in Physics*, Free Press, New York, 2016.

20. M. J. Nye, *Before Big Science. The Pursuit Of Modern Chemistry and Physics, 1890–1940*, Twayne Publishers, New York, 1996, pp. 28–56.

21. M. Faraday, *Experimental Researches in Electricity*, R & JE Taylor, London, 1839, vol. 1, p. 265.

22. D. Knight, *Atoms and Elements*, Routledge, New York, 1967.

23. From Kekulé's recollections at the celebration in 1890. Quoted in A. J. Rocke, *Image and Reality. Kekulé, Kopp, and the Scientific Imagination*, University of Chicago Press, Chicago, 2010, p. 194.

24. Kekulé's speech on the occasion was summarized in a contemporaneous publication and later included in a biography: R. Anschutz, *August Kekulé. Vol 2, Abhandlungen, Berichte, Kritiken, Artikel, Reden*, Verlag Chemie, Berlin, 1929, pp. 939, 942.

25. A. Kekulé, On some points of chemical philosophy, *The Laboratory*, 1867, **1**, 303–306; reprinted in R. Anschutz, *August Kekulé. Vol 2, Abhandlungen, Berichte, Kritiken, Artikel, Reden*, Verlag Chemie, Berlin, 1929.

26. J. P. Perrin, Discontinuous Structure of Matter, *Nobel Prize in Physics*, 1926. Online at www.nobelprize.org/prizes/physics/1926/perrin/lecture/.

27. W. C. Röntgen, Ueber eine neue Art von Strahlen, *Ann. Phys.*, 1898, **300**, 12–17.

28. The term electron had been introduced in 1891 to describe the unit of electric charge. It was applied to Thomson's corpuscles after his paper was published.

29. It's actually 1/1836 the mass of a proton.

30. J. J. Thomson, On the structure of the atom: an investigation of the stability and periods of oscillation of a number of corpuscles arranged at equal intervals around the circumference of a circle; with application of the results to the theory of atomic structure, *Philos. Mag.*, 1904, **7**, 237–265.

31. In a lecture in 1936, quoted in E. N. Do Casta Andrade, *Rutherford and The Nature Of The Atom*, Doubleday, New York, 1964, p. 111.

32. J. J. Thomson, Cathode rays, *Philos Mag*, 1897, **44**, 293–316; E. Rutherford, The scattering of α and β particles by matter and the structure of the atom, *Philos. Mag.*, 1911, **21**, 69–688; N. Bohr, On the constitution of atoms and molecules, *Philos. Mag.*, 1913, **26**, 1–24; E. Schrödinger, An undulatory theory of the mechanics of atoms and molecules, *Phys. Rev.*, 1926, **28**, 1049–1070; J. Chadwick, The existence of a neutron, *Proc. R. Soc. London, Ser. A*, 1932, **136**, 692–708.

33. Quoted in G. Holton, *The Advancement of Science, and Its Burdens*, Cambridge University Press, Cambridge, 1986, p. xvi.

34. E. Rutherford and F. Soddy, The radioactivity of thorium compounds. I. An investigation of the radioactive emanation, *J. Chem. Soc.*, 1902, **81**, 321–350.

35. Quoted in M. Howorth, *Pioneer Research on the Atom. Rutherford and Soddy in a Glorious Chapter of Science; the Life Story of Frederick Soddy*, New World Publications, London, 1958.

36. M. Longair, Rutherford and the Cavendish Laboratory, *J. R. Soc. N. Z.*, 2021, **51**, 444–466.

37. F. Joliot and I. Curie, Artificial production of a new kind of radio-element, *Nature* , 1934, **133**, 201–202.

38. This was an observation that could have been made earlier by the Cavendish or by the Berkeley cyclotron if the equipment had been configured appropriately. M. Hiltzik, *Big Science. Ernest Lawrence and the Invention That Launched the Military-Industrial Complex*, Simon & Schuster, New York, 2015, pp. 123–128.

39. S. Weinberg, *The Discovery Of Subatomic Particles*, Cambridge University Press, Cambridge, 2nd edn, 2003.

40. Attributed to Rutherford in J. D. Bernal, *The Social Function of Science*, George Routledge, London, 1939, p. 9. The exact occasion on which he may have said it has never been identified.

41. Quoted in A. S. Eve, *Rutherford. Being the life and letters of the Right Hon. Lord Rutherford, O.M.*, Cambridge University Press, Cambridge, 2013, p. 183. Letter from Rutherford to Hahn, December 22, 1908.

42. Quoted in A. Hermann, *The Genesis Of Quantum Theory*, The MIT Press, Cambridge, MA, 1971, p. 23. Letter from Planck to Robert Williams Wood, October 7, 1931.

43. The anecdote is widely quoted in this form, but may be a popularization. The source may be a recollection in a lecture that Planck gave in 1924 according to a letter from Friedrich Katscher in *Sci. Am.*, 1996, **274**(2), 10.

44. M. Kumar, *Quantum: Einstein, Bohr and The Great Debate About The Nature Of Reality*, Icon Books, London, 2008; A. Lightman, *The Discoveries: Great Breakthroughs In 20th-Century Science, Including The Original Papers*, Vintage, New York, 2006, pp. 3–15.
45. Planck expressed the assumption in the equation $E = h\nu$, where E (the quantum) is the product of a constant, h, now known as Planck's constant, and ν is the frequency of the radiation. This is probably the second-most famous equation in physics.
46. The anguish of uncertainty that was to be created by quantum mechanics is well summarized by the quotations from Turing and Schrödinger at the beginning of the Physical Sciences section.
47. T. H. Levere, *Transforming Matter. A History Of Chemistry From Alchemy To The Buckyball*, Johns Hopkins University Press, Baltimore, 2001, pp. 171–175.
48. L. Pauling, *The Nature of the Chemical Bond and the Structure of Molecules and Crystals; An Introduction to Modern Structural Chemistry*, Cornell University Press, NY, 1939.
49. The Heitler–London theory shows how the wave functions of two atoms join together to form a covalent bond.
50. Pauling's model is called the valence bond (VB) model. Subsequently it was somewhat replaced by the molecular orbital (MO) model, a more powerful theory that visualizes bonds in the geometry of the molecule, proposed by Robert Mulliken, who was awarded a Nobel Prize in 1966. However, Pauling refused to include molecular orbital theory, even in later editions of his book when MO was replacing VB. This was often felt to hold back the field. See S. Shaik, D. Danovich and P. C. Hiberty, Valence bond theory—its birth, struggles with molecular orbital theory, its present state and future prospects, *Molecules*, 2021, **26**, 1624–1648.

Bohr's Quantum: 1926–

"HOW WONDERFUL THAT WE HAVE MET WITH A PARADOX. Now we have some hope of making progress," said Niels Bohr with characteristic irony.[1] The quantum view of the atom began with Bohr. Coming from Denmark, he worked with Rutherford, who although famously skeptical about theoreticians, supported him. Max Planck had previously introduced the idea of the quantum, and Einstein had extended it from matter to light, but Bohr's proposal that electrons can adjust their orbits only by quantum leaps was revolutionary. The next revolution was Schrödinger's description of electrons as waves instead of particles. This led to recognition of the wave–particle duality. The rest of the 20th century was marked by the realization that the apparently fundamental particles of the nucleus, the proton and the neutron, each actually consisted of a series of sub-particles. The wave–particle duality created a baffling series of paradoxes. Following revolutions tended to be driven by theoreticians, and massive team efforts were required to identify the sub-particles experimentally.

The Frontiers of Science
By Benjamin Lewin
© Benjamin Lewin 2026
Published by the Royal Society of Chemistry, www.rsc.org

Timeline for Defining Sub-atomic Particles

1900 — Thomson discovers electrons (1897)
Planck introduces quanta (1900)

1905 — Einstein proposes photons are quantal (1905)
Einstein explains Brownian movement by atoms (1905)

Perrin calculates size of atom (1908)
Millikan measures charge on electron (1909)
1910 — Rutherford discovers nucleus (1911)

Bohr proposes quantum theory (1913)

1915 — } FIRST WORLD WAR

1920 — Rutherford discovers proton (1919)

1925 —

Schrödinger treats electrons as waves (1926)

Heisenberg proposes Uncertainty Principle (1927)

Dirac proposes existence of antimatter (1928)

1930 — Pauli proposes neutrino (1930)

Chadwick discovers neutron (1932)
Cockcroft & Walton split the atom (1932)

Enrico Fermi splits uranium (1934)

1935 —

Pauling defines bonds between atoms (1939)
Lise Meitner describes fission (1939)

1940 — } SECOND WORLD WAR

1945 — Manhattan Project concludes with atom bomb (1945)

A Dialog between Dmitri Mendeleev and Niels Bohr

Dmitri Mendeleev (1834–1907) spent his career in St. Petersburg. When writing a textbook of chemistry, he realized that he could organize elements into groups with related properties. This became the periodic table.

Niels Bohr (1881–1962) was a Danish physicist, working in Copenhagen after a period in England when he developed the concept that electrons can exist only at specific quantum levels. This was the start of quantum mechanics.

Dmitri Mendeleev: Good day, Dr. Bohr. It's a pleasure to meet you. It came to me in a dream that I should organize elements based on their properties. My periodic table has revolutionized chemistry and predicted the existence of new elements.

Niels Bohr: Good day, Professor Mendeleev. It's an honor to engage in a conversation with the architect of the periodic table. Your contributions to chemistry provided a remarkable framework for understanding the relationships between elements. This laid the groundwork for our exploration of atomic structure.

Dmitri Mendeleev: Your work on the structure of the atom has brought about a revolution in our understanding of the fundamental building blocks of matter. How do you perceive the relationship between the periodic table and the atomic model?

Niels Bohr: As I peered into atomic structure, I was able to explain the patterns in the periodic table by arranging electrons in discrete energy levels. My atomic model was an attempt to bring a new level of understanding to the periodicity of elements.

Dmitri Mendeleev: In my time, the periodic table reflected observations. Its predictive power was a testament to the empirical foundations of science. By providing an explanatory theory, your quantum theory is the ultimate validation.

Niels Bohr: The transition from classical physics to quantum mechanics was a profound shift. Classical physics described the macroscopic world well, but at the atomic level, it fell short. Quantum mechanics, with its probabilistic nature and wave-particle duality, emerged as a more accurate framework for describing the behavior of particles at the atomic and subatomic scales.

Dmitri Mendeleev: The challenge to deterministic notions prompts us to consider the philosophical implications of such uncertainty at the foundational level of matter.

Niels Bohr: Quantum mechanics challenges our classical intuitions about a deterministic universe. Heidelberg's uncertainty principle suggests inherent limits to our precision in simultaneously measuring certain pairs of properties. It raises questions about the nature of reality and the role of observation in shaping that reality. But it provides a statistical framework for predicting the behavior of ensembles of particles. While we may not predict the exact path of a single particle, we can make probabilistic predictions about the behavior of a group of particles. Those predictions have been verified. I rest my case.

BOHR'S QUANTUM

Sometimes I've believed as many as six impossible things before breakfast,[2] The Red Queen, 1872.

True, the Standard Model does explain a very great deal. Nevertheless it is not yet a proper theory, principally because it does not satisfy the physicists naive faith in elegance and simplicity. It involves some 17 allegedly fundamental particles and the same number of arbitrary and tunable parameters, such as the fine-structure constants, the muon-electron mass ratio and the various mysterious mixing angles,[3] Sheldon Glashow, 1991.

Driven by equations, physics of the 20th century started with theories that were developed as mathematical tools, but which then turned out to represent physical reality. Planck and Einstein were reluctant to accept the physical reality of the quantum. Dirac predicted the positron because it was an alternative solution to an equation. With their belief in the power of mathematics, the ancient Greeks would have loved this outcome.

The 20th century began with the atom as a discrete unit, revealed through a series of discoveries to consist of a nucleus containing protons and neutrons, surrounded by electrons in circular orbits. It was revolutionary to argue that the nucleus could be disintegrated, and to propose discontinuities in the electron orbits. But even though the atom was no longer indivisible, the proton, neutron, and electron were still regarded as particles.

The transition to regarding the structure of the atom as based on probability and statistics rather than discrete particles came when Erwin Schrödinger proposed that electrons could be treated as waves surrounding the nucleus as a cloud of probability. As the name of the concept, wave–particle duality, shows, this was not a revolution in replacing one paradigm by another. Rather it was a recognition that there are two ways of looking at the properties of sub-atomic particles. Each is valid in certain circumstances.

All of these concepts, of course, were actually developed as sets of equations. Visualizations such as 'waves' or 'clouds' are attempts to explain in simple language the working of the equations. (The Schrödinger equations, for example, describe the probability of finding a charge at any particular point in space.)

As the equations become more complicated, translating into language becomes more difficult.

Bertrand Russell put it very well when he said, "Ordinary language is totally unsuited for expressing what physics really asserts, since the words of everyday life are not sufficiently abstract. Only mathematics and mathematical logic can say as little as the physicist means to say. As soon as he translates his symbols into words, he inevitably says something much too concrete."[4] (This becomes even more difficult in cosmology, which I discuss in Chapter 17.) Attempts to express the wave–particle duality led to a series of paradoxes (discussed in the next chapter).

The concept that electrons can be treated as waves was proposed in 1924. Louis de Broglie (1892–1987) and his brother (Maurice) broke with the aristocratic family's tradition of diplomatic service when they became physicists. (Their father was the Duc de Broglie.) Louis became a theoretical physicist. He proposed that any particle (and in particular the electron) can be treated as a wave whose wavelength is inversely proportional to its momentum. His insight was so outrageously theoretical that the citation for his Nobel Prize in 1929 stated that, "no single known fact supported this theory."[5]

Erwin Schrödinger (1887–1961) then modified Bohr's quantum theory by describing the behavior of electrons within the atom in terms of waves.[6] Schrödinger was born in Vienna and went to university there. He then held positions at Stuttgart, Breslau, and Zurich. He published what are now known as the Schrödinger equations in 1926, when he was 39, a relatively late age for a physicist to make such novel contributions. (Many of the major contributions in the period were being made by physicists in their twenties, to the point at which the field was sometimes called Knabenphysik [boys' physics].)

In 1927, Schrödinger succeeded Max Planck in Germany's most important academic position as Professor in Berlin. When Hitler came to power in 1933, he moved to Austria, and then escaped in 1938 after the Anschluss (German invasion), ending up in Dublin. A noted womanizer, he somewhat offended the straight-laced Irish by living in a ménage à trois with two women, his wife and a long-time mistress. Schrödinger shared a Nobel Prize in 1933 with Paul Dirac, an English physicist who (among other things) discovered the existence of anti-matter.

Schrödinger lecturing in Graz, Austria in 1937.

Possibly by Heinz Reuter. Courtesy of the University of Vienna. Used under fair dealing.

Heisenberg developed the same theory as Schrödinger, but using a different mathematical method. Werner Heisenberg (1901–1976) started as a Privatdozent (able to teach but without a Chair) at Göttingen in 1924. He spent six months with Niels Bohr in Copenhagen in 1924–1925, returned to Göttingen, and then obtained a position in Copenhagen in 1926. He returned to Leipzig to a Chair in 1928.

Heisenberg used matrix algebra to formulate quantum mechanics. Schrödinger used differential equations to represent probability. The papers were published simultaneously. It took more than a year to realize that they presented the same analysis in different mathematical forms.[7] Heisenberg obtained the 1932 Nobel Prize for physics. It was announced at the same time as Schrödinger's prize. The two physicists were rewarded for creating quantum mechanics.[8]

Schrödinger was as influential in biology as he was in physics. When he turned his attention to biology, he gave a series of lectures under the title *What is Life?*, published as a short book in 1944. As noted in Chapter 10, and discussed in more detail in Chapter 19, this was enormously influential in attracting

The most brilliant group of scientists ever assembled together at one time? The participants in the 5th Solvay Conference on Electrons and Photons in 1927.[9]

physicists to enter biology in order to tackle the question of the nature of genetic material.

The concentration of physics into a small number of leading scientists is emphasized by the participants at the 5th Solvay conference on 'Electrons and Photons,' held in Brussels in 1927. Out of 29 participants, 17 either had already earned, or would in future obtain, a Nobel Prize. A photograph shows a surprisingly elderly group of men (Marie Curie was the only woman), considering that physicists usually make their major discoveries when young. The most remarkable feature was how compact physics was at the time, and how well connected the leading scientists were. This isn't often considered in accounts of how science functions.

There is also a lineage in which Nobel Prizewinners beget Nobel Prizewinners: not literally—although there are several cases where fathers gained Nobel Prizes and so did their sons a generation later—but because many Nobel Prizewinners previously worked in the laboratories of Nobel laureates.[10] The series of discoveries shows a concentration of people and places: for example, Thomson to Rutherford to Chadwick (and others), in

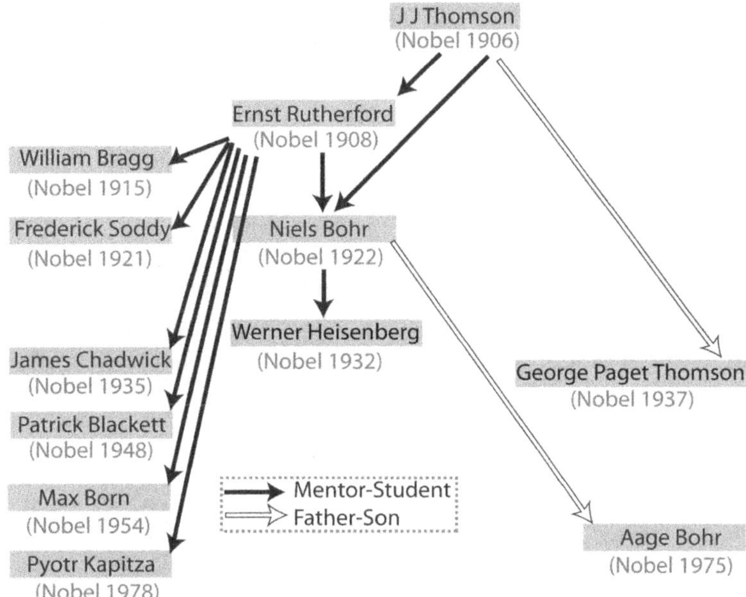

A lineage of Nobel Prizewinners in atomic physics started with J. J. Thompson in 1906 and lasted for more than half a century. It included both mentor–student and father–son relationships.

the Cavendish Laboratory and at Manchester. The importance of the discoveries was recognized in short order, with Nobel Prizes following after delays of only 3–9 years. This is something of a contrast with the period from the second half of the 20th century, when the importance of discoveries tended to be more disputed (especially in biology) and the delay before universal recognition could be much longer.

With the discovery of the neutron, the nucleus was defined as comprising 3 types of elementary particles. The revolution came with the realization that these in turn could be described in terms of other (sub)particles. "The discovery of many new types of elementary particle [in the second half of the twentieth century] not only increased the catalog of particles—they also instigated a revolution in what is meant by a particle being elementary," says Nobel Prizewinner Steven Weinberg.[11]

Developed through the 1920s and 1930s, quantum field theory describes all particles in terms of quanta—combinations of field energy and momentum. All known particles are combined into the Standard Model, which is intended to be a complete description of the nucleus (see the box below). However, the revolution is a work in progress. It's unclear if it will ever really be completed.[12]

A crisis began as early as 1927, with the discovery that some energy appeared to be missing when the energy emitted by a radioactive radium nucleus was totaled. This contradicted the first law of thermodynamics—that energy is conserved. The missing energy led to a spirited debate over the next few years as to whether the first law was going to hold when electrons were emitted from or captured by a nucleus.[13] Physicists were forced to consider a possible exception to the law, but were very unhappy about it. "A radical departure from this principle would imply strange consequences," Niels Bohr said.[14]

The solution was to introduce another particle. Wolfgang Pauli (1900–1958), an Austrian physicist working in Zurich, proposed

Wolfgang Pauli lecturing at the Niels Bohr Institute in Copenhagen in 1930. Pauli is slightly blurred because he tended to rock back and forth when talking. At the Institute, an instrument that oscillated was called a Pauli.

Reproduced with permission from the American Institute of Physics.

the neutrino in 1930. He called it "a desperate way out." (He proposed the idea in a letter to other physicists rather than a formal paper.[15] The world of atomic physicists was very close knit.) The neutrino has no charge. It's created when a neutron turns into a proton plus an electron plus a neutrino. The mass of the neutrino is only 10^{-5} the mass of an electron. This explains the missing energy. The neutrino was not detected directly until 1955.

Pauli gave his name to two principles. The Pauli exclusion principle refers to his paper in 1925 defining the quantum properties of electrons in an atom. He was awarded the Nobel Prize in Physics for it in 1945. "I was unable to give a logical reason for the exclusion principle or to deduce it from more general assumptions. I had always the feeling and I still have it today, that this is a deficiency," Pauli said in his Nobel acceptance speech.[16] Pauli could be biting about work he considered was not up to standard. "This is not even wrong," he famously said of one paper.[17] The second principle is the Pauli effect, describing his belief that his presence caused equipment to malfunction. It was formalized as "a functioning device and Wolfgang Pauli may not occupy the same room."[18]

Paul Dirac lecturing. Dirac was famous for the elegance of his science—Einstein described the Dirac equation as "the most logically perfect presentation of quantum mechanics"—and the messiness of his life. There are many stories about Dirac's offbeat comments on science and other matters. Niels Bohr called him "the strangest man."

Reproduced with permission from the American Institute of Physics.

Another surprising discovery came with the attempts of Paul Dirac (1902–1984), a theoretical physicist in Cambridge, to reconcile quantum mechanics with the theory of relativity. (Dirac was known for his reclusive nature. He was taciturn to the point that his colleagues named a unit of speech after him: a dirac was one word per hour.[19]) The Dirac Relativistic Wave Equation in 1928 accounted for the behavior of atomic electrons at the velocity of light. However, the equation had two solutions: one for a particle with positive energy; and one for a particle with negative energy. But energy could not be negative!

Dirac solved the dilemma by proposing the existence of the positron, a particle identical to the electron but with positive instead of negative charge.[20] The positron was detected in 1932 by an American experimental physicist, Carl Anderson (1905–1991), in the form of tracks of cosmic rays that went in the opposite direction from electrons.[21] This alternation of predictions from theoretical physicists and confirmations from experimental physicists marks the progress of physics through the 20th century.

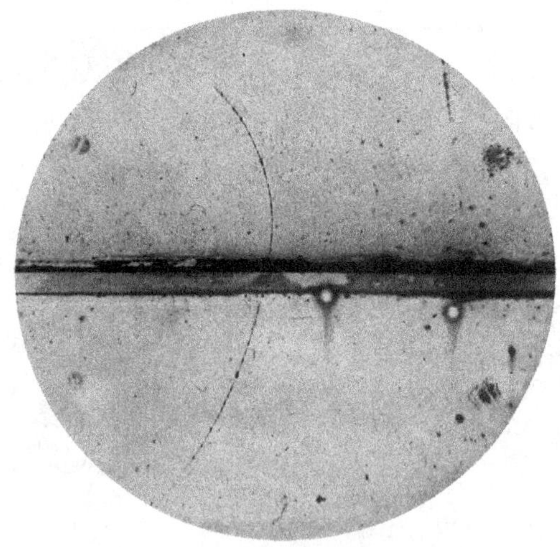

The positron was detected in a cloud chamber. It entered at lower left and was deflected by a magnetic field perpendicular to the image. The direction of the curved path shows it was positively charged. The length of the path after it was slowed by a lead plate (horizontal line in center) is 10× greater than a proton would show.

Carl Anderson with the cloud chamber in 1931. The actual chamber was 17 cm in diameter and 3 cm thick, inserted between copper coils about 1 m in diameter. It required about 10% of the power required for the entire Caltech campus and could only be run at night. The apparatus weighed 2 tons.

Courtesy of Caltech Archives and Special Collections.

The discovery of the positron was an accident. Anderson was famously skeptical of theory. "I was not familiar in detail with Dirac's work. I was too busy operating this piece of equipment to have much time to read his papers ... The discovery of the positron was wholly accidental," Anderson said.[22] He was using a cloud chamber to measure the energy of cosmic rays.

The cloud chamber had been invented in completely different circumstances. Studying light effects on the Ben Nevis mountain in 1894, a Scottish physicist, Charles Wilson (1869–1959), created a chamber to study cloud formations and optical effects in moist air. Ions provided foci for the condensation of water drops. By 1911, Wilson had developed the cloud chamber in which a sealed chamber was saturated with water vapor. Then the chamber was expanded so that air cooled and water vapor condensed. When a charged particle passes through the chamber, it leaves a track as water vapor condenses on the ions.

The cyclotron was invented in 1932 by Ernest Lawrence at the University of California, Berkeley. The 60-inch cyclotron at Berkeley was the biggest in the world in 1939. It's a huge contrast with the latest particle accelerator, the synchrotron at CERN (see Chapter 11).

U.S. National Archives and Records Administration.

The cloud chamber became a standard tool for particle physics. By applying a magnetic field, it's possible to deduce the mass and the charge of the particle from its deflection. The positron was detected because it moved in the opposite direction from an electron (meaning it was positively charged), but had a similar mass to an electron. (The name positron was actually introduced by the editor of the journal where the article was published.)[23]

The path through the nucleus was not painless in the thirties. Some important laboratories made embarrassing mistakes. The cyclotron at Berkeley was the most powerful accelerator in the world. Ernest Lawrence was using it to bombard targets with deuterons, the nuclei of deuterium, the heavy isotope of hydrogen. This was extremely productive because the extra weight of deuteron (consisting of one proton plus one neutron) gave it much greater power than a proton to disintegrate the target. In fact, bombarding several different types of targets produced very fast protons, so energetic that Lawrence concluded that the deuteron itself must be unstable and was disintegrating. This was a mistake. The result was actually due to contamination of the target with deuterons that were released under the bombardment.[24]

Defining the structure of the atom moved steadily from a small number of discrete particles into smaller and smaller entities and interactions between them. After the proton and neutron were defined, it turned out that they are themselves made of smaller particles. More types of particles were discovered through the 1930s and 1940s, and then many more when synchrotrons were developed in the 1950s.

The results led to the question: what is an elementary particle? The basic view is that a particle is *composite* if its properties can be explained by combinations of other particles: otherwise it is *elementary*. The description of subatomic particles now is derived from a mix of theoretical and experimental physics.

The Standard Model classifies all known elementary particles and describes three fundamental forces responsible for interactions between them. Some of the particles were at first proposed on theoretical grounds to make the Standard Model work. Experiments to detect them followed later. So far 37 elementary particles have been detected. Extensions of the Standard Model

The Standard Model

Elementary Subatomic Particles

Fermions consist of 6 types of quarks (which make up protons and neutrons and interact by strong forces) and 6 types of leptons (including the electron and the neutrino).

The 12 types of gauge bosons include 8 gluons, which have no mass and mediate strong forces, and 3 bosons with charges that mediate weak forces.

Types of Forces

The electromagnetic force involves interactions between charged particles. It's carried by photons. Quantum electrodynamics (QED) describes charged particles in terms of the emission and absorption of photons.

The strong force involves interactions between uncharged particles. It's carried by gluons (particles with no mass). Quantum chromodynamics (QCD) was constructed by analogy to QED and describes how gluons transmit force between quarks (and particles built of quarks).

The weak force is carried by specific elementary particles, the W and Z bosons, and operates over very short distances (less than the diameter of a proton). It allows quarks to change from one type to another.

---

have proposed other elementary particles that have not, however, been detected (at least yet).[25]

Protons and neutrons are composite particles, consisting of quarks. Electrons and neutrinos are leptons. Quarks and leptons are the two major groups of particles, together known as fermions. They interact with one another by transferring energy by means of another type of particle, bosons. They acquire their mass by moving through the Higgs field (consisting of Higgs bosons; see Chapter 11).

What holds the nucleus together? It was a mystery for decades why the nucleus did not fly apart because protons repel one another. The particles interact by three types of force: electromagnetic, strong, and weak (see box). The 'strong' force between uncharged particles is the basic counter to repulsion between protons.

The Standard Model is theoretically self-consistent (and many of its parameters have been verified to an excruciating number of decimal places). However, to the frustration of physicists, it has been impossible to extend it to include an explanation of gravity. Physicists are still producing models, but from the outside it appears very much as though things are stuck.

NOTES AND REFERENCES

1. Quoted in R. Moore, *Niels Bohr: The Man, His Science, & the World They Changed*, Hodder & Stoughton, London, 1967, p. 196.
2. L. Carroll, *Through the Looking-Glass, and What Alice Found There*, MacMillan, Philadelphia, 1871.
3. S. Glashow, *The Charm of Physics*, Touchstone Books, New York, 1991, p. 268.
4. B. Russell, *The Scientific Outlook*, George Allen & Unwin, London, 1931, p. 85.
5. Presentation Speech by Professor C.W. Oseen, Chairman of the Nobel Committee for Physics of the Royal Swedish Academy of Sciences, on December 10, 1929. Online at www.nobelprize.org/prizes/physics/1929/ceremony-speech.
6. This is a simplistic translation of the actual format, in which Schrödinger introduced eigenvalues (a mathematical formulation) to describe quanta.
7. G. Gamow, *Thirty Years That Shook Physics: The Story of Quantum Theory*, Dover Publications, New York, 1966, p. 3.
8. A. Lightman, *The Discoveries: Great Breakthroughs In 20th-Century Science, Including The Original Papers*, Vintage, New York, 2006, pp. 190–202.
9. Top row: A. Piccard, E. Henriot, P. Ehrenfest, E. Herzen, Th. De Donder, E. Schrödinger, J. E. Verschaffelt, W. Pauli, W. Heisenberg, R. H. Fowler, L. Brillouin. Middle row: P. Debye, M. Knudsen, W. L. Bragg, H. A. Kramers, P. A. M. Dirac, A. H. Compton, L. de Broglie, M. Born, N. Bohr. Bottom row:

I. Langmuir, M. Planck, M. Curie, H. A. Lorentz, A. Einstein, P. Langevin, Ch. E. Guye, C. T. R. Wilson, O. W. Richardson.

10. H. Zuckerman, *Scientific Elite: Nobel Laureates In The United States*, Free Press, New York, 1977.

11. S. Weinberg, *The Discovery of Subatomic Particles*, Cambridge University Press, Cambridge, 2nd edn, 2003.

12. M. Dine, *This Way to the Universe: A Theoretical Physicist's Journey to the Edge of Reality*, Dutton, New York, 2022.

13. A. D. Franklin and A. D. Marino, *Are There Really Neutrinos?*, CRC Press, Boca Raton, FL, 2nd edn, 2021, pp. 58–62.

14. N. Bohr, Faraday Lecture: Chemistry and the Quantum Theory of Atomic Constitution, *J. Chem. Soc.*, 1932, **135**, 349–384.

15. W. Pauli, Letter to a physicists' gathering at Tübingen, December 4, 1930, in *W. Pauli, Collected Scientific Papers*, ed. R. Kronig and V. Weisskopf, Interscience, New York, 1964, vol. 2, p. 1313.

16. www.nobelprize.org/uploads/2018/06/pauli-lecture.pdf.

17. Often embellished as "This isn't right. It isn't even wrong," R. Peierls, *Biographical Memoirs of Fellows of the Royal Society, Wolfgang Pauli*, 1960, vol. 5, p. 186.

18. This may seem humorous, but Pauli took it seriously, and even corresponded with Jung about a possible basis for the effect. G. Gamow, *Thirty Years That Shook Physics: The Story of Quantum Theory*, Dover Publications, New York, 1966 p. 64; C. G. Jung and W. E. Pauli, *The Interpretation of Nature and the Psyche*, Routledge & Kegan Paul, London, 1952.

19. G. Farmelo, *The Strangest Man: The Hidden Life Of Paul Dirac, Mystic Of The Atom*, Basic Books, New York, 2009, pp. 89, 120.

20. G. Gamow, *Thirty Years That Shook Physics: The Story of Quantum Theory*, Dover Publications, New York, 1966, p. 4.

21. C. D. Anderson, The positive electron, *Phys. Rev.*, 1933, **43**, 491–494.

22. Quoted in M. Dine, *This Way to the Universe: A Theoretical Physicist's Journey to the Edge of Reality*, Dutton, New York, 2022.

23. A few months later, Patrick Blackett and Giuseppe Occhialini at the Cavendish Laboratory used a more sophisticated cloud chamber to obtain positron tracks. They commented that Dirac's hypothesis predicted "a time of life for the positive electron that is long enough for it to be observed in a cloud chamber but short enough to explain why it had not been discovered by other methods." Blackett was awarded a Nobel Prize in 1948. P. M. S. Blackett and G. Occhialini, Some photographs of the tracks of penetrating radiation, *Proc. R. Soc. London, Ser. A*, 1933, **139A**, 699–727.

24. M. Hiltzik, *Big Science: Ernest Lawrence and the Invention That Launched the Military-Industrial Complex*, Simon & Schuster, New York, 2015, pp. 105–121.

25. Some particles that were thought to be elementary when they were discovered later turned out to be composite, such as the meson (1947), pion (1964), kaon (1968), which are made up of quarks.

Heisenberg's Uncertainties: 1927–

UNTIL HEISENBERG PROPOSED THE UNCERTAINTY PRINCIPLE in 1927, it was an open question whether models for the structure of matter reflected actual reality or were useful fictions for operational purposes. There wasn't any sense in either case that there might be parameters it was impossible to determine. The uncertainty principle says that it is impossible to determine simultaneously the location and momentum of an electron. While it applies to a specific situation, 'uncertainty' is a good description for the entire approach of quantum theory. Schrödinger's equations show that electrons behave simultaneously as waves and particles. The double-slit experiment, when a beam of light or electrons passes through two parallel slits, shows that the act of observation influences the behavior of the wave–particle duality. The ability of a pair of 'entangled' electrons (or other particles) to influence each other instantaneously appears to exceed the speed limit of light. Paradoxes of quantum behavior may even extend to biological situations. Intuition or common sense are no guide to quantum behavior.

The Frontiers of Science
By Benjamin Lewin
© Benjamin Lewin 2026
Published by the Royal Society of Chemistry, www.rsc.org

Timeline for Defining the Atom
by Quantum Theory

1900 — Planck introduces quanta (1900)

Einstein proposes photons are quantal (1905)

1910

Bohr proposes quantum theory of atom (1913)
Gilbert Lewis arranges electron orbits into quantum shells (1916)

1920 — de Broglie proposes particles can be treated as waves (1924)
Schrödinger treats electrons within atom as waves (1926)
Gilbert Lewis proposes term 'photon' for unit of light (1926)
Heisenberg proposes Uncertainty Principle (1927)
Dirac's equations predict existence of positron (1928)
1930 — Pauli proposes existence of neutrino (1930)
Pauling defines chemical bonds by quantum mechanics (1931)
Yukawa proposes strong forces for proton-neutron interaction (1935)
EPR paper challenges quantum mechanics (1935)
Schrödinger develops cat-in-box thought experiment (1935)
1940

} SECOND WORLD WAR

1950

Reines and Cowan detect neutrinos (1956)
Everett proposes multiverse theory (1957)
1960

Bell supports theory of entanglement (1964)
Higgs proposes new boson (1964)
Gell-Mann proposes quarks (1964)
Weinberg introduces weak interactions (1967)
1970

Clauser and Freedman confirm entanglement (1972)
Pais and Treiman introduce term *Standard Model* (1975)

1980

1990

First quantum computer developed with 2 qbits (1998)
2000

2010

CERN identifies Higgs boson (2012)

Loopholes excluded in experimental test of entanglement (2015)

2020 — Google claims quantum supremacy (2019)

A Dialog between J. J. Thomson and Werner Heisenberg

J. J. Thomson (1856–1940) was a classical physicist whose work at Cambridge on cathode rays led to the discovery of electrons in 1897. He viewed the atom as a solid body embedded with electrons.

© Oxford Science Archive/Heritage Images/Science Photo Library.

Werner Heisenberg (1901–1976) was a German atomic physicist who worked with Niels Bohr in Copenhagen to develop quantum mechanics in 1925. He returned to Leipzig in Germany in 1928, and remained in Germany.

AIP Emilio Segrè Visual Archives.

Thomson: Ah, Professor Heisenberg, it's an honor to finally meet. I must say, I find your theories on uncertainty rather… unsettling. It wasn't so long ago that we believed in finding the precise positions and velocities of particles. We were sure that nature could indeed be known fully.

Heisenberg: And I, in turn, am humbled to speak with the discoverer of the electron! I understand, Professor Thomson, that your generation fought hard to reveal a structure to the atom. But over time, it turned out that measuring certain qualities simultaneously is impossible. It's not a matter of precision in our instruments but an inherent limit—a boundary set by nature.

Thomson: Fascinating, though troubling. Does this uncertainty suggest that reality itself is somehow unknowable? It sounds as if you're suggesting that the very act of observation disturbs the thing we are observing.

Heisenberg: Precisely! That is the core of quantum mechanics. You see, the electron does not exist as a solid object in a fixed place until we measure it. Until that moment, it's more accurate to say it exists as a probability wave. It's not just difficult to measure both position and momentum. It's fundamentally impossible to know both with absolute certainty.

Thomson: But, Professor Heisenberg, if electrons do not exist as discrete particles until we observe them, does this imply that they are somehow… less real? In my work, I attempted to show that these particles were fundamental building blocks of matter—real and measurable entities that constitute the very fabric of the universe. Modern physics is based on explaining interactions between real objects.

Heisenberg: We are dealing with potential reality. The observer plays an active role in defining certain properties of a particle. Before we observe them, particles exist in a superposition of states—many possibilities, rather than a single reality. When we measure, the wave function "collapses," so to speak, into a single outcome.

Thomson: In my time, we believed that reality exists independently of our observation. Our measurements only reveal it, not determine it. The idea that uncertainty is an intrinsic feature of nature is puzzling. It seems the more we uncover, the less we truly understand. You speak of reality not as something we can touch but as something we shape through our observations.

Heisenberg: Exactly! And yet it's not mystical; it's mathematical. Quantum mechanics, for all its strangeness, allows us to predict the outcomes of experiments with remarkable precision. Science, I believe, is not merely a tool for understanding, but a constant reminder of our limits, our humility before the unknown. I rest my case.

HEISENBERG'S UNCERTAINTIES

Those who are not shocked when they first come across quantum theory cannot possibly have understood it,[1] Niels Bohr, 1952.

When I use a word it means just what I choose it to mean— neither more nor less,[2] Humpty Dumpty, 1872.

"It was as if the ground had been pulled out from under one, with no firm foundation to be seen anywhere," Einstein felt, when he heard about Planck's proposal for the quantum.[3] He then went on to pull out the ground from under everyone else with his 1905 paper proposing that light should be treated in the same way as matter: as consisting of discrete particles.[4] The paper derived an equation to explain the photoelectric effect: that electrons are emitted from a metal surface when light is shone on it. This interaction between light and matter cannot be explained by classical physics, where light is treated as an electromagnetic wave.[5]

1905 was Einstein's *annus mirabilis*. Albert Einstein (1879–1955) was completely unknown in 1905, when he published four papers that solved the two major problems that had dominated physics for the second half of the 19th century. What are atoms? And what is electromagnetic radiation? The papers were as remarkable for their style as for their content. They started from propositions (rather than a review of experimental evidence), and cited very little previous work.

The proposal that light exists as quanta was in his first paper on the photoelectric effect. The second paper proposed an equation to explain Brownian motion (the random motion of molecules in a gas or fluid). This was the final straw in convincing physicists that atoms existed (see Chapter 13). The third paper introduced special relativity and established a universal constant (c) for the speed of light. Further developing special relativity, it was the fourth paper that derived $E = mc^2$ to define the equivalence of mass and energy.

There were actually *no* references to previous work in the third paper, introducing relativity. All but unprecedented in a scientific contribution, this might lead to the conclusion that in fact there were no precedents. Yet Einstein consistently said that he regarded relativity as evolutionary, not revolutionary. "I want to emphasize that this theory [relativity] has no speculative origin, it rather owes its discovery only to the desire to adapt theoretical physics to observable facts as closely as possible. This is by no means a revolutionary process but merely the natural development of a trail that can be traced through the centuries[6]... The four men who have laid the foundation of physics on which I have been able to construct my theory are Galileo, Newton, Maxwell, and Lorentz.[7]

Einstein pointed in the opening of his first paper, on the nature of light, to the "profound formal difference" between perceptions of matter (particulate) and the view of light (continuous). Einstein regarded his first paper as the one that was most innovative. "The first [paper]... deals with radiation and the energy properties of light and is very revolutionary."[8] One impetus for the paper was a discrepancy Einstein had seen between Planck's theory and Maxwell's equations.[9] (Einstein was cautious in the way he discussed this in the paper, probably as the result of discussions with his friend Michele Besso who toned down the original draft.[10]) Not only did Planck accept the paper, but he became Einstein's friend and supporter. However, he remained opposed to the quantal theory of light for at least a decade. (By contrast, Planck was one of the few physicists not to dismiss the theory of special relativity at the time.)

Einstein submitted his four papers to *Annalen der Physik*, then the leading research journal for physics, published from Berlin, with Max Planck as the Editor. Indeed, Einstein was lucky in that Max Planck was known to have a policy of publishing papers with which he disagreed so long as they had no obvious flaws. In the modern era, it would be unheard of for a journal to publish multiple papers from an unknown author proposing to overthrow the status quo (see Chapter 24).

This makes it all the more striking that when Einstein submitted his papers he held no academic position, but was working in the patent office in Bern, Switzerland. He came from a secular Jewish family in Germany. His parents moved to Italy when he was 15, but he stayed at high school in Munich, and then moved to Switzerland to complete his education, first in high school, then at the Zurich ETH (polytechnic). He acquired Swiss citizenship and a position at the Patent Office, while working on his Ph.D.

(Einstein obtained his Ph.D. at the start of 1906. The only other case that comes to mind for such significant work written by someone without a Ph.D. is Watson and Crick's proposal for the double helix of DNA in 1953, but that was because Crick had been delayed in getting his Ph.D. by the Second World War.) Einstein became a professor in Zurich only in 1909, and moved to a position in Berlin in 1913. He was the supreme theoretician, working alone and publishing most of his papers as sole author.[11] (He only ever had one graduate student.)

Einstein was completely unknown to Max Planck in 1905, so when one of Planck's students, Max von Laue, went on vacation to Switzerland, he visited Bern to meet the author of these papers. von Laue later recollected the unexpected nature of the meeting. "I looked him up in the Patent Office. In the general waiting room an official said to me: 'Go down the corridor and Einstein will come out and meet you.' I followed his instructions but the young man who came to meet me made so unexpected an impression on me that I did not believe he could possibly be the father of the relativity theory. So I ignored him and we only finally met when he came back to the waiting room."[12] von Laue, who obtained a Nobel Prize in 1914 for inventing X-ray crystallography (the method of analyzing protein structures for the next century), struck up what was to be a life-long friendship with Einstein.

After his papers were published, Einstein began to correspond with many of the famous physicists of Europe, but his first meetings in person did not happen until he was invited by Max Planck to give the plenary lecture at a major conference in Salzburg in 1909.[13] First he dismissed one of the longstanding principles of electromagnetic radiation. "It was necessary to assume that... it is essentially the luminiferous ether that mediates the propagation of light... However, today we must regard the ether hypothesis as an obsolete standpoint." Then he made a comment that was prescient, although well ahead of its time. "The next stage in the development of theoretical physics will bring us a theory of light that can be understood as a kind of fusion of the wave and emission [particle] theories of light."[14]

Although the existence of the ether had been disproven by the Michelson–Morley experiment in 1887 (see Chapter 9), the result was more or less ignored because it did not fit with current thinking. In fact, the idea of the ether was so embedded that Einstein's dismissal of it was regarded as bold, if not revolutionary. (It's unclear to what extent the Michelson–Morley experiment played a role in his thinking. Indeed, as the paper cited no earlier work, it gave no information about previous influences. Einstein's recollections on the subject varied widely.[15])

Einstein's proposal that light consisted of quanta was not really accepted for another couple of decades, when the quanta were named as photons. (In 1913, four leading German scientists said, "[Einstein] may sometimes have missed the target in his speculations, as, for example, in his hypothesis of light quanta."[16]) Photons were detected experimentally in 1923. They have energy and momentum but no mass. The discovery was rewarded with a Nobel Prize in 1927.

Even in 1921, when Einstein was recognized as one of the world's great scientists and awarded his Nobel Prize, the citation pointedly honored his work on the photoelectric effect without mentioning quanta. The opposition was based on the substantial body of evidence supporting the wave theory of light. Planck was among the doubters. American experimental physicist Robert Millikan, whose Nobel Prize in 1923 was in part for verifying

Einstein lecturing to a crowded room in Paris in 1922.

Einstein's equation for the photoelectric effect, recollected later (in 1949): "I spent ten years of my life testing that 1905 equation of Einstein's, and, contrary to all my expectations I was compelled in 1915 to assert its unambiguous experimental verification in spite of its unreasonableness since it seemed to violate everything that we knew about the interference of light... [It] seemed... to be a straight return to the corpuscular theory of light which had been completely abandoned... since around 1800."[17] This was a classic demonstration of scientific methodology, where attempts to disprove a hypothesis failed, to leave the hypothesis standing.

The transition from classical physics to quantum physics was painful for many of the participants. When theory runs ahead of experiments, physicists are sometimes reluctant to believe the real-world implications of their theory. It's an occupational hazard for theoretical physicists to create models based on mathematical formulae, and then be brought to question whether the situations predicted by those models exist in reality. The dichotomies and paradoxes of the quantum approach only multiplied during the 20th century.

Planck and Einstein believed that quanta were mathematical solutions to the problems they were addressing, but were reluctant to believe they reflected physical reality. Einstein's public reputation is dominated by $E = mc^2$ and his description of relativity. "God does not play dice with the universe" is his famous comment about quantum physics. However, he was deeply involved in quantum theory, although at the end of his career he effectively withdrew from it.[18] Yet he said that, "I have thought a hundred times as much about the quantum problems as I have about Relativity Theory."[19] Ultimately, he came to the view that the uncertainty predicted by equations reflected a failure in human comprehension rather than actual uncertainties in reality.

Perhaps the difficulties in expressing the mathematical concepts of quantum physics in words start with Heisenberg's Uncertainty Principle. While he was working with Niels Bohr in Copenhagen in 1927, Werner Heisenberg established that it is impossible to know both the position and speed of a particle. The better the position is defined, the less accurate is information about speed, and *vice versa*. (Heisenberg originally used the term "Ungenauigkeit," which means imprecision, but this rapidly became "uncertainty" in English.) Heisenberg returned to Germany, to Leipzig, in 1928. The blot on his career that he was a leading figure in German efforts to produce an atomic bomb during World War II is usually glossed over.

The issue of whether theory represents reality or is simply a useful mathematic construction is especially pointed with the uncertainty principle. Heisenberg attributed the development of the theory to an idea he had as the result of a conversation with Einstein, who had rejected the view that the next theory should describe observable quantities. Einstein said, "It's the theory that decides what can be observed."

Heisenberg recollected, "I suddenly remembered my conversation with Einstein... We had always said so glibly that the path of the electron in the cloud chamber could be observed. But perhaps what we really observed was something much less. Perhaps we merely saw a series of discrete and ill-defined spots through which the electron had passed... The right question should therefore be: Can quantum mechanics represent the fact that an electron finds itself approximately in a given place and that it moves approximately with a given velocity, and can we make these approximations so close that they do not cause experimental difficulties? A brief calculation... showed that one could indeed represent such situations mathematically, and that the approximations are governed by what would later be called the uncertainty principle of quantum mechanics."[20]

The reality of reality is where physics merges into metaphysics and then into science fiction.[21] Are the uncertainties of quantum physics a true reflection of reality—is it really the case that an electron is a smear of probabilities—or is this an illusion reflecting our inability to analyze the system? The 'Copenhagen interpretation' was so-named because Niels Bohr developed it in a series of lectures in the 1920s. It proposes that subatomic particles exist only in probabilistic states until they are forced into a single state by an act of observation. ('Observation' does not necessarily imply human intervention. It can be any interaction with the environment.) Werner Heisenberg used the phrase 'potential reality' to describe the power of quantum equations.[22]

(There is a counterpart in mathematics in the question: does mathematics proceed by invention or discovery? In other words, is a theorem an intellectual construction that is useful but does not necessarily represent the real world, or is it revealing truths that previously existed independently of the mathematician? The extreme of the second position would be Plato's view that *only* mathematics can reveal the real world.[23] Einstein famously captured the conundrum when he asked, "How can it be that mathematics, being after all a product of human thought which is independent of experience, is so admirably appropriate to the objects of reality?"[24,25] There's concern about the extreme dependence of physics upon mathematics coupled with the search for elegance or beauty. "This mathematical take-over of physics has its dangers, as it could tempt us into realms of thought which embody mathematical perfection but might be far removed, or even alien to, physical reality."[26])

The Solvay conference of 1927 (see Chapter 14) was a turning point. Heisenberg described it as "the completion of quantum physics," but another participant, Paul Langevin, famously said it was "where the confusion of ideas reached its peak." The situation was summarized when Paul Ehrenfest wrote on the blackboard some lines from Genesis describing the tower of Babel. "And the Lord said, let us go down, and there confound their language that they may not understand one another's speech." The big question was how to reconcile quantum theory and reality. A century later, it is no better resolved.

The debate as to whether we are uncovering physical reality or making useful models for it goes further into the question of the inevitability of discovery. If experiments (or theories) represent reality, they are in a sense inevitable. Many cases of simultaneous discovery show that discoveries are made when the time is right, when technology and context come together. But if theories are models, operationally useful but not necessary representing physical reality, one model might be as good as another. Indeed, the example of Heisenberg's matrix and Schrödinger's wave mechanics shows that entirely different ways of visualizing the world can have equally valid predictive value.

∞ ∞ ∞

The paradox inherent in the wave–particle duality became apparent with later variants of the famous double-slit experiment originally designed by Thomas Young in 1801. A light projects through two parallel slits in a plate on to a screen. This produces an interference pattern—comparable to waves meeting when two stones are dropped into a pond at different places at the same time. This was taken as evidence for the wave theory of light.

However, the varying density of the pattern suggests that light is hitting the screen at discrete points, that is, like particles. More recent versions of the experiment include detectors at each slit. It turns out that any individual photon passes through only one slit. So light is behaving simultaneously as a particle (with a discrete passage through one slit) and a wave (propagating through both slits). If a photon is detected, however, it does not generate an interference pattern. Observation has destroyed the experiment. Electrons show the same behavior if the experiment is performed with an electron gun instead of a light source.[27,28] The effect of measurement is described as causing the wave function to *collapse*.

The wave–particle duality is at the heart of the paradoxes of quantum physics. Richard Feynman called it the "deep mystery,"

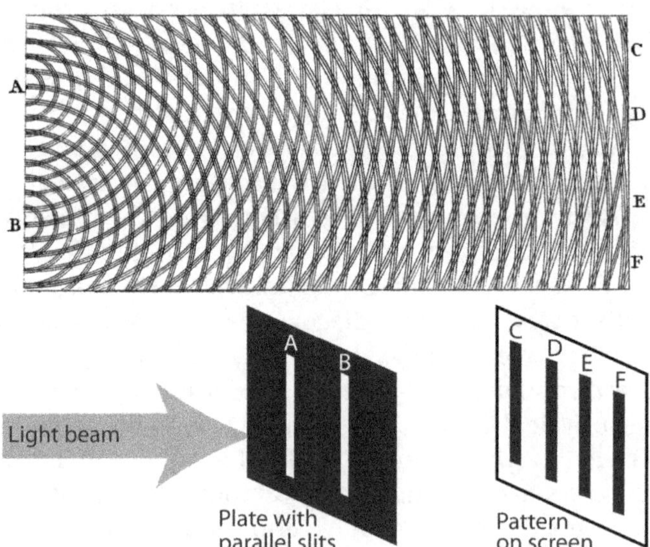

Young's original sketch to explain the interference of light waves (top). The experimental design is shown below (with a simplified version of the interference pattern).

and took the view that all the paradoxes of quantum physics follow from this one experiment.[29] Feynman intensified the paradox with a thought experiment: suppose you could fire just a single electron at the screen.

He suggested it would create an interference pattern. The experiment later became possible by using an electron gun with extremely low energy.[30] As electrons pass successively through the screen, an interference pattern forms. Electrons are somehow passing though both slits simultaneously and interacting with themselves. The ability to be in two places (actually, all possible places) at one time is called *superposition*.

Do we always want to follow where the equations lead? Mathematical constructions can produce the most amazing—and often enough incredible—results when interpreted in everyday terms. One problem with quantum theory comes from the demonstration that the act of observation changes the situation. Schrödinger's famous thought experiment (see the box below), in which he showed that the effects of applying quantum theory to the survival of a cat would lead to absurdity, illustrates the limits of everyday applications. "One can even set up quite ridiculous cases," Schrödinger said.[31]

Einstein's suspicion of the implications of quantum theory led to a thought experiment. Named for its authors, Albert Einstein, Boris Podolsky, and Nathan Rosen, the EPR paper, as it became known, was a 4-page article in 1935 that led to another paradox.[32] The basic argument of the paper was that quantum mechanics is missing some "elements of reality," which must mean it was incomplete.[33]

The paper imagined a pair of particles that interact briefly and then move off in different directions. According to Heisenberg's uncertainty principle, either the position or momentum of a particle could be measured (but not both). But if the momentum of one particle is measured, the momentum of the other particle can be calculated (because the total momentum of the pair is constant). Similarly, if the position of one particle is measured, the position of the other can be calculated from their distance apart.

The implication is that both position and momentum are 'real' properties of each particle. This contradicts the uncertainty principle. Niels Bohr, the creator of quantum theory, reacted indignantly that quantum theory was complete.[34] (Bohr's reply

Schrödinger's Cat

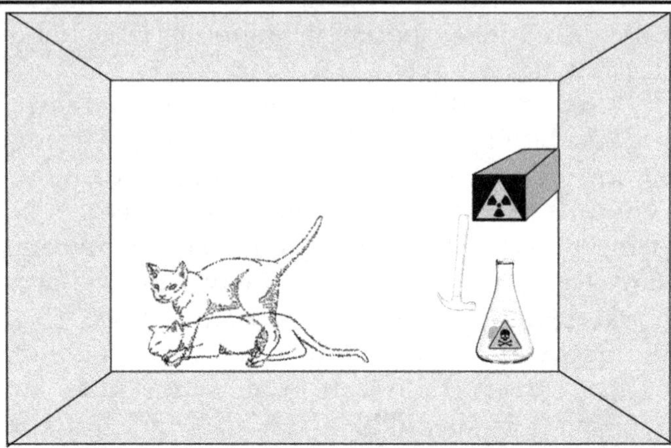

Schrödinger's famous thought experiment imagined a cat in a closed box that contains a radioactive atom. When the atom decays, it triggers a mechanism causing poison to be released, killing the cat. The probability that the atom will decay is 50% in a given period of time. If the probability is influenced by the observer, the atom must remain in the combined state of half decay/half no decay until the box is opened. But the cat cannot be dead and alive at the same time! The point is that therefore the probability of quantum change cannot be driven by the observer alone. An alternative view in terms of the multiverse (the theory that there many universes) is that when the atom decays, there is a bifurcation. In one branch the cat dies; in the other branch the cat lives. You find out which branch you are in when you open the box.

was published under the exact same title as the EPR paper.) The Copenhagen school is generally considered to have won the argument. The debate between Einstein and Bohr demonstrates that science is not driven by logic alone—or at least, the logic is not necessarily inescapable. The difference in their positions was driven to some extent by their more general beliefs about the nature of Nature.

An even more problematic paradox is that a change introduced in one particle's position or momentum must *instantaneously* be reflected in a change in the other particle, even if they are far apart. This implies that information can move between the particles faster than the speed of light! The technical term is that the phenomena are 'non-local,' meaning that the quantum state of one particle cannot be described independently of the quantum state of the other particle. The basis of Einstein's quarrel with quantum mechanics now was that he did not believe phenomena could be non-local.

The interaction between the pair of particles is now known as *entanglement*. (Schrödinger introduced the term, originally *verschränkung* in German.) Einstein actually called it "spooky action at a distance."[35] One explanation that would avoid the need for instantaneous communication was that the particles might carry 'hidden variables' that determine their responses to perturbations.

Everyone's instinctive explanation was that the response was predetermined in some way we did not (yet) understand. John Stewart Bell, who wrote a famous paper about it, explained the situation. "Dr. Bertlmann likes to wear two socks of different colours. Which colour he will have on a given foot on a given day is quite unpredictable. But when you see that the first sock is pink you can be already sure that the second sock will not be pink. Observation of the first, and experience of Bertlmann, gives immediate information about the second. There is no accounting for tastes, but apart from that there is no mystery here. And is not the EPR business just the same?"[36]

Bell was an Irish physicist working at CERN in Switzerland on accelerator design, but he showed in a theoretical paper in 1964 that, if a correlation between particles were due to hidden variables, certain 'inequalities' would result.[37] This is known as Bell's Theorem or Bell's Inequalities. It's a mathematical formulation showing that measurements of a quantum property would give different results depending on whether it was predetermined or due to a response in real time. This opened the way to an experimental test, although no one was very interested at the time.[38]

The first experiment to test the theory showed there were no inequalities. This means the correlations exist in the present. Physicists were so reluctant to believe this that they spent a lot of time looking for loopholes in the experiments, until finally they were satisfied that all loopholes had been excluded.[39] Bell summarized the situation. "The reasonable thing just doesn't work."[40]

Entanglement has been confirmed experimentally by measuring parameters such as polarization or spin of entangled particles. When one particle changes, so does the other—instantaneously. The paradox of how this can happen without exceeding the speed of light remains a challenge, but as a practical matter, it raises the question of whether it may become possible to use entangled states for communication. Entanglement appears to work also at a much higher scale than atomic particles, in fact, the latest experiments show it functions in the context of a fiber-optic network.[41] The Nobel Prize for physics in 2022 was awarded to a group of three physicists for "pioneering quantum information science."

Entanglement is perhaps the most striking example ever recorded of an experimental result that defies theory. We simply lack any basis for understanding the phenomenon. To anyone but a theoretical physicist (and to some physicists!) it defies all reason. It's definitely a bold concept to conceive a new technology based on a phenomenon we do not understand, and which indeed is impossible according to the theory of general relativity. Usually when you get an absurd result, you either question the result or revise the theory, but today physicists are living in the schizophrenic state of accepting both entanglement and relativity.

There have always been difficulties in understanding forces that act at a distance with no visible intermediaries. Magnetism appeared to be magical. Kepler could not bring himself to visualize gravity as an attraction between bodies in spite of his observations and calculations. Descartes believed that inertia should cause bodies to move in a straight line. In order to explain circular motion, he was forced to invoke vortices in the ether. Newton called his own theory of gravity an 'absurdity'.[42]

The basic problem in extending the Standard Model is that quantum mechanics cannot explain gravity. This is sometimes called the quantum gravity problem. General relativity treats space and time as continuous. They can be curved and distorted by the presence of matter and energy. All other forces are quantal in nature. At one time it was proposed that gravity was due to a quantum particle, the graviton. On even less certain ground, a quantum counterpart for time has been called a chronon. But today it's regarded as unlikely that gravity or time can be quantized.

∞ ∞ ∞

Extending the quantum view of the world to biology was perhaps not what Schrödinger had in mind when he wrote his book, *What is Life?*, but probably it would not have surprised him. Schrödinger's view that biology might provide new insights in physics was extraordinarily stimulating, but led to the conclusion that the laws of classical physics were sufficient to explain biological phenomena. More recently, however, quantum effects have been implicated in molecular biology.

Quantum physics is a world in itself. To work at the atomic level, it's necessary to explain subatomic particles in terms of wave–particle duality, and to view the construction of the atom as a matter of probability. We have to try to understand effects such as entanglement or superposition or electron tunneling (see box). Yet none of this is required to use the laws of physics or chemistry at a higher level. We do not need to invoke quantum mechanics in order to use the laws of gravitation or to understand why salt dissolves in water. At this level, quantum effects collapse and we are dealing with reliable, predictable situations that can be described by classical physics. The technical name for this is *decoherence*.[43]

To say that biological systems are noisy and messy is a simplified way of describing the fact that they do not maintain the isolated environment necessary to maintain quantum effects. You might say that biological systems are intrinsically decoherent. In a sense, they always provide the equivalent of an observation that causes quantum effects to collapse when they are measured.

---

*Three Inexplicable Effects of Quantum Physics
that Might be Relevant to Molecular Biology*

Entanglement describes two particles that are generated with linked properties; for example, two electrons are generated with a total spin of zero. If the spin of one of them changes, the other changes instantaneously to maintain total zero spin, no matter how far apart they are.

Superposition is the ability of a quantum system to be in multiple states at the same time until it is measured. In the double slit experiment, a single photon goes through both slits, but more than that, it simultaneously takes every possible trajectory.

Quantum tunneling describes the ability of an electron or other particle to pass through an energy barrier although it does not have enough energy to overcome the barrier. It's an essential part of nuclear fusion. (Treated as a wave, some of it is reflected from the barrier but some passes through.)

There are, however, some situations in which it seems that biological phenomena might be explained by directly invoking quantum actions. It's fair to say that in none of these cases has a mechanism been demonstrated, but it does not seem that explanations in terms of classical physics will work.[44] The challenge in each case is to explain how the system protects quantum effects against decoherence in the biological environment. The best examples are instances where enzymatic systems in the cell directly manipulate electrons: respiration and photosynthesis. The involvement of magnetism in migration is intriguing but not so well based.

It's not obvious to see a connection between nuclear fusion in stars and biological systems. The gravitational force of a star forces two hydrogen nuclei (protons) together so they fuse. (This is the first step in generating helium.) But there is a problem. As the protons get closer together, the repulsion between them creates an energy barrier. This is overcome indirectly by quantum tunneling, a process in which a subatomic particle passes through a barrier without having enough energy to go over it. (Think of it as tunneling underneath the barrier.) The decoherence of a living cell ought to prevent tunneling, but honing in at an even finer level, in effect within the molecular structure of the enzyme, the environment is more controlled.[45]

The respiratory chain is a series of enzymes that are a crucial part of the process for generating energy in the living cell. In effect, the chain passes an electron along a series of enzymes. The electron loses energy at each transfer. The energy is transferred to power the cell. The puzzle is that the system works much faster than expected. This may be explained by quantum tunneling.

Another example where the internal structure of an enzyme protects against decoherence is photosynthesis. Chlorophyll absorbs photons provided by sunlight. The chlorophyll is surrounded by a light-harvesting complex, which transfers the energy of the photons to electrons, which are transported in a wavelike form called excitons. The excitons have to find their way to the reaction center. The issue is whether they get there by a random walk or by a more directed process.[46] It's controversial whether the efficiency of the process (it appears to harvest 95% of the energy of the photons) might be explained by maintaining quantum coherence.[47]

A wide range of species can migrate reliably over vast distances. There is indirect evidence that magnetism is involved because magnets interfere with the ability to migrate. Light is necessary as well as magnetism. However, the energetics of the potential interaction with the Earth's magnetic field seem to be several orders of magnitude too low to be explained by classical chemistry or physics. This led to suggestions that quantum effects might be involved, mediated by a protein called cryptochrome. This has the ability to move one electron from a pair in a light-sensitive donor molecule to a recipient.

The theory is that the electron in the recipient remains entangled with its partner in the donor. Because they are sensitive to magnetic fields, this is used for direction finding.[48] Cryptochromes are present in the retinas of many migrating species, but the mechanism by which they might be involved in using light and magnetism to find direction remains controversial.[49] The verdict is out on whether and to what extent quantum effects are important in biological systems.

∞ ∞ ∞

While current paradoxes identify problems in quantum physics, the nature of the revolution needed to progress to the next stage is not at all obvious. *Dreams of a Final Theory* was the title of Nobel Prizewinner Steven Weinberg's book in 1992 describing attempts to unite quantum mechanics and relativity.[50] Three decades later, this is still a dream. It leaves physics in the grip of two conflicting paradigms. It will require more than a revolution to resolve the situation.

Quantum mechanics and general relativity are both consistent in their own sphere. It's certainly far from unknown for there to be competing theories to explain known data—science is replete with examples from ancient Greece to the present. But the situation in physics today is unprecedented, with two theories, quantum mechanics and relativity, each self-consistent in its own sphere, but incompatible with the other. Quantum mechanics explains the behavior of subatomic particles and atoms, while general relativity explains the behavior of gravity on a cosmic scale. Entanglement describes one of the major incompatibilities between them. It seems fair to ask whether physics is stuck, as attempts at alternative models have failed to resolve the problem (see Chapter 16). Have we reached the limits of our understanding?

At the start of the 20th century, physicists suffered from a malaise that everything interesting had been discovered (see Chapter 13). Ironically this was just before the great spurt of activity that led to quantum theory. At this time, in 1923, Einstein expressed an expectation that gravitation and electromagnetism should be combined into a single theory.[51] Einstein was the most famous scientist of the 20th century, but by the end of his career, other physicists had ceased to pay him much attention, because they did not believe in his search for a 'unified field theory'.

So far, a unified theory has yet to be achieved. Yet in spite of failures over the past half-century to unite quantum theory and relativity into a 'Theory of Everything' (as it is now described), physicists today do not seem to suffer from lack of confidence. In fact, they are somewhat indignant about suggestions that science may have reached an end.[52]

The division between theoretical analysis and experiments makes for significant differences between physics and other sciences. The contrast between the brevity of theoretical papers making paradigm shifts and the scale of the experimental apparatus needed to investigate them is a vivid demonstration of the difficulties in proceeding from theoretical physics to experimental confirmation.

Many of the major advances of physics since the definition of the atom as electrons, protons, and neutrons, have come from theory (later confirmed by experiments). Theoretical physicists work on the basis of known data, but theory can get out of synch with experiments. Sometimes theory has outrun experiments by decades. The situation can be uncomfortable when theory is inconsistent with experiment. By contrast, almost all the major advances in biology, in fact perhaps all, have been based on experimental observations. Theory has followed as an explanation of the experiments. Paradigms in physics have been overthrown by either theory or experiment; but in biology only by experiment.

This difference between theoretical and experimental science can show in the chain of logic leading to a new theory. In experimental science, theory is presented as a logical consequence

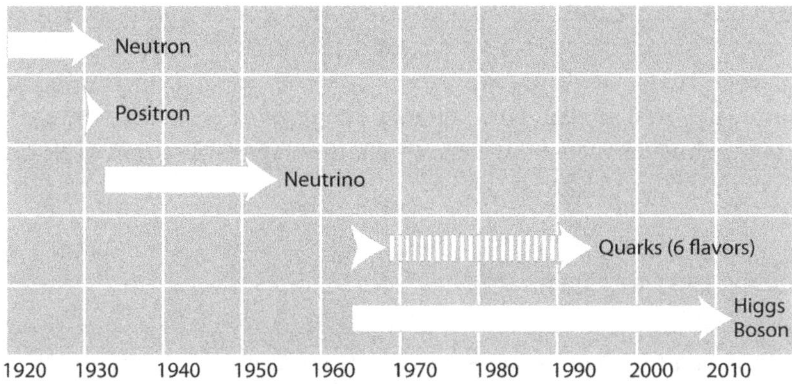

The delay between predicting an atomic particle theoretically and detecting it experimentally has lengthened as the particles have become 'more' elementary and the equipment needed to detect them becomes larger. (The dotted line for quarks indicates that the first of the predicted types of quark was detected quickly, but it took longer to detect the others.)

of the experiment. A common format would be to say, "Our results can be explained by supposing..." There is no such obligation in theoretical physics, where new theories can arise as logical consequences of experiments or previous theories, but sometimes without any apparent precedent out of the blue.

Steven Weinberg calls physicists who produce new theories out of the blue: "magician-physicists, who do not seem to be reasoning at all but who jump over all intermediate steps to a new insight about nature... Planck was a magician in inventing his 1900 theory of heat radiation, and Einstein was playing the part of a magician when he proposed the idea of the photon in 1905... The papers of magician-physicists are often incomprehensible. In this sense, Heisenberg's 1925 paper was pure magic... I have never understood Heisenberg's motivations for the mathematical steps in his paper."[53]

Is there any equivalent to the 'final theory' in other sciences? Since chemistry diverged from physics, it has become a more mundane science and does not have such lofty ideals. Biology tackles the nature of humanity, from origins and evolution to consciousness. Perhaps understanding consciousness is biology's equivalent to physics' final theory. The ideal for a reductionist would be to explain consciousness in terms of known functions, such as neuronal interactions. This would ultimately bring it into the realm of the laws of physics and chemistry, just as the final theory in physics would unite quantum mechanics and relativity. In that sense, molecular biology is the ultimate biology.

NOTES AND REFERENCES

1. According to an account of a conversation Heisenberg had with Wolfgang Pauli and Niels Bohr in June 1952. See W. Heisenberg, *Physics and Beyond: Encounters and Conversations*, translated by A. J. Pomerans, Harper & Row, New York, 1971, p. 206.

2. In L. Carroll, *Through the Looking-Glass, and What Alice Found There*, MacMillan, Philadelphia, 1871.

3. *Albert Einstein Philosopher-Scientist*, ed. P. A. Schilpp, MJF Books, New York, 1949, p. 45.

4. A. Einstein, On a heuristic viewpoint concerning the production and transformation of light, *Annalen der Physik*, 1905, **17**, 132–148. Translation online at einsteinpapers.press.princeton.edu/vol2-trans/100.

5. Classical electromagnetism predicts that light consists of continuous waves that transfer energy to electrons. The electrons would then be emitted when they accumulate enough energy. The emission should depend on the intensity of the light. But experimental results show that emission requires light to exceed a certain frequency and does not depend on intensity or duration of exposure. Einstein explained this by arguing that the light should be treated as a beam of photons.

6. From a lecture at King's College, London, June 13, 1921. Reproduced in *The Collected Papers of Albert Einstein. Volume 7: The Berlin Years: Writings, 1918–1921* (English translation supplement) (Princeton University Press, Princeton), p. 238. Online at einsteinpapers.press.princeton.edu/vol7-trans/254.

7. Einstein Sees End of Time and Space, *New York Times*, April 4, 1921.

8. Quoted in A. D. Stone, *Einstein and the Quantum: The Quest of the Valiant Swabian*, Princeton University Press, Princeton, 2nd edn, 2015, p. 78.

9. A. D. Stone, *Einstein and the Quantum: The Quest of the Valiant Swabian*, Princeton University Press, Princeton, 2nd edn, 2015, pp. 82–92.

10. A. D. Stone, *Einstein and the Quantum: The Quest of the Valiant Swabian*, Princeton University Press, Princeton, 2nd edn, 2015, pp. 85–86.

11. Einstein's papers are available (in English translation where appropriate) online at einsteinpapers.press.princeton.edu/vol2-trans.

12. C. Seelig, *Albert Einstein: A documentary biography*, translated by M. Savill, Staples Press, London, 1956 p. 78; C. Seelig, *Albert Einstein; eine dokumentarische Biographie*, Europa Verlag, Zurich, 1954, pp. 92–93.

13. A. D. Stone, *Einstein and the Quantum: The Quest of the Valiant Swabian*, Princeton University Press, Princeton, 2nd edn, 2015, pp. 152–155.

14. A. Einstein, *On the Development of Our Views Concerning the Nature and Constitution of Radiation*, originally published in German, translated by A. Beck, *The Collected Papers of Albert Einstein, the Swiss Years: Writings, 1900–1909*, Princeton University Press, Princeton, 1989, vol. 2, pp. 379–394.

15. G. Holton, Einstein, Michelson, and the "Crucial" Experiment, *Isis*, 1969, **60**, 132–197.

16. The four scientists were Max Planck, Emil Warburg, Heinrich Rubens, and Walther Nernst. They were nominating Einstein for the Prussian Academy of Sciences. Quoted in A. País, *Subtle Is The Lord: The Science and Life Of Albert Einstein*, Oxford University Press, New York, 1982, p. 382.

17. R. A. Millikan, Albert Einstein on his seventieth birthday, *Rev. Mod. Phys.*, 1949, **21**, 343–345.

18. A. D. Stone, *Einstein and the Quantum: The Quest of the Valiant Swabian*, Princeton University Press, Princeton, 2nd edn, 2015.

19. Quoted in A. País, *Subtle Is The Lord: The Science and Life Of Albert Einstein*, Oxford University Press, New York, 1982, p. 9.

20. W. Heisenberg, *Physics and Beyond: Encounters and Conversations*, translated by A. J. Pomerans, Harper & Row, New York, 1971, pp. 77–78.

21. Philosophers have two terms. *Realism* supposes that science directly represents reality. *Instrumentalism* takes the position that theories are mathematical models, and it does not matter whether they correspond to reality, simply that they explain the observations.

22. W. Heisenberg, *Across The Frontiers*, Harper & Row, New York, 1975, pp. 83–84.

23. R. Penrose, *The Emperor's New Mind—Concerning Computers, Minds, and The Laws Of Physics*, Oxford University Press, Oxford, 1989, pp. 96–97.

24. In a lecture in 1921, published as A. Einstein, *Sidelights on Relativity*, Dutton & Co., New York, 1922. Online at www.gutenberg.org/files/7333/7333-h/7333-h.htm.

25. The theme was taken up later in an influential paper from Eugene Wigner, awarded the Nobel Prize in Physics in 1963 for his work on symmetry in atomic physics, when he argued that mathematics was 'unreasonably effective.' E. Wigner, The unreasonable effectiveness of mathematics in the natural sciences, *Commun. Pure Appl. Math.*, 1960, **13**, 1–14.

26. M. Atiyah, Pulling the Strings, *Nature*, 2005, **438**, 1081–1082.

27. S. Frabboni, *et al.*, The Young-Feynman two-slits experiment with single electrons: Build-up of the interference pattern and arrival-time distribution using a fast-readout pixel detector, *Ultramicroscopy*, 2012, **116**, 73–76.

28. Even larger entities display the same behavior. Y. Fein, *et al.*, Quantum superposition of molecules beyond 25 kDa, *Nat. Phys.*, 2019, **15**, 1242–1245.

29. R. Feynman, *The Character Of Physical Law (Messenger Lectures, 1964)*, The MIT Press, Cambridge, MA, 1967, p. 145.

30. A. Ananthaswamy, *Through Two Doors at Once: The Elegant Experiment That Captures the Enigma of Our Quantum Reality*, Dutton, New York, 2018.

31. E. Schrödinger, The Present Status of Quantum Mechanics, *Die Naturwiss*, 1935, **23**, 1–26. Translation published in *Proc. Am. Phil. Soc.*, 1980, 124, 323–338.

32. This was a rare exception in which Einstein was not a sole author. In fact, he did not write the paper himself, and he was not happy with the

presentation, as he thought its main point had been "buried by the erudition." See R. Deltete and R. Guy, Einstein and the EPR, *Philos. Sci.*, 1991, **58**, 377–391.

33. A. Einstein, B. Podolsky and N. Rosen, Can quantum-mechanical description of physical reality be considered complete?, *Phys. Rev.*, 1935, **47**, 777–780.

34. N. Bohr, Can Quantum-Mechanical Description of Physical Reality Be Considered Complete?, *Phys. Rev.*, 1935, **48**, 696–702.

35. R. Moore, *Niels Bohr: The Man, His Science, & The World They Changed*, Hodder & Stoughton, London, 1967, p. 196.

36. J. S. Bell, *Speakable and Unspeakable in Quantum Mechanics*, Cambridge University Press, Cambridge, 1987, p. 139.

37. J. S. Bell, On the Einstein Podolsky Rosen Paradox, *Physic*, 1964, **3**, 195–290.

38. C. Bernhardt, *Quantum Computing for Everyone*, The MIT Press, Cambridge, MA, 2020, pp. 72–87; J. Bernstein, *Quantum Profiles*, Princeton University Press, Princeton, 1991, pp. 7–8, 74–77.

39. A paper proposed in 1969 that Bell's Inequalities could be tested: Clauser, *et al.*, Proposed Experiment to Test Local Hidden Variable Theories, *Phys. Rev. Lett.*, 1969, **23**, 880–884. The first experimental test followed in 1972: S. J. Freedman and J. F. Clauser, Experimental test of local hidden-variable theories, *Phys. Rev. Lett.*, 1972, **28**, 838–941. There have been many papers since. They were considered to have possible loopholes until a paper in 2015 B. Hensen, *et al.*, Loophole-free Bell inequality violation using electron spins separated by 1.3 kilometres, *Nature*, 2015, **526**, 682–686.

40. Quoted in C. Bernhardt, *Quantum Computing for Everyone*, The MIT Press, Cambridge, MA, 2020, p. 84.

41. R. Riedinger, *et al.*, Remote quantum entanglement between two micro-mechanical oscillators, *Nature*, 2018, **556**, 473–477.

42. "That gravity should be innate inherent & [essential] to matter so that one body may act upon another at a distance through a vacuum without the mediation of any thing else by & through which their action or force [may] be conveyed from one to another is to me so great an absurdity that I believe no man who has in philosophical matters any competent faculty of thinking can ever fall into it. Gravity must be caused by an agent {acting} constantly according to certain laws, but whether this agent be material or immaterial is a question I have left to the consideration of my readers." Letter from Isaac Newton to Richard Bentley, 5 February 1692/3. 189.R.4.47, ff. 7-8, Trinity College Library, Cambridge. Online at www.newtonproject. ox.ac.uk/view/texts/normalized/THEM00258.

43. Quantum coherence describes the correlations between the physical properties of atomic particles caused by their wave-like nature. Decoherence describes the loss of these correlations due to averaging out the distribution at higher levels.

44. I have drawn in this account on the reviews by J. McFadden and J. Al-Khalili, The Origins of Quantum Biology, *Proc. R. Soc. A*, 2018, **474**(2220), 20180674, and by A. Marais, *et al.*, The Future of Quantum Biology, *J. R. Soc. Interface*, 2018, **15**, 20180640.

45. J. J. Hopfield, Electron transfer between biological molecules by thermally activated tunneling, *Proc. Natl. Acad. Sci. U. S. A.*, 1974, **71**, 3640–3644.
46. G. D. Scholes, *et al.*, Lessons from nature about solar light harvesting, *Nat. Chem.*, 2011, **3**, 763–774.
47. P. Ball, Is Photosynthesis Quantum-ish?, *Phys. World*, 2018, **31**, 44–48.
48. T. Ritz, S. Adem and K. Schulten, A model for photoreceptor-based magnetoreception in birds, *Biophys. J.*, 2000, **78**, 707–718.
49. Cryptochromes were discovered as blue-light receptors in plants.
50. S. Weinberg, *Dreams of a Final Theory: The Scientist's Search for the Ultimate Laws of Nature*, Vintage, New York, 1992.
51. "The mind striving after unification of the theory cannot be satisfied that two fields should exist which, by their nature, are quite independent. A mathematically unified field theory is sought in which the gravitational field and the electromagnetic field are interpreted only as different components or manifestations of the same uniform field, the field equations where possible no longer consisting of logically mutually independent summands." In A. Einstein, *Fundamental ideas and problems of the theory of relativity*, 1923. Lecture delivered to the Nordic Assembly of Naturalists at Gothenburg. Online at www.nobelprize.org/prizes/physics/1921/einstein/lecture/.
52. Horgan makes the case that science has come to an end, but it is scarcely convincing that physics, cosmology, evolutionary biology, neurobiology, *etc.*, should all simultaneously have come to a halt. Physicists are either amused or irritated by the idea, according to their disposition. J. Horgan, *The End Of Science*, Basic Books, New York, 2015.
53. S. Weinberg, *Dreams of a Final Theory: The Scientist's Search for the Ultimate Laws of Nature*, Vintage, New York, 1992.

CHAPTER 16

Einstein's Universe: 1905–

"I CANNOT SERIOUSLY BELIEVE IN IT [QUANTUM THEORY] because the theory cannot be reconciled with the idea that physics should represent a reality in time and space, free from spooky actions at a distance [spukhafte Fernwirkungen]," said Albert Einstein.[1] Einstein is usually regarded as the greatest scientist of the 20th century. Coming from obscurity, he revolutionized several fields with the simultaneous publication of papers on topics ranging from the quantum nature of light to the equivalence of mass and energy. "Einstein's Revolution" is at the least an ambiguous term as it could apply to more than one groundbreaking discovery. It might be fair to say Einstein created as many paradoxes as he solved. The longest-lasting paradoxes have been the consequences of the theory of general relativity, where investigations even today of the nature of the universe produce inconclusive results. Ironically, Einstein did not really participate in the later developments of relativity, partly because he was reluctant to believe some of the implications of his own model. It remains impossible to reconcile relativity and quantum mechanics. Paradox piles upon paradox.

The Frontiers of Science
By Benjamin Lewin
© Benjamin Lewin 2026
Published by the Royal Society of Chemistry, www.rsc.org

Timeline for Key Events in Relativity

1890 — Michaelson and Morley cannot detect ether drift (1887)

1900 — Poincaré derives relativistic formula for adding velocities (1902)
Lorentz discusses transformations of spacetime coordinates (1904)
Einstein proposes special relativity (1905)
Einstein proposes equivalence of mass and energy ($E=mc^2$) (1905)

1910 — Einstein proposes general relativity (1915)
Karl Schwarzschild publishes solution to Einstein field equations (1916)
Einstein predicts gravitational waves (1916)
Einstein introduces cosmological constant (1917)
1920 — Eddington detects bending of light at eclipse (1919)

Lemaître proposes expanding universe (1927)
1930 — Hubble shows red shift depends on distance away (1929)
Einstein abandons cosmological constant (1931)
Cockcroft and Walton confirm $E=mc^2$ (1932)

Einstein and Rosen describe solution for calculating gravitational waves (1937)

1940

1950

1960 — Quasars and pulsars are analyzed by theory of relativity (1963)

1970 — Emission of gravitational waves implied by properties of double star system (1974)

1980

1990 — Cosmological constant resurrected because of accelerating expansion (1998)
Hubble space telescope finds first Einstein ring (1998)
2000

2010 — LIGO detects gravitational waves (2016)

2020

A Dialog between Newton and Einstein

Isaac Newton (1642–1726) brought the Scientific Revolution to its culmination by formulating laws of physics governing general phenomena that previously had been explained only on an *ad hoc* basis.

Albert Einstein (1879–1955) triggered the transition from classical to modern physics by replacing the idea of fixed forces between objects with the concept of spacetime functioning by relativity.

Newton: Greetings, Herr Professor Einstein. I've heard of your revolutionary ideas regarding space, time, and the very fabric of the universe. It seems you've built upon the groundwork I laid centuries ago.

Einstein: Sir Isaac, an honor to engage in conversation with the mind that laid the foundations of classical physics. I have built upon the principles you so brilliantly formulated, particularly in the realm of gravity. Our ideas about the fundamental relationships of matter have evolved.

Newton: Indeed, gravity—the force that binds all matter and dictates the motion of celestial bodies. I described it as a force acting at a distance. What brings you to reconsider this concept?

Einstein: My theory of general relativity offers a new perspective. Rather than a force, I describe gravity as the curvature of spacetime caused by the presence of mass and energy.

Newton: Curvature of spacetime? An intriguing notion, but how does this curvature manifest itself? Surely time is absolute and true and flows from its own nature equably without regard to anything external? And what becomes of my concept of a force acting between masses?

Einstein: In relativity theory, massive objects, such as planets and stars, warp the fabric of spacetime around them. Other objects, following the curvature of this warped spacetime, experience what we perceive as gravity. It's a departure from the idea of a force transmitted through space.

Newton: It's a great departure, indeed. But how does your theory account for the precise predictions made by my laws of motion and gravitation? Are they not sufficient for understanding the cosmos?

Einstein: Your laws are remarkably accurate within the limits of certain conditions, Sir Isaac. However, at the extremes of speed and gravity, such as near massive celestial bodies or at speeds close to that of light, they show their limitations.

Newton: Limits, you say? I deliberately did not propose any hypothesis to explain gravitational force. What advantage is gained by abandoning the simplicity of force? The abstraction of curved spacetime is an absurdity beyond gravity!

Einstein: Your laws are wonderful, Sir Isaac, and they remain accurate in most practical situations. But we need deeper understanding when we move closer to the speed of light. General relativity allows us to grasp the nature of the cosmos in ways your laws cannot. I rest my case.

EINSTEIN'S UNIVERSE

For us believing physicists, the separation between past, present and future has only the meaning of an illusion, albeit a persistent one,[2] Albert Einstein, 1955.

Relativity is full of counter-intuitive paradoxes. If you move towards an object that is itself moving, perception of its speed is affected by whether it is moving towards you or away from you. But relativity theory says that perception of the speed of light is the same irrespective of the observer's position or movement. Nothing can move faster than the speed of light. Yet electrons can be connected in such a way that information appears to travel between them instantaneously. Sometimes it seems that science is moving back towards the ancient view of mysterious forces acting beyond our understanding.

At the start of the Scientific Revolution, knowledge of the universe was confined to the solar system, supposed in effect to comprise a compact set of concentric spheres. By the end of the 17th century, the solar system was seen as a part of the universe, which extended out to the stars. The Milky Way was a concentration of stars, about as far away as we could see. This was a major advance, but there was more to come. (Today the Milky Way is thought to have about 100 billion stars.)

Before the invention of the telescope, the lack of technology limited astronomers to observations with the naked eye. The Scientific Revolution in astronomy followed directly from Galileo's construction of a telescope. But telescopes of the day were limited by technology to relying on relatively small glass lenses. The production of large parabolic mirrors in 1721 made it possible to advance to a new level of scale. It was during the Industrial Revolution that astronomers were able to turn to cosmology.

Our perception of the universe expanded enormously as the result of William Herschel's observations with a telescope that he assembled in 1774. Only a handful of nebulae were known until Herschel published his paper, *One Thousand New Nebulae*, in 1786.[3] Two years previously, he had published *On the Construction of the Heavens*, which replaced the idea of "fixed stars" with the concept that stars were dispersed throughout the universe as far as the telescope could see.[4] Nebulae were clusters of stars forming from condensing gas. Far from being unchanging,

Herschel's Telescope

William Herschel (1738–1822) came from a musical background in Hanover in Germany. When he was 19, he was sent with his brother to take refuge in England after Hanover was defeated in battle with France in 1757. William earned his living as a musician, at first as an oboist and then as an organist in Bath. He started to observe the stars in 1766. Working with his sister Caroline, who joined him in 1772, he began to observe the night sky at a resolution well above anything that had been achieved before.[6] Caroline also had a career as a soprano, but later became an astronomer in her own right, discovering several comets. This was still the era when amateurs could master an area of science.

Herschel's early discoveries were a mix of misinterpretations and inspired leaps. Observing the surface of the Moon, he leapt to the conclusion that the craters were evidence of artificial constructions, made by 'Lunarians.' This attracted considerable scepticism! The discovery that made his name was the observation of what at first appeared to be a new comet in 1781. But it was round and lacked a tail. Further analysis over the next few months showed that it was a new planet, the seventh to be discovered in the solar system, which Herschel called Uranus. He was awarded the Copley Gold Medal by the Royal Society for the discovery.

Herschel's 'Great Forty' telescope was constructed between 1785 and 1789 with a 48-inch concave metal mirror. It was the largest telescope in the world until it was dismantled in 1840. The Astronomer Royal, Nevil Maskelyne, admitted that his telescopes at Greenwich could barely resolve Uranus. "Mr Herschel is undoubtedly the most lucky of Astronomers in looking accidentally at the fixt stars with a 7 foot reflecting telescope magnifying 227 times to discover a comet, which if he had magnified only 100 times he could not have known from a fixt star."[7] Herschel was most indignant at the suggestion that his discovery had been accidental.

the universe was an act of continual creation. Herschel called the clusters, "the Laboratories of the universe." By the time he had counted up to 1000 nebulae, he considered the possibility that some were very distant, beyond the Milky Way.[5] This was the start of cosmology as a modern science.

The very existence of astronomy depends on the ability of light to cross space. By analogy with sound waves traveling through the air, or ripples spreading on a pond, light waves were thought to require a medium in which to propagate. "In space no one can hear you scream," is the iconic tagline of the 1979 sci-fi film *Alien*. This is because sound cannot propagate if there is no medium. By comparison, this is why it was so revolutionary when Einstein abandoned the ether as a medium of propagation for light.

The ether had been a fixed feature of classical physics, but the Michelson–Morley experiment to detect it at the end of the 19th century failed conspicuously, as I describe in Chapter 9. Without the ether, there is no universal frame of reference, which means that motion can only be measured relative to some other object: that's why it's a theory of relativity. The theory of special relativity deals with 'spacetime'—the relationship between the speed (of light), mass, time, and space, in the absence of gravity.

Einstein's papers were in a sense the foundation of the great theoretical tradition in 20th century physics as they started from postulates rather than experimental observations.[8] The two basic postulates for special relativity are that the laws of physics are the same for all observers,[9] and that the speed of light in a vacuum is always the same (which implies the absence of the ether). (The dilemma, prior to Einstein, was that Maxwell's equations describing electromagnetic radiation required the speed of light to be invariant. This was hard to equate with its supposed passage through the ether.)

The basic consequences of special relativity are counter-intuitive. Mass, length, and time are not absolute but depend on the relative motions of observers. Moving near the speed of light, time slows down, space contracts (objects become smaller), and objects gain mass. This was a revolution in the full sense of the term. It suddenly replaced classical physics with an entirely new way of looking at the universe.

The conception of relativity received mixed reactions. When the paper was submitted to the University of Bern in 1907 to support Einstein's application to be a Privatdozent (allowed to teach students), it was rejected with the comment, "I don't understand it at all."[10] On the other hand, Max Planck became an advocate for the paper, although he admitted there was no proof for or against it. "I find it more appealing," he said in justification.[11] In fact, the only one of Einstein's four papers in 1905 that won rapid acceptance was the paper confirming atomic theory. Photons and relativity remained in the realm where personal value systems were more determinative than the equations.

$E = mc^2$ is surely the most famous equation in all of science. Even people whose eyes glaze over at the very thought of an equation know what it means. Einstein described the equivalence of mass and energy as the "most important upshot" of special relativity.[12] The first experimental test of $E = mc^2$ came in 1932, when Cockcroft and Walton at the Cavendish Laboratory used an accelerator to bombard lithium with protons (see Chapter 13). The mass of lithium plus a proton was converted into the mass of two α-particles (helium nuclei) plus kinetic energy.

This wasn't a direct proof. The experiment (conducted for other purposes) simply assumed the validity of the equation in calculating its results. All the same, it's often taken to be the first experimental validation of the equation (others followed along similar lines). Cockcroft and Walton were awarded the Nobel Prize for physics in 1951. A definitive test of the equation had to wait until 2005, when it was confirmed within 0.4 parts per million.[13]

Einstein himself offered two criteria for a scientific theory. It should not have been falsified by empirical facts; and it should be simple. "The supreme goal of all theory is to make the irreducible basic elements as simple and as few as possible without having to surrender the adequate representation of a single datum of experience."[14] (This is often popularly restated as "Everything should be made as simple as possible, but not simpler.") This is the equivalent of Dirac's call for a principle of mathematical beauty (see Chapter 13). It reprises the theme that laws of physics should be elegant.

Einstein believed in the importance of speculation and was prepared to consider the possibility that 'facts' might be wrong or misleading. He was asked by Ilse Rosenthal-Schneider, a doctoral student in Berlin, what his reaction would have been if Eddington's observations of the eclipse at Principe in 1919 had not confirmed the theory of relativity (see below). Einstein replied, "I could just feel sorry for the good Lord. The theory is correct."[15]

General relativity extended the revolution to include gravity. Completed in 1915, like special relativity it was based on empirical reasoning and thought experiments. In a lecture on how he created the theory of relativity, Einstein later recollected, "I was sitting on a chair in my patent office in Bern. Suddenly a thought struck me: If a man falls freely, he would not feel his weight. I was taken aback. . . This led me to the theory of gravity."[16]

Newton had defined rules describing the operation of gravity, but could not explain the nature of the gravitational force. General relativity replaces Newtonian gravity, in which the force of attraction between two masses is proportional to the product of their masses, with the view that large masses warp the shape of the space around them. This controls how matter moves in that space. The key to developing the theory was the realization that spacetime is not flat, but is curved. The Einstein field equations of 1915 relate the curvature of spacetime to mass, energy, and momentum. Put into simplistic terms, think of the Earth as skating round the interior rim of a bowl created by the gravity of the Sun.

Here there was an immediate experimental test to support the theory. It had been known since 1859 that Mercury (the nearest planet to the Sun) has an orbit that is not quite regular. General relativity explains that this is due to the warping of space near the Sun. It exactly predicts the deviation.

Einstein's theory of special relativity was published during the First World War at a time when international cooperation in science had broken down. Einstein was in Berlin, maintaining contact with other physicists through the neutral Netherlands, in particular with astronomers Willem de Sitter and Paul Ehrenfest. In spite of the war, it was Sir Arthur Eddington in Cambridge who

Einstein did not meet Eddington until a visit to England in 1921. On a later visit, in June 1930, he spent a week with Eddington at Cambridge.

The photograph was taken in the garden of the Observatory, possibly by Winifred Eddington.

There were two expeditions to measure star positions at the eclipse of 1919, one to Sobral, Brazil and one to Principe, Africa. All the equipment had to be transported, creating difficulties in calibration. The photograph shows the instruments for Sobral.

Reproduced from Space, Time and Graviation: An Outline of the General Relativity Theory (1920) with permission from Cambridge University Press.

proselytized the theory of relativity among his fellow physicists. Eddington had been aware of the theory since it was published, but became really involved when he received details in a series of letters from de Sitter in 1916.[17]

Dramatic support for the theory came from a solar eclipse in 1919. General relativity predicts that light should be bent by gravity. The eclipse made it possible to photograph stars whose light passed close to the Sun during the blackout. Comparison with photographs taken six months earlier, when the stars were in locations where their light passed farther from the Sun, would reveal whether there was any difference.

It was not straightforward to organize observations of the eclipse, as they were possible only in restricted locations. Together with the Astronomer Royal, Frank Dyson, who raised the funds, Eddington organized the expeditions to measure bending of light during the eclipse. There were two expeditions, one to

Glass positive photograph of the solar eclipse taken May 29, 1919 at Sobral in Brazil. Asterisks mark the stars whose positions were measured.

Reproduced from https://www.eso.org/public/images/potw1926a/, under the terms of the CC BY 4.0 license https://creativecommons. org/licenses/by/4.0/.

Sobral, in Brazil, and the other to Principe, an island off the west coast of Africa.

These were not easy measurements to make. The deflection expected from the theory of general relativity is very small, 1.7 arcsec (about $0.0004°$). Newton's theory of gravity had already predicted that a massive body such as the Sun would deflect light. But it predicted a smaller effect with a difference of about 0.8 arcsec. Tiny changes in focal length, other mechanical deformations, or changes in temperature, all can have effects of comparable magnitude to the 2-fold margin for supporting relativity.

Earlier attempts had produced inconclusive results. The British expedition to Principe in 1919 was the first to produce positive results, but recently there has been a controversy as to whether in fact it did succeed.[18] The two critical photographs at Principe produced a deflection of 1.61 ± 0.45 arcsec. They are in better agreement with relativity than the alternative, but the evidence is not very strong given the high error bar.

The Hubble telescope was built by NASA and the ESA (European Space Agency). Launched in 1990, it extends observations well beyond Earth-bound astronomy and examines the spectrum from ultraviolet to infrared.

NASA.

When the results were reported as supporting the theory, the resulting publicity made Einstein an international celebrity. Yet everyone—even physicists—found it difficult to understand relativity. Even J. J. Thomson (discoverer of the electron, who had become President of the Royal Society) commented that, "Perhaps Einstein has made the greatest achievement in human thought, but no one has yet succeeded in stating in clear language what the theory of Einstein's really is."[19] In 1920, Eddington wrote a semi-popular book to explain relativity to the public, *Space Time and Gravitation*.

During the 6-minute total eclipse on Principe, Eddington photographed star positions. This required only simple equipment. The more sophisticated analyses of the present era, examining the bending of light around galaxy masses or black holes, require vastly more complicated equipment. This can be done only from space. The Hubble space telescope is a demonstration that the cost and facilities required for research into astronomy now can be provided only at governmental level.

∞ ∞ ∞

One consequence of the bending light of light by gravity predicted by general relativity is that when light passes by an extremely large mass (like a galaxy cluster or a black hole) it is diverted and forms an arc. This is known as an Einstein Ring. This image from the Hubble telescope supports the theory. (For a picture of a black hole see Prologue.)

Reproduced from https://esahubble.org/images/potw1814a/, under the terms of the CC BY 4.0 license https://creativecommons.org/licenses/by/4.0/.

Although by now obvious as one of the major scientific achievements of the century, relativity was never acknowledged with a Nobel Prize. Einstein's prize was in fact awarded in spite of considerable opposition. Not only did the award avoid mentioning quanta, it never cited relativity. Einstein had been nominated for the prize since 1910. In 1921, when he was a leading contender, the committee decided not to award any prize. Their grounds were that no nomination was worthy of it! The next year, they awarded two prizes, the 1921 (retroactively) to Einstein, and the 1922.

Records show intense opposition from some members of the committee to any award to Einstein for relativity, based on spurious scientific arguments. There have certainly been other cases where misguided opposition to a scientific advance has delayed recognition, but here the arguments went beyond reason. Antisemitism played a significant part. This factor has been somewhat hidden in historical accounts of the process (and notably in the Nobel Committee's own account).[20]

Pulling out the ground from underneath one's feet is a fair description of the effects of the introduction of relativity. It took more than two decades for general relativity to be accepted. Revolution is really inadequate to describe the effects of general relativity on either physicists or laymen. Not only was all of classical physics turned upside down, but there was a perpetual series of dichotomies and paradoxes that were simply impossible to assimilate. Attempts to understand this series of mind-bending conundrums move into philosophy and metaphysics. We are back close to where we started.

Part of the reason for delays in accepting the new paradigms may have been that they were, at first, theoretical. When a new theory is based on experiments, first the experiments are scrutinized for errors in methodology, and then the interpretation is examined. When theory precedes experiments, there is far more scope for argument. The acceptance of special and general relativity required experiments to produce supporting data. Indeed, the jury is not yet entirely in.

Quantum mechanics and general relativity both obey uniform laws, even if we can't completely define them or relate the two situations. It becomes more confused when physicists try to go back to understand the origins of the universe. Paul Dirac made a comment in 1939 that seems prophetic in view of the difficulties encountered in cosmology today. "At the beginning of time the laws of Nature were probably very different from what they are now. Thus we should consider the laws of Nature as continually changing with the epoch, instead of as holding uniformly throughout space-time."[21]

In its first draft, the theory of general relativity predicted that the universe must be expanding, because gravity would cause a static universe to contract. Einstein was not happy with the exclusion of a static universe. He introduced a term into his equations, the cosmological constant, to make it possible. This was an example of 'fine-tuning' (see Chapter 11). Einstein was never reconciled to the cosmological constant. He said in 1923 that it was "gravely detrimental to the formal beauty of the theory.[22]... Since I have introduced this term I had always a bad conscience... I am unable to believe that such an ugly thing should be realized in nature."[23] (Here is the concept of beauty again.) According to cosmologist George Gamow, Einstein regarded the cosmological constant as his "biggest blunder."[24] In any event, relativity started out with a static universe.

This was a case of being fooled by the data. The universe appeared to be static at the time, but that was about to change. However, a lasting consequence was that Einstein became mistrustful of the need to modify theory to accommodate facts. Together with his reluctance to accept the uncertainties of quantum mechanics, this led to his becoming cut off from developments in his later years.[25]

∞ ∞ ∞

In the first half of the 20th century, the Mount Wilson observatory near Los Angeles had the biggest telescope in the world. Astronomer Edwin Hubble (1889–1953) was there in the 1920s. Due to family pressure, he became a lawyer after graduating from the University of Chicago, but then took up his true love of astronomy. His career was interrupted by the First World War, when he worked on ballistics in the army. He went to Mount Wilson at the end of the war in 1919.

The expansion of the universe beyond known bounds started in 1924, when Hubble measured the luminosity of stars known as Cepheids. The first Cepheid was in the Andromeda area. It was 900 000 light years away, $3\times$ the estimate for the diameter of the Milky Way.[26] This meant that Andromeda must be outside the Milky Way. We now know it is the nearest external galaxy to the Milky Way.[27]

Before Hubble's observations, our universe was effectively restricted to the Milky Way. (There is a parallel here with Tycho Brahe's demonstration that a comet must be farther away than the Moon, which expanded our view of the cosmos beyond the sublunar area; see Chapter 4.) The impact of Hubble's discovery was immediate. When Harlow Shapley, head of the Harvard College Observatory, received a letter from Hubble describing the result, he turned to a colleague and said, "Here is the letter that has destroyed my universe."[28]

The next breakthrough came in 1929 when Hubble measured the distances to even more distant nebulae. It was already known that distant objects display a red shift, a change in the light spectrum indicating that they are moving away from us. Hubble obtained an amazing result: the extent of the red shift (indicating the speed with which the object is receding) increases with distance away from us. The farther away the object is, the faster it is receding![29]

Hubble's paper presented his observations without much interpretation—he was another experimentalist who was skeptical of theory. However, unknown to Hubble, Georges Lemaître had previously calculated from Hubble's earlier results that the speed with which a galaxy retreated from us should be proportional to its distance away. Lemaître (1894–1966) was a Belgian priest and astronomer who had been a student of Arthur Eddington. He constructed a model in 1927 to show that the universe was expanding. In fact, Russian astrophysicist Alexander Friedmann had provided the same solution in 1922, but he was interested in the mathematics, not physical reality. No one knew about his results either. The 1920s–1940s (and sometimes even later) were a period when the theoreticians and experimentalists continually talked past one another.

Lemaître had published in an obscure journal, so his results were largely unknown.[30] He tried to discuss his conclusions with Einstein at the Solvay Conference in 1927 (see Chapter 14), but Einstein dismissed him. "Vos calculs sont corrects, mais votre physique est abominable" (Your calculations are correct but your physical insight is terrible), Einstein said.[31]

<hr>

Components of the Universe

<hr>

The universe was created 13.7 billion years ago by the Big Bang.

The universe is 5% visible matter, consisting of atomic particles (as defined on Earth) organized into different structures.

Most of the universe is dark, with 72% dark energy and 23% dark matter. They are called dark because their properties are unknown.

A nebula is a cloud of gas and dust in space. Nebulae vary from regions where new stars are being formed to those that have the remnants of dead stars.

A galaxy is a large grouping of stars and other materials, held together by gravity. Galaxies vary from a few million to more than a trillion stars.

The Milky Way is a large galaxy, containing more than 100 billion stars. It is part of a group of galaxies. The Solar System is about a third of the way from the galactic center of the Milky Way.

<hr>

More missed opportunities to put things together! It's a great contrast with the practice of science in the era of the instant communication of the internet. Actually, the failure of theorists and experimentalists to talk to one another might have been as well, because Lemaître's prediction held only for objects more than 100 million light years apart. Hubble's initial measurements went out to about 6 million light years. In fact, the data look terrible. A straight line through them shows a relationship between speed of movement and distance, but the scatter is so great as to cast doubt on it.[32] It was a leap of inspiration to propose a proportional relationship.

Two years later, Hubble made measurements going out to 120 million light years that produced a convincing straight-line relationship. By then, Lemaître's results had come to light (he had sent a copy of his paper to Eddington, who realized its importance) and everything was put together in the interpretation that the universe is expanding.[33] Although this is a keystone of current cosmological theory, neither Hubble nor Lemaître was ever awarded a Nobel Prize.

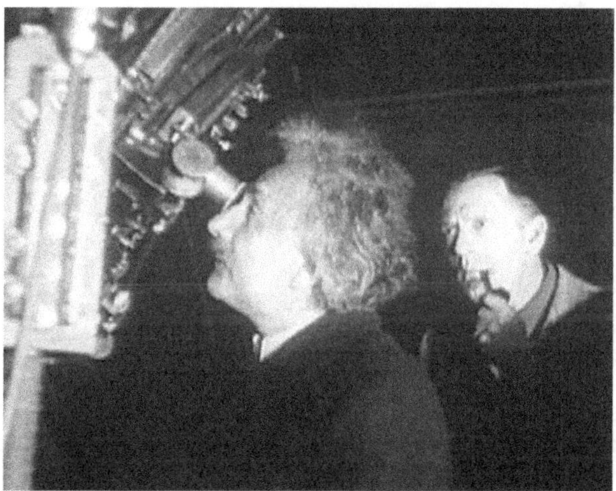

Einstein looking into the eyepiece of the 100-inch Hooker telescope at Mt. Wilson Observatory, with Edwin Hubble in January 1931. The photograph was posed for the photographers.

Reproduced with permission from Caltech Archives.

The results made a stir: Einstein visited Hubble at the Mt. Wilson observatory in 1931. When he accepted the theory of the expanding universe that year, Einstein set the cosmological constant at zero (essentially taking the view that it was, after all, superfluous). Today it's thought to have a small positive value. In fact, the cosmological constant has become an important factor in current cosmology. Very small changes in its value have drastic effects on the models for construction of the universe (see Chapter 17). Einstein's blunder may not have been a mistake after all. (This does not disguise the fact, however, that it remains essentially a fudge factor, accepted because it makes the theory work, but without a sound theoretical basis.[34])

The impression that Einstein's formulation of special relativity was a bolt out of the blue was enhanced by the way his papers cited little prior work. But what would have happened if Einstein had never published his papers? Mathematician Henri Poincaré (1854–1912) was approaching the idea of relativity, and in 1905 considered the possibility of discarding the ether. Physicist Hendrik Lorentz (1853–1928) developed the equations in 1904, now known as the Lorentz transformations, that relate different frames in spacetime. It's unknowable whether Poincaré or Lorentz would have made the leap to integrate the equations into a system of relative spacetime, but sooner or later, the ether would have been discarded and a theory of relativity would have emerged.[35] Einstein himself remarked that, "There is no doubt, that the special theory of relativity, if we regard its development in retrospect, was ripe for discovery in 1905."[36]

NOTES AND REFERENCES

1. In a letter from Einstein to Max Born, March 3, 1947, reproduced in A. Einstein and M. Born, *Born-Einstein Letters, 1916–1955: Friendship, Politics and Physics in Uncertain Times*, Macmillan, London, 1971, p. 158.
2. Letter of condolence to family of Michael Besso in 1955. Translated from German in H. D. Zeh, *The Physical Basis of the Direction of Time*, Springer, Berlin, 2001, p. 199.
3. W. Herschel, Catalogue of one thousand new nebulae and clusters of stars, *Philos. Trans. R. Soc. London*, 1786, **76**, 1–47.
4. W. Herschel, On the construction of the heavens, *Philos. Trans. R. Soc. London*, 1785, **75**, 13–266.
5. R. Holmes, *The Age of Wonder: How the Romantic Generation Discovered the Beauty and Terror of Science*, Harper, New York, 2009, pp. 60–124.

6. Conventional telescopes worked on refraction, with two lenses, one at each end of the tube. Newton had invented the reflecting telescope, in which a concave mirror concentrates starlight. Herschel's first reflecting telescope had a 6-inch diameter mirror that he made himself by polishing metal.

7. Quoted in R. Holmes, *The Age of Wonder: How the Romantic Generation Discovered the Beauty and Terror of Science*, Harper, New York, 2009, p. 74.

8. A. Lightman, *The Discoveries: Great Breakthroughs In 20th-Century Science, Including The Original Papers*, Vintage, New York, 2006, pp. 60–71.

9. Assuming they are traveling at constant velocities (meaning there is no acceleration).

10. G. Holton, *The Advancement Of Science, and Its Burdens*, Cambridge University Press, Cambridge, 1986, p. 71.

11. G. Holton, *The Advancement Of Science, and Its Burdens*, Cambridge University Press, Cambridge, 1986, p. 70.

12. A. Einstein, What is the Theory of Relativity. Letter to London Times, November 28, 1919.

13. S. Rainville, *et al.*, World Year Of Physics: A Direct Test of $E = mc^2$, *Nature*, 2005, **438**, 1096–1097.

14. In The Herbert Spencer Lecture, delivered at Oxford, June 10, 1933. Reprinted in A. Einstein, On the Method of Theoretical Physics, *Philos. Sci.*, 1934, **1**, 163–169.

15. "Do könnt' mir halt der liebe Gott leid tun. Die Theorie stimmt doch." Quoted in I. Rosenthal-Schneider, *Reality and Scientific Truth. Discussions with Einstein, Von Laue, and Planck*, Wayne State University Press, Detroit, MI, 1980, p. 74.

16. In a lecture in Kyoto in 1922. A. Einstein, How I Created the Theory of Relativity, translated by Y. A. Ono, *Physics Today*, 1922, **35**, 45–47.

17. J. Stachel, Eddington and Einstein, in *The Prism of Science*, ed. E. Ullmann-Margalait, Riedel, Dordrecht, 1986, pp. 225–250, p. 227.

18. There were 3 photographs from Principe. Two showed the deflection predicted by general relativity and were taken to confirm the theory. The third, and the photographs from Sobral, were discarded because they were judged to be out of focus. This led to an accusation that Eddington had massaged the data. J. Earman and G. Glymour, Relativity and eclipses: the British Eclipse expeditions of 1919 and their predecessors, *Hist. Stud. Phys. Sci.*, 1980, **11**, 49–85. This resulted in popular reports that another scientific proof was based on fudged data. Indeed, this is now the dominant view. That analysis has been debunked by G. Gilmore and G. Tausch-Pebody, The 1919 eclipse results that verified general relativity and their later detractors: a story re-told, *Notes Rech.*, 2022, **76**, 155–180. In fact, all of the data were available publicly and the reasons for discarding the results were stated.

19. D. Brian, *Einstein: A Life*, John Wiley, New York, 1996, p. 101.

20. R. M. Friedman, The 100th Anniversary of Einstein's Nobel Prize: Facts and Fiction, *Ann. Phys.*, 2022, 534.

21. Letter of condolence to family of Michael Besso in 1955. Translated from German in H. D. Zeh, *The Physical Basis of the Direction of Time*, Springer, Berlin, 2001, p. 199.

22. I. Lorentz *et al.*, *The Principle of Relativity*, Dover, New York, 1923, p. 189.

23. In a letter from Einstein to Lemaître, September 1947. Einstein Collection. box 6, Boston University.

24. G. Gamow, *My Worldline*, Viking Press, New York, 1970, p. 44.

25. D. Bodanis, *Einstein's Greatest Mistake: A Biography*, Mariner Books, New York, 2016.

26. Current estimates are that the Milky Way has a diameter of about 100 000 light years and Andromeda is about 2 million light years away.

27. Hubble called Andromeda an extragalactic nebula; it was only later it was realized it was actually a galaxy.

28. Quoted in B. J. T. Jones, V. J. Martinez and V. L. Trimble, *The Reinvention of Science: Slaying the Dragons of Dogma and Ignorance*, World Scientific Publishing, Singapore, 2023, p. 270. For general accounts of Hubble's discovery see Jones *et al.*, pp. 259–274 and M. Bartusiak, *The Day We Found the Universe*, Pantheon Books, New York, 2009, pp. 199–224.

29. N. A. Bahcall, Hubble's Law and the Expanding Universe, *Proc. Natl. Acad. Sci. U. S. A.*, 2015, **112**, 3173–3175.

30. G. Lemaître, Un Univers homogène de masse constante et de rayon croissant rendant compte de la vitesse radiale des nebuleuses extra-galactiques, *Ann. Soc. Sci. Bruxelles*, 1927, **47**, 49–59. Republished as G. Lemaître, A homogeneous universe of constant mass and increasing radius accounting for the radial velocity of extra-galactic nebulae, *Gen. Relativ. Gravitation*, 2013, **45**, 1635–1646.

31. Quoted in A. Deprit, Monsignor Georges Lemaître, in *The Big Bang and Georges Lemaître*, ed. A. Barger, Reidel, Dordrecht, 1984, p. 370.

32. E. Hubble, A relation between distance and radial velocity among extra-galactic nebulae, *Proc. Natl. Acad. Sci. U. S. A.*, 1929, **15**, 168–173.

33. A. Lightman, *The Discoveries: Great Breakthroughs In 20th-Century Science, Including The Original Papers*, Vintage, New York, 2006, pp. 230–245.

34. P. J. E. Peebles, *Cosmology's Century. An Inside History of Our Modern Understanding of the Universe*, Princeton University Press, Princeton, 2020, p. 3.

35. It's controversial to what extent Einstein might have known of Poincaré or Lorenz's work. There are opposing views in E. T. Whittaker, *A History of the Theories of Aether and Electricity: Volume 2 The Modern Theories 1900–1926*, Nelson, London, 1953 and G. Holton, *Thematic Origins of Scientific Thought: Kepler to Einstein*, Harvard University Press, Cambridge, 1973, for example. It's in any case a striking difference compared to the practice of science today that until relatively recently in the 20th century researchers did not necessarily know of prior relevant work (especially concerning division between theoreticians and experimentalists; see, for example, Chapter 17).

36. Quoted in M. Born, *Physics in My Generation*, Pergamon Press, London, 1956, p. 194.

Lemaître's Big Bang: 1931–

"WE COULD CONCEIVE THE BEGINNING OF THE UNIVERSE IN the form of a unique atom, the atomic weight of which is the total mass of the universe," said Georges Lemaître.[1] The Big Bang is one of the scientific concepts that has taken hold in the popular imagination. It originated when Lemaître, a Jesuit and physicist, proposed that the universe had originated in a primeval atom. Ironically, the theory was not named by its supporters, but by Fred Hoyle, an opponent who favored the alternative theory that the universe is in a steady state that has existed for ever. The models make different predictions, because a Big Bang leaves relics of the original explosion that can be detected in the form of cosmic radiation and gravitational waves. While some of these predictions have been confirmed, there are many discrepancies. The universe is expanding too fast. There are gravitational effects that cannot be explained by the existence of matter as we know it. In fact, matter can account for only 5% of the universe. And the likelihood that the universe would develop conditions propitious to support life is infinitely small. Efforts to reconcile these results with what we can infer of the Big Bang seem to come closer to science fiction than describable reality.

The Frontiers of Science
By Benjamin Lewin
© Benjamin Lewin 2026
Published by the Royal Society of Chemistry, www.rsc.org

Timeline for Key Discoveries in Cosmology

1900

Einstein proposes special relativity (1905)

1910

Einstein proposes general relativity (1915)
Einstein introduces cosmological constant (1917)

1920

Erwin Hubble detects red shift (1924)

Lemaître proposes expanding universe (1927)

1930

Hubble shows red shift depends on distance away (1929)
Lemaître proposes universe originated in primeval atom (1931)

Fritz Zwicky proposes existence of dark matter (1933)

1940

Hans Bethe shows nuclear fusion generates energy inside stars (1938)

Gamow formalizes model later known as Big Bang (1946)
Hoyle establishes principle of nucleosynthesis in stars (1946)

Hoyle proposes steady state for origin of universe (1948)

1950

Hoyle introduces term 'Big Bang' (1950)

1960

Penzias & Wilson detect cosmic background radiation (1964)
Everett proposes multiverse theory (1965)
Bell Burnell and Hewish detects pulsars (1967)
John Wheeler introduces term 'black hole' (1967)

1970

Penrose & Hawking characterize collapse of black holes (1969)
Gabriele Veneziano introduces equations that become string theory (1969)
Black holes detected by several researchers (1971)
Carter proposes anthropy (design of universe to support life) (1973)
Stephen Hawking postulates black holes emit radiation (1974)

1980

Five superstring theories proposed (1985)

1990

Hubble Space Telescope launched (1990)

Dark energy inferred from accelerating expansion of universe (1998)

2000

2010

LIGO detects gravitational waves (2016)

2020

Event Horizon Telescope photographs black hole (2019)

A Dialog between William Herschel and Georges Lemaître

William Herschel (1738–1822) was a German-born British astronomer who together with his sister Caroline invented more powerful telescopes leading to the discovery of clusters of stars that expanded the known universe.

© DeAgostini/Getty Images.

Georges Lemaître (1894–1966) was a Belgian astronomer who was also ordained as a priest in the Catholic Church. He was Professor of astrophysics at the Catholic University of Leuven.

Archives de l'Université catholique de Louvain, BE A4006 FG LEM-603, © ESA.

Herschel: I have spent countless nights observing the heavens, Monsignor Lemaître. The vast star systems, the nebulous regions—all of them seemed like eternal fixtures. I have often wondered if the universe has existed indefinitely.

Lemaître: I understand. That was the prevailing thought for centuries. A static, eternal universe is a comforting idea. But our understanding has evolved. My studies in mathematics and physics have led me to another conclusion.

Herschel: You do not believe the universe is eternal?

Lemaître: I came to believe that the universe had a beginning—a singular, primordial event. What I call the "primeval atom," or what others might now call the Big Bang. I theorized that all matter, energy, and even space itself, were once compressed into a single, dense point, which then expanded.

Herschel: Extraordinary! I have concluded that the universe continuously changes, but you suggest that it had a specific origin? A moment when it came into being?

Lemaître: Precisely. I believe that the universe is expanding. I developed this idea after studying Einstein's equations, which hinted at a dynamic universe. Observations by astronomers, such as Erwin Hubble, later supported the idea. If we reverse the process, we come to the origin of the universe.

Herschel: A magnificent theory, but I must wonder: what came before this primeval atom? And why did it expand? Do we not need a cause for this great event?

Lemaître: These are indeed profound questions. As a man of faith and science, I do ponder the ultimate cause, but scientifically, I must admit we do not know what preceded the Big Bang. It's not clear if "before" even holds meaning when time itself was bound to that moment. The mathematics grows hazy at such depths.

Herschel: Your ideas make the universe a grand, living thing, continuously growing and changing. There is something poetic, even divine, in a universe with a beginning and an unfolding purpose. How does this equate with your position as a man of faith within the Church?

Lemaître: There are those who believe that the theory aligns closely with the doctrine of Creation, that it represents God's will bringing the cosmos into existence from nothingness. However, I believe that scientific theories should remain neutral with respect to theology. The model is based on observations and mathematics. It should neither confirm nor contradict theology, for it addresses a separate dimension of truth. It should be judged solely by the criteria of science. Belief is to theology as data are to science. I rest my case.

LEMAÎTRE'S BIG BANG

The new cosmology will probably turn out to be philosophically even more revolutionary than relativity or the quantum theory, although at present one can hardly realize its full implications,[2] Paul Dirac, 1939.

If the rate of expansion one second after the Big Bang had been smaller by even one part in a hundred thousand million million, it would have recollapsed before it reached its present size. On the other hand, if it had been greater by a part in a million, the universe would have expanded too rapidly for stars and planets to form,[3] Stephen Hawking, 2010.

The Big Bang doesn't actually mark the beginning of our universe; it marks the end of our theoretical understanding,[4] Sean Carroll, 2016.

The universe has either existed forever or was produced by a discrete moment of creation. The idea that the universe began with the explosion of a single very dense particle—the 'primeval atom'—originated in 1931 with Georges Lemaître. Lemaître's proposal was made in a very short letter to *Nature*, less than one page.[5] It's a bit like the primeval atom itself—its impact has been out of all proportion to its size.

Pope Pius XII took the position at a meeting of the Pontifical Academy of Sciences in 1951 that the concept of a sudden act of creation was consistent with the Church's view of God as the creator. This was not at all Lemaître's position. He wanted his hypothesis to be judged solely as a theory in physics. Supposedly Lemaître later met with Pius XII to argue that physics and religion were independent realms of thought. At a conference in 1958, he said: "As far as I can see, such a theory remains entirely outside any metaphysical or religious question. It leaves the materialist free to deny any transcendental Being... For the believer, it removes any attempt at familiarity with God..."[6]

The primeval atom developed into the Big Bang, advocated by George Gamow, originally a Russian astrophysicist, who worked with Alexander Friedman in Leningrad. He defected to the United States in 1934, and helped to develop Lemaître's idea into a more formal model in the 1940s. It was controversial. British

Robert Millikan, Georges Lemaître and Albert Einstein after Lemaître's lecture at the California Institute of Technology in January 1933.

astronomer Fred Hoyle regarded it as pseudoscience, describing it as "an irrational process that cannot be described in scientific terms." He coined the term Big Bang as a description in a talk he gave on the BBC radio in 1949, but it has stuck.[7] He denied later that it was intended to be pejorative.

Hoyle was a great popularizer of science (and beyond that an author of science fiction, including the BBC television series A for Andromeda in the 1960s.) He was a leading proponent of the steady state theory, which proposed that the universe has always existed and expands steadily. The laws of physics would be unchanging. Attempts to test the different predictions made by the 'steady state' and 'Big Bang' theories were inconclusive through the 1950s, creating another of those periods in cosmology when practitioners were free to choose which theory to believe.[8]

Hubble's observations on the red shift (see previous chapter) were the impetus for the Big Bang theory. If the universe is expanding in such a way that distant galaxies recede at a speed proportional to their distance away, what happens if you extrapolate backward by reversing the expansion? This suggests an origin in a single particle—Lemaître's primeval atom—whose disintegration generated the universe. Going back to the beginnings, Gamow supposed that the density and temperature of the primordial particle would start nuclear fusion after the Big Bang. This would create atomic nuclei. The process would quickly stop as heat was released and the universe cooled down.

The model predicts that the universe started as a dense mass of atomic subparticles, so energetic they were moving close to the speed of light. After about 10^{-6} sec, energies dropped to the point at which protons and neutrons could form. They combined to form deuterium and helium nuclei after 3 minutes. After about 400 000 years, electrons combined with atomic nuclei to form hydrogen, helium, and lithium.

Heavier elements were formed much later within stars by a fusion process (called nucleosynthesis, which requires the gravitational pressure created by a massive body). Ironically, Fred Hoyle discovered nucleosynthesis in stars because he did not believe the Big Bang could have provided appropriate conditions. He was right about that, but wrong to dismiss the Big Bang as a result. The matter of the universe still consists of about 70% hydrogen. Our understanding of how the basic units of chemistry—the elements—were created therefore depends upon cosmology.[9]

As early as 1948, two physicists at the Applied Physics Laboratory at Johns Hopkins University predicted in a one-page note in *Nature* that the heat of the Big Bang should have radiated throughout the universe. It should have cooled down with cosmic expansion over around 14 billion years to 5° Kelvin or –267 °C. It's a sort of shockwave or echo of the Big Bang, effectively a cooled fossil of the first release of energy. Current theory says that it should be detectable as waves in the microwave frequency.[10]

The detection of the waves is another story of missed opportunities. Canadian astrophysicist Andrew McKellar detected waves at 2.3° above absolute zero in 1941, but there was no context at the time to interpret them. A few years later, a team led by George Gamow predicted the existence of waves at a slightly higher temperature—but they were theoretical physicists and did not know about the earlier experimental results. Early in 1964, Robert Dicke, a cosmologist at Princeton, predicted that cosmic microwave background radiation (CMB) should exist at a temperature of about 10° above absolute zero, again without knowing about any of the earlier work!

Dicke was about to start an experimental search for the CMB when he heard from Arno Penzias and Robert Wilson at Bell Labs in New Jersey that they had detected peculiar signals when they were calibrating a giant microwave antenna—at first they thought it was random noise or equipment failure.[11] At one point, they thought the problem was due to pigeons that had nested in the antenna, but the signal continued after they removed the pigeons (and their droppings). This happy accident turned out to be the CMB.[12]

The waves are an echo of the Big Bang. Now they are widely regarded as a refutation of the steady state theory. (Although it is still possible to produce alternative theories arguing, for example, that the universe is alternately expanding and contracting.) Penzias and Wilson's paper reporting the CMB carried an account of the experiments but little interpretation.

"A possible explanation for the observed excess noise temperature is the one given by Dicke *et al.* in a companion letter in this issue," they said in introducing their results. Dicke's paper

The Bell Labs facility where CMB was detected is now the National Historical Landmark of the Holmdel Horn Antenna. Made of aluminum, it is 15 m long with an aperture of 6 m × 6 m. The elevation wheel, which is used to turn the antenna, is 10 m in diameter.

offered the expanding universe as one possible explanation, among others. Penzias and Wilson were awarded the Nobel Prize in 1978, but none of the theorists who had predicted the existence of the waves were included in the award!

One difficulty with the original model for an expanding universe was that, because the distances of remote galaxies were underestimated, the Universe appeared to be a mere 2 billion years old. Such a young age contrasted with geological evidence that the Earth is 3–5 billion years old. The problem was resolved in the 1950s when new calculations gave the universe a much older age. It's now thought to be about 13.7 billion years.

∞ ∞ ∞

The Big Bang was a singularity—a point at which the laws of physics break down. Or at least, a point at which they may have been different. The level of energies during the Big Bang was several orders of magnitude greater than anything since.

The best evidence for singularities within our present universe comes from black holes—stars that have collapsed into masses so dense that no light can escape from them. In a paper in 1965 that was only 3 pages long, Roger Penrose calculated that all sufficiently massive stars eventually collapse.[13] The basis of the calculation is that gravitational collapse creates a point of no return after which further collapse cannot be avoided. This forms a 'horizon surface' that traps light internally where spacetime breaks down—the black hole results from light's failure to escape.

The idea that an object of sufficient mass could trap light actually goes back to the 18th century. John Michell argued in 1783 that if a body is sufficiently massive, "all light emitted from such a body would be made to return to it by its own power of gravity."[14] While at the front in World War I, Karl Schwarzschild derived an equation from Einstein's theory of general relativity to show how a massive star could collapse to a tiny radius. Known only in theory, such stars were variously called frozen stars, dark stars, or collapsed stars.

They were regarded as theoretical oddities. The term 'black hole' was introduced by American astrophysicist John Wheeler in an exchange at a conference in 1967, about observations made a couple of years before. Of course, it's impossible to see a black hole directly, because it emits no light. So the presence of a black hole has to be inferred from its effects on the environment around it, as predicted by the theory of relativity. One of these effects is the generation of X-rays.

Cygnus X-1 was detected because of a pattern of flickering X-rays, but it was not visible at any other wavelength. The closest visible object was a star whose orbit suggested it was part of a binary system. To explain the orbit, Cygnus X-1 would have to have a very large mass, but the pattern of X-ray emissions indicated it must be very small. The contradiction identified a black hole.

The first images of black holes were published in 2019 (see Prologue). Penrose shared a Nobel Prize for Physics in 2020 with the leaders of the teams that identified a massive black hole at the center of the Milky Way by observing the motions of stars around it. Other black holes have been detected since.

Black holes were regarded originally as the ultimate destroyers of information: it can go in, but it cannot get out. Stephen Hawking developed the idea that in fact their behavior is quite the opposite: information is stored on the horizon surface (the point where the singularity occurs and light cannot get out). In effect, they are holograms. This creates another conundrum. Black holes eventually disintegrate, which should mean all that holographic information is lost. Yet according to quantum theory, such loss of information is impossible.[15] There is, of course, no direct evidence to bear on the question.

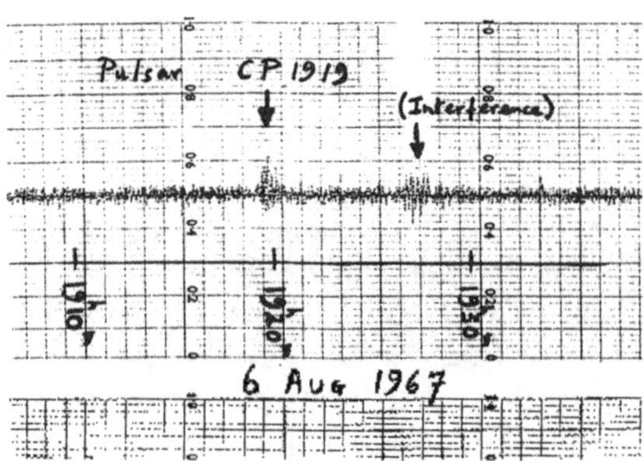

The first trace of a pulsar, from Jocelyn Bell's chart. It took an acute eye to see that it might be distinguished from interference.

Reproduced with permission from Jocelyn Bell Burnell.

If collapsing stars are massive but not sufficiently massive, they may become pulsars. When a massive star collapses, protons and electrons are crushed into neutrons. At certain masses, the neutrons stop the collapse, and these neutron stars emit radio waves.[16] The pattern of the radio waves gave rise to the name pulsar. (For larger masses the collapse continues, generating a black hole.)

The credit for the discovery of the radio waves became controversial. British astrophysicist Antony Hewish was awarded a Nobel Prize in 1974 for discovering pulsars, but the Nobel Committee did not acknowledge Jocelyn Bell. She had actually made the observation while a postgraduate student (and reportedly had difficulties convincing her supervisor of its reality).

The first signals, detected in August 1967, were barely distinguishable from interference. By November, regular pulses, repeating every 1.337 seconds, and lasting for milliseconds, were detected. The possibility that the signal might emanate from an extra-terrestrial intelligence led to the pulsar being named (semi-facetiously) as LGM1, for Little Green Man 1. However, it was soon realized that it could not be an artificial signal. The parameters of the signal indicated that the emitting object must be relatively small—a condensed neutron star.

The theory of relativity says that the interactions of moving massive objects, such as two pulsars or a collision between black

Jocelyn Bell connecting the instrument at the Mullard Radio Astronomy Observatory at Cambridge.

One of LIGO's gravitation wave interferometers. The arms are each 2.5 miles long, constructed of concrete to protect the steel vacuum tubes inside.

holes, should disrupt spacetime by creating gravitational waves that propagate away from the source. They are "ripples in spacetime." The practical problem is that the waves decrease in amplitude as they travel. By the time waves from a distant system reach Earth, measurements may need to detect distances 1000 times smaller than the atom. It was a technological triumph when LIGO (Laser Interferometer Gravitational-Wave Observatory) used huge assemblies of vacuum tubes placed at distant locations on Earth to detect gravitational waves created by the collision of two black holes 1.3 billion light years away.

∞ ∞ ∞

The working model for the origin of the universe says that after the Big Bang, as the cosmic gas cooled down, denser regions condensed into stars; then gravity pulled the stars into galaxies. There are some important discrepancies between observations and the model. Some old galaxies are too large.[17] Different ways of measuring the rate with which the universe is expanding (called the Hubble constant) are inconsistent.[18] And some galaxies are moving much too fast. Do these discrepancies mean there something wrong with the theory?

Everything in the universe is moving: stars move within galaxies, and galaxies move within clusters, *etc.* The concept that matter as we know it cannot account for the universe comes from astronomer Fritz Zwicky's observation at Caltech in 1933 that galaxies in the Coma cluster move much faster than expected. The speed implies they contain 50× more mass than can be accounted for by totaling the masses of all the galaxies in the cluster.[19,20]

To explain the anomaly, Zwicky, a Swiss astronomer who had emigrated to the Unites States in 1925, proposed that the extra mass is *dunkle Materie.* Now called dark matter, its basis is unknown, and it is undetectable to us.[21] It was not really obvious what to make of this, and Zwicky himself said in 1957 that, "It's not certain how these startling results must ultimately be interpreted."[22]

Not much attention was paid until observations in 1970 showed that gas clouds orbiting the Andromeda galaxy also move at speeds implying the existence of a great amount of undetectable mass.[23] The possibility that it might be normal matter in some diffuse form that we cannot detect is excluded by calculations of how much matter should have been created by the Big Bang.

We now have a measure quantitatively of the extent of the problem. Looking at the ripples the CMB creates, it seems that only 5% of the universe consists of normal matter (made of protons, neutrons, electrons, and neutrinos: sometimes called baryonic matter). Another 23% behaves like matter in its effects on gravitation. This is the 'dark matter,' basically inferred only from its gravitational effects.[24] No one has the faintest idea about its physical makeup.[25] The other 72% is not matter, at least as we can recognize it, and is now called 'dark energy.'[26] It's even more difficult to understand than dark matter. Cosmologists are embarrassed that we do not really know what the universe is made of.

The expansion of the universe appears to be accelerating.[27] This is an inference based on observations that supernovas that explôded when the universe was only two thirds of its present age are fainter than they should be at a constant rate of expansion.[28] Dark energy is thought to be the force that's responsible. The shape of the universe is flat—not concave or convex—which requires a certain density of energy and matter. Known matter does not provide enough. The cosmological constant in Einstein's equations is a form of energy that allows for expansion. Given an appropriate value it makes the universe flat. So it's thought that it may correspond to dark energy.[29]

The gap between data and theory is unprecedented. Kepler deduced the laws of planetary motion, describing elliptical orbits around the Sun. They were purely descriptive, but accurate. Newton deduced that the orbits result from gravitational attraction. Even without any explanation for action at a distance, their predictions are accurate. Einstein formulated the concept of curved spacetime. Its predictions for how mass bends light were confirmed. At each stage, theory explained the data.

But now the discrepancies between data and theory argue for an adjustment of one or the other. There are too many different lines of data to argue that the data are incorrect. Dark matter and dark energy are the means of restoring consistency with the theory of relativity. But we have no means of detecting dark matter or dark energy. Their existence is an inference from inconsistencies of data with theory.

Accumulating data can be reconciled with the theory of relativity only by *ad hoc* assumptions, such as adjusting the cosmological constant (whose introduction into the theory was *ad hoc* in the first place). Some attempts to maintain the theory seem desperate, such as proposing that dark energy occupies an extra dimension (this is a variant of string theory, which I discuss below). In historical terms, the introduction of increasing numbers of *ad hoc* assumptions to reconcile data with theory is the mark of a collapsing paradigm.

It's fair to say that there is confidence in the concept of the Big Bang, but understanding of the following events that led to the creation of the universe has been somewhat shaken by the discrepancies between data and theory. Cosmologists expect that understanding dark energy will need a major revolution—a paradigm shift and a half, if you like.

Spacetime is a four-dimensional entity (three dimensions of space and one of time), but there is a sense in which space and time are arbitrary dimensions. Rotating the dimensions is a trick that physicists use to make calculations. Stephen Hawking went further. "One could take the attitude that quantum gravity and indeed the whole of physics is really defined in imaginary time. It's simply a consequence of our perception that we interpret the universe in real time."[30] Thomas Hertog, who worked with Hawking, adds, "The story of the universe, the theory holds, is that once upon a time there was no time." This leads to the model that there was a singularity. History cannot be extended beyond it.[31]

∞ ∞ ∞

Cosmology is in a revolution, but it's an incomplete revolution—and no end is in sight. Actually it might be appropriate to say revolutions in the plural, as there is no end to the concepts, some extraordinarily clever, some quite fantastic, trying to resolve the conundrums.

One attempt to reach a Theory of Everything was the construction in 1969 of string theory, which proposes that every object in the universe consists of vibrating strings of energy. String theory requires extra dimensions beyond the 4 dimensions of spacetime. They are supposed to be so tiny that we never detect them.[32] String theory became superstring theory in 1985 by introducing the concept of supersymmetry. This says that each of the particles in the Standard Model has a partner with a 'spin' that differs by half of a unit. Five superstring theories were then combined into a single theory, called M-theory.

To say that string theory is controversial would be an understatement. The theory intends to reconcile quantum physics with relativity, but it's impossible to test it. This has led to a debate about scientific realism *versus* empiricism. Proponents of string theory admit that it was originally an "exotic fantasy," but see it now as the "the only promising candidate for the construction of a truly unified theory."[33] The structure and scale of the strings (they are so small that they approach the point at which gravitational force is close to the strong force within the nucleus) makes it impossible to conceive any way to verify their existence.

Applying Karl Popper's criterion of falsifiability raises the question as to whether this is science. Indeed, many eminent physicists believe that string theory resembles pseudoscience rather than science.[34] A call that we should change the standards for judging science led to a cry to "defend the integrity of physics."[35] The essence of the argument for string theory is that it provides a unique explanation to which there is no alternative. Refusing to consider it because of some arbitrary principle (falsifiability) is deemed nonscientific.[36] On the other hand, it's impossible to know there is no alternative theory.

Science would have gone down many deep rabbit holes in the past if theories had been adopted because there seemed to be no alternative. The deep question is whether the ability of physics to explain fundamental subatomic forces has reached a point at which traditional scientific principles break down (indeed, perhaps like the laws of physics themselves at the origin of the universe).

Because you can't perform experiments in cosmology, our knowledge rests on interpreting observations. Modeling by simulations replaces the perturbations of the system that would be involved in an experimental process. This is a powerful approach, but suffers from the limitation that a simulation can do no more than demonstrate plausibility.[37] And, of course, we must ask if it's possible that, like Ptolemy's geocentric model, a simulation of the universe might have high predictive value but not in fact reflect reality?

One simulation that gave an interesting result suggested how supernovae might originate. The simulation was run because of a requirement when the test ban treaty was being negotiated in 1963. It was necessary to be able to distinguish cosmic flashes from the radiation flash that would result from detonating a nuclear weapon in the atmosphere. The simulation showed how to discriminate between them. It demonstrated that when a small star collapses (because its center has run out of fuel), it generates a shockwave, with the outer layers shooting out at close to the speed of light. This explains the origin of the sudden explosion of light that marks a supernova.[38]

While there is general acceptance of a Big Bang, this has become something of a generic theory, with many inconsistencies yet to be explained, and many possible alternatives. There comes a point at which physics, at least as we know it today, cannot account for the properties of the universe. Physics becomes metaphysics or even philosophy. The steady state and Big Bang theories of the universe are equally difficult to place in conceptual terms of human understanding.

It's unknowable what the laws of physics might have been at the time of the Big Bang and whether and how they have changed. The fact that we can deduce consequences of the Big Bang that can be interpreted according to the present laws of physics at least argues that the laws have been in effect since very soon after the Big Bang. But in the early stages, any sort of accident might have 'frozen' the form of the universe. Here is a limitation on our ability to define the course of events according to the laws of science, because unpredictable random events may have selected one outcome over another.

∞ ∞ ∞

Is the universe stable? This appears to be the case, not in the sense that it's unchanging, but insofar as it obeys a reliable set of physical laws. Gravity ensures that the apple always falls down from the tree. If sometimes it flew upwards, life would be very different, in fact it would probably be impossible. Until the establishment of 'chaos theory,' science was based on the principle that, given a defined starting point, the laws of physics would govern progress towards an end point or equilibrium. A small change in the initial conditions would produce a small change in the final state.

"Does the Flap of a Butterfly's Wings in Brazil Set off a Tornado in Texas?" was the question proposed by Edward Lorenz, a meteorologist at MIT, in a talk he gave in 1972.[39] Lorenz had discovered in 1963 that a very small change in his starting parameters caused a very large change in the direction of the air stream predicted by his meteorological model. In fact, the change was discovered because Lorenz used an approximation, putting 0.506 instead of the full value of 0.506127, into his equation. He did not think this would matter until he saw the effect.

The technical name for a large change in output resulting from small changes in input is "sensitive dependence on initial conditions," but this is now universally known as the "butterfly effect." Chaos seemed to be a shocking new idea when it was proposed, but amazingly it was actually conceived by the great mathematician Henri Poincaré. He predicted in 1905 that, "It may happen that small differences in the initial conditions produce very great ones in the final phenomena. A small error in the former will produce an enormous error in the latter. Prediction becomes impossible, and we have the fortuitous phenomenon."[40]

It would be easy to dismiss the application to the real world as an error in the model, but it became clear over the next few decades that a similar model applies to other situations, such as turbulence in liquids, or transitions between solid and liquid, or between liquid and gaseous states.[41] The properties of a solid form are predictable, the properties of a liquid form are predictable—but the transition point is a state of chaos.[42]

The key feature is that the system is deterministic—it can be described by equations—but the initial conditions evolve in a way that never repeats itself, so the state of the system at any moment is not predictable. Lorenz described this as, "When the present determines the future, but the approximate present does not approximately determine the future."[43] Chaos theory would be a major threat to the existence of science as we practice it today if it were not for the fact that it applies only in certain, special situations.

∞ ∞ ∞

It was very likely chaos after the Big Bang. (In this, modern physics shows a striking parallel to the ancient view of the world, as I discuss in Chapter 1.) Current calculations suggest that the period of chaos started at 10^{-43} seconds after the Big Bang and lasted for 10^{-36} seconds.[44] After that, the laws of physics, as we know them, would have applied. But if things had been different during the 10^{-36} seconds of chaos, the universe would be quite different today—and life would not exist.

Taking a typically dispassionate view, molecular biologist Jacques Monod argued in 1971 in his marvelous book *Chance and Necessity* (the title gives away the theme) that human existence is due to a freak accident. "The universe was not pregnant with life...Our number came up in the Monte Carlo game."[45]

To the extent that our universe may have depended on an accident, the philosophers might be right in questioning reality. The laws of physics are what they are because we are here to observe them in this particular universe. If the Big Bang had come out differently, the laws might be different—and we would not be here to observe them. This does not, however, invalidate the fact that the predictability of the laws of physics refutes any relativistic view of science.

Theoretical reconstructions of the origin and evolution of the universe point to the extreme unlikelihood of the development of the conditions for supporting life. The tiniest of adjustments at the beginning would have meant that carbon, essential for all living forms, would never have been created. At each subsequent stage, it seems that random events moved in the direction of ultimately creating conditions favorable for the evolution of life.[46] The view that this cannot be mere accident leads to a position rather reminiscent of the Middle Ages: the universe was created with a clockwork-like mechanism ensuring that afterwards everything worked perfectly (see Chapter 6).

This view has been formalized into the anthropic principle: our observations of the universe are limited by the fact that they could be made only in a universe capable of developing intelligent life.[47] This seems like a circular argument. Indeed, it was widely regarded as nonsense when it was first proposed in 1973. It has become part of cosmology, although there are many dissenters.[48] David Gross, who shared the Nobel Prize for physics in 2004, regarded the anthropic principle like a virus. "Once you get the bug you can't get rid of it," he complained at a conference on the Future of Cosmology in 2003.[49]

The anthropic principle gained more traction with the concept of the multiverse: that there could be many, many universes, only some (probably a tiny minority) with the possibility of supporting life.[50] It says that we see only our own universe and have no means of detecting the others. The theory takes its force from the thought that any one of many very tiny changes in cosmological principles would make our universe completely inhospitable for life. The debate comes down to whether the laws of physics by mere chance allow life or whether they have been determined by the requirement that intelligent life is possible.[51] This can come dangerously close to intelligent design.

The multiverse is also known as the 'many worlds' theory. One theory for the origin of multiple universes is that they occur by a succession of Big Bangs. (One wonderfully far-fetched idea is that our universe is in fact the inside of a black hole. Science is never far from science fiction in cosmology.) An extreme version of multiverse theory holds that a quantum event triggers a branch: a separate universe is created for each outcome.[52]

Returning to the question of illusion, we know that our brain interprets its inputs from the surroundings—it does not, as it were, carry a scale model of what we see. Could it be that 'reality' is in fact a clever construction that the brain produces from sensory input? Are we living in a computer simulation? While this may be a science fiction fantasy, it's not entirely removed from the laws of physics.[53] Has the joke replaced reality?

String or superstring theory, the idea of the multiverse, anthropic or otherwise, and holographic black holes take us to the far edges of science. Since there is no possible way to test such theories—indeed they make no predictions at all that can be tested—do they qualify formally as pseudoscience? Indeed, there is some concern in the community that by mixing science and fiction they may damage the view of science as a search for objective truth. Of course, if the laws of physics have changed with time, and there is no way to deduce what the original laws might have been, investigations of the origin of the universe also become pseudoscience.

It might not seem obvious why cosmology should depend on particle physics. On the one hand, we have the vast expanse of the universe; on the other, the nano-world (well, really much, much smaller than nano) of the atom. Nobel Prizewinner Abdus Salam, whose Prize in 1979 was for unifying weak and electromagnetic interactions in the nucleus, explains: "Early cosmology [from 10^{-43} seconds until 10^{-12} seconds after the Big Bang] has become synonymous with particle physics. This is because phase transitions, which separate one era of cosmology from another, are also the mechanisms through which... [forces are converted]. The fact that these transitions occur at high temperatures... that are unlikely ever to be realized with man-made accelerators, makes the early universe and cosmology attractive to the experimental particle physicist as providing the only laboratory which will, at least indirectly, be able to test our theories."[54]

Applying the laws of physics to a cosmological situation depends on the starting conditions. For example, eclipses can be predicted from the known movements of the Earth and the Moon from a given starting point. The starting point is called more technically the boundary condition. One difficulty in cosmology is that the boundary conditions for the origin of the universe are impossible to define if the laws of physics may have been different then.

This parallels in a way the difference between science and society. The boundary condition separating science from society is the belief in objective validation of facts. That's equivalent to a 'law' of science that does not apply anywhere else in the intellectual (or political) world. Does cosmology cross the boundary when we follow the equations to the point of interpreting them in terms of everyday life?

To be sure, there are no *data* to distinguish between models that the existence of life is due to accident or purpose, that there is one universe or many universes, and even whether reality actually exists. It's a matter of philosophy, or to use a simpler term, belief.

Before the attempt to merge quantum mechanics with general relativity, cosmology had the self-contained quality that typifies science. But has it now returned to its origins in crossing the boundaries to become a blend of science, philosophy, religion, and (sometimes) science fiction?

NOTES AND REFERENCES

1. G. Lemaître, The Beginning of the World from the Point of View of Quantum Theory, *Nature*, 1931, **127**, 706.
2. Lecture when accepting the James Scott Prize, Feb 6, 1939. Published as P. Dirac, The Relation between Mathematics and Physics. *Proc. R. Soc. (Edinburgh)*, 1939, **59**, 122–129. Online at www.damtp.cam.ac.uk/events/strings02/dirac/speach.html
3. S. Hawking, Quantum cosmology, in *The Nature of Space and Time*, ed. S. Hawking and R. Penrose, Princeton University Press, Princeton, 2010, p. 89.
4. S. Carroll, *The Big Picture. On the Origins of Life, Meaning, and the Universe Itself*, Dutton, New York, 2016.
5. G. Lemaître, The beginning of the world from the point of view of quantum theory, *Nature*, 1931, **127**, 706.
6. Quoted in A. Deprit, *Monsignor Georges Lemaître*, in *The Big Bang and Georges Lemaître*, ed. A. Berger, Riedel, Dordrecht, 1984, pp. 363–392, p. 388.
7. "These theories were based on the hypothesis that all the matter in the universe was created in one Big Bang in a particular time in the remote past." F. Hoyle, Continuous creation, *The Listener*, 1949, **41**, 567–568.
8. P. J. E. Peebles, *Cosmology's Century. An Inside History Of Our Modern Understanding Of The Universe*, Princeton University Press, Princeton, 2020, pp. 6–8.

9. H. Y. McSween Jr and G. R. Huss, *Cosmochemistry*, Cambridge University Press, Cambridge, 2010.

10. For a history of the events leading to the discovery of the expanding universe and of cosmic background radiation, see R. Evans, *The Cosmic Microwave Background, How it Changed Our Understanding of the Universe*, Springer, Cham, Switzerland, 2015.

11. A. Penzias and R. Wilson, *Astrophys. J.*, 1965, **142**, L419.

12. A. Lightman, *The Discoveries: Great Breakthroughs In 20th-Century Science, Including The Original Papers*, Vintage, New York, 2006, pp. 409–426.

13. R. Penrose, Gravitational collapse and space-time singularities, *Phys. Rev. Lett.*, 1965, **14**, 57–59.

14. J. Michell, On the Means of Discovering the Distance, Magnitude, &c. of the Fixed Stars, in Consequence of the Diminution of the Velocity of Their Light, in Case Such a Diminution Should be Found to Take Place in any of Them, and Such Other Data Should be Procured from Observations, as Would be Farther Necessary for That Purpose, *Philos. Trans. R. Soc. London*, 1783, **74**, 35–57.

15. T. Hertog, *On the Origin of Time*, Bantam, New York, 2023, pp. 210–249.

16. A. Hewish, *et al.*, Observation of a rapidly pulsating radio source, *Nature*, 1968, **217**, 709–713.

17. The oldest galaxies, formed soon after the Big Bang, should be relatively small. However, results from the new James Webb Space Telescope in 2022 identify some very large galaxies that appear to have originated early (about 600 million years after the Big Bang). I. Labbé, *et al.*, A population of red candidate massive galaxies ~600 Myr after the Big Bang, *Nature*, 2022, **616**, 266–269.

18. Two different ways of measuring the Hubble Constant produce a difference that is greater than the apparent error limit in either method. (One method utilizes the CMB, the other looks at distant stars.) P. Shah, P. Lemos and O. Lahav, A Buyer's Guide to the Hubble Constant, *Astron. Astrophys. Rev.*, 2021, **29**, 9.

19. The basic principle is that the inverse law of gravitational attraction (returning to Newtonian physics) says, for example, that more distant planets orbit the Sun more slowly than closer planets. In the same way, star clusters that orbit galaxies should decrease in velocity when they located farther from a galaxy. But moving out from the center of a galaxy, stars and clusters move much faster than predicted. This means there must be more mass than is apparent.

20. Based on the then-current value for the Hubble constant that was about $8\times$ too high, Zwicky's original estimate was actually for a $400\times$ discrepancy.

21. F. Zwicky, Die Rotverschiebung von extragalaktischen Nebeln, *Helv. Phys. Acta*, 1933, **6**, 110–127.

22. F. Zwicky, *Morphological Astronomy*, Springer-Verlag, Berlin, 1957, p. 132.

23. V. C. Rubin and W. K. Ford, Rotation of the andromeda nebula from a spectroscopic survey of emission regions, *Astrophys. J.*, 1970, **159**, 379–403.

24. P. J. E. Peebles, *Cosmology's Century. An Inside History Of Our Modern Understanding Of The Universe*, Princeton University Press, Princeton, 2020, p. 4.

25. Proposals based on string theory led physicists to search for WIMPs (weakly interacting massive particles) in the 1970s, and today they are searching for 'axions,' but without success.

26. E. Gates, *Einstein's Telescope: The Hunt for Dark Matter and Dark Energy in the Universe*, Norton, New York, 2009, pp. 3–30.

27. E. Gates, *Einstein's Telescope: The Hunt for Dark Matter and Dark Energy in the Universe*, Norton, New York, 2009, pp. 67–89.

28. The supernovae were observed by two groups, one led by Saul Perlmutter at the Lawrence Berkeley National Laboratory in California, the other by Adam Reiss and Brian Schmitt at the Johns Hopkins Space Telescope Science Institute in Baltimore. Perlmutter, Schmitt, and Reiss were awarded the Nobel Prize in physics in 2011.

29. E. Gates, *Einstein's Telescope: The Hunt for Dark Matter and Dark Energy in the Universe*, Norton, New York, 2009, pp. 196–224.

30. Quoted in T. Hertog, *On the Origin of Time*, Bantam, New York, 2023, p. 98.

31. T. Hertog, *On the Origin of Time*, Bantam, New York, 2023, p. 76.

32. Depending on the flavor of string theory, there may be 9, 10, or 11 dimensions in total.

33. R. Dawid, *Scientific Realism in the Age of String Theory*, Physics and Philosophy, 2007, Philsci-archive.pitt.edu/3584/1/latex-scientific_realism_in_the_age_of_string_theory(new).pdf.

34. P. Woit, *Not Even Wrong: The Failure of String Theory and the Continuing Challenge to Unify the Laws of Physics*, Jonathan Cape, London, 2006, pp. 180–181.

35. G. Ellis and J. Silk, Defend the Integrity of Physics, *Nature*, 2014, **516**, 321–323.

36. S. Carroll, Falsifiability, in *This Idea Must Die: Scientific Theories That Are Blocking Progress*, ed. J. Brockman, Harper Perennial, New York, 2015.

37. A. Pontzen, *The Universe in a Box: A New Cosmic History*, Jonathan Cape, London, 2023.

38. A. Pontzen, *The Universe in a Box: A New Cosmic History*, Jonathan Cape, London, 2023, p. 88.

39. Quoted in E. Lorenz, *The Essence of Chaos*, University of Washington Press, Seattle, 1993, pp. 179–182.

40. H. Poincaré, *Science and Method*, translated by F. Maitland, Thomas Nelson, London, 1905.

41. J. Gleick, *Chaos: Making a New Science*, Penguin Books, New York, 2nd edn, 2004.

42. The term 'chaos' was first used in a paper by mathematician James Yorke in 1975: T.-Y. Li and J. A. Yorke, Period Three Implies Chaos, *Am. Math. Monthly*, 1975, **82**, 985–992.

43. Quoted in C. M. Danforth, *Chaos in an Atmosphere Hanging on a Wall. Mathematics of Planet Earth*, 2013, Online at mpe.dimacs.rutgers.edu/2013/03/17.

44. K. Gelfert and E. M. Adilson, (Non)Invariance of Dynamical Quantities for Orbit Equivalent Flows, *Commun. Math. Phys.*, 2010, **300**, 411–433.

45. J. Monod, *Chance and Necessity*, Vintage Books, New York, 1971, pp. 145–146.

46. Astronomer Martin Rees argues that change in any one of six key numbers describing the parameters of atomic structure and interactions would have made formation of the universe impossible. See M. Rees, *Just Six Numbers: The Deep Forces That Shape the Universe*, Basic Books, New York, 2000.

47. First proposed by Brandon Carter in 1973 at a symposium, and published in the proceedings. B. Carter, Large number coincidences and the anthropic principle in cosmology, *IAU Symposium 63: Confrontation of Cosmological Theories with Observational Data*, Reidel, Dordrecht, 1974, pp. 291–298.

48. L. Susskind, *The Cosmic Landscape: String Theory and the Illusion of Intelligent Design*, Hachette Book Group, New York, 2005.

49. D. Overbye, Zillions Of Universes? Or Did Ours Get Lucky? New York Times, 2003, October 28.

50. T. Hertog, *On the Origin of Time*, Bantam, New York, 2023, pp. 26–29, 181–182.

51. L. Susskind, *The Cosmic Landscape: String Theory and the Illusion of Intelligent Design*, Hachette Book Group, New York, 2005.

52. S. Hossenfelder, *Existential Physics: A Scientist's Guide to Life's Biggest Questions*, Viking, New York, 2022, pp. 105–124.

53. S. Hossenfelder, *Existential Physics: A Scientist's Guide to Life's Biggest Questions*, Viking, New York, 2022, pp. 51–78.

54. A. Salam, *Unification Of Fundamental Force: The First Of The 1988 Dirac Memorial Lectures*, Cambridge University Press, Cambridge, 1990, pp. 72–73.

Science and War

"WHEN YOU SEE SOMETHING THAT IS TECHNICALLY SWEET, you go ahead and do it and you argue about what to do about it only after you have had your technical success. That is the way it was with the atomic bomb," said Robert Oppenheimer.[1] Scientists have often put their knowledge to use in war. Far from reluctant, Archimedes, Leonardo da Vinci, and Galileo seemed if anything eager to test their theories in the theater of war. Like anyone else, scientists identify with the society in which they live, so German scientists were involved in the war effort in the first and second world wars just as much as scientists in the opposing democracies. It's a truism that war is a crucible. This is as true of science as anything else. War is certainly responsible for technological advances, and sometimes also for scientific advances. Science was in the background of war efforts until the Second World War. After the war, some of the scientists became famous, notably Alan Turing for the analysis of intelligence, and Robert Oppenheimer and others for the atom bomb. Since science became an acknowledged tool of war in more recent times, it has become more secret, and public knowledge of relevant scientific developments has receded.

The Frontiers of Science
By Benjamin Lewin
© Benjamin Lewin 2026
Published by the Royal Society of Chemistry, www.rsc.org

Timeline for Involvement of
Great Scientists in War

200 BCE — Archimedes designs war machines (215 BCE)

1500 — Leonardo da Vinci becomes military engineer (1500)

Mathematician Niccoló Tartaglia founds ballistics (1537)

1600 — Galileo offers telescope to navy (1610)

Galileo defines projectile motion (1638)

1700 —

1800 —

Alfred Nobel invents dynamite (1867)

1900 — Fritz Haber develops poison gas (1915)
Watson-Watt invents radar (1935)
Turing develops cryptography (1942)
Oppenheimer leads Manhattan project (1945)
Heisenberg leads German effort to produce atom bomb (1945)

2000 —

A Dialog between Archimedes and Leonardo da Vinci

Archimedes (*c.* 287–212 BCE) was a polymath, living in Syracuse, Sicily, famous not only as a mathematician but also as an inventor who used mathematics to develop engineering principles.

Leonardo da Vinci (1452–1519) was a polymath ranging from arts to sciences. He was a painter and sculptor in the Italian Renaissance, an architect and engineer who designed amazing objects, and an experimentalist in science.

© GeorgiosArt/Getty Images.

Archimedes: Good day, Leonardo. It's an honor to meet a mind as versatile as yours. Your artistic and scientific endeavors have left an indelible mark on the world.

Leonardo: Thank you, Archimedes. I must say, your contributions to mathematics and physics, particularly in the realm of mechanics, have inspired countless thinkers through the ages. What brings us together today?

Archimedes: I've often pondered the role of science in the affairs of men, particularly in times of conflict. It seems our innovations, whether in mathematics or art, can be harnessed for both constructive and destructive purposes. What are your thoughts on the usefulness of science in war?

Leonardo: A profound question, Archimedes. I, too, have grappled with the dual nature of knowledge. Science, art, and invention can enhance civilization, but in the wrong hands, they can become tools of destruction. Take, for instance, my designs for war machines—innovations that, while born of curiosity, could be wielded as instruments of harm.

Archimedes: Indeed, I am reminded of your sketches for war machines, such as the tank and the machine gun. It's a curious paradox—our endeavors to explore the boundaries of knowledge can inadvertently empower conflict.

Leonardo: While I created those designs to demonstrate my engineering and artistic skills, I did so with the understanding that their use would be beyond my control. War, unfortunately, has been a constant companion to humanity. Innovations are often driven by the necessity of survival.

Archimedes: In my own time, during the Siege of Syracuse, I applied my knowledge of geometry and mechanics to design ingenious war machines to defend the city. I may not have been able to find a lever to move the world, but I was able to hold off the Romans.

Leonardo: Your defenses in Syracuse were legendary, Archimedes. The "Archimedes Screw" for irrigation, your clever use of mirrors to focus sunlight on enemy ships—it demonstrated the power of applying scientific principles to the challenges of war. Yet, I can't help but wonder if our creations inadvertently perpetuate the cycle of violence.

Archimedes: The same principles that allow us to envision machines of war can be applied to better the human condition. It's a delicate balance—one that society, not just individuals, must navigate. But in Syracuse we were forced to defend ourselves against the Romans. Scientists must use their knowledge to defend democracy against tyranny. I rest my case.

SCIENCE AND WAR

When men are engaged in war and conquest, the tools of science become as dangerous as a razor in the hands of a child of three,[2] Albert Einstein, 1940.

However distinguished, intelligent and practical scientists may be, they cannot be so well qualified to decide what the needs of the nation are, and their priorities, as those responsible for ensuring that those needs are met,[3] Victor Rothschild, 1971.

War (sadly) has advanced science; and science has certainly advanced war. (Although the so-called 'science of war' is not science at all, but simply a more organized and systematic approach to war.) The history of the involvement of science in war has often involved the diversion of scientific genius from fundamental to more practical problems, which, however, seemed to become equally fascinating to the participants. It would be reasonable to expect war to drive applied science, developing known principles into technological advances, but it has also led to fundamental advances in basic science.

Archimedes (*c.* 287–212 BCE) is almost as well known for his military inventions as for shouting "Eureka" when he jumped out of his bath. Archimedes was a notable figure in Syracuse—so notable that during the siege of Syracuse there were orders not to harm him. However, he was killed by a soldier when the Romans finally took the town, giving them command of Sicily during the second Punic War.

The Romans had a formidable range of weapons, including a floating siege tower with grappling hooks, and scaling ladders lowered with pulleys from ships on to the city walls. Archimedes used his knowledge of mathematics to devise defensive measures that held the Romans at bay for months. He appears to have been as much an inventor as a pure mathematician. Unfortunately, there are no contemporaneous accounts of his work. The earliest date from 70 years after his death. It's difficult to distinguish legend from reality, but the extent of the legends is a tribute to Archimedes' reputation as a genius.

Archimedes established the principle that a body placed in water is buoyed up by a force equal to the weight of water that it displaces. Described in his treatise, *On Floating Bodies*, this is known today as Archimedes' principle. The principle was famously used to determine whether a crown made of gold had been adulterated with silver (which would have lowered the density and changed the displacement of water). Actually, it's not quite clear whether the displaced volume could have been measured precisely enough directly to determine the difference in density. Galileo developed a hydrostatic balance (allowing the weight of two objects to be compared in water) according to Archimedes' principle. He thought this was probably what Archimedes had actually done.[4]

Two stories about Archimedes' inventions in the siege of Syracuse are probably apocryphal. He is supposed to have invented a 'claw' consisting of a crane-like arm with a large grappling hook that could be dropped on an attacking ship to lift it out of the water and sink it. And he was described as using parabolic

The legends of Archimedes' inventions at the siege of Syracuse spread far and wide. The title page of Opticae Thesaurus, *a Latin edition in 1572 of Ibn al-Haytham's* Book of Optics *from c. 1021 CE, had an engraving to illustrate Archimedes' alleged use of mirrors to burn the invading ships.*

mirrors to focus a 'heat ray' on the ships that would burn them. Recent reconstructions suggest that the claw could have worked, but any effects of the mirrors would have been marginal.

"Give me a place to stand on, and I will move the Earth," Archimedes is supposed to have said.[5] This was certainly the principle used to construct the weapons mounted on the city walls that threw projectiles at the invading ships. Archimedes defined the law of the lever and pulley, which was the basic principle behind the catapult. Archimedes provides an early example of the application of scientific method, as he appears to have used experiments to test his theories. We have no means of knowing whether he regarded his military efforts as an opportunity to test his theories or as a distraction from theoretical matters. At all events, the catapults of the ancient world were far more effective than the equivalent artillery weapons of the medieval period, an example of the loss of technology.[6]

Leonardo da Vinci (1452–1519) designed a great number of machines, many useful for war. One of his designs, which he stated was based on Archimedes' ideas, was for a cannon fired by steam. "The Architronito is a machine of fine copper, an invention of Archimedes, and it throws iron balls with great noise and violence."[7] (Architronito means 'thunder of Archimedes'.) Leonardo made a wonderful drawing of how the machine should work. It's been suggested that Archimedes could have used such a device to fire projectiles, perhaps filled with flammable materials, at the Roman fleet. One problem here is generating pressurized steam. This could have been achieved by using parabolic mirrors.[8] Reconstruction experiments at MIT suggest that making a steam cannon might have been possible.[9] At any event, it's more likely than using parabolic mirrors aimed at the ships.

Leonardo's designs included catapults, a giant crossbow, a tank, and a machine gun, but often could not be built because of the lack of appropriate technology. Because there was continuous war between the independent city-states of Tuscany, Leonardo was employed by his patrons as a military engineer. Indeed at one point he listed the categories of military engineering in which he was proficient as an inducement to hire him.

Calculating trajectories of projectiles was a military preoccupation that recruited the greatest mathematicians of their eras: Archimedes, Leonardo, and Galileo. Indeed, the need to

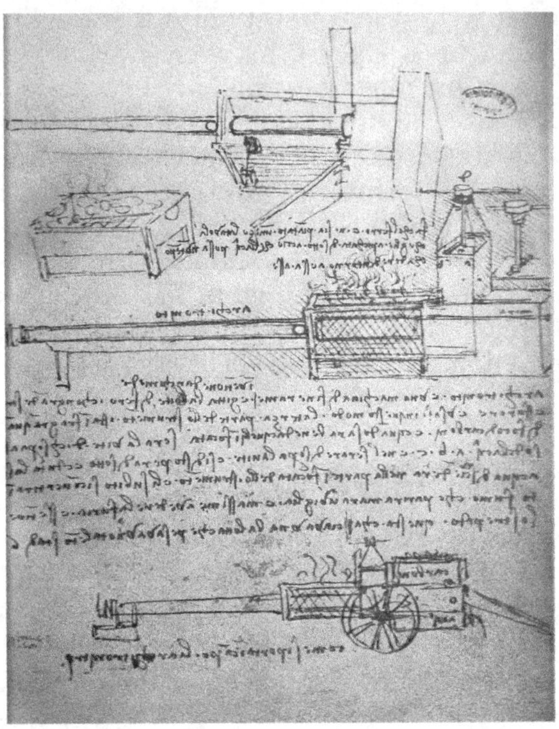

A page from Leonardo's notebook showing the design for a steam cannon, with the description in mirror handwriting.

calculate the paths of projectiles has driven scientific developments from the medieval to the present. At the start of the Second World War, one limiting factor for artillery was the need to produce firing tables. These factor in all the conditions that might affect the path of a projectile, so that the right angle of elevation can be used. They were produced using mechanical calculators by a large staff, mostly women, who were known as 'computers.' (The same term had been used previously for a group of women cataloguing the night sky at Harvard, see Chapter 23.) The Army decided to speed up the process by developing a digital electronic computer that would allow a firing table to be calculated in 100 seconds instead of 1 month.[10] This led to the production of ENIAC, one of the first computers of the modern era (see Chapter 23).

Whether originating from military preoccupations or from his other intellectual interests, a recently discovered manuscript shows

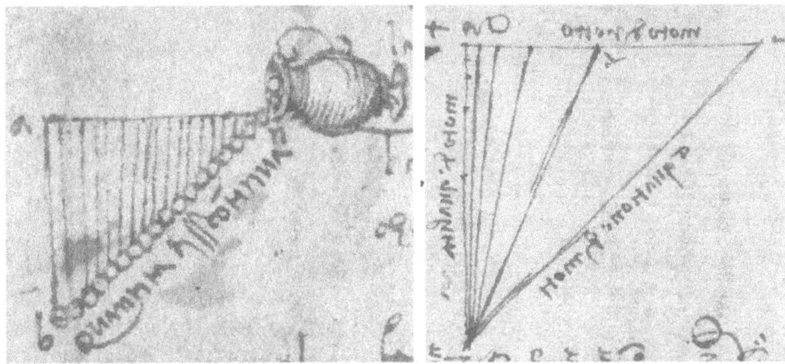

Leonardo illustrated his experiment on motion with grains of sand falling as the jar moves horizontally (left). The diagram on the right shows the line formed by the sand, depending on the relationship between vertical motion (due to gravity) and horizontal motion (due to movement of the jar). When the two are equal, the line is the hypotenuse of a right-angled triangle with equal sides, marked on the diagram as 'equationi di moto' (equivalence of motion).[13]

Reproduced with permission from The British Library.

that Leonardo became interested in gravity. A century before Galileo, he deduced that it represented a form of acceleration.[11] From experiments rolling balls on inclined surfaces, Leonardo worked out that "a weight that descends freely in every degree of time acquires… a degree of velocity." He distinguished between *moto natural* (falling under gravity) and *moto directo* (directed motion).

Leonardo designed an experimental situation to compare *moto natural* with *moto directo* by allowing sand to fall (*moto natural*) from a jar that was moving horizontally (providing *moto directo*). When the horizontal acceleration equals the acceleration due to gravity, the particles of sand line up along the hypotenuse of a right-angled triangle where the vertical axis (*moto natural*) and the horizontal axis (*moto directo*) are equal. Simulation of the experiment shows that it could have been used to make a remarkably close estimate of the force of gravity.[12]

Given the politics of the late medieval era, the patrons who could afford to support the research of the philosophers were often engaged in war. It was natural that they should recruit them for developing military technology. In fact, Galileo did not need recruiting.

Soon after he obtained the chair of mathematics at the University of Padua in 1592, he opened a workshop in his house where he taught military fortifications and built instruments for the military.[14]

Galileo's designed a military compass, produced by an instrument maker who lodged in his house. It was useful for surveying and drawing plans, especially for attacking a fortress. It could be used as a quadrant, a device used in conjunction with firing tables that allowed calculation of the elevation needed for the cannon, depending on the distance of the target and size of artillery. It was also equipped with scales for commercial calculations, such as converting currency. Most of these functions were previously known and could be performed by earlier instruments, but the great success of Galileo's compass was its versatility in combining so many functions. His workshop also produced other compasses, including the surveying compass, which was used to help construct tunnels for undermining a target fortress.

Galileo's most famous use of his first telescope, constructed in 1609 (see Chapter 4), was in astronomy. But he immediately realized its military implications and offered it to the Doge of Venice for use in the army and navy. He was rewarded with lifetime tenure at the university. By 1617, he had taken over the development of binoculars (originally regarded as of such military importance that they were considered as a state secret). He was involved in developing a great variety of other military devices, from manufacturing gunpowder to improving cannons.

Galileo's first thoughts on motion preceded his career as a military engineer. He wrote a treatise, *De Motu*, on the subject at Pisa from 1589–1592 but never published it, possibly because he changed his mind about the character of motion. The last chapter of *De Motu* has the title, *"Why Objects Projected by the Same Force Move Farther on a Straight Line the Less Acute are the Angles they Make with the Plane of the Horizon."*[15] This accepted the conventional wisdom of the time, but was contradicted by his later conclusions.

After a decade of experience with military matters, in 1604 Galileo turned his attention back to motion (see Chapter 5). In fact, together with the mathematician Guidobaldo del Monte, he had obtained data in 1592 suggesting that a projectile follows a symmetrical parabolic course. This was evidently inconsistent

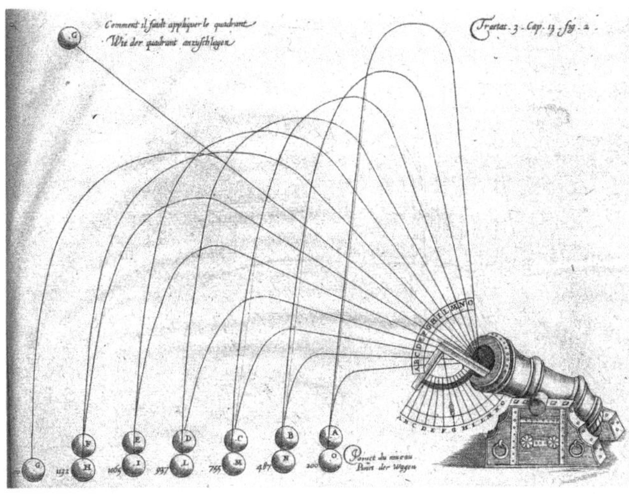

A diagram from Spanish artillery captain Diego Ufano's textbook in 1613 shows calculations for distance traveled depending on the angle of projection.

with the analysis in *De Motu,* but Galileo did not follow it up theoretically for another decade (see Chapter 5). He was now in a position to develop his theory of fall, which he completed in the period 1604–1610 (although it was not published until much later), before he turned his attention fully to astronomy.[16] His practical experience as a military engineer may well have been a spur to development of the theory.

The English Civil War (1642–1650) was over, and the Restoration of the Monarchy had occurred (1660), by the time Isaac Newton came to prominence. Newton was the only one of the major figures involved in development of the theory of gravity not to be involved in military matters. One can only speculate on what he might have brought to bear.

Well versed in mathematics, and trained as a Captain of Artillery, Napoleon had a grasp of scientific matters that far exceeded that of the usual political leader. While this might have led to an expectation of significant involvement of scientists in military matters during the Napoleonic Wars, in fact Napoleon had a greater effect upon the (civilian) development of science.[17] He encouraged innovation and extended teaching of science across the universities. His expedition to Egypt in 1798–1801 included 500 scientists (paid on a scale from Savant 1st class to Savant 10th class). Surprisingly, Napoleon paid little attention to the possible employment of science for war.[18]

∞ ∞ ∞

Scientists were recruited on both sides for the military effort in the First World War. Fritz Haber directed his Institute of Physical Chemistry and Electrochemistry in Berlin to focus on developing poison gas from the outbreak of war. The institute reported directly to the Supreme Army Command from 1916. It employed more than 1500 people. This was applied rather than pure research. Not a glorious moment in the scientific record, it did not lead to any fundamental insights.

Haber also discovered the reaction of nitrogen from air with hydrogen under pressure, to form ammonia. This was useful in munitions production. More positively, the ammonia can be used to make nitric acid, which can be used to produce fertilizers (still a useful method).[19] (Haber enthusiastically promoted the use of poison gas. He was not regarded as a war criminal, however, but in 1919 was awarded a Nobel Prize for his work on nitrogen. This was widely regarded as an outrage at the time, yet today there is a Fritz Haber Institute in Berlin as part of the Max Planck Society.)

The First World War became known as "the chemists' war."[20] Germany had been ahead of any other country in linking chemistry to industry in the second half of the 19th century, but employment of chemists to design explosives or poison gases occurred on both sides in the war. Chemists were involved in designing high explosives and securing essential supplies for them, such as toluene (for TNT) and acetone (used in the production of cordite). French Nobel Prizewinners were involved at high levels, with Marie Curie directing mobile radiology units, Jean-Baptiste Perrin involved in work on acoustics, and Louis de Broglie directing radio transmissions.

Physicists were involved in both basic and applied developments. Astronomer Charles Nordmann, Nobel Prizewinning physicist Lawrence Bragg, and medical researcher Lucien Bull developed an acoustic system to pinpoint enemy artillery positions. Nobel Prizewinners Ernest Rutherford and Robert Millikan worked on acoustic methods to detect submarines.

Eventually, in 1976, the acoustic methods were developed into ultrasound for medical purposes.

For the most part, scientists who worked on the war effort were diverted from primary research to more technological developments. War was a spur for developing technology, which in turn was subsequently useful in science. However, it's a moot point whether the First World War more advanced or retarded basic scientific development (because scientists were diverted from original research).

Employment of scientists for research for military purposes was sporadic in the First World War. Rutherford made a plea to the government to use scientists for research, but it was largely ignored. More scientists were sent to fight in the trenches than undertook research. The Second World War was more of an incubator for science as well as technology. Scientists were actively recruited to use their knowledge for the war effort.

War has been responsible more for directly developing science and technology in physics and chemistry than in biology. The development of penicillin and its production on a mass scale is an exceptional example of a development in biology directly driven by the needs of war.[21] Serendipity, or the Law of Unexpected Consequences, led to other major medical advances as the result of priorities set in the Second World War.

Research into mustard gas was triggered in the Second World War because it had been used in the First World War. The basic intention was to find some protection against the gas. Alfred Gilman and Louis Goodman at Yale University were in charge of the project because they had previously shown that mustard gas was cytotoxic (killed cells). Lymphoid tissues were especially sensitive. One of their colleagues obtained an extraordinary result: "We gave the mustard to this one lone mouse which had a fairly advanced tumor [a lymphoma]. After just two administrations of the compounds the tumor began to soften and regress... to such an extent that we could no longer palpate it." This lone mouse led to the use of chloroethylamine derivatives (the agent of mustard gas) to treat lymphoma in human patients, and indeed to the field of cancer chemotherapy.[22] (Another case of lethal wartime agents becoming useful in peacetime is the development from nerve gases to insecticides—a similar principle, but here with the target of insects rather than people.)

The use of cortisone to treat inflammatory diseases has an even more far-fetched origin. During the Second World War, a rumor developed among the allies that the Luftwaffe was using an adrenal extract to help pilots to fly at higher altitudes, because the extract counteracted the effects of oxygen deprivation. As a result, it became a priority to identify and isolate the active component. The compound, cortisone, was not actually obtained until 1948, at the Merck drug company. Its first clinical use to treat rheumatoid arthritis followed in 1949.[23] (Research projects have a life of their own once they have started. Sometimes this is productive, sometimes not.)

Of course, clinical usage did not come out of the blue. It had been known since the 1920s that jaundice alleviated the symptoms of rheumatoid arthritis. There had been unsuccessful efforts to isolate the responsible compound, known as 'Substance X.' One hypothesis was that it might be produced by the adrenal glands. It was already known that an adrenal extract could be used to treat Addison's disease, in which adrenal function is deficient. The connection was not actually very convincing. It owed quite a lot to a personal friendship between Philip Hensch, a physician with a research interest in rheumatoid arthritis, and Edward Kendall, a chemist trying to isolate the active component from adrenal extract, both at the Mayo Clinic. But then Substance X turned out to be identical with cortisone.

∞ ∞ ∞

Splitting the atom to produce the atom bomb is the most famous example of the application of science for war, but the war also led to radar, jet propulsion, and computers. The Second World War is sometimes called the Physicists' War. There were also major medical advances, especially the introduction of plasma for blood transfusions, as well as the exploitation of penicillin as an antibiotic. It might be fair to say that the Second World War blurred the boundary between science and technology, setting the scene for the rest of the 20th century.

The path to nuclear fission started in 1934 when Enrico Fermi bombarded uranium with neutrons. Fermi (1901–1954) started as a theoretical physicist. His first major contribution, in 1926, was the development together with Paul Dirac of rules to explain the behavior of particles that obey the Pauli exclusion principle

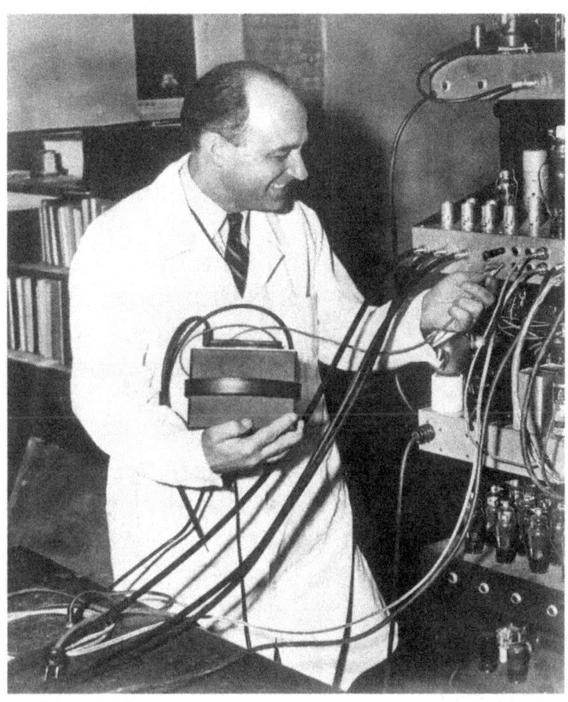

Enrico Fermi working on the atomic pile c. *1942.*

© *Mondadori Portfolio/Getty Images.*

(which governs the quantum behavior of subparticles in the atom). (The class of particles including electrons, protons, and neutrons are called fermions after him.) This led to a position in Rome, where he assembled a group of young physicists. In 1934, he proposed a theory to explain beta decay (when electrons are emitted from a nucleus), which famously was rejected by *Nature* as "too remote from physical reality." The theory was correct. Fermi became known as the "Pope of Physics."[24]

Fermi now turned to experimental physics. Bombarding uranium with neutrons resulted in the release of radioactivity, but there was a curious discrepancy in that the result depended on the material surrounding the equipment. Fermi discovered the cause when he placed a block of paraffin wax between the neutron source and its target: radioactivity increased more than $100\times$. Asked why he had thought to do that, he just said, 'intuition.'[25] The cause of the effect was that the paraffin wax had slowed down the neutrons because it is rich in hydrogen.

The release of radioactivity from the uranium suggested that new radioactive elements had been created. The implication was that they were transuranic—they would be the first elements discovered with greater atomic weights than uranium. Fermi was awarded the Nobel Prize for Physics in 1938 for discovering "new radioactive elements beyond uranium.' But this was a mistake. In fact, the uranium had split into two lighter elements. (A proposal in Italy to call the new element Mussolinium was thwarted, but would not have been so inappropriate since it was, after all, ephemeral.)

Working in Berlin in 1938, Otto Hahn and Fritz Strassman found isotopes of barium among the decay products of uranium. They thought it was impossible that a tiny neutron could cause the uranium to divide. Hahn wrote to Lise Meitner, with whom he had worked before she was forced to flee Germany. Meitner together with her nephew Otto Frisch worked out that uranium could split into two atoms, which together would have less mass than the uranium nucleus. The difference is released as energy according to $E = mc^2$.

When Frisch told Niels Bohr of the results, Bohr banged his head and exclaimed, "Oh what idiots we have all been."[26] Meitner and Frisch published a paper in *Nature* in 1939,[27] and Frisch called the process 'fission.' (Hahn and Strassman published a separate paper.[28]) Hahn was awarded a Nobel Prize in 1944 (at a time when Germany was working to develop an atomic bomb, although Hahn never worked on the project, but continued his work on isotopes). Meitner and Frisch were not acknowledged.

Fission of the uranium nucleus releases neutrons. This creates the possibility of a chain reaction, as each disintegration of a nucleus releases enough neutrons to cause the disintegration of more nuclei. The process occurs with U^{235}, which is unstable. This is why it can propagate a chain reaction, compared to U^{238}, which is stable. By the time the Second World War broke out, the process was well enough understood to consider its use for making a bomb. The effort was triggered after Albert Einstein (then at the Princeton Institute of Advanced Study) signed a letter to President Roosevelt in 1939 endorsing the possibility. Because of the situation in Italy, Fermi had emigrated to the

United States immediately after receiving his Nobel Prize in 1938. He became one of the driving forces in the Manhattan Project.

This was really the point at which science became permanently entangled with the military. Until then, scientists had been recruited in time of war, effectively by their patrons. Before the First World War, this was generally to apply their existing expertise. During the First World War, they might well be directed to work on other matters related to their expertise. After the Second World War, research for the military, both basic and applied, was sequestered. It became possible to make a career in science solely in military matters. As such work is often classified, it is hard to describe.[29]

It's a paradox that, before the Second World War, Germany was the leader in theoretical physics, followed by England. The United States scarcely showed any interest. Asked in 1927 if he would like to visit the United States, Paul Dirac said, "There are no physicists in America."[30] Yet it was the United States that developed atomic weapons.

Robert Oppenheimer was an exception in the United States. After his undergraduate degree from Harvard, he went to Göttingen and obtained his Ph.D. in 1927 under Max Born. He also published several important papers with Born. After the oral for the Ph.D. was over, one of the examiners was heard to say, "I'm glad that is over. He was on the point of questioning me."[31]

Recognized for his brilliance and breadth, Oppenheimer spent time at Harvard, Caltech, and Zurich before taking a position at the University of California, Berkeley. This was not the most obvious place, but Oppenheimer expressed his rational: "I thought I'd like to go to Berkeley because it was a desert. There was no theoretical physics."[32] Of course, Berkeley was strong in experimental physics: it was the home of the cyclotron (see Chapter 14). By the 1930s, the situation in the United States was changing with the arrival of Jewish nuclear physicists driven out of Europe by Nazism.

When the Manhattan project was started, Oppenheimer was recruited in 1942 to run it. By the end of the war, the United States was a leader in physics. After the war was over, in 1947 Oppenheimer became the Director of the Institute for Advanced Studies at Princeton. It was a sad result of the new connection between politics and science that the anti-Red hysteria of the fifties led to the revocation of his security clearance in 1954, which effectively ended his career in nuclear physics.[33] Oppie, as Oppenheimer was universally known, was by all accounts one of the most brilliant theoretical physicists of his generation, but (until the Manhattan project) he never stayed very long with any specific project.

Perhaps the most direct connection that war forced with science and technology was the development of the computer. The need to make complex calculations of trajectories for artillery firing tables drove the need to develop a way to make calculations faster than could be achieved manually, as mentioned above. (I discuss the development of computers in Chapter 23.) When more powerful computers were required to make the calculations necessary for controlling nuclear explosions, Johnny von Neumann at Princeton was instrumental in the development in 1951 of MANIAC (Mathematical and Numerical Integrator and Computer). (It was constructed in spite of local opposition from the faculty.)

At the end of the Second World War, President Roosevelt asked Vannevar Bush, the Director of the Office of Scientific Research and Development, "What can be done... to make known to the world as soon as possible the contributions which have been made during our war effort to scientific knowledge?"[34] It would not have been surprising if the experience of the war had led to a demand for the government to control the direction of research, but Bush's response was a call to support basic research. In *Science, the Endless Frontier,* he argued: "As long as... scientists are free to pursue the truth wherever it may lead, there will be a flow of new scientific knowledge to those who can apply it to practical problems in Government, in industry, or elsewhere." Public funding then massively expanded, bringing science into the modern era.

NOTES AND REFERENCES

1. *In the Matter of J. Robert Oppenheimer, Transcript of Hearing Before Personnel Security Board*, 1954, United States Atomic Energy Commission, Washington DC, Reprinted by Cornell University Press, 2001.
2. Quoted in D. E. Rowe and R. J. Schulmann, *Einstein on Politics: His Private Thoughts and Public Stands on Nationalism, Zionism, War, Peace, and the Bomb*, Princeton University Press, Princeton, 2007, p. 470.
3. V. Rothschild, *A Framework for Government Research and Development by Lord Rothschild*, HMSO, 1971.

4. L. Fermi and G. Bernardini, *Galileo and the Scientific Revolution*, Basic Books, New York, 1961, pp. 21–22.

5. As quoted by Pappus of Alexandria (*c*. 290–350 CE) in *Synagoge*, Book VIII (a series of eight books from *c*. 340 CE).

6. L. Russo, *The Forgotten Revolution: How Science Was Born in 300 BC and Why It Had to Be Reborn*, Springer, Berlin, 2003, p. 109.

7. Bibliothèque de l'Institut de France, Manuscrits de Léonard de Vinci, Ms 2173.

8. C. Rossi, Archimedes' Cannons Against The Roman Fleet?, in *The Genius of Archimedes — 23 Centuries of Influence on Mathematics, Science and Engineering*, ed. N. Palladino, S. A. Paipetis and M. Ceccarelli, Springer, New York, 2010, pp. 113–131.

9. designed.mit.edu/gallery/data/2011/homepage/experiments/steamCannon/ ArchimedesSteamCannon.html.

10. W. Isaacson, *The Innovators: How a Group of Hackers, Geniuses, and Geeks Created the Digital Revolution*, Simon & Schuster, New York, 2015, pp. 72–73.

11. M. Gharib, C. Roh and F. Noca, Leonardo da Vinci's Visualization of Gravity as a Form of Acceleration, *Leonardo*, 2023, **56**, 21–27.

12. M. Gharib, C. Roh and F. Noca, Leonardo da Vinci's Visualization of Gravity as a Form of Acceleration, *Leonardo*, 2023, **56**, 21–27.

13. Interpretation according to M. Gharib, C. Roh and F. Noca, Leonardo da Vinci's Visualization of Gravity as a Form of Acceleration, *Leonardo*, 2023, **56**, 21–27.

14. M. Valleriani, *Galileo Engineer*, Springer, Dordrecht, 2010.

15. P. Damerow, G. Freudenthal and J. Renn, *Exploring the Limits of Preclassical Mechanics: A Study of Conceptual Development in Early Modern Science: Free Fall and Compounded Motion in the Work of Descartes, Galileo, and Beekman*, Springer Verlag, New York, 2nd edn, 2004, pp. 141–157.

16. P. Damerow, G. Freudenthal and J. Renn, *Exploring the Limits of Preclassical Mechanics: A Study of Conceptual Development in Early Modern Science: Free Fall and Compounded Motion in the Work of Descartes, Galileo, and Beekman*, Springer Verlag, New York, 2nd edn, 2004, pp. 180–237.

17. The fact that science was not regarded as a direct factor in the war at this time was shown by Napoleon's award of a prize in 1808 to Humphry Davy in England for his scientific achievements; and Davy went to Paris to collect it in the middle of the Napoleonic wars. See Chapter 9.

18. A. Roberts, *Napoleon and Wellington*, Phoenix Press, London, 2002, p. 48.

19. The connection is that fertilizers can be used to make bombs.

20. M. Freemantle, *The Chemists' War: 1914–1918*, Royal Society of Chemistry, London, 2014.

21. Fleming discovered penicillin in 1929 but it was not exploited until 1939. Driven by the wartime situation, Ernst Chain and Howard Florey at Oxford first isolated it and then developed a scheme for mass production.

22. S. L. Smith, *Toxic Exposures: Mustard Gas and the Health Consequences of World War II in the United States*, Rutgers University Press, New Brunswick, NJ, 2017, p. 104.

23. J. Le Fanu, *The Rise and Fall of Modern Medicine*, Basic Books, New York, 2nd edn, 2012, pp. 27–39.

24. G. Segrè and B. Hoerlin, *The Pope of Physics: Enrico Fermi and the Birth of the Atomic Age*, Henry Holt, New York, 2016.

25. G. Segrè and B. Hoerlin, *The Pope of Physics: Enrico Fermi and the Birth of the Atomic Age*, Henry Holt, New York, 2016, p. 103.

26. R. Rhodes, *The Making of the Atomic Bomb*, Simon & Schuster, New York, 1986, p. 261.

27. L. Meitner and O. Frisch, Products of the Fission of the Uranium Nucleus, *Nature*, 1939, **143**, 471–472.

28. O. Hahn and F. Strassmann, Concerning the Existence of Alkaline Earth Metals Resulting from Neutron Irradiation of Uranium, *Naturwissenschaften*, 1939, **27**, 11–15; translation in H. G. Graetzer, Discovery of Nuclear Fission, *Am. J. Phys.*, 1964, **32**, 9–15.

29. Secrecy is usually regarded as incompatible with the practice of science. Sociologists sometimes find examples to the contrary, but these are usually short-term, with researchers trying to protect their position before they publish. In classified research, of course, secrecy is a one-way process: scientists engaged in classified research have access to the rest of science. It's an interesting question whether classified research would proceed more effectively if it were open to external criticism. For a review of the conduct of secret research, see B. Balmer, *Secrecy and Science. A Historical Sociology of Biological and Chemical Warfare*, Ashgate, Farnham, 2012.

30. Quoted in R. Monk, *Robert Oppenheimer: A Life Inside the Center*, Anchor Books, New York, 2013, p. 145.

31. Quoted in R. Monk, *Robert Oppenheimer: A Life Inside the Center*, Anchor Books, New York, 2013, p. 137.

32. Quoted in R. Monk, *Robert Oppenheimer: A Life Inside the Center*, Anchor Books, New York, 2013, p. 169.

33. K. Bird and M. J. Sherwin, *American Prometheus: The Triumph and Tragedy of J. Robert Oppenheimer*, Vintage Books, New York, 2007.

34. Letter from Roosevelt to Bush, November 17, 1944. Online at www.nsf.gov/about/history/nsf50/vbush1945_roosevelt_letter.jsp.

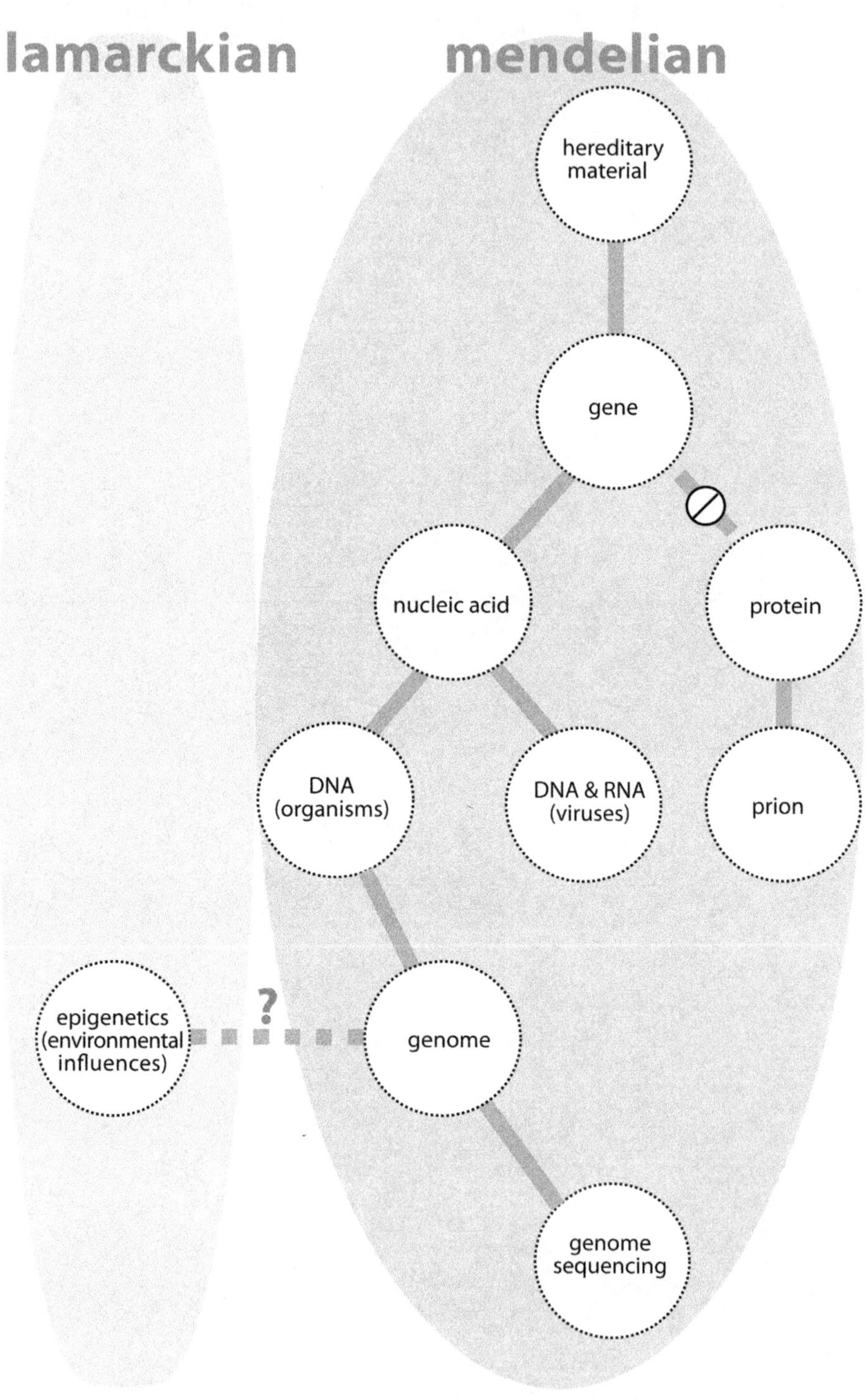

lamarckian
mendelian
hereditary material
gene
nucleic acid
protein
DNA (organisms)
DNA & RNA (viruses)
prion
genome
epigenetics (environmental influences)
?
genome sequencing

Biological Sciences

Only with the discovery of the double helix and the ensuing genetic revolution have we had grounds for thinking that the powers held traditionally to be the exclusive property of the gods might one day be ours, J. D. Watson, 1968.

CHAPTER 19

Crick's Double Helix: 1953–

"ALMOST ALL ASPECTS OF LIFE ARE ENGINEERED AT THE molecular level, and without understanding molecules we can only have a very sketchy understanding of life itself," said Francis Crick.[1] The most important discovery of the 20th century in biology, the model for the double helix of DNA required structural information from Maurice Wilkins and Rosalind Franklin, and key insights about structure and genetics from Watson and Crick. The concept that DNA is a code-script carrying information permeates all biology. It was followed by the most fertile period for discovery in the history of biology. Many brilliant minds were involved. Perhaps it is an over statement to call this Crick's Revolution, but if any single figure was a focal point, it was Francis Crick. While he was rarely subsequently involved in experiments, he was a constant presence, a rare instance of a decisive role for a theoretician in biology. He continued to generate theories for the rest of his life, finally turning from molecular biology to the intractable problem of consciousness.

The Frontiers of Science
By Benjamin Lewin
© Benjamin Lewin 2026
Published by the Royal Society of Chemistry, www.rsc.org

Timeline for Major Discoveries in Molecular Biology

1920

1930 Griffiths finds 'transforming principle' that
transfers hereditary characteristics between bacteria (1928)

1940

Avery shows that transforming principle is DNA (1944)

1950 Hershey & Chase show only viral DNA enters bacterium (1952)
Watson and Crick propose model of double helix (1953)
Kornberg isolates enzyme that copies DNA (1955)
Crick proposes Central Dogma (1957)
Meselson & Stahl identify mechanism of DNA replication (1958)

1960 Brenner and others discover messenger RNA (1961)
Jacob & Monod propose model for gene regulation (1961)
Holley sequences small RNA (80 bases) (1965)
Nirenberg and others define genetic code (1966)

1970 Temin & Baltimore discover reverse transcription (1970)

Fiers sequences first gene (400 bases) (1972)

Roberts & Sharp discover split genes (1977)

1980 Cech and Altman discover ribozymes (1981)
Prusiner discovers prions (1983)

Sinsheimer proposes human genome sequencing (1985)
Gilbert describes RNA World theory for origin of life (1986)
Human genome project starts (1988)

1990

2000

First draft sequences of human genome (2001)

2010

Church & Doudna publish CRISPR gene editing (2013)

2020 Completion of human genome sequences (2022)

A Dialog between Friedrich Miescher and Oswald Avery

Friedrich Miescher (1844–1895) was a Swiss biochemist. Trying to purify material from the nucleus of the cell, he found a new type of compound, containing phosphorus and nitrogen.

Oswald Avery (1877–1955) was an American biochemist, working at the Rockefeller University, who discovered that DNA is the genetic material of a bacterium.

© Ann Ronan Pictures/Print Collector/ Getty Images.

Friedrich Miescher: Good day, Dr. Avery. It's an honor to meet a fellow pioneer who first showed that DNA is the component of nuclein. Is it not a sad comment that neither of our works were appreciated in our own time?

Oswald Avery: Thank you, Dr. Miescher. It's a pleasure to meet the scientist who first isolated the genetic material. Your discovery of nuclein laid the groundwork for my own research.

Friedrich Miescher: Indeed, it has been a fascinating journey. I set out to determine the chemical basis for life. I obtained precipitates that could not be dissolved in water, acetic acid, dilute hydrochloric acid, or in solutions of sodium chloride. So this material could not belong to any of the hitherto known proteins. I could not have anticipated this material would be identical with the transforming principle in bacteria.

Oswald Avery: We came from another direction, because we knew that the transforming principle carried hereditary information. We sought to understand its nature, so that was parallel to your efforts to define the chemical basis for nuclein.

Friedrich Miescher: I anticipated that an entire family of such phosphorus-containing substances, which differ slightly from one another, would reveal itself, but I never thought that this family of nuclein bodies would prove tantamount in importance to proteins. I was unable to make a firm connection with the transmission of heredity.

Oswald Avery: We had never imagined that there was DNA in bacteria. It opened up a new era in biology by providing a molecular foundation for understanding heredity. You were trying to find proteins involved in heredity and isolated nuclein. I was trying to discover how bacteria cause pneumonia and I ended up discovering DNA! It's a testament to the unpredictable nature of scientific progress and the unity of science. I rest my case.

CRICK'S DOUBLE HELIX

The ultimate aim of the modern movement in biology is in fact to explain all biology in terms of physics and chemistry,[2] Francis Crick, 1966.

The discovery in 1953 that DNA is a double helix was one of the most fundamental observations ever made in any area of science.[3] For the first half of the 20th century, genetics was something of a black box. We knew that genes carry heredity, and that mutations represent changes in those genes, but it was impossible to understand the underlying basis in terms of the properties of physical molecules. After 1953, it was possible to interpret all genetics, and indeed the characteristics of every organism, in terms of its DNA sequence. And the fact that all living organisms have DNA as their genetic material validates the theory of evolution beyond anything achieved by fossil records.

The importance goes beyond the facts themselves, and their implications within genetics and biology, into establishing a reductionist manifesto for biology. All features of animals (including humans) can be explained in terms of the laws of physics and chemistry. This settled the question raised by physicist Erwin Schrödinger in 1944 in his influential book, *What is Life?*, when he considered the possible nature of the genetic material.[4]

"We must be prepared to find [the structure of living matter] working in a manner that cannot be reduced to the ordinary laws of physics," Schrödinger said.[5] The prospect that defining the nature of the genetic material might uncover new laws in addition to those already known in physics and chemistry was an epochal event, bringing many physicists into biology.[6] Of course, with the discovery of the double helix it turned out that new laws were not needed after all.

The concept that DNA carries *information* allows the biology of an organism to be analyzed in terms of how that information is expressed. In the brave new world of today, we can change the properties of an organism directly by introducing changes in its DNA sequence. Perhaps it's a little far-fetched to compare the difference between the pre- and post-DNA eras to the difference between astronomy in the ancient world and astronomy after the telescope was invented, but there is something of the same change in perspective.

Throughout the first half of the 20th century, the prevailing paradigm was that the genetic material must be protein. Proteins were known to have complex three-dimensional structures. It was assumed that only protein could have the complexity to specify their formation. Proteins are constructed from 20 different types of amino acids. The types and order of the amino acids in any particular protein enable it to fold into its specific structure. DNA, by contrast, is a nucleic acid that has only 4 types of subunit. It was therefore not at all obvious how it might be able to give rise to proteins.

Ironically, research to suggest that genetic material is protein, and work to exclude nucleic acid, both came from the Rockefeller University in New York, then a powerhouse of American science, where ultimately Oswald Avery showed that DNA is the genetic material.

It created a worldwide sensation when Wendell Stanley at the Rockefeller announced in 1935 that he had crystallized tobacco mosaic virus.[7] He believed the material consisted of protein. This led to the view that the infectious ability might depend on an autocatalytic function in which the protein activated itself. Showing that infection could be caused by crystallized material was taken to support the reductionist view that disease could be produced by inorganic material. The idea caught the imagination, although actually the fact of crystallization is more or less irrelevant.[8]

Much of the debate was based on the erroneous view that the crystallized material was exclusively protein, although within a year of Stanley's report it was shown that it also contained 6% nucleic acid.[9] None of this prevented Stanley from sharing a Nobel Prize in 1946. The question of whether small amounts of undetected material might be the true active agent was to bedevil attempts to find the basis of heredity or infection through the rest of the century.

Part of the problem with nucleic acid was that over the first years of the 20th century, chemist Phoebus Levene at the Rockefeller concluded (incorrectly) that DNA has a simple repeating structure made from the same set of four subunits. Thinking in terms of a structural basis for the genetic material, this reinforced the widely held view that DNA could not be complex enough.

When Oswald Avery (1877–1955) at the Rockefeller demonstrated that the material responsible for transferring hereditary properties between bacteria was DNA, there was widespread reluctance to believe the result. The most common criticism was that there must be a loophole—an undetected protein must be responsible.

Avery never received a Nobel Prize, because the Nobel committee was not convinced of the importance of his discovery before he died in 1955. "It's to be regretted that he did not receive the Nobel Prize. By the time dissident voices were silenced, he had passed away," the official history records.[10]

Driven by the belief that DNA was the genetic material, and that its structure must therefore be important and informative, Watson and Crick set out to build a model. Working at the Cavendish Laboratory at Cambridge University, neither had experience in the field. Their approach was closer to theoretical than to experimental science (at least insofar as they conducted no experiments themselves). There were several false steps along the way. However, when they saw a photograph of DNA, taken using X-ray crystallography by Ray Gosling, a student with

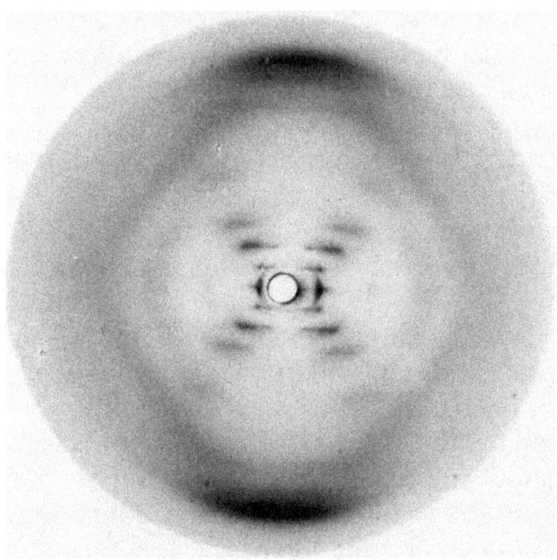

The famous photograph 51 showing the X of reflections.
King's College London Archives, KDBP/1/1.

Rosalind Franklin in Paris. Her laboratory obtained the critical data to show a double helix. Would she have been awarded the Nobel Prize if she had not died prematurely?

Reproduced from https://en.m.wikipedia.org/wiki/File:Rosalind-franklin-in-paris_crop.jpg, under the terms of the CC BY-SA 4.0 license, https://creativecommons.org/licenses/by-sa/4.0/.

Rosalind Franklin at King's College, London, they realized that the structure must be helical. "The instant I saw the picture my mouth fell open and my pulse began to race. ... the black cross of reflections which dominated the picture could only arise from a helical structure," Watson recollected.[11] The photograph passed into history as the iconic photograph 51.

When Watson and Crick built their model in 1953 to fit with the parameters implied by the X-ray crystallography, everything fell into place. The model was widely (but not universally) accepted. They were awarded a Nobel Prize in 1961 (together with crystallographer Maurice Wilkins, who was in charge of the department at King's College).

One reason for the difference between resistance to Avery's result compared with acceptance of Watson and Crick's model was that the absence of a structure for DNA had made it hard to see just how DNA could function as the genetic material. Watson and Crick in effect rephrased the paradigm. Instead of functioning as a structure, DNA functioned as information. Even though it took another decade to work out how that information

Maurice Wilkins with a camera he developed for X-ray diffraction. In this era, experimenters still built their own equipment.

is interpreted, the concept provided a framework for thinking about the issue, and designing experiments to test the theory. Before the model of the double helix, there had been a blank. Nature may not, after all, abhor a vacuum—but scientists do.

"Chance favors the prepared mind," Louis Pasteur said. The discovery of DNA is a wonderful example. Watson and Crick were prepared because they thought about both heredity and chemistry. Constructing the double helix model also required knowledge of crystallography. Watson and Crick between them (with the help of Jerry Donahue, a chemist in the laboratory) combined insights from all three disciplines. With the increasing specialization of science, a comparable effort involving so few people would be most unlikely today.

It turned out subsequently that William Astbury, one of England's best-known crystallographers, working at the University of Leeds, had obtained an X-ray diffraction photograph of DNA in 1951 that was very similar to photograph 51. But Astbury

Francis Crick and James Watson walking in Cambridge in 1953.

CSHS.

never published the photograph. From his background in protein structure, he may simply have thought it was too uninteresting in structural terms.[12]

The double helix model has a backbone made from alternating sugar and phosphate residues, with 'bases' on the inside. There are four types of base. Each base consists of a ring structure made from carbon and nitrogen. The bases consist of two types of single-ringed structures (pyrimidines), cytosine and thymine, and two types of double-ringed structures (purines), adenine and guanine. Watson and Crick knew from the previous work of Erwin Chargaff (a biochemist in New York) that the amount of adenine in DNA always equals the amount of thymine, and the amount of guanine always equals the amount of cytosine.

The genius of the model is that the bases form two types of pairs, each consisting of a purine paired with a pyrimidine, G-C and A-T, held together only by hydrogen bonds. Whereas the covalent bonds that link atoms into molecules require a lot of energy to break, hydrogen bonds require much less energy. This allows the two strands of the double helix to separate, in effect by pulling each base pair apart into its individual bases. Then each

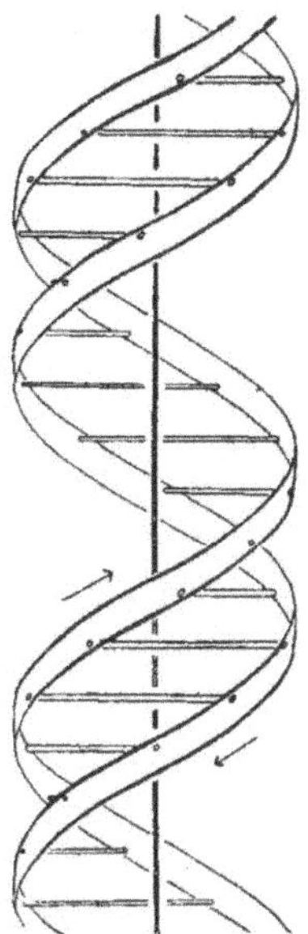

The original illustration of the structure of the double helix for the paper in Nature in 1953 was drawn by Odile Crick.

Reproduced from https://doi.org/10.1038/171737a0 with permission from Springer Nature, Copyright 1953.

base in the separated strands can pair with a new partner, specified by the pattern of hydrogen bonding, to produce two identical daughter duplexes.

The famous understatement that concluded Watson and Crick's paper foresaw how it might replicate itself. "It has not escaped our notice that the specific pairing we have postulated immediately suggests a possible copying mechanism for the genetic material."[13]

It's unlikely that there will be another major insight made in a comparable way to the discovery of the double helix. It had a greater component of theoretical science than is usual for biology, admittedly with a foundation in the X-ray diffraction—but the crystal structure wasn't 'solved' as such. The model was built on the principle of string and sealing wax—in fact, first efforts used paper cutouts—until the workshop was persuaded to make metal parts. The model did not emerge from testing a hypothesis as such, but from the perspective that the structure of DNA was a prize because—whatever it was—it would be essential to understanding heredity. A series of models that were wrong (in some cases wildly wrong) were tested until Watson and Crick found one that stood up to the data.

The model was so striking that it became the prevailing paradigm almost instantaneously. This was not a revolution—that should have taken place when Avery's paper was published in 1944—but it provided a basis for the new paradigm: DNA is the genetic material. Once again, an insight into how things might work was required for the data to be assimilated. Accordingly, when

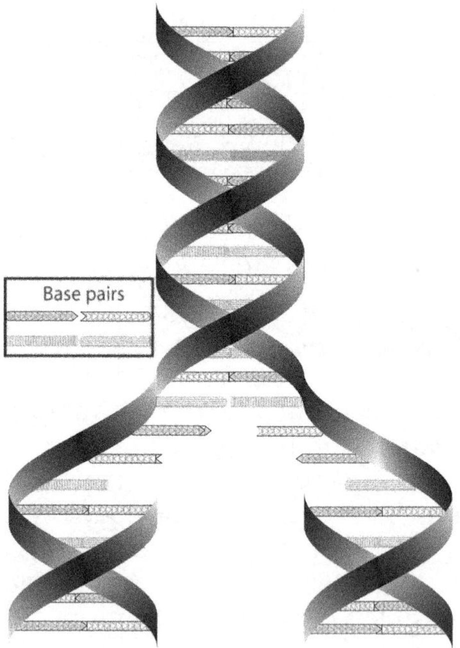

Base pairs are in the center of the backbone of the double helix. In order to replicate, the duplex (at the top) unwinds into separate strands (in the middle) and each base now pairs with a new partner to generate two identical double helices (at the bottom).

Heinz Fraenkel-Conrat showed in 1956 that RNA was the genetic material of TMV (tobacco mosaic virus), the result was accepted immediately.

RNA is another nucleic acid, or to be more precise, a class of nucleic acids. It has two chemical differences from DNA: it has the sugar ribose whereas DNA has deoxyribose (this means that RNA has a hydroxyl, –OH group, on the sugar ring at a point when DNA has only a –H atom); and it has the base uracil instead of thymine.[14] It's single-stranded, although many RNAs have double-stranded regions. Its role was unknown when the double helix model was proposed for DNA. Within the cell, RNAs are made by copying DNA, using the same principles involved in DNA replication (the technical term for the process is 'transcription'). Over the next few years, it turned out that several types of RNA are involved in interpreting the sequence of DNA. And more recently a variety of other roles have been found for new classes of RNAs.

The era of molecular biology started with the transition from thinking that the role of the genetic material was structural to realizing that it carries a *code*. The relationship between the sequence of DNA and the sequence of proteins is called the genetic code. The next decade was spent working out the nature of the code and the apparatus that is responsible for executing it.

There's a gap in the argument here. A protein's function is determined by its three-dimensional structure. But the genetic code gives a linear sequence of amino acids. How do we get from one to three dimensions? Working at the NIH in Bethesda, Maryland, Christian Anfinsen showed in a classic study in 1961 that when a protein is denatured (treated to cause it to lose its structure), it can refold into its original structure. The discovery that sequence carries the information necessary to fold into the correct structure won Anfinsen the Nobel Prize in 1972.[15]

The worldview at the start of this period was governed by a set of paradigms that followed from the model for replication and expression of the double helix:

- DNA is the universal genetic material.
- RNA can be the genetic material in viruses (as of 1956).
- All infectious agents have genomes of nucleic acid.
- DNA is transcribed into RNA, but RNA cannot generate DNA.
- All enzymes (the catalytic agents of the cell) are proteins.

Except for the extension of the role of DNA as genetic material to include the RNA of viruses, all of the subsidiary paradigms were overthrown during the last three decades of the 20th century.

∞ ∞ ∞

The Central Dogma ruled molecular biology from its formulation by Francis Crick in 1957 to its overthrow in 1970. What Crick said exactly was that, "once 'information' has passed into protein *it cannot get out again.*"[16] If any single phrase captures the spirit of molecular biology—indeed of the basis of life—it is the Central Dogma. It states the basis of heredity and excludes heresies such as Lamarck's idea that acquired characters can be inherited.

The process of reading a sequence in DNA to generate a protein sequence passes through an intermediary of RNA (this is called messenger RNA). (The technical term for reading the messenger RNA into protein is 'translation.') As the process was worked out, the Central Dogma was transmogrified into the vernacular, "DNA makes RNA makes protein." This was taken to mean that it was impossible to reverse the process at either stage. That wasn't, of course, exactly what Crick had said, but it became the dominant paradigm.

At the University of Wisconsin, Howard Temin had spent the 1960s arguing that RNA tumor viruses (they have RNA as the genetic material and can cause tumors in animals) function by copying themselves into a double strand of DNA. Temin later described the reception of his work as, "Essentially ignored, if not derided."[17] But in 1970, he was able to show directly that this was precisely how the virus worked. David Baltimore at MIT made the same discovery at the same time, and the two papers were published side by side in *Nature*.

There was an argument over whether this was truly a paradigm shift. After *Nature* published a commentary under the title *Central Dogma Reversed,*[18] Crick replied that this was a misreading of his formulation, which had never precluded transfer between nucleic acids.[19] While technically correct, this missed the point that everyone had assumed for 20 years that "DNA makes RNA makes protein." A paradigm is what people believe, not necessarily what is stated in the literature. In any case, that paradigm was well and truly overthrown.

∞ ∞ ∞

Yet this reinforced, if anything, the paradigm that the genetic material can only be nucleic acid, and that infectious agents must therefore have genomes of nucleic acid. Scrapie is an infectious neurological disease of sheep. When Stanley Prusiner at UCSF (University of California, San Francisco) set out to isolate the infectious agent, he tried to purify its nucleic acid. After failing to find any nucleic acid in the preparation, he was forced to conclude in 1983 that the infectious agent was protein. He called it a *prion*, for proteinaceous infectious particle.[20]

The results were received with indifference, and sometimes hostility, based on the argument that the real agent must surely be an undetected nucleic acid in the preparation. History was repeating itself. This was an ironic reversal of the arguments when DNA was identified as the genetic material: that the real agent might have been an undetected protein in the preparation. In the absence of a plausible mechanism, it was easier to reject the results than think the unthinkable.

The protein that causes scrapie is responsible for a series of neurodegenerative diseases in sheep, deer, cows, and humans. It's infectious between as well as within species, and was responsible for the outbreak of "mad cow disease," when people suffered lethal neurodegenerative lesions after eating meat from cows with the disease. A mutation in the gene coding for the prion is responsible for the inherited disease CJD (Creutzfeldt–Jakob disease) in humans.

The disease is caused when, instead of folding into its normal structure as a single molecule, the protein forms an abnormal structure from an aggregate of many misfolded protein molecules. Once an abnormal structure has formed, newly synthesized proteins are recruited to it to join the misfolded structure. Prusiner was the sole recipient of the Nobel Prize for Physiology or Medicine in 1997, but even after that, there was still criticism that the Nobel committee would be embarrassed when the real (nucleic acid) agent was found.[21,22]

The mechanism of the prion's infectious property is only slowly emerging. The prion protein has a central core consisting of two α-helices (the α-helix is a common feature of protein structure). The infectious form has converted the α-helices to a flat structure known as β sheets. The sheets are sticky—Velcro would be a good analogy—which is how they recruit more prion proteins to join them. The disease may be caused by the accumulation of long fibers (β-amyloid) made from the sheets. There are still many details we don't understand, but the principle has been established to the point at which it is now accepted that the abnormal protein can be infectious.

Proteins are divided into several groups, including structural, hormonal, transport, antibodies, and enzymes. The enzymes are catalysts—agents that speed up chemical reactions. The structure of an enzyme has a *catalytic center* that brings together the reacting components. In an echo of the old view that only proteins could have the complexity needed to be the genetic material, it seemed obvious that only proteins could have the structural complexity to be enzymes.

Wrong again! Trying to purify the enzymes that modified the structures of certain RNA molecules, Sidney Altman at Yale and Tom Cech at the University of Colorado found that they could not separate the activity from the RNA. They were forced to conclude that the RNA itself must have the catalytic activity. Cech called an RNA with catalytic activity a ribozyme.[23] Altman and Cech shared the Nobel Prize for Chemistry in 1989.

This was the forerunner of an entire new insight into molecular biology and evolution. There are many ribozymes, each with the characteristic property that the RNA can fold into a structure forming a catalytic center. This lends support to the "RNA World Hypothesis"—the idea that life may have originated with RNA rather than DNA.[24] It's plausible that an RNA with a primitive catalytic activity was at the origin of life. DNA could have emerged later as a means to store information for the system. (Is there an analogy here with the development of computers? The first computers functioned on the basis of information input from the exterior, for example, in the form of punch cards. It was a key transition into the modern era when they were provided with internal memory.)

Ribozymes were accepted immediately as reality. The difference from the resistance to prions was that the discovery of this catalytic activity was associated with a mechanism. As Tom Cech puts it, "Splicing [the reaction the RNA catalyzes] is a chemical reaction and saw bonds being made. When you see something like that, it gives a handle on the system that is undeniable."[25]

∞ ∞ ∞

The period from 1953 to 1966 was spent defining the genetic code—the relationship between the sequence of DNA and the sequence of proteins—and the means of its execution. The code is triplet: a series of 3 bases in nucleic acid codes for each amino acid in protein. As there are 64 possible triplet combinations, the code is redundant. More than one triplet may code for the same amino acid. The code also has triplets for starting and ending protein sequences.

The code was worked out in microorganisms, where a protein is coded by a continuous sequence of triplets, running from a start to a stop signal. It was assumed that, since DNA is the universal genetic material, this colinearity would also be universal, so a continuous sequence in nucleic acid would always represent a continuous sequence in protein. That turned out to be wrong when groups at Cold Spring Harbor in New York, and at MIT, working with adenovirus, showed in 1977 that a gene is not necessarily continuous.

A gene is expressed by making an RNA copy of the DNA. This messenger RNA has the same sequence as one strand of the DNA. Genes of higher organisms (and their viruses) can be *interrupted*, consisting of a series of *exons* coding for proteins, alternating with *introns* that have no coding function. Colinearity between the nucleic acid and protein sequences is created by *splicing* the sequences of the exons together. (This process occurs in the messenger RNA. The introns are thrown away.) Rich Roberts from Cold Spring Harbor and Phil Sharp from MIT shared a Nobel Prize in 1993.

Perhaps because the evidence for splicing was partly visual (consisting of electron micrographs directly showing the interruptions in the gene), it was accepted immediately. Not only was it accepted, but by the end of 1977, tens of laboratories were working to define interrupted genes in other systems. The mechanism of splicing was worked out only later, but one impetus for accepting the results was that they explained several puzzling features about the way genes are expressed. No one had previous understood the discrepancy that a messenger RNA in bacteria has a length corresponding to the protein it codes for, but the equivalent RNA in higher organisms is often very much larger. It's reduced to an appropriate length by the process of splicing.

∞ ∞ ∞

The relationship between experiment and theory may be clearer in physics, but the history of paradigm shifts in molecular biology over the second half of the 20th century follows the same principles. Sir Arthur Eddington, who obtained key data to support Einstein's theory of relativity, expressed a much-quoted view in 1934. "It's a good rule not to put overmuch confidence in a theory until it has been confirmed by observation. . . It's also a good rule not to put overmuch confidence in the observational results that are put forward until they have been confirmed by theory."[26] Everyone accepts that theories must be confirmed by experiment, but the reverse part is regarded as problematic. However, it captures an essential truth about the scientific attitude. There can be difficulties in accepting observations until there is some theoretical framework that makes sense of them.

A discrepancy between theory and data may result in reluctance to accept the theory or renewed questioning of the data. There is a telling difference between the paradigm shifts of physics and biology in the 20th century, because physics functions by laws whereas biology functions more by examples.

Theory tends to rule supreme in physics. Newton's theory of gravity was accepted, even though there was no means of understanding force at a distance, because data conformed with the predictions of the laws. When neutrinos appeared to defy Einstein and move faster than the speed of light, the data were questioned because they appeared to contradict the cold equations (see Chapter 25).

Data tend to rule supreme in biology. All known enzymes were protein—until the ribozyme was found. All known infectious agents were based on nucleic acid—until the prion was found. Physics functions on laws, but molecular biology has been based more on the black swan theory—a generalization is disproven if a single example is found that's inconsistent with one of its fundamental postulates.

NOTES AND REFERENCES

1. F. Crick, *What Mad Pursuit: A Personal View Of Scientific Discovery*, Basic Books, New York, 1988, p. 61.

2. F. Crick, *Of Molecules and Men*, University of Washington Press, Seattle, 1966.

3. As expressed graphically by Watson (see the beginning of the Biological Sciences section).

4. E. Schrödinger, *What is Life? The Physical Aspect of the Living Cell*, Cambridge University Press, Cambridge, 1944.

5. E. Schrödinger, *What is Life? The Physical Aspect of the Living Cell*, Cambridge University Press, Cambridge, 1944, p. 81.

6. Schrödinger viewed the chromosome as a 'code-script' and drew an analogy with Morse code (see ref. 4).

7. W. M. Stanley, Isolation of crystalline protein possessing the properties of tobacco-mosaic virus, *Science*, 1935, **81**, 644–645.

8. The debate about origins was reminiscent of the 19th century debate whether crystallization might account for the formation of organic material (see Chapter 21).

9. L. E. Kay, W. M. Stanley's Crystallization of the Tobacco Mosaic Virus, 1930–1940, *Isis*, 1986, 77, 450–472.

10. H. N. Schuck, *Nobel: The Man and His Prizes*, Elsevier, New York, 1962.

11. J. D. Watson, *The Double Helix*, Weidenfeld and Nicolson, London, 1968, pp. 167–168.

12. Astbury had previously (in 1938) published a model for DNA that identified it as a repeating structure with bases sticking out from a backbone every 3.4 Å.

13. J. D. Watson and F. H. C. Crick, Molecular structure of deoxypentose nucleic acids, *Nature*, 1953, **171**, 737–738.

14. Uracil and thymine are very similar: thymine has an extra methyl (CH_3) group compared to uracil.

15. We now know that the system is not quite so simple, and that accessory proteins (called chaperones) assist and influence folding, but it remains true that the structure is determined by the sequence. The ability of the AlphaFold software to predict structure from sequence is the ultimate validation (see Chapter 23). (The example of prions, discussed later, is an interesting exception.)

16. F. H. C. Crick, On protein synthesis, *Symp. Soc. Exp. Biol.*, 1958, **12**, 138–163.

17. Quoted in J. Crewdson, *Science Fictions*, Little, Brown, New York, 2002, p. 6.

18. J. Tooze, Central Dogma Reversed, *Nature*, 1970, **226**, 1198–1199.

19. F. Crick, Central Dogma of Molecular Biology, *Nature*, 1970, **227**, 561–563.

20. M. P. McKinley, D. C. Bolton and S. B. Prusiner, A protease-resistant protein is a structural component of the scrapie prion, *Cell*, 1983, **35**, 57–62.

21. S. B. Prusiner, *Madness and Memory*. Yale University Press, New Haven, CT, 2014.

22. That noted authority on science, the *New Yorker*, even called it 'pathological science,' meaning that it was a delusion. R. Rhodes, Pathological Science, *New Yorker*, 1997, December 1, p. 54.

23. With typical understatement, he said, "[It] is not an enzyme but has some enzyme-like characteristics...we call it a ribozyme, an RNA molecule that

has the intrinsic ability to break and form covalent bonds." See K. Kruger, *et al.*, Self-splicing RNA: autoexcision and autocyclization of the ribosomal RNA intervening sequence of Tetrahymena, *Cell*, 1982, **31**, 147–157.
24. M. Yarus, *Life from an RNA World: The Ancestor Within*, Harvard University Press, Cambridge, MA, 2011.
25. Discussion with Tom Cech, February 2022.
26. A. Eddington, *New Pathways in Science: Messenger Lectures*, Cambridge University Press, Cambridge, UK, 1945.

CHAPTER 20

Watson's Genomes: 1990–

"I WOULD ONLY ONCE HAVE THE OPPORTUNITY TO LET MY scientific life encompass a path from double helix to the three billion steps of the human genome," said Jim Watson.[1] The sequencing of the human genome was a *tour de force*. Controversial at the outset for its potential to change the nature of biology, the project gained acceptance and impetus when Watson became director of the Human Genome Project. A vast team effort, bringing the intrusion of Big Science into biology, it was a revolution in style as well as substance. The pace with which the human, and other, genomes have been sequenced has astounded even the participants. While no single person could therefore be described as responsible, it is in a sense the culmination of the idea inherent in the double helix model that the sequence of DNA is important because it carries information. The real revolution is that DNA sequencing has become a routine tool not only in science but also in medicine.

The Frontiers of Science
By Benjamin Lewin
© Benjamin Lewin 2026
Published by the Royal Society of Chemistry, www.rsc.org

Timeline for the Development of Genome Sequencing

1975
Sanger invents dideoxy DNA sequencing method (1977)
Maxam & Gilbert invent base modification method for DNA sequencing (1977)

1980

1985
Human genome sequencing proposed at Santa Cruz meeting (1985)
Opposition to HGP expressed at Cold Spring Harbor meeting (1986)
DOE and NIH begin funding Human Genome Project (HGP) (1988)
HUGO founded to support international efforts of HGP (1988)

1990
Sequencing begins (1990)

Five-year target for HGP achieved one year early (1994)

1995

2000
Sequence of fruit fly genome (2000)
First working draft of human genome (2001)
Sequnce of mouse genome (2002)

2005
Sequence of chimpanzee genome (2005)
23andMe founded (2006)

2010
Sequence of Neanderthal (2010)

Sequence of Denisovan human (2012)

2015

2020

Final human genome sequence (2022)
Human pangenome (47 diverse individuals) 2023
2025

A Dialog between Jean-Baptiste Lamarck and Gregor Mendel

Jean-Baptiste Lamarck (1744–1829) was a French military man who later became a zoologist and developed a theory of evolution. Part of the theory was that hereditary characters could be acquired in response to the environment.

Gregor Mendel (1822–1884) was an Augustine monk who studied plant breeding to improve agriculture. He proposed that heredity depends on discrete particles (genes), in contrast to previous ideas that heredity depends on blending.

Lamarck: Good day, Abbot Mendel. What a pleasure it is to converse with the "father of modern genetics." Your laws of inheritance have had a profound impact on our understanding of heredity.

Mendel: Thank you, Mr. Lamarck. I am very pleased to have the opportunity of discussing your unconventional ideas of heredity. I must say, your ideas on the inheritance of acquired traits are somewhat of a break with genetics.

Lamarck: Indeed, I proposed that traits acquired during an organism's lifetime could be passed on to future generations. Your laws of inheritance, on the other hand, emphasize the transmission of traits through discrete units. How do you see the relationship between our respective theories?

Mendel: The distinction lies in the mechanisms we propose. While your theory suggested a direct inheritance of acquired traits, my laws of inheritance emphasize the role of discrete hereditary units, which are now called genes. These units are separation between the traits acquired during an individual's lifetime and the traits inherited genetically.

Lamarck: But I ask whether this separation is artificial? Your laws emphasize the stability of inherited traits, whereas mine suggest a more dynamic interaction between an organism and its environment. In your experiments with pea plants, did you ever consider the possibility that acquired traits might influence inheritance?

Mendel: My experiments focused on traits that are stable and can be reliably observed. The principles I proposed, now known as Mendelian inheritance, form the basis of modern genetics. The discovery of DNA as the genetic material and the understanding of gene transmission provide a basis for my laws. By contrast, your theory is a speculation without any experimental basis—much like the situation in heredity before I performed my experiments.

Lamarck: Some recent research has suggested the potential for environmental factors to influence gene expression and, in turn, be inherited. Epigenetics may validate my work after a century of criticism! Do you not think it challenges the strict separation between acquired traits and genetic inheritance?

Mendel: It's true that the field of epigenetics has revealed a complex interplay between genes and the environment. While my laws provide the foundations, ongoing research continues to uncover the nuances of how environmental factors can influence gene expression *via* interactions with the underlying DNA sequence. But that is not the same as changing the pattern of inheritance! Epigenetics is to genetics as alchemy is to chemistry. My laws stand unchallenged. I rest my case.

WATSON'S GENOMES

The question is often asked whether after the Human Genome, will our life be the same? Yes, it will be because the Human Genome is only a telephone book; but try to find out from a city telephone directory how the city works. There is need for a theory. It is the wrong approach to collect the data, stuff them into the computer, and wait for the emerging results. This is not artificial intelligence, rather, it is artificial stupidity. We have to have a theory.[2] Sydney Brenner, 2003.

Completed just after the turn of the Millennium, the sequence of the human genome is one of the great accomplishments of science. Ah—but is it science or is it technology?

It had taken a century to get from the concept of the gene to the idea that the identity of a species lies in its DNA sequence. Heredity had been a mystery in the 19th century. Gregor Mendel (1822–1884) established the basis for genetics when he started his work on the garden pea in 1854 at the monastery in Brno (now in the Czech Republic). But no one knew about the work until it was rediscovered in 1900. Genetics then became a science, really the first quantitative sub discipline in biology. Experiments principally on microorganisms and fruit flies established the existence of the gene as a discrete entity subject to sudden change (mutation).

The idea that genes are located on chromosomes dates from soon after the rediscovery of Mendel in 1902. Friedrich Miescher (1844–1895) had identified nuclein in 1869 as a component of the nucleus. A crude extract, it contained both proteins and nucleic acid.[3] Once it was purified, the nucleic acid was identified as DNA. Yet no one thought it was the genetic material. When Oswald Avery showed in 1944 that DNA is the genetic material of a bacterium, he was surprised. "Who could have guessed it? This type of nucleic acid has not to my knowledge been recognized in *Pneumococcus* before," he wrote in a letter to his brother.[4]

It took the structure of the double helix to bang home the importance of DNA. When it was realized that the significance of DNA lies in its sequence, the idea of being able to determine that sequence was no more than a fantasy. It had been difficult enough to sequence proteins. When Fred Sanger, working at the University of Cambridge, started to analyze insulin in 1945, it was not even certain at the time that a protein would have a unique sequence. Working painstakingly fragment by fragment, Sanger determined the sequences of the two protein chains in insulin. He received a Nobel Prize in 1958.

Sanger turned his attention to DNA sequencing in 1973. With remarkable modesty for someone who won two Nobel Prizes, Sanger once said of himself, "Unlike most of my scientific colleagues, I was not academically brilliant."[5] His work is one of the most remarkable examples of how scientific curiosity can drive technology, which in turn feeds back into science. He once expressed this by saying that "anytime you get technical development that's two to threefold or more efficient, accurate, cheaper, a whole range of experiments opens up."[6]

The idea of sequencing the whole genome, and in particular the human genome, arose from an overt proposal to increase the

Fred Sanger in his laboratory in 1969. A big contrast with the automated sequencing centers of the next century.

Reproduced with permission from MRC Laboratory of Molecular Biology.

scale of research in biology. Robert Sinsheimer at the University of California, Santa Cruz started the ball rolling with a meeting in 1985. "Biology had always been 'small science.' I wondered if there were scientific opportunities in biology that were being overlooked, simply because we were not thinking on an adequate scale," he said.[7] (Ironically, Sinsheimer's own work had been on the smallest scale possible, on a virus that infects bacteria.)

The idea was discussed the following year at a meeting at Cold Spring Harbor on the Molecular Biology of *Homo sapiens*, when concerns were expressed that the project might compete for resources rather than complement existing efforts. People were worried that the introduction of 'big science' could change the nature of 'small science.'

Until now, biology had been funded on a relatively small scale, largely through grants to PI's (Principal Investigators) who ran individual laboratories. Competition to obtain grants was based

---

DNA Sequencing

The relative simplicity of DNA, with 4 types of base compared with the 20 types of amino acids in proteins, made an approach comparable to protein sequencing problematic for anything other than rather short sequences. Analysis by breaking an RNA into smaller pieces that were then analyzed separately was used by Robert Holley in 1965 to sequence a small (80 base) RNA. Walter Fiers used the same method in 1972 to sequence a longer stretch of a virus gene (just under 400 bases). Followed four years later by the entire viral genome, this was more or less at the limit.

Sanger developed a method in 1977 in which DNA could be copied using a precursor that blocks the reaction specifically at one of the four bases. The technique involves repeating the reaction four times, once for each type of base. All the newly synthesized chains start at the same position, but end at positions corresponding to the base that blocks the reaction. Separating the chains on the basis of their length identifies which base lies at each end position. This can be 'read' into a sequence. Another technique, based on breaking the chain at specific bases, was developed by Alan Maxam and Walter Gilbert at Harvard University. Sanger and Gilbert shared a Nobel Prize in 1980.

The length of DNA that could be analyzed in one stretch increased. The technology moved from visual reading of the gels to automation. The first automated DNA sequencers came on the market in 1986. Sequencing is now into third-generation techniques, using nanopore methods. The cost has dropped dramatically. The speed has reached a point at which one recent research paper reported the results of sequencing 10 000 individual human genomes.

---

on a system of reviewing proposals by committees at the National Institutes of Health. Sequencing the human genome was far too large a project to be managed in such a way. The DOE (Department of Energy) competed with the NIH (the major source of biology science funding in the United States) to run the project. (The DOE's interest in a biology project was triggered by the responsibility it had been given for assessing the dangers of radiation after the atom bombs were dropped in 1945.)

The idea of such a large-scale project in biology was, to say the least, controversial. Most of the opponents were eventually won over, but it's fair to say that the project went ahead not so much because of a consensus in the scientific community, but because its proponents were more skillful at mustering political support than its opponents.[8] Efforts to block the project continued for the first few years after it was approved. By contrast, the Superconducting Super Collider (the SSC), a larger project to construct an accelerator for high-energy particle physics, was initially approved, but later killed on grounds of cost, although it was largely supported by the physics community.[9] Politics rather than science played the major role in both cases.

With Jim Watson appointed as the first Director of the Human Genome Project (HGP), operating out of NIH, the program started with $17 million in 1988.[10] (James Wyngaarden, the Director of NIH, who made the appointment, said, "I had an A list and a B list, and Watson was the only name on the A list.")[11] The first five years required about $400 million of funding.[12] When the project formally began in 1990, at several dedicated genome centers, it had an annual budget of $87 million.[13] (This was not solely a government-sponsored effort, as some of the technical and conceptual advances that were required came from work sponsored by large philanthropic organizations such as the Wellcome Trust.) HUGO (human genome organization) became an international effort, with sequencing centers in Europe (mostly the U.K.) as well as the United States. This was a completely new way of managing science in biology, although it was actually small in scale compared to some projects in physics.

There were debates about the best technical means of sequencing whole genomes. There was also a division between the public effort (HUGO) and a private effort by a company, Celera, established by Craig Venter, formerly a sequencer at the NIH.

This led to spirited, not to say rather contentious, debates as to whether anyone could own and patent a sequence. One result of the competition was that the first sequences to be published (in 2001) were labeled as 'draft' sequences, as they covered 94% rather than the complete genome. The sequence was completed in 2022.

The cost of sequencing fell dramatically to the point at which commercial services now offer to sequence anyone's genome. You can have your own genome analyzed by a commercial service for under $200. This makes it possible to determine risk factors for diseases with a known genetic basis. There was much concern about the possibility that sequencing might be used to discriminate against people, for example, who had specific risk factors, but so far these fears have not been realized. Improved diagnosis or treatment for specific diseases has been making its way slowly into medicine, but this is just the beginning.

The big surprise in the human genome is the small number of genes. When the project started, estimates of the total number were based on the average size of the gene and the assumption that most of the genome would code for proteins. A simple division gave a number around 100 000. Actually, there are just less than 20 000 genes coding for protein. This is not very different from the number of genes in the mouse or the worm *C. elegans.*

It had been taken more or less for granted that more complicated organisms would have more genes. This is another

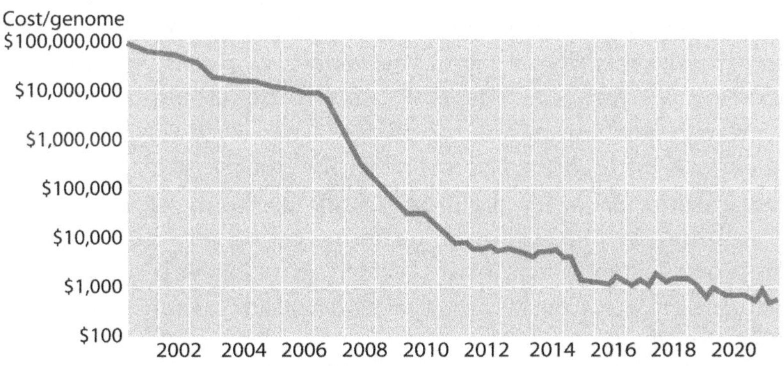

The cost of sequencing a human genome has decreased from $100 million in 2000 to under $1000 today. Data from National Human Genome Research Institute.

example of an assumption that was so common to the field that it was never questioned, even though there was never any evidence to support it.

The constitution of the human genome turns out to be very different from what we had expected before the genome project was completed. Instead of being dominated by genes coding for proteins, only about 1% codes for protein, another 5% for known regulators—and 94% has no known function! This was so mystifying that at one point the term 'junk DNA' was introduced to express the fact that we had no way of accounting for its function. We now believe that some part of it plays essential regulatory functions, although the quantity remains puzzling. (It's uncanny that 94% of the human genome has no known function, and 95% of the universe consists of something that we cannot identify. I discuss the situation in physics in Chapter 17.)

Comparing the human genome with the chimpanzee genome suggests that differences in protein sequences are unlikely to be responsible for the distinction between humanity and its near relatives.[14] The sequences are very closely related, with 1.2% variation between equivalent sequences, and another 3% of differences from rearrangements, insertions, and deletions.[15] Indeed, comparing protein sequences suggested as long ago as 1975 that differences between proteins are minor: many proteins are identical. The average human protein differs by only one amino acid from its chimpanzee counterpart.[16]

In addition to genes coding for proteins, the complete human genome sequence includes 27 000 genes coding for RNAs—meaning that an RNA copy of the gene is transcribed, but it does not code for protein. These RNAs are divided into at least 14 different types. Some, such as those involved in the execution of the genetic code, have been known for a long time. Others have been discovered only recently. Many of them are involved in controlling the expression of protein-coding genes. Perhaps the difference between humans and chimpanzees will lie in the way their protein-coding genes are controlled rather than in the genes themselves.

But how to identify the crucial differences? The sequencing of other hominoid genomes, from Neanderthals and Denisovans, may offer a route. Triangulating features that are common to the three hominoids but absent from chimpanzee is one intriguing possibility. Comparing differences between humans and Neanderthals or Denisovans may hone in on details that make *H. sapiens* distinctive.

The human genome sequence has changed the face of biology. It's a means not an end, but marks a transition from small-scale biology led by individuals to large-scale research led top-down by committees. Papers reporting genome sequences have tens and sometimes hundreds of authors. Most of the authors are specialists in some aspect who do not necessarily understand the other aspects. One of the early genome sequence papers, on the plant Arabidopsis, assigned each author to one of 14 specific roles, such as comparisons with other genomes, analysis of specific parts of the chromosome, cellular organization, metabolism, response to pathogens, and so on.[17] This is a different way of doing science.

One objective of the genome project is to define the essence of being human. The Zoonomia Consortium takes a broader view and tries to define the essence of being mammalian by sequencing genomes of many mammalian species.[18] (There are just fewer than 7000 known mammalian species.) There is also a project called the Vertebrate Genomes Project, which has the objective of sequencing all of the 70 000 known species of vertebrates.[19]

Sequences representing functions that are common (for example, found in monkeys as well as humans) are expected to be conserved, that is, to change much less across species. Conserved sequences are only a small proportion (almost 11%) of the human genome. Half of the conserved sites have completely unknown functions.[20] Regions that have been evolving more rapidly are likely to be involved in speciation. It's true of both types of sites that only a small proportion (20%) code for proteins. This suggests that protein sequences as such are not the driving force in evolution. The highly evolved sites tend to be near genes that code for proteins involved in developmental or neurological processes.

Machine learning (artificial intelligence) has been used to identify sites that may be important but whose sequences have changed beyond the bounds of what can be identified by conventional sequence analysis.[21] Sites (presumed to be regulatory because they do not code for proteins) that have changed specifically in humans but that are conserved in other mammals potentially could contribute to unique human traits. This reinforces the view that regulation of genes may be just as, or even more, important than evolving new proteins. The role of accidents in evolution is emphasized by an increase in the rapidity of evolution around the time of the extinction of the dinosaurs.[22]

Sequencing a genome has become a technological project. In one sense, the data provide another tool that allows hypotheses to be formulated and tested. But the availability of the sequences of many individual human genomes has led to attempts to replace the search for individual genes and their functions by looking for correlations on the basis of comparing the sequences. This is the approach of big data that I discuss in Chapter 11.

Analyzing the sequence database requires a different skill set from that of the experimentalists, dividing biology in a way analogous to the division in physics between theoreticians and experimentalists. In fact, performing sequence analysis is a completely different task from analyzing the data, where the importance lies not in the raw sequence but in the algorithms that are devised to annotate it. It also requires a different mindset. Instead of examining individual data points, it's necessary to come to grips with the algorithmic analysis of the whole genome (sometimes aided by machine-learning techniques). This is now a specialized task in itself. It may be difficult for scientists who were 'classically' trained to assess this new style of research critically.

The sequences of the human and other genomes have certainly revised the existing paradigms. Before genome sequence analysis, it was assumed that most of the genome coded for proteins. It was a fair assumption that the specificity of each species would be found in the proteins. Now it turns out there are fewer protein-coding genes than we had expected, there is a much more important role for RNAs in controlling gene expression, and there are many regulatory elements in the genome with undefined roles. The essence of humanity is likely to be regulatory rather than coding.

Physicists have been wrestling with the changes resulting from big science for much longer than biologists. As long ago as the 1960s, Alvin Weinberg, who directed the Oak Ridge Laboratory, with an annual budget of $100 million per year, expressed

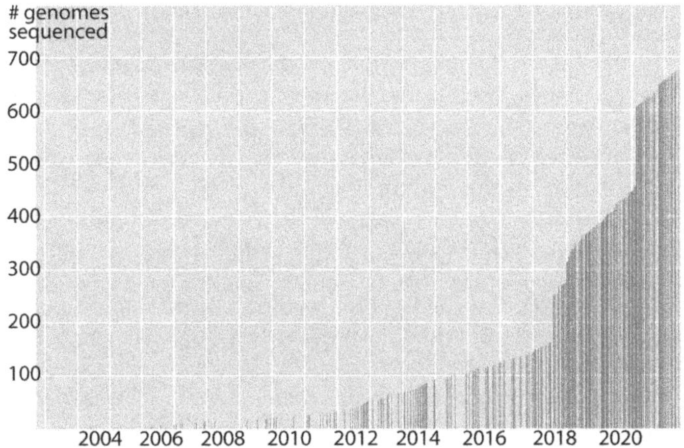

The number of mammalian genomes that has been sequenced has increased dramatically since 2018.[23] By the end of 2022, it corresponded to almost 10% of mammalian species.

concern about the effects of big science in reducing intellectual innovation.[24] (In fact, he introduced the term.) The question is even more pertinent today than it was in the last century.

At least for biology, there is a major transition in the way of doing science. The transition from small groups to large groups has been evolving over several decades, but the genome sequencing projects move the scale by an order of magnitude. Compared to only a couple of decades ago, this feels more like 'huge science' than 'big science.' Not only has the scale changed enormously, but the way of thinking about the data, how to analyze it and how to advance to the next stage, has changed. This is a revolution as much in style as in content.

NOTES AND REFERENCES

1. Quoted in R. Cook-Deegan, *The Gene Wars: Science, Politics, and the Human Genome*, W.W. Norton & Co., New York, 1994, p. 161.
2. Quoted in I. Hargittai and M. Hargittai, *Candid Science VI: More Conversations With Famous Scientists*, Imperial College Press, London, 2006, p. 39.
3. R. Dahm, From discovering to understanding: Friedrich Miescher's attempts to uncover the function of DNA, *EMBO Rep.*, 2010, **11**, 153–160.
4. Quoted in R. J. Dubos, *The Professor, the Institute, and DNA*, The Rockefeller University Press, New York, 1976, pp. 218–219.

5. F. Sanger, Sequences, sequences, and sequences, *Annu. Rev. Biochem.*, 2008, **57**, 1–28.

6. www.genome.gov/sites/default/files/media/files/2019-05/david_schlessinger_transcript.pdf.

7. R. L. Sinsheimer, The Santa Cruz workshop, *Genomics*, 1989, **5**, 954–956.

8. For a triumphalist account of how big science succeeded, see D. J. Galas, *et al.*, Lessons from the human genome project: a critical scientific effort that almost didn't happen illustrates the need for a rigorous but flexible process to evaluate large-scale transformative research proposals, *Issues Sci. Technol.*, 2017, **33**, 57–62.

9. D. J. Kevles, Big Science and Big Politics in the United States: Reflections on the Death of the SSC and the Life of the Human Genome Project, *Hist. Stud. Phys. Biol. Sci.*, 1997, **27**, 269–297.

10. $12 million came from new funds and $5 million was diverted from existing programs: see R. Cook-Deegan, *The Gene Wars: Science, Politics, and the Human Genome*, W.W. Norton & Co., New York, 1994, p. 144.

11. Quoted in D. J. Kevles, Big Science and Big Politics in the United States: Reflections on the Death of the SSC and the Life of the Human Genome Project, *Hist. Stud. Phys. Biol. Sci.*, 1997, **27**, 269–297.

12. R. Cook-Deegan, *The Gene Wars: Science, Politics, and the Human Genome*, W.W. Norton & Co., New York, 1994, p. 176.

13. Two thirds controlled by NIH, almost all of which came from new appropriations.

14. Chimpanzee Sequencing and Analysis Consortium, Initial sequence of the chimpanzee genome and comparison with the human genome, *Nature*, 2005, **43**, 69–87.

15. The most dramatic rearrangement is the fusion of chimpanzee chromosomes 2A and 2B into the single human chromosome 2, explaining why chimpanzees have 48 chromosomes while humans have 46.

16. Based on comparison of protein sequences, see M. C. King and A. C. Wilson, Evolution at two levels in humans and chimpanzees, *Science*, 1975, **188**, 107–116.

17. A formalized system, called CRediT (Contributor Roles Taxonomy), describes specific roles that can be attributed to authors. They include Conceptualization, Data curation, Formal Analysis, Funding acquisition, Investigation, Methodology, Project administration, Resources, Software, Supervision, Validation, Visualization, Writing – original draft, Writing – review & editing (see credit.niso.org).

18. N. S. Upham and M. J. Landis, Genomics expands the Mammalverse, *Science*, 2023, **380**, 358–359; I. G. Romero, Seeing Humans through an Evolutionary Lens, *Science*, 2023, **380**, 360–361.

19. A. Rhie, *et al.*, Towards complete and error-free genome assemblies of all vertebrate species, *Nature*, 2021, **592**, 737–746.

20. M. Christmas, *et al.*, Evolutionary constraint and innovation across hundreds of placental mammals, *Science*, 2023, **380**, eabn3943.

21. I. M. Kaplow, Relating enhancer genetic variation across mammals to complex phenotypes using machine learning, *Science*, 2023, **360**, eabm7993.

22. N. M. Foley, *et al.*, A Genomic Timescale for Placental Mammal Evolution, *Science*, 2023, **380**, eabl8189.
23. Data from N. S. Upham and M. J. Landis, Genomics expands the Mammalverse, *Science*, 2023, **380**, 358–359.
24. A. M. Weinberg, Impact of Large-scale Science on the United States, *Science*, 1961, **134**, 161–164.

CHAPTER 21

Science and Medicine

"IN OUR AGE NOTHING HAS BEEN SO DEGRADED AND THEN wholly restored as anatomy," said Andreas Vesalius in his book of 1543. If any single figure typified the transition from the Roman learning of Galen to modern knowledge, it would be Vesalius. After it became possible to practice human dissection, Vesalius placed knowledge of human anatomy on a scientific basis. Yet it was a century later before William Harvey described the circulation of the blood, and the modern era fully began. Treatments for diseases remained based on alchemy, following the work of Paracelsus (who died just before Vesalius published his book). Medical practice remained almost entirely pragmatic until it began to acquire a scientific basis in the 19th century. There was no single transition, perhaps not even a single century, that can be pointed to as the time of revolution. However, there has certainly been a transition from the medieval era, when a physician was more likely to kill you than to cure you, to the present era, when diagnosis and treatment are based on increasingly scientific tests and methodology.

The Frontiers of Science
By Benjamin Lewin
© Benjamin Lewin 2026
Published by the Royal Society of Chemistry, www.rsc.org

Timeline for Scientific and Technological Advances in Medicine

1750 — James Lind shows citrus fruits prevent scurvy (1747)

1760

1770

1780

1790

Edward Jenner develops vaccination for smallpox (1796)

1800

1810

René Laennec invents stethoscope (1816)

1820 — James Blundell performs blood transfusion (1818)

1830

1840

William Morton uses nitrous oxide as anesthetic (1846)

1850 — Ignaz Semelweis shows infection causes puerperal fever (1847)

John Snow identifies cholera as infection (1854)

Louis Pasteur identifies germs as cause of disease (1857)

1860

Joseph Lister develops antiseptic methods (1867)

1870

1880 — Vaccines for cholera, anthrax, and rabies (1879-1882)

Robert Koch discovers TB bacillus (1882)

1890 — First contact lenses developed (1887)

Wilhelm Röntgen discovers X-rays (1895)

1900 — Felix Hoffman develops aspirin (1899)

Karl Landsteiner classifies blood groups (1901)

1910 — Paul Dudley White introduces ECG (electrocardiograph) (1901)

Ehrlich releases salvarsan as cure for syphilis (1910)

1920 — Edward Mellanby shows rickets due to lack of vitamin D (1921)

Insulin first used to treat diabetes (1922)

1930 — Alexander Fleming discovers penicillin (1928)

Albert Szent-Györgyi discovers vitamin C (ascorbic acid) (1932)

1940 — Karl Theodore Dussik introduces medical ultrasound (1942)

Selman Waksman discovers streptomycin antibiotic (1943)

1950 — John Hopps invents cardiac pacemaker (1950)

Joseph Murray performs first organ (kidney) transplant (1954)

1960

1970

Robert Ledley invents CAT-scans (1975)

1980 — First MRI (1977)

First test-tube baby born (1978)

1990 — Restrictions on smoking (1990)

2000 — Robotic surgery introduced (2000)

2010

Genome profiling for cancer (2018)

2020 — First mRNA vaccine (COVID) (2020)

A Dialog between Galen and William Harvey

Galen (129–216 CE) was a Greek physician, working in Rome, who established the foundations of medical practice based on diagnosing the symptoms of the patient.

© NLM/Science Source/Science Photo Library.

William Harvey (1578–1657) was a prominent English physician. As well as specializing in anatomy based on human dissection, he studied embryogenesis in animals.

Galen: Greetings, Dr. Harvey. I have heard of your theories on the circulation of blood. I must say, they challenge the principles I laid down regarding the humoral theory. In my time, we understood the circulatory system as a system of channels filled with air and vital spirits, not an unending circuit of blood. How do you defend a departure from the ancient understanding of the body and its humors?

Harvey: Greetings, Galen. It's fascinating to meet the physician who established the basis of medicine in the West for a thousand years. I arrived at my conclusions through systematic observation and dissection. By examining the beating heart and studying the flow of blood through arteries and veins, I observed a consistent pattern that could only be explained by the continuous circulation of blood. My work, *De Motu Cordis*, presented detailed evidence in 1628 supporting this concept.

Galen: Dissection and observation were foundations in my own work. I wonder if the continuous circulation you propose adequately explains the complexities of the humors and vital spirits that were central to our understanding of the body's functioning in my time. Is there not a risk that by ignoring the humors you may understand the functions of parts of the body without understanding the whole?

Harvey: Humoral theory was all very well in your era, Galen. However, my observations focused on the physical movement of blood and its vital role in carrying nutrients and oxygen throughout the body. This paved the way for a more detailed understanding of physiological processes.

Galen: Mechanistic explanations are well and good, but the humoral theory accounts for the delicate balance of the body's fluids and their influence on temperament and health. Your focus on the mechanical aspects neglects the broader understanding of bodily functions and the interconnectedness of the humors. How did you approach the challenges of convincing your peers about the continuous circulation, especially in an era where beliefs were deeply ingrained?

Harvey: I faced considerable skepticism initially, Galen. In fact, I faced greater skepticism than you did! Many people were entrenched in traditional beliefs about the ebb and flow of blood. However, as the evidence supporting continuous circulation became more compelling, the scientific community gradually accepted this new understanding. The humors are to the circulation as demons are to humans. I rest my case.

SCIENCE AND MEDICINE

> *From the earliest times, medicine has been a curious blend of superstition, empiricism, and that kind of sagacious observation which is the stuff out of which ultimately science is made... Medicine was constituted in the days of the priest-physicians of Egypt and Babylonia of the same three strands.... The proportions have, however, varied significantly; an increasingly alert and determined effort, running through the ages, has endeavoured to expel superstition, to narrow the range of empiricism, and to enlarge, refine, and systematize the scope of observation... The general trend of medicine has been away from magic and empiricism and in the direction of rationality,*[1]
> Abraham Flexner, 1925.

In another hundred years, medicine of the early 21st century may look as primitive then as medicine of the 19th century does now. Looking back, there are points at which one would be simultaneously amazed by the availability of treatments, but horrified at the primitive nature of basic knowledge. In spite of all the talk of 'personalized medicine' (and some spectacular individual successes), today we still are forced to treat the whole body for ailments that specifically affect particular organs. This will seem *so* primitive in a hundred years. The real revolution is yet to come.

Medicine has certainly become more scientific with an array of tests producing clearer results than basing diagnoses solely on symptoms. Some people worry that new doctors have lost the diagnostic ability, but some physicians have another view. When I remarked to an internist that doctors seem to have lost the art of diagnosis and tend instead to rely on a lot of (very expensive) tests, he said, "Yes, but in the old days, a lot of those diagnoses would have been wrong—in fact we have no idea how many."

Medicine was well developed in the ancient Near East by 2000 BCE. Doctors were divided into those who practiced empirical medicine, based on symptoms, and those who offered prayers to drive out the demons that were believed to cause the disease. Pharmaceutical treatments were based on plant preparations: a mixture of alcohol, honey, and myrrh was used as a disinfectant. Wounds were washed before they were treated.

A thousand years later, Hippocrates (*c.* 460–370 BCE) established medicine as a profession in ancient Greece. Little is actually known of his life, although he is thought to have come from a medical family. Most knowledge relates to the Hippocratic Corpus, a collection of about 70 early medical works in Alexandria, dating from around 420 BCE, probably written by his students and followers.[2] Its most famous document contains the Hippocratic Oath.

Hippocrates' reputation as the "father of medicine" reflects his role in the transition to the realization that diseases have natural, as opposed to supernatural, causes. Epilepsy was called the 'sacred disease' because it was often attributed to an act of the gods, but Hippocrates had another view. "I do not believe that the so-called sacred disease is any more divine or sacred than any other disease. It has its own specific nature and cause, but because it is completely different from other diseases, men through their inexperience and wonder at its peculiar symptoms have believed it to be of divine origin."[3]

Medicine was based on the view that health depended on the 'humors,' the balance between blood, phlegm, yellow bile, and black bile. Treatment was based on enabling the body to restore the proper balance between the humors. The famous Roman physician Galen (129–216 CE) took up the theory of the humors, in which each humor has a pair of qualities: hot–wet, hot–dry, cold–dry, and cold–wet. Classifying illnesses in terms of balance between humors, he then applied treatments to restore the balance. After he became physician to the gladiators, his treatment of their wounds was so successful that he reduced deaths by ten-fold.

Galen came from a wealthy family in Pergamum (today Bergama in Turkey), the second most important cultural center after Alexandria. Pergamum had become part of the Roman Empire, and in 162 CE, Galen moved to Rome. Regarding himself as a philosopher as well as physician, and wealthy enough that he did not need to charge patients, Galen became a society doctor. He had a great success in treating several well-known patients whose doctors had previously failed to cure them. (This did not make him popular with the medical profession.) He became the Court physician, taking care of several Emperors. As a scholar as well as physician, he maintained a large library, a pharmacopoeia of drugs, and surgical instruments.[4] He ran a clinic at his house. Galen based diagnoses on taking a history based on extended observation of the patient's symptoms, and he kept case notes. He was notable for reading the patient's pulse. (He wrote four books *On the Diagnosis of Pulses.*)

Galen extended anatomy by practicing dissection, first on apes, and then on pigs.[5] Knowledge of human anatomy was limited by a ban on dissection.[6,7] Galen studied the spinal cord and achieved some understanding of the nervous system connected to it. He was notable for introducing experiments to prove his point, for example, tying off or cutting the laryngeal nerve in pigs to show that vocal cords were responsible for making sound. Hippocrates may have been the first physician to reject supernatural causes for disease, but Galen was the first physician to introduce an experimental approach to medicine.

Galen was a pagan, but in the monotheistic tradition, so his work did not clash with Christianity. With the rise of Christianity, the attitude that illness came from offending the gods was replaced by the belief it was caused by original sin. Miracles were regarded as evidence for the effectiveness of religious as opposed to secular healing. There was basically no progress in medicine in the West for more than a thousand years after Galen.

Medicine in the Middle Ages was more advanced in Arabia than in Europe (I discuss science in Arabia in Chapter 3). One of

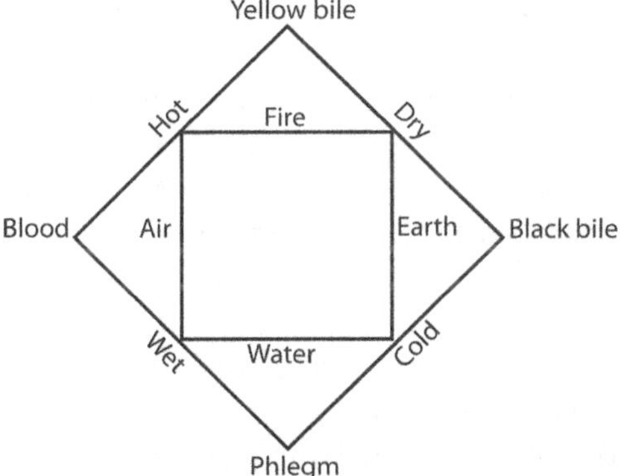

Hippocrates' scheme for the four humors in On the Nature of Man. *Each of the humors is associated with two of the properties of hot, dry, wet, cold. They were also linked to the seasons, and by the Roman era to Aristotle's elements of air, fire, earth, and water (see Chapter 7).*

the most influential physicians of the medieval era was Abū Bakr al-Rāzī (865–925 CE), known as Rhazes in the West. His early career is uncertain, but he appears to have turned to medicine in his thirties. Head of the hospital established in Baghdad after the foundation of the Abbasid caliphate, he wrote a large number of books, following Galen, but including an unpublished manuscript entitled, *Doubts about Galen*.

His major work was *Kitab al-Hawi* (The Comprehensive Book on Medicine).[8] Translated into Latin in the 13th century, it became an influential 23-volume series of books.[9] Rhazes was famous for distinguishing between smallpox and measles, and wrote a book on it that was reprinted many times in the 16th and 17th centuries.[10] He performed what may have been the first clinical trial in medicine. In European eyes, he was the Arabian equivalent of Galen, but in Arabia he was regarded with more suspicion as a philosopher who failed to conform to all the precepts of Islam.

Avicenna (980–1037 CE) was the most famous of the medical scholars in Arabia (in Persia). He wrote the *al-Qanun* (Canon of Medicine), five volumes summarizing Greek and Islamic medicine and biology. Translated into Latin in the 12th century, it became a standard textbook for medical education in Europe until the 17th century.

∞ ∞ ∞

It was not until the 10th–11th centuries that medical learning as such began to revive in the West.[11] By the 12th century, the works of Galen were among those being translated in Spain from Arabic into Latin. However, it was not until the 14th century that direct translations from the Greek into Latin became widely available.[12] One effect of the translations was that many of Galen's qualifications were replaced by certainty. Of course, with literacy largely restricted to the clergy, dissemination was limited. Monasteries were the major source of medical knowledge until the rise of the universities in the 13th and 14th centuries, when it became possible to obtain a medical degree (with a qualification to practice).

By the 14th century, reactions to the plague showed a split between the religious approach and the beginnings of a scientific attitude. Belief in divine intervention clashed with realization that quarantine measures were needed because a disease was contagious.[13] This was a modification of Galen's opus, which had not allowed for contagion.

By the 15th century, Bologna was the leading center for medicine in Europe, and Italian medical schools were at the forefront of medical training. Education was theoretical, based largely on the works of Hippocrates or Galen. Astrology remained an important part of medical training until the end of the 17th century. Casting horoscopes was a routine procedure in predicting the course of an illness. A distinction between medicine and surgery was emphasized by the Hippocratic oath, which includes, "I will not use the knife... but I will give place to such as are craftsmen therein."

One of the common references for astrology was an image intended to guide physicians by showing how the universe was reflected in the human body. Zodiac Man is a male body showing which body part is associated with each of the 12 signs of the zodiac. The principle was that a part of the body should not be treated by bloodletting when the Moon is associated with its zodiac sign. The idea is ancient, but by the medieval era was associated with Ptolemy. "Draw not blood from that member, whilst the Moon is in a Sign representing the same."[14] This represented an important principle of medical understanding through the 16th century. Physicians were required to check the position of the Moon before performing any surgical procedure.[15]

Pharmaceutical treatment was driven by alchemy. Paracelsus (1493–1541) was both the most important physician since Galen, and the leading alchemist of the era. (I discuss his role in alchemy in Chapter 7.) Born as Theophrastus Bombastus von Hohenheim, he Latinized his name to Paracelsus in 1592.[16]

Paracelsus traveled widely, started a practice in Strasbourg, and then successfully treated two notable patients in Basel, one of whom was Erasmus, the famous humanist. He obtained a position at the University, but soon was regarded as a rebel for abandoning Hippocrates and Galen. "We shall not [advance medicine] by strictly adhering to the rules of the ancients." His criticism was so vehement—he burned famous books, including Galen's and Avicenna's, in the market square—that he needed to leave Basel, and he became itinerant. He was hostile towards the medical profession. "You have entirely deserted the path indicated by Nature, and built up an artificial system, which is fit for nothing but to

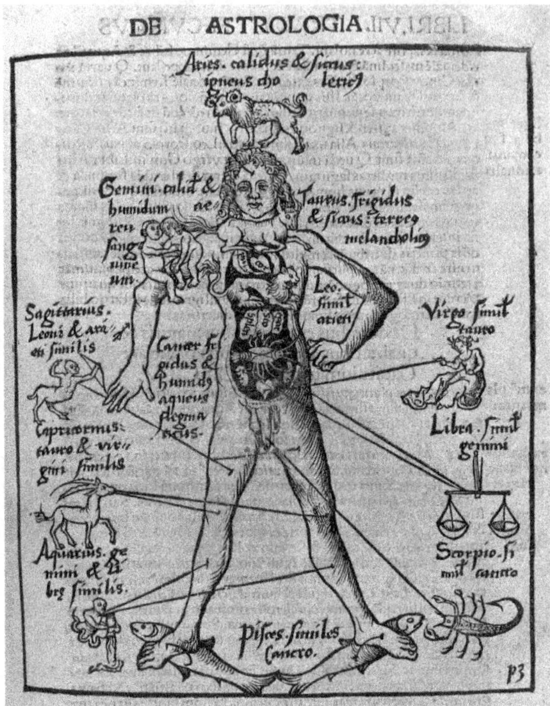

Zodiac Man was a common illustration in works from the 13th through 16th centuries. This copy is from the encyclopedia Marga- rita Philosophica of Gregor Reisch (c. 1467–1525), published in Freiburg in 1503. The encyclopedia was a standard reference work of the period, with chapters devoted to each part of the trivium and quadrivium, and other topics.

swindle the public and to prey upon the pockets of the sick."[17] As a result, he was often prevented from practicing. Probably his main offense against the establishment was to believe in observation of patients on the grounds that diseases had external causes.

Much ancient pharmaceutical knowledge had been lost in the medieval era, and many ingredients were not available. Paracelsus created his own medicines. He described each illness as caused by a specific part of the body. Instead of trying to restore balance of the humors by applying the opposite pair of characters, he applied chemicals resembling the offending humor. Paracelsus was an alchemist par excellence, using procedures such as distillation, calcination, and sublimation (see Chapter 7).[18]

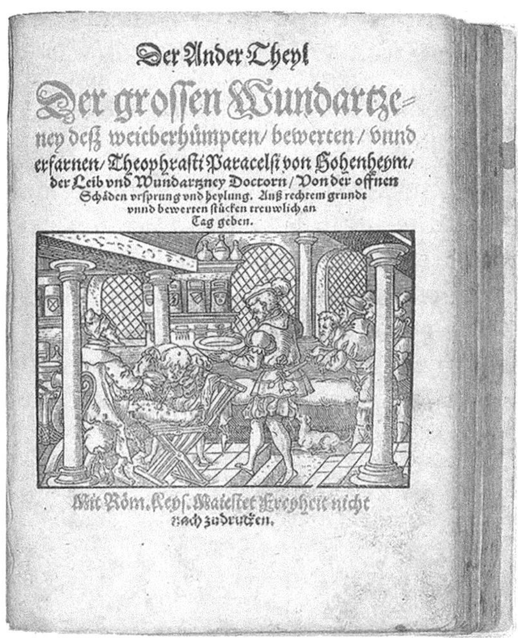

An illustration of surgery from Der Großen Wundartzney, *first published in 1536, consisting of two parts: The doctor's body and wound remedy for all wounds, stings, shots, burns, animal bites, broken bones; and On the origin and healing of open sores. Like his other books of the period, it was published under Paracelsus' real name, Theophrastus Bombastus von Hohenheim.*

Sometimes called the "father of toxicology," Paracelsus called his approach 'iatrochemistry.' His belief that disease was caused by external agents attacking the body, not by supernatural forces, meant that disease could be treated through chemical remedies. However, illustrating the dichotomy of the age, he also believed that the *tria prima* of salt, sulfur, and mercury possessed spiritual properties. He published at least a dozen books during his lifetime, and more were published posthumously. He was scathing about pharmacists who merely ground up ingredients to make complicated (and expensive) preparations consisting of many compounds. Alchemy remained important in medicine until late in the 18th century.

∞ ∞ ∞

Galen dominated medicine until Andreas Vesalius (1514–1564) published the results of his studies in dissection at Padua in *De Humani Corporis Fabrica Libri Septem* (On the Structure of the Human Body in Seven Books) in 1543—the same year that Copernicus' magnum opus was published (Chapter 4). Dissection appears to have started in 1315 in Bologna, and others had preceded Vesalius without, however, forcing any revision of Galen. Vesalius is regarded as the father of modern anatomy. Anatomical studies became important to artists in the 15th century, as witnessed by Leonardo's marvelous drawings.[19]

Coming from three generations of a wealthy medical family, Vesalius was offered the chair of surgery and anatomy at the

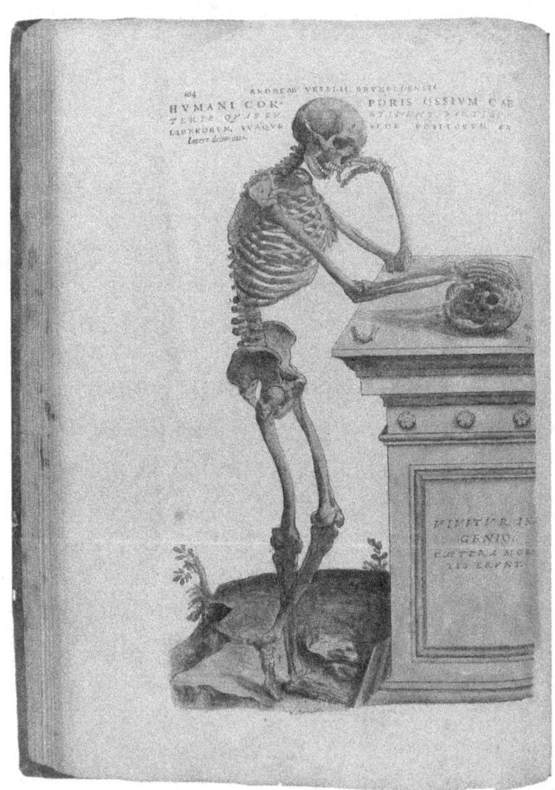

De Humani Corporis Fabrica, which Vesalius published in 1543, was illustrated with many detailed drawings. This famous drawing shows a skeleton contemplating a tomb, which bears the motto "Genius lives on, all else is mortal."

University of Padua as soon as he graduated from the University of Leuven. In addition to his anatomical studies, he was also revising the Latin translations of Galen. He presented *Fabrica* as an updated version of Galen, pointing out that Galen had actually based his analysis on animals, not humans. (Previously no one had paid any attention to the implications.) This was the first time in more than a thousand years that anyone in the West had questioned Galen's analysis. The importance of *Fabrica* extends beyond the study of anatomy. It was one of the foundations of the view that knowledge should be based on empirical observations rather than ancient writings.

Vesalius came to a sad end. Following the careers of his grandfather and father, he left the University of Padua soon after the publication of *Fabrica*, and became a physician at the court of Emperor Charles V, and later his successor King Philip II of Spain. Although *Fabrica* was highly successful (including pirated copies), Vesalius was roundly criticized by the followers of Galen. The sharpest criticism came from his former teacher, Jacobus Sylvius in Paris, who even went so far as to argue that the human body had changed in the period since Galen. For reasons that have never been clear, Vesalius left the court for a pilgrimage to Jerusalem.[20] While there, he was invited to rejoin the University of Padua, but he died following a shipwreck on the way back.

The final blow to Galenism came from Harvey's discovery of the circulation of the blood. According to Galen, there were two separate systems for distributing blood. Arterial blood absorbed air in the lungs and moved to all parts of the body. Venous blood came from the liver and moved through the veins to body parts as needed. Blood was used up: it did not recirculate. Galen supposed that the blood seeped through invisible channels from the right to left ventricle in the heart. His influence was so strong that the first anatomists claimed to have observed the (nonexistent) channels.

William Harvey (1578–1657) was at the top of his profession in England, where he was Physician to the King. After graduating in medicine from the University of Padua, he returned to England, where he lectured at St. Bartholomew's hospital in London. Insisting on the importance of observation and experimentation, he based his theory of circulation of the blood on comparative dissections of a variety of animals.

In his book *De Motu Cordis*, published in 1628, he pointed to a series of problems with Galen's theory. He then proposed that the heart functions as a pump, sending blood from the right to the left of the heart *via* the lungs. Using ligation as an experimental method, he showed the effects of preventing movement of the blood. This was perhaps the first real direct impact of scientific method on medicine. Limited by his era, however, Harvey thought the heart's action was due to some sort of vitalistic pulsation. Even though Descartes argued in *Discours* in 1637 that the action must be mechanical, Harvey stuck to his view in 1660 that it represented a property of the soul.[21]

Circulation of the blood was actually proposed in Arabia three centuries earlier. Indeed, not only the concept of circulation, but also the mixing of blood with air in the lungs, was suggested by Ibn al-Nafis (1213–1288), Chief of Physicians in Egypt from 1260–1277. His works were translated into Latin in 1547. Ibn al-Nafis explicitly rejected the existence of Galen's invisible pores in the heart.[22] It's unclear whether the work of Ibn al-Nafis had any impact in Europe,[23] but in 1553, Michael Servetus described pulmonary circulation in a book, *Christianismi Restitutio*. He was accused of heresy (only incidentally for rejecting Galen's authority) and burned at the stake in Geneva.

It was only in 1661 when the microscope was developed that it became possible actually to see the vessels connecting veins with arteries. And it was only subsequently when the properties of air were investigated that it was realized that oxygen combines with blood in the lungs. This was the decade when medicine moved from its old basis in Galenism to its new basis in scientific experimentation. This was part of the Scientific Revolution. Many advances in medicine were now driven by scientific discovery.

Yet this was more an attitude than reality for the first half of the 18th century. There was much emphasis on 'philosophical' (for which read 'scientific') theorizing, there was an atmosphere of enthusiasm, but not as much as might be expected in the way of tangible medical advances during the Enlightenment.[24] Although there were significant advances in anatomy, there was less progress in treating disease. There was a debate as to whether diseases were caused by 'bad air' or contagion. Scientific progress was offset by belief in magical remedies.

∞ ∞ ∞

The major medical problems at the start of the 19th century were caused by infectious diseases. Smallpox was epidemic, common childhood diseases were often lethal, dysentery was common, and venereal disease was rampant in the military. By the end of the century, many infectious diseases had succumbed to improvements in public health or to advances in medicine.

Inoculation (also known as variolation) against smallpox had become a folk-practice in the 18th century. It involved treating a scratch on the skin with pus from a smallpox victim. The idea was to induce the disease at such a low level it did not leave pockmarks, but did confer immunity. After Edward Jenner (1749–1823) noticed that cowpox gave milkmaids immunity to smallpox, his introduction of vaccination in 1796 made for a safer treatment. This was a medical advance fueled by science.[25]

A cartoon in 1806 from a campaign by Dr Rowley and Dr Moseley against vaccination suggested that vaccination might cause births of half-cow, half-human babies.

Jenner had been apprenticed when only 14 years old to a local surgeon (in the area of the Cotswolds to the west of London), became a surgeon after training at a London hospital, and then moved back to the Cotswolds.

Jenner's new procedure was called vaccination from vacca— Latin for cow—and the name for cowpox, vaccinia. Opposition to vaccination was strong, and just as irrational as today's anti-vaxxer movement. Vaccination gradually replaced variolation, which became illegal in Britain in 1840.

Jenner looked for a place to publish his results in 1796. He turned to the Royal Society, where he had been elected a Fellow in 1788, but they refused, saying that he "should be cautious and prudent... and ought not to risk his reputation by presenting to the learned body anything which appeared so much at variance with established knowledge, and withal so incredible."[26] Jenner published his results privately in a monograph two years later.[27] (The rejection might have been based on the view that Jenner tended to publish prematurely. There had been a kerfuffle about an earlier paper, in which he reported that the cuckoo lays its eggs in other birds' nests. This was received with incredulity until it was verified somewhat later by photography.)

Cholera became epidemic in Western Europe in 1830. There was no medical treatment. Here the solution was technological: introducing a sewage system. Public opposition reduced the impact the system would have had after it was introduced in 1848. The demonstration in 1854 by John Snow (1813–1858) that cholera was spread by contaminated water led to improvements (again against some resistance) in the supply of clean water. One of Snow's arguments was that workers at a brewery were rarely infected because they drank beer instead of water.

Coming from a poor background, John Snow became a surgeon-apothecary apprentice in 1827, and a decade later entered medical school. He became eminent in anesthesia and childbirth, rising to a point at which he treated Queen Victoria with chloroform during the births of her later children. His origins, together with exposure to cholera outbreaks during his early work in medicine, gave him a lifelong interest in epidemiology. He became a skeptic of the Miasma theory that cholera was due to transmission of bad air.

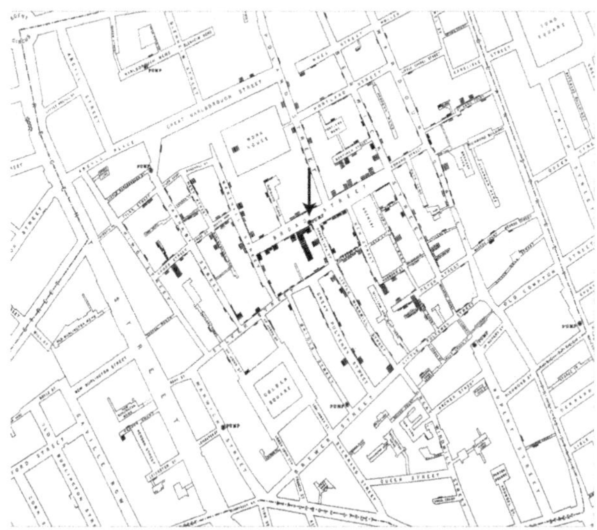

Snow's map of the cholera outbreak in 1854 showed the cluster of cases (black squares) centered on the Broad Street pump (arrow added).

When Snow investigated a cholera epidemic that broke out in 1854 in the area of Soho in London, he found that cases were concentrated in people who lived close to the pump for drinking water in Broad Street.[28] After Snow persuaded the authorities to remove the pump handle, the epidemic subsided. Here was proof that cholera was due to an infection carried by the water, not to transmission of bad air. It was a case where scientific method drove technological improvement. (The bacterium responsible for cholera, *Vibrio cholerae*, was actually grown in 1854, but remained unknown because it was reported in an obscure Italian journal. Robert Koch isolated it in 1884.)

One of the first examples of applying scientific method actually preceded Snow's work by a century, when James Lind (1716–1794) conducted a clinical trial to find a cure for scurvy in 1747.[29] A physician in the Royal Navy, he compared the effects of 6 different treatments on 12 sailors on a naval vessel. He correctly concluded that citrus fruits were effective, but because of difficulties with storage and preservation, failed to make the link to a curative regime. He followed up with further studies with

greater numbers of patients and revised his treatise on the results. (It took almost half a century before the British Navy
introduced a regime that eliminated scurvy.)

Lind's trial was successful because the results were clear even
though the numbers were very small. Subsequent improvements
in methodology included using placebos to compare with the
proposed active ingredient in 1863, double blind trials in 1943
(when the experimenter does not know which patients receive
which regime), and randomized trials in 1946 (when patients are
assigned to regimes on a statistical basis).[30] The use of statistical
techniques to analyze the results of clinical trials became
steadily more sophisticated after their introduction at the start of
the 20th century.

The great English statistician Ronald Fisher introduced the
null hypothesis as a crucial feature in designing experiments in
1935.[31] The null hypothesis says that there is no statistical difference between two sets of data. If the hypothesis is falsified, it
means there is a significant difference. This is now a crucial
element in clinical trials. Fisher's definition of the null hypothesis came one year after Karl Popper introduced the concept of
falsifiability (see Chapter 10).

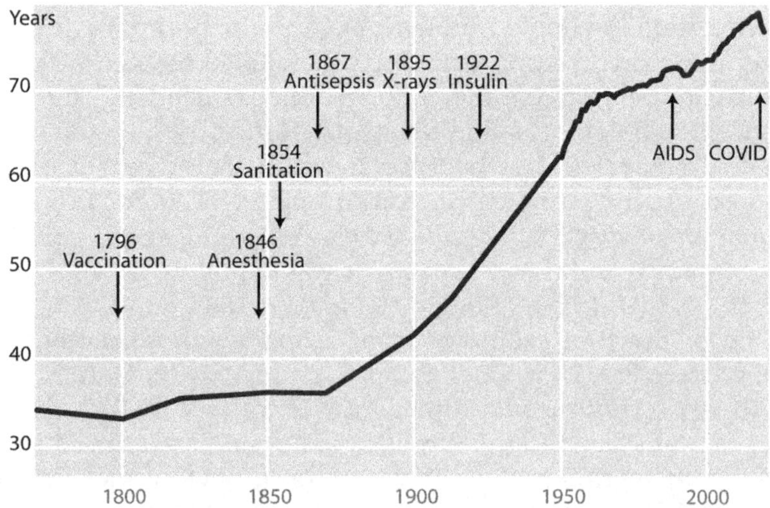

*Life expectancy has increased steadily since the mid-nineteenth
century.*[33]

The application of scientific method to medicine may have been more important prior to the 19th century than any actual discovery in science. The combination of scientific methodology with mathematical analysis comprised a revolution in medicine at least equaling the effects of science itself. Medicine has become more scientific even where it does not rest on scientific advance as such.

Life expectancy began to rise sharply in the second half of the 19th century. The slow start to the rise shows the delay before new developments take effect. It's hard to quantitate the effects of vaccination *versus* improvements in sanitation in reducing mortality due to infectious diseases. Vaccination almost eliminated smallpox in England between 1850 and 1900. Cholera and typhoid were very much reduced in the same period by improvements in public sanitation as the use of experimental methods demonstrated the route of contagion. The rise in hospitals equipped with laboratory facilities also had a significant effect. Chlorination of water may have saved more lives than penicillin.[32]

∞ ∞ ∞

It was a continuing controversy in biology during the 18th and 19th centuries whether living organisms could be accounted for solely in terms of physics and chemistry or are distinguished by an additional principle—a *vis vitalis* (vital force). Vitalism continued to be a force in spite of the successes of science. Its persistence led to the famous Reymond-Brücke oath in 1842; actually signed by Emil du Bois-Reymond, Hermann Helmholtz, Ernst Brücke, and Karl Ludwig, all at the University of Berlin. They were in the department of Johannes Peter Müller, the leading exponent of vitalism.

Allegedly signed in their own blood, the oath read: "No other forces than the common physical chemical ones are active within the organism. In those cases which cannot at the time be explained by these forces one has either to find a specific way or form of their action by means of physical mathematical method, or to assume new... chemical–physical forces inherent in matter, reducible to the force of attraction and repulsion."

Friedrich Wöhler (1800–1882), a German chemist who bracketed organic and inorganic chemistry, is often attributed with destroying vitalism in chemistry in 1828 when he showed in Berlin that urea (an organic compound) could be synthesized from

Three Ancient Ideas That Survived Far Too Long

Vitalism

Vitalism holds that living organisms have a vital principle (*élan vitis* or *vis vitalis*) that distinguishes them from nonliving material. It became standard philosophy by the time of Aristotle, when it was called the pneuma (breath of life). In spite of the move to mechanistic analysis during the Scientific Revolution, vitalism remained controversial even during the Enlightenment and through the 19th century. Proving that organic substances could be synthesized inorganically dented the idea that living organisms have a non-physicochemical force. Yet a leading textbook on physiology by Johannes Peter Müller argued for the existence of a 'soul.' Today, in science at least, vitalism is considered pseudoscience. The organism is governed by the laws of physics and chemistry.

Spontaneous Generation

Aristotle established the theory of spontaneous generation, that living creatures can arise from nonliving matter. He explicitly defined it as a property that animals share with plants. It was regarded as a fact until the first experiments to exclude it, by Francesco Redi in 1668. There was a continuing controversy until the issue was regarded as settled two hundred years later by Pasteur's experiment in 1859. Of course, now we know that all living organisms develop from genomes of DNA, so the question has, as it were, been kicked back to the issue of how the first organisms arose.

Preformationism

Preformationism is the theory that organisms develop from miniature versions of themselves. Either the egg or the sperm contains a homunculus that develops into the entire organism. Aristotle thought that the female provides the material and the male directs the form of development. Leeuwenhoek's discovery of spermatozoa in 1677 was taken to support preformationism, with a division between spermists (the sperm contains the homunculus) and ovists (the egg contains the homunculus). One advantage of the theory was that no mysterious forces were required to generate the organism: the homunculus simply had to grow. The decisive experiments were performed only at the end of the 19th century, when Hans Driesch showed that separate cells of the early sea urchin embryo each had the capacity to develop into a full organism (according to preformationism each should by then have represented only specific body parts).

---

inorganic components (by heating ammonium cyanate).[34] "Without doubt, urea and this crystalline substance, or ammonium cyanate, if one can so call it, are absolutely identical compounds."[35]

With the eventual defeat of vitalism, the reductionist manifesto led, implicitly if not explicitly, to the view that physiology could be explained in terms of the laws of physics and chemistry. Yet the concept of vitalism survived in the background to the point at which Francis Crick could say as recently as 1965, "I also suspect that many workers in [molecular biology] and related fields have been strongly motivated by the desire, rarely actually expressed, to refute vitalism."[36] The issue may have been settled in science, but still resonates in society. Indeed, it can impact science when religious beliefs lead to bans on lines of research. (I discuss current controversies in Chapter 26.)

The cell theory proposed by Theodor Schwann (1810–1882) in 1838 made its impact on medicine when Rudolf Virchow (1821–1902) proposed in a lecture series in Berlin in 1858 that all diseases are due to disturbances in cells.[37] Schwann qualified as a physician in 1834, but instead of practicing medicine, worked with his mentor, Johannes Peter Müller, in physiology. He was led to the cell theory by observing through the microscope that plant and animal cells share a similar organization.

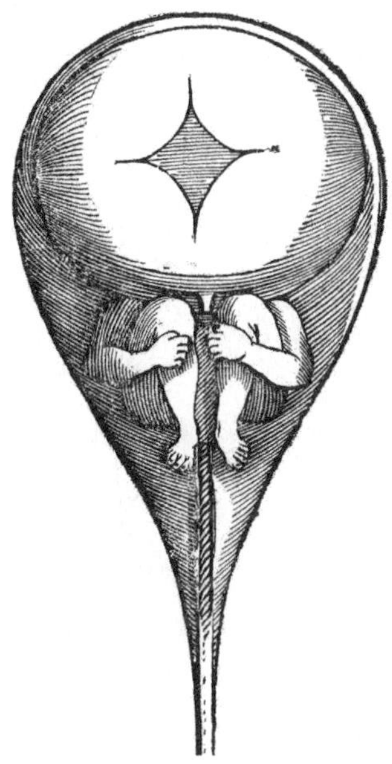

It was more imagination than observation when microscopist Nicolaas Hartsoeker depicted a homunculus in sperm in 1694 with the description "How the new generation is formed."

He published his observations in a scientific journal in 1838, followed by a famous monograph in 1839.

Schwann concluded that living organisms consist of cells, but was unable to take the next step of identifying their origins. His indecision was summarized in the foreword to the book. "A common principle underlies the development of all the elementary subunits of all organisms, much as the same laws govern the formation of crystals despite their differences in shape."[38] This referred to an old idea that crystallization might provide a model for the formation of organic material.

The next step was left to Virchow, who is often credited for stating the doctrine *omnis cellula e cellula* ("every cell is derived from a [preexisting] cell") in 1855. Actually, it had been stated previously in 1825 by François Raspail (1794–1878). (Another

Converting Scientific Advances into Medicine

	Discovery	Technology	
1718	Folk-practice of inoculation introduced into Europe	Edward Jenner introduces vaccination for smallpox	1796
1628	William Harvey discovers circulation of blood	James Blundell performs the first successful transfusion of human blood	1818
1800	Humphrey Davy discovers properties of nitrous oxide	William Morton uses nitrous oxide for anesthesia in dentistry	1846
1838	Theodor Schwann introduces cell theory	Rudolf Virchow proposes that diseases are due to malfunctions in cells	1858
1857	Louis Pasteur shows microorganisms spoil wine and milk and develops germ theory of disease	Joseph Lister uses carbolic acid as antiseptic	1867
		Robert Koch defines criteria for infectious agents	1890
1895	Wilhelm Röntgen discovers X-rays	First X-rays of patients	1896
1838	Raffaele Piria extracts salicylic acid from willow	Felix Hoffman synthesizes aspirin (acetyl salicylic acid)	1899
1921	Banting and Best discover insulin	Insulin used to treat diabetes	1922
1901	Karl Landsteiner discovers ABO blood groups	Norman Bethune develops blood transfusion	1936
1928	Alexander Fleming discovers penicillin	Penicillin becomes general purpose antibiotic	1943
1998	Sodium channels in neurons responsive to pain	Blocking specific channel gives non-addictive opioid	2025

demonstration of the difference between science today, when communication is all but instantaneous, and the 19th century, when important observations could rest in obscurity because of the lack of communication.)

Virchow was another student of Johannes Peter Müller, about a decade after Schwann, but he went into medicine. Virchow was ahead of his time in believing in clinical observations and the application of scientific method. He described leukemia as a disease of white blood cells and worked out an explanation for the formation of blood clots. Virchow's book, *Cellular Pathology*, published in 1858, encouraged physicians to think about diseases in terms of cells.

The cell theory is a demonstration that "history is written by the victors" can apply in science as well as history. Matthias Jakob Schleiden (1804–1881), who published with Schwann, is often credited as co-founder of the cell theory, although he mistakenly proposed that cells arise from nuclei. Schleiden did his best to suppress the work of Raspail. Virchow, admittedly a polymath with substantial accomplishments, downplayed the work of Raspail and Robert Remak (1815–1865), who was probably the first to demonstrate decisively that cells arise only from cells. Nationalism and antisemitism played a significant role in establishing the claims to priority that triumphed.[39]

In spite of the Scientific Revolution, in spite of the progress made in physics and chemistry, biology was still primitive enough that Aristotle's view of the origin of life persisted into the second half of the 19th century. Aristotle argued that life could arise from nonliving material if the material contained 'pneuma' (vital spirit). Some experiments as early as the 17th century suggested that in fact living organisms were needed for situations where life had appeared to arise spontaneously. Other experiments appeared to exclude the requirement (usually because of inadequate sterilization procedures). The situation remained so controversial that the Académie des Sciences in Paris offered a prize for resolution of the issue.

Louis Pasteur (1822–1895) won the prize for his experiments showing that sterilization prevented the growth of microorganisms. However, there was a continuing debate about germ theory *versus* spontaneous generation between Pasteur and his principal opponent, Félix Pouchet. Pouchet presented a paper in 1858 claiming that if hay was sterilized by heating, exposed to oxygen, but separated from atmospheric air by mercury, microorganisms appeared *de novo*. The experiment was essentially an attempt to exclude the possibility that atmospheric air contains micro-organisms that are responsible for putrefaction.

Pasteur's experiment to exclude spontaneous generation in 1860 modified Pouchet's protocol by boiling a solution of sugar and yeast in water in a 'swan-necked' flask. The neck of the flask was constructed so as to prevent atmospheric air from entering. The contents remained pristine, whereas a control that had not been heated rapidly developed mold. When the neck of the flask from a boiled sample was broken off, atmospheric air entered, and mold developed rapidly. This suggested that the mold was produced by microorganisms in the air.[40]

Pasteur's result was one of the major scientific advances of the 19th century, but the comparison with Pouchet's experiments was less decisive than it seemed. Each was able to criticize the other's experiments. Pasteur suggested that Pouchet had not sterilized his equipment well enough. Pouchet suggested that

Pasteur's famous swan-neck flask.

Reproduced from https://commons.wikimedia.org/wiki/File:Swan-necked_flask_used_by_Pasteur._Wellcome_M0012521.jpg, under the terms of the CC BY 4.0 license, https://creativecommons.org/licenses/by/4.0/.

Pasteur's procedures had destroyed the 'vegetative forces' responsible for spontaneous generation.[41] The result was not so clear-cut in scientific terms as it was regarded by society, where opinion came down firmly on Pasteur's side. This may have been influenced by the situation in France, where the Catholic Church was resurgent. Spontaneous generation was seen as an atheistic denial of the divine role in creating life.[42]

But Pasteur provided strong inferential evidence for the germ theory by showing that the potential of atmospheric air to sponsor growth of microorganisms decreased with elevation (because the concentration of microorganisms declines with the density of the air). Even so, he might have had a nasty shock if he had directly repeated Pouchet's experiments, because hay contains microorganisms that are exceedingly difficult to kill by normal sterilization procedures.[43] Of course, Pasteur was supported by subsequent developments in medicine during the second half of the 19th century.

The similarities between fermentation and putrefaction led Pasteur to the involvement of microorganisms. Having proved that microorganisms are responsible for fermentation, he went on to argue that "life takes part in the work of death in all its phases."[44] Subsequently Pasteur developed vaccines for several

Louis Pasteur in his laboratory, in a contemporaneous painting in 1855.

diseases, chicken cholera (1879),[45] anthrax (1881), and rabies (1885). The triumph of cell theory and refutation of spontaneous generation marked the start of cell biology as a science.

Pasteur is one of the most revered figures in French history. His accomplishments ranged widely across chemistry, biology, and medicine. Perhaps not surprisingly, given how many of his conclusions went against conventional wisdom, he could be a somewhat pugnacious figure, determined to protect his reputation. Perhaps for that reason he specified that his laboratory notebooks should not be made public, but when eventually they came into the public domain, it seemed there were significant discrepancies with the published results.[46]

Opinion is polarized as to whether this was scientific genius in knowing which results to accept or reject, or 'deception' or

A heroic view of Pasteur in 1886 on the cover of the satirical republican magazine Le Don Quichotte. *The caption says, The Angel of Inoculation.*

'fraud' in misrepresenting the data. The debate is a great example of a clash of cultures. Historians of science have universally leapt to the defense of the criticism; scientists have deplored the attack on Pasteur. Be that as it may, virtually all of Pasteur's experimental results have since been vindicated. It's nonetheless fair to say that his reputation has been embellished by myth making, not least from his family.

The greatest figures of the 19th century in medical microbiology were Louis Pasteur and Robert Koch (1843–1919). As Pasteur was French and Koch was German, and given that the two countries went to war in 1870, perhaps it's not surprising that their rivalry was not always friendly. Starting his studies while working as a physician, Koch identified the microbial basis of anthrax in 1876. After being appointed to a government position in Berlin, he identified the bacterium responsible for tuberculosis in 1882, and found the bacterium for cholera in 1884. Koch then developed what are now known as Koch's postulates, establishing the standard—which still holds—for demonstrating that a particular microorganism causes a specific disease.[47]

Koch's discoveries of bacteria for tuberculosis and cholera, and Pasteur's success with vaccination, changed thinking (albeit slowly) about appropriate treatments or preventative regimes. Following Koch, there were systematic attempts to identify microbes ranging from viruses to protozoa as the causal agents for various diseases. One of the great successes was the identification in 1905 of the bacterium responsible for syphilis, followed in 1906 by the Wassermann blood test as a diagnostic. Then in 1909, the immunologist Paul Ehrlich (who after working with Koch moved with his own institute to Frankfurt) tested 900 different compounds for their effect on sleeping sickness, caused by trypanosomes (protozoan microbes). The result was the discovery of Salvarsan, an arsenic-based compound. This was first drug that killed the syphilis bacterium (in humans). It's another demonstration that the results of research, even tightly directed research, are unpredictable.

To be sure, it was not all smooth sailing once Koch's postulates had been accepted. There was a myriad of reports falsely implicating germs in all sorts of diseases. But the discovery of microorganisms, leading to the germ theory of disease, followed by the introduction of antiseptic practices, were major advances directly driven by science. Even before Joseph Lister (1827–1912) introduced the use of carbolic acid in 1867, it was clear that hand washing would reduce mortality rates. Working at the Vienna Maternity Hospital in 1847, Ignaz Semmelweis (1818–1865) noticed a striking difference in the mortality rate in a clinic attended by physicians compared with a second clinic attended only by midwives. Mortality in the physicians' clinic was 98.4 per 1000 births, while in midwives' clinic the rate was 36.2 deaths per 1000.

Semmelweis deduced that the difference was because the physicians came straight from postmortem exams to the maternity ward where they examined the patients; the midwives did neither. When he introduced hand-washing (with bleaching powder) mortality fell to 12.7 per 1000 births. This amounted to the use of full scientific method with a controlled trial, yet the work was almost completely ignored.[48] It was only after another 20 years, when Joseph Lister independently introduced antiseptic methods, that mortality generally began to drop. Lister's practices may have been accepted because he gave them a scientific explanation based on the germ theory of disease.[49]

∞ ∞ ∞

Medicine became more scientific during the 19th century, but there were contrasting attitudes about the appropriate approach. One was increased focus on observations of patients, repeated and reported so that general rules for diagnosis and treatment could be deduced. A book in 1844 introduced the concept of 'medical science.'[50] In 1865, Claude Bernard (1813–1878), a physiologist in Paris who made many important discoveries including the role of the pancreas, emphasized the need for experimental results. "The true sanctuary of medical science is the laboratory."[51] He is often credited for the move to integrate medical practice with laboratory science.

The state of medicine a century ago is revealed by a book by William Osler, sometimes called the Father of Modern Medicine. Originally graduating in medicine in Toronto, Osler trained with Virchow in Germany, and then moved through Philadelphia, followed by Johns Hopkins in Baltimore, before moving to Oxford University in 1905. First published in 1892, successive editions of his 1200-page book dominated medical thinking for the first decades of the 20th century.[52] It marks the transition to 'evidence-based' medicine. Osler was the foremost advocate of basing diagnosis on symptoms. The book discusses virtually every known disease, its symptoms, diagnosis, and treatment. A contrast with current medicine is the much greater reliance today on laboratory tests. The nature of 'evidence' has extended from symptoms to scientific measurements.

Major medical advances in the 20th century owe as much to technology as to science. The discovery of insulin and the discovery of penicillin were major scientific advances, as was the development of more vaccines, including vaccines against polio and (most recently) COVID. But the ECG, the pacemaker, defibrillation for the heart, dialysis for the kidney, organ transplants (admittedly dependent on advances in immunology), CAT scanner, robotics in surgery, all are technological developments. Of course, science and technology are more closely connected today than they were, say, during the Industrial Revolution.

The use of stem cells to regenerate specific tissues that have developed malfunctions is one of the most direct applications of science to medicine. Today more than 100 clinical trials are underway to test the potential of stem cells for 'regenerative medicine.'

Science has had its major impact on medicine in the treatment of infectious diseases. Virtually all bacterial diseases can be treated with antibiotics. Science has been less effective with viral diseases. Vaccination is effective against many serious diseases, but we lack effective agents to counteract sporadic infections by other viruses. Although there have been some remarkable successes—the COVID vaccine stands out as an example—there have also been some devastating failures, the lack of an AIDS vaccine being perhaps the most prominent.

I am also inclined to the view that, whether because of lack of perspicacity or failure to commit resources, we have failed to take full advantage of the technology. Why cannot there be a PCR-based, virtually instant, assay for all known infectious diseases based upon the DNA sequence of the agent?[53] Some bacterial diseases are still diagnosed by much slower means, such as growing the agent in a Petri dish!

Completing the sequence of the human genome raised hopes for a revolution in treating inherited conditions. The view in

2001 was optimistic. "By 2020, the impact of genetics on medicine will be even more widespread. The pharmacogenomics approach for predicting drug responsiveness will be standard practice... New gene-based 'designer drugs' will be introduced to the market for diabetes mellitus, hypertension, mental illness, and many other conditions... It's likely that every tumor will have a precise molecular fingerprint determined... and therapy will be individually targeted to that fingerprint."[54] This promise has been fulfilled only in a relatively small number of cases. Indeed, a decade later, although still relentlessly optimistic, Francis Collins, Director of the National Institutes of Health, admitted, "The consequences for clinical medicine, however, have thus far been modest."[55]

Early attempts at human gene therapy were unsuccessful. In fact, by the time the human genome project was under way, there was general disillusionment in the community.[56] The first real prospect of treating a major disease systematically came with the approval in 2024 of two procedures for treating sickle cell anemia. The disease is caused by a defective gene for hemoglobin (the protein of red blood cells). The treatment consists of inserting a functional gene into the patient's cells.[57]

There has been more impact on diagnosis than treatment. This is especially true in the case of pregnancy, where it's possible to scan for most major inherited diseases that are due to single genetic traits. However, more complex cases, where multiple genes are involved, remain refractory to diagnosis. Some of these diseases may have both an inherited component and an infectious component, which creates a complexity beyond the capacity of present technology.[58] There has been progress in identifying more cancers as due to specific cellular or molecular causes, leading to cures in an increasing number of cases.

More than 6000 inherited human diseases are caused by a mutation in a single gene.[59] Probably there are many more that have not yet been identified. Another 600 or so genes cause susceptibility to a complex disease or infection. Panels can be created to allow screening for multiple possible targets at one time. There is no reason why this approach cannot ultimately be extended to all genes where mutations are known to cause human diseases. The problem, however, is that the sequencing approach has not proved so successful at identifying mutations at sites that are outside a gene but affect its function.[60]

∞ ∞ ∞

In spite of the successes from placing medicine on a scientific basis, there is a powerful movement of 'alternative medicine,' rejecting conventional, which is to say science-based, treatments. Alternative medicine takes us right into pseudoscience by making sweeping claims unsubstantiated by the data, sometimes based on beliefs that defy the laws of physics and chemistry.

We should distinguish alternative medicine as such from 'traditional medicine,' which tries to derive treatments from folk remedies. It's not always an easy distinction to draw because often enough terms such as alternative, complementary, or traditional medicine are used interchangeably. There are some famous examples of modern drugs derived from folk remedies. The bark of the willow tree contains salicylic acid, which is the active ingredient in aspirin. An old Chinese practice (recorded around 300 CE!) to treat malaria used sweet wormwood, from which Tu Youyou isolated artemisinin, gaining a Nobel Prize in 2015.

These dramatic successes do not imply that other folk remedies will necessarily be effective, but there are intensive efforts to understand the basis for treatments developed in traditional medicine.[61] There's a huge literature reporting trials of traditional remedies or attempts to isolate active compounds, but no statistics to identify what proportion of these efforts is successful.[62]

Perhaps the least credible part of alternative medicine is homeopathy. This originated with Samuel Hahnemann (1755–1843), a physician who qualified in Leipzig and Vienna. Horrified by a belief that treating patients often made them worse rather than better, he abandoned medical practice in 1784. "I could not conscientiously treat the unknown morbid conditions of my suffering brethren by these unknown medicines, which being very active substances, may... so easily occasion death, or produce new affections and chronic maladies... I renounced the practice of medicine."[63]

Hahnemann's belief in homeopathy was triggered by a report that cinchona (the bark of a Peruvian tree) was effective against malaria. (It contains quinine, which is one of the old treatments for malaria.) When he tried taking cinchona, he discovered that it induced malarial-like symptoms. He decided that a treatment inducing symptoms in a healthy person could cure the same symptoms in a person who was ill. This became the principle known as "like cures like," proposed in a paper in a medical journal in 1796. Reminiscent of the ideas introduced by Paracelsus in the medieval era when he replaced balancing the humors by treating with like humors (see above), it has no better basis in fact.

To avoid the problem of over-dosage, Hahnemann introduced serial dilution. He introduced the term homeopathy in 1810 in his book, *The Organon of the Healing Art,* which went through six editions by 1842.[64] The concepts of serial dilution and 'dynamization' were refined in successive editions. By the 6th edition, Hahnemann specified that "simple dilution results in nothing but water," but dynamization, putting 1 drop of solution into 100 drops of rectified wine spirit and shaking 100 times, creates "a spirit-like medicinal force."

This dilution is repeated multiple times. In fact, at the end of the process, there are actually no molecules left of the original active substance! The specification of the exact extent of dilution and the number of repetitions sounds scientific, but it's not at all clear how Hahnemann conceived the process. Homeopathic preparations often carry the description 30X or 30C, which refers to the series of successive dilutions. The original preparation is diluted 10× in an 'X' series and 100× in a 'C' series, and the resulting solution is diluted the same amount, and so on, 30 times. While it's credible that low dosage treatments could provoke a response, it's not credible that dilution to zero molecules of active substance can have any effect. The laws of physics and chemistry say that it's impossible for the treatments to have any effect based on their content. (If homeopathic treatments have any results, it's probably due to a placebo effect.) Homeopathy is used by about 2% of the population (compared to 1% for acupuncture and 20% for herbal remedies).[65]

Claims for alternative medicine are often supported by supposedly scientific results. Sometimes these are spurious, simply made up out of thin air. Sometimes a so-called trial has been done, usually without any controls.[66] Controlled studies have to date shown no effects for any homeopathic treatments, aside from placebo effects.[67] One statistician, who was involved in testing claims for alternative medicine at a center of the National Institutes of Health, calls it "snake oil science."[68] But total spending on alternative medicine in the United States run at $30 billion per year, including $3 billion on homeopathy.[69]

The knee-jerk reaction against current medical practice that leads to belief in homeopathy is an impediment to wider acceptance of the necessary role of science in medicine. It's unfortunate (I would like really to use a stronger term) that homeopathic preparations are sold in pharmacies alongside genuine medicines, because this gives them an impression of legitimacy they certainly should not have.

It's noteworthy that crackpot theories of medicine often try to validate themselves by claiming a scientific basis. In this context, 'scientifically proven' is one of those phrases, like 'with all due respect,' that actually means exactly the opposite of what the words state. But it shows the extent to which people who reject science actually believe that the public perception is that science speaks to truth.

Medicine has an increasingly scientific basis. The transition to personalized medicine will require a detailed understanding of several disciplines—molecular biology, cell biology, immunology to start. At that point, medicine really will become scientific.

NOTES AND REFERENCES

1. A. Flexner, *Medical Education. A Comparative Study*, Macmillan, New York, 1925.
2. L. Totelin, Hippocratic Corpus, *Oxford Classical Dictionary*, Oxford University Press, Oxford, 2021, DOI: 10.1093/acrefore/9780199381135.013.8525.
3. Sacred Disease, attributed to Hippocrates, possibly written later. Quoted in J. Longrigg, *Greek Medicine: from the Heroic to the Hellenistic Age. A Source Book*, Routledge, London, 1998.
4. Much of this was destroyed in a fire in 192 CE.
5. S. P. Mattern, *The Prince of Medicine: Galen in the Roman Empire*, Oxford University Press, Oxford, 2013.
6. Dissection was banned in ancient Greece. It appears to have first been practiced in Alexandria in the 3rd century BCE, but was later banned again. D. C. Lindberg, *The Beginnings of Western Science: The European Scientific*

Tradition in Philosophical, Religious, and Institutional Context, Prehistory to A.D. 1450, University of Chicago Press, Chicago, 2007, p. 119.

7. Herophilos (335–280 BCE) was a leading physician in Alexandria, who developed the science of anatomy based on dissecting human cadavers, but his work has been completely lost. L. Russo, *The Forgotten Revolution: How Science Was Born in 300 BC and Why it Had to Be Reborn*, Springer, Berlin, 2003, pp. 143–158.

8. Originally this comprised his working notes, not intended for publication, but they were collected and published after his death.

9. S. Tibi, Al-Razi and Islamic medicine in the 9th century, *J. R. Soc. Med.*, 2006, **99**, 206–207.

10. Such as *A Discourse on the Small-pox and Measles. By Richard Mead, To Which is Annexed, a Treatise on the Same Diseases, by the Celebrated Arabian Physician Abubeker Rhazes*, Alex Donaldson, Edinburgh, 1763.

11. N. G. Siraisi, *Medieval and Early Renaissance Medicine*, University of Chicago Press, Chicago, 1990.

12. S. P. Mattern, *The Prince of Medicine: Galen in the Roman Empire*, Oxford University Press, Oxford, 2013, pp. 279–289.

13. V. Nutton, Medicine in Medieval Western Europe, 1000–1500, in L. I. Conrad *et al.*, *The Western Medical Tradition: 800 BC to AD 1800*, Cambridge University Press, Cambridge, 1995, pp. 139–205.

14. Ptolemy, Saying XX in the *Centiloquium* (100 sayings).

15. We may not be impressed today with the prescription, but is it so different from today's principle of biodynamic agriculture that plants should not be pruned at the full Moon? The basic idea is the same, that the influence of the Moon might lead to uncontrolled release of material.

16. This probably meant 'above Celsus' (25 BCE–50 CE), who wrote a medical encyclopedia, *De Medicina*.

17. Quoted in F. Hartmann, *The Life of Paracelsus*, Kegan Paul, London, 1896, p. 166.

18. Although most medicines came from plants, Paracelsus believed in the use of mineral remedies, in particular based on sulfur, salt, and mercury. The theory was that these were appropriate for treating sulfurous, saline, and mercurial diseases. This was why traditional alchemical techniques, involving use of a furnace, were required. Antimony was at the base of several remedies.

19. A. Wear, Medicine in Early Modern Europe, 1500–1700, in L. I. Conrad *et al.*, *The Western Medical Tradition: 800 BC to AD 1800*, Cambridge University Press, Cambridge, 1995, pp. 215–361.

20. There are stories that he was denounced as a heretic and that he was atoning for dissecting a person who turned out still to be alive.

21. A. Wear, Medicine in Early Modern Europe, 1500–1700, in L. I. Conrad *et al.*, *The Western Medical Tradition: 800 BC to AD 1800*, Cambridge University Press, Cambridge, 1995, p. 339.

22. "When the blood in the right chamber has been refined, it must reach the left one where the vital spirit originates. Now there is no communication between these two cavities, …not even invisible pores, like Galen has

postulated." Quoted in K. Rothschuh, *History of Physiology*, Krieger, New York, 1973.

23. A. Majeed, How Islam changed medicine, *BMJ*, 2005, **331**, 1486–1487; S. S. Amr and A. Tbakhi, Ibn al-Nafis: discoverer of the pulmonary circulation, *Ann. Saudi Med.*, 2007, **5**, 385–387.

24. R. Porter, The Eighteenth Century, in L. I. Conrad *et al.*, *The Western Medical Tradition: 800 BC to AD 1800*, Cambridge University Press, Cambridge, 1995, pp. 371–475.

25. S. Riedel, Edward Jenner and the history of smallpox and vaccination, *Proc. (Bayl. Univ. Med. Cent.)*, 2005, **18**, 21–25.

26. The paper was never formally submitted, but Joseph Banks, the President of the Royal Society, rejected it on the basis of the opinions of two experts who thought it was premature, although putting things together, his letter appears to have gone beyond the bounds of their comments. See M. Bennet, *War Against Smallpox: Edward Jenner and the Global Spread of Vaccination*, Cambridge University Press, Cambridge, 2020, p. 72; D. Baxby, Edward Jenner's Unpublished Cowpox Inquiry and the Royal Society: Everard Home's Report to Sir Joseph Banks, *Med. Hist.*, 1999, **43**, 108–110.

27. A current review in the *Royal Society Proceedings* recalls the history without giving details of the aspersion cast on Jenner's work: R. A. Weiss and J. Esparza, The prevention and eradication of smallpox: a commentary on Sloane (1755) An account of inoculation, *Philos. Trans. R. Soc., B*, 2015, **370**, 2014037820140378.

28. "The result of the inquiry then was that there has been no particular outbreak or increase of cholera, in this part of London, except among the persons who were in the habit of drinking water of the pump-well," J. Snow, *On the mode of transmission of cholera*, John Churchill, London, 1855, Reprinted by UCLA Fielding School of Public Health, 2001. Online at: www.ph.ucla.edu/epi/snow/snowbook3.html.

29. U. Tröhler, Lind and scurvy: 1747 to 1795, *J. R. Soc. Med.*, 2005, **98**, 519–522.

30. A. Bhatt, Evolution of clinical research: a history before and beyond James Lind, *Perspect. Clin. Res.*, 2010, **1**, 6–10.

31. R. Fisher, *The Design of Experiments*, Oliver and Boyd, Edinburgh, 1935.

32. S. Pinker, *Enlightenment Now: The Case for Reason, Science, Humanism, and Progress*, Viking, New York, 2018, p. 64.

33. Data are for Europe, but the graph is similar for all continents, with variations due to local conditions. From ourworldindata.org/life-expectancy.

34. A caveat is that cyanate is not purely inorganic as it is produced in various metabolic pathways. However, the conclusion stands.

35. F. Wöhler, On the artificial production of urea, *Ann. Phys. Chem.*, 1828, **88**, 253–256.

36. F. H. C. Crick, Recent Research in Molecular Biology: Introduction, *Br. Med. Bull.*, 1965, **21**, 183–186.

37. The lectures were published as a book in 1859 in German, with an English translation following in 1860: R. Virchow, *Cellular pathology as based upon physiological and pathological histology twenty lectures delivered in the*

Pathological Institute of Berlin during the months of February, March, and April, 1858, John Churchill, London, 1860.

38. T. Schwann, *Microscopical Researches into the Accordance in the Structure and Growth of Animals and Plants*, Sander'schen Buch, Berlin, 1839. Translated by H. Smith for the Sydenham Society, London, 1847.

39. For a review see H. Harris, *The Birth of the Cell*, Yale University Press, New Haven, CT, 2nd edn, 2000, pp. 94–137.

40. N. Roll-Hansen, Revisiting the Pouchet-Pasteur Controversy Over Spontaneous Generation: Understanding Experimental Method, *Hist. Philos. Life Sci.*, 2018, **40**, 1–28.

41. From a modern perspective, this seems a weak retort, but belief in 'vegetative forces' was not so unreasonable in mid-nineteenth century.

42. According to this view, there was really nothing to choose between the theories of Pasteur and Pouchet on strict scientific grounds, and the 'fix was in' with the committees that investigated the matter stacked with establishment figures supporting Pasteur. For the view that the science was less important than sociological factors see J. Farley, *The Spontaneous Generation Controversy from Descartes to Oparin*, Johns Hopkins University Press, Baltimore, 1977.

43. Heat-resistant spores of *Bacillus subtilis*.

44. W. F. Bynum, *Science and the Practice of Medicine in the Nineteenth Century*, Cambridge University Press, Cambridge, 1994, pp. 127–132.

45. Not related to the human disease, although it has the same name.

46. G. L. Geison, *The Private Science of Louis Pasteur*, Princeton University Press, Princeton, 1995.

47. The four postulates were formulated by Koch and Loeffler in 1884 and published by Koch in 1890. They are: The microorganism must be found in abundance in all organisms suffering from the disease but should not be found in healthy organisms. The microorganism must be isolated from a diseased organism and grown in pure culture. The cultured microorganism should cause disease when introduced into a healthy organism. The microorganism must be re-isolated from the inoculated, diseased experimental host and identified as being identical to the original specific causative agent.

48. I. Loudon, Ignaz Phillip Semelweiss' studies of death in childbirth, *J. R. Soc. Med.*, 2013, **106**, 461–463.

49. W. F. Bynum, *Science and the Practice of Medicine in the Nineteenth Century*, Cambridge University Press, Cambridge, 1994, p. 134.

50. E. Bartlett, *An Essay on the Philosophy of Medical Science*, Lea & Blanchard, Philadelphia, 1844.

51. C. Bernard, *Introduction a la médecine expeérimentale*, 1865; C. Bernard, *An Introduction to the Study of Experimental Medicine*, translated by H. C. Greene, Dover, New York, 1957, p. 146.

52. W. Osler, *The Principles and Practice of Medicine*, Appleton & Co., New York, 1892.

53. PCR is a technique that identifies nucleic acid sequences.

54. F. Collins and V. McKusick, Implications of the Human Genome Project for Medical Science, *JAMA*, 2001, **285**, 540–544.

55. F. Collins, Has the revolution arrived?, *Nature*, 2010, **464**, 674–675.

56. Le Fanu sets this in a wider context of the failure of two approaches, genetic analysis and epidemiology, both trying to "replace the chance discovery of new drugs and the empiricism of technological innovation by… the elucidation of the causes of diseases that would lead either to rational forms of treatment or the prevention of common illnesses." See J. Le Fanu, *The Rise and Fall of Modern Medicine*, Basic Books, New York, 2nd edn, 2012, pp. 309–535.

57. Details of the two treatments differ, but the common approach is treating the patient's own blood stem cells so they produce a functional hemoglobin.

58. J. Le Fanu, *The Rise and Fall of Modern Medicine*, Basic Books, New York, 2nd edn, 2012, pp. 406–413.

59. www.omim.org/statistics/geneMap, accessed February 2024.

60. H. H. Ropers and C. D. van Karnebeek, Rare diseases: human genome research is coming home, *Cold Spring Harbor Mol. Case Stud.*, 2022, **8**(2), a006210.

61. One survey identified 100 drugs derived from plants that have been used as folk remedies. D. S. Fabricant and N. R. Farnsworth, The value of plants used in traditional medicine for drug discovery, *Environ. Health Perspect.*, 2001, **109**, 69–75.

62. WHO (World Health Organization) stated in 2023 that "around 40% of pharmaceutical products today draw from nature and traditional knowledge," but provided no evidence for this assertion.

63. T. S. Bradford, *The Life and Letters of Dr. Samuel Hahnemann*, Forgotten Books, London, 2012.

64. S. Hahnemann, *Organon of Medicine*, 1842, translated by J. Kunzli, A. Naude and P. Pendleton, Houghton Mifflin Company, Boston, 6th edn, 1982.

65. A survey for 2007 showed 21.7% of the population use herbal remedies and megavitamin therapy; 25.6% for meditation, breathing, yoga *etc.*; 15.5% for chiropractic therapy, massage *etc.*; 1.7% for homeopathy; 1.1% for acupuncture; 3.5% for a variety of other treatments. These numbers do not include faith healing. P. M. Barnes *et al.*, Complementary and Alternative Medicine Use Among Adults and Children: United States. *National Health Statistics Report No 12*, December 10, 2008.

66. B. Goldacre, *Bad Science: Quacks, Hacks, and Big Pharma Flacks*, Farrar, Straus and Giroux, New York, 2008.

67. A. Shang, Comparative study of placebo-controlled trials of homoeopathy and allopathy, *The Lancet*, 2005, **366**, 726–732; House of Commons Science and Technology Committee report: Evidence Check 2: Homeopathy. Fourth Report of Session 2009-10; NHMRC (National Health and Medical Research Council, Australia), Information Paper. Evidence on the effectiveness of homeopathy for treating health conditions, 2015.

68. R. B. Bausell, *Snake Oil Science: The Truth About Complementary and Alternative Medicine*, Oxford University Press, Oxford, 2007.

69. R. L. Nahin *et al.*, Expenditures on Complementary Health Approaches: United States, 2012, *National Health Statistics Reports*, #95 June 22, 2016. www.cdc.gov/nchs/data/nhsr/nhsr095.pdf.

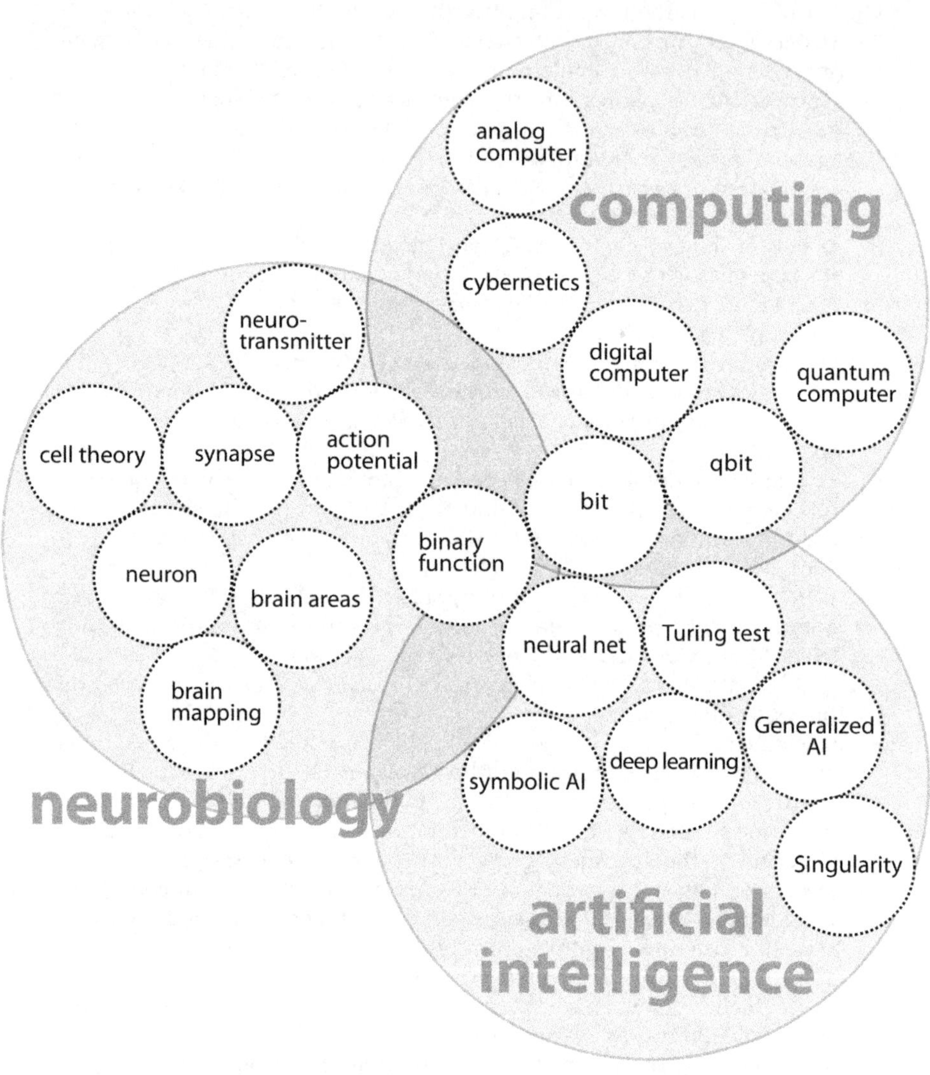

analog computer
computing
cybernetics
neuro-transmitter
digital computer
quantum computer
cell theory
synapse
action potential
qbit
bit
neuron
brain areas
binary function
brain mapping
neural net
Turing test
Generalized AI
neurobiology
symbolic AI
deep learning
Singularity
artificial intelligence

Mind and Brain

If there were machines that bore a resemblance to our bodies and imitated our actions as closely as possible for all practical purposes, we should still have two very certain means of recognizing that they were not real humans. This first is that. . . it is not conceivable such a machine should produce arrangements of words so as to give an appropriately meaningful answer to whatever is said in its presence, as the dullest of men can do. Second, even though some machines might do some things as well as we do them, or perhaps even better, they would inevitably fail in others, which would reveal that they are not acting from understanding, René Descartes, 1637.

There lies the frontier, still almost as impassable for us as it was for Descartes. Not until that barrier has been passed will dualism cease to be a force, and to that extent a truth, in the lives of all of us. We today are no less in the habit of differentiating between brain and mind than they were in the eighteenth century. Objective analysis obliges us to see that this seeming duality within us is an illusion. Jacques Monod, 1971.

Cajal's Neuron Doctrine: 1898–

"[NEURONS ARE] MYSTERIOUS BUTTERFLIES OF THE SOUL, the beating of whose wings might one day, who knows, reveal the secrets of mental life," said Ramón y Cajal,[1] in a poetic turn of phrase that captures the following century of research. The 20th century started with the neuron doctrine and ended with neural nets. Neurons are so completely different in appearance from other cell types, that previously it was not even clear whether they were in fact cells. By proposing the neuron doctrine, that neurons are discrete, independent cells, Cajal ushered in a revolution. Over the course of the twentieth century, different types of neurons were distinguished, including sensory neurons (which communicate with one another) and motor neurons (which signal to muscle). Neuron signaling was shown to take place by chemical or electrical transmission. The concept developed that the brain functions by means of the network of connections between its neurons. The model of the neuron as an input–output device was the spark for the development of A.I. The biology of the brain as seen today is truly the result of Cajal's revolution.

The Frontiers of Science
By Benjamin Lewin
© Benjamin Lewin 2026
Published by the Royal Society of Chemistry, www.rsc.org

Timeline for Completion of
the Neuron Doctrine

1870

Gerlach proposes reticular network for nervous system (1871)
Golgi introduces staining technique (1873)

1880

1890

Cajal proposes neurons are independent cells,
consisting of body, axon, and dendrites (1888)
Cajal proposes Law of Dynamic Polarization of neurons (1892)

1900

Sherrington proposes that synapse connects neurons (1897)

1910

Brodmann maps human brain into 52 areas (1909)

1920

Loewi & Dale show chemical transmission at synapse (1921)

1930

Dale shows neuron releases same transmitter at all axon terminals (1935)

1940

Hodgkin & Huxley identify action potential (1939)

McCulloch & Pitts propose binary model for neuron (1943)
Eccles proposes electrical transmission (1947)
Hebb shows coincidental firing strengthens connections (1947)

1950

Roberts & Awapara identify GABA as inhibitory transmitter in brain (1950)
Katz shows quantal release of acetylcholine (1952)
Palay & Palade analyze synapse by electron microscopy (1955)

1960

Furshpan & Potter find electrical transmission (1957)

1970

Neher & Sakman develop patch clamp technique (1976)

1980

Brenner maps all neuronal connections in worm brain (1986)

1990

2000

2010

Human Connectome Project begins mapping connections in brain (2016)

2020

Connections mapped in fruit fly brain (2024)

A Dialog between Ramón y Cajal and Warren McCulloch

Ramón y Cajal (1852–1934) was a Spanish physician and scientist who defined the neuron doctrine around 1890, proposing that neurons are cells whose connections account for brain function.

Legado Cajal-CSIC.

Warren McCulloch (1898–1969) was an American neuropsychologist who collaborated with Walter Pitts to propose in 1943 that neurons function as binary input–output devices, and can be connected to form a neural net.

Cajal: Dr. McCulloch, it's an honor to converse with a pioneer in the field of cybernetics. Your work with Warren Pitts approaches the neuron from a completely new perspective.

McCulloch: Thank you, Dr. Cajal. Your detailed studies of neuronal morphology and connectivity were the inspiration for our work.

Cajal: Histology allows us to unveil the intricate architecture of the nervous system. However, I understand that your work views neurons as information-processing units? How does this converge with their physiological functions?

McCulloch: Histology explains how one neuron contacts another, but it does not explain how the network functions. By treating neurons as input–output devices, we can decipher the algorithms by which neural circuits work.

Cajal: Structure–function relationships are at the core of my investigations. Through detailed anatomical studies, I aim to unravel the connections and arrangements of neurons. The morphology of neurons, with their dendrites and axons, provides crucial insights into the transmission of information within the brain.

McCulloch: Your Law of Dynamic Polarization, that neurons fire in one direction only, was a major insight into function. We have extended it by arguing that neurons function in a binary manner.

Cajal: We know that there is plasticity in the nervous system. How does your computational perspective accommodate the dynamic nature of neural networks?

McCulloch: Neural plasticity is a cornerstone of our understanding. Computational models enable us to simulate how synaptic strengths change over time. By changing the strengths of connections, we aim to capture the dynamic nature of learning and memory, and to develop artificial intelligence.

Cajal: Fascinating. I understand that it now seems the number of neurons in the human brain is somewhat less than we had thought. Is there a problem that the complexity of the neuronal network does not seem to be great enough? Computational systems are even less complex!

McCulloch: We have moved from individual neurons to the network. Our computational models have achieved great success in approaching an artificial intelligence. They are becoming more complex every day.

Cajal: Technological advancements such as imaging techniques and molecular tools continue to push the boundaries of what we can uncover about the brain. I foresee a more comprehensive understanding of neural circuits, perhaps even an understanding of consciousness. But we still need to understand the function of the neuron itself! I rest my case.

CAJAL'S NEURON DOCTRINE

[This book] is designed, in the first place, to investigate the fundamental laws of those operations of the mind by which reasoning is performed. It's unnecessary to enter here into any argument to prove that the operations of the mind are in a certain real sense subject to laws, and that a science of the mind is therefore possible.[2] George Boole, 1854.

Most of the complexity of a human neuron is devoted to maintaining its life-support functions, not its information-processing capabilities. Ultimately, we will be able to port our mental processes to a more suitable computational substrate. Then our minds won't have to stay so small.[3] Ray Kurzweil, 2005.

Aristotle thought the brain was a radiator—no more than a cooling agent for the really important organ: the heart. He described the *sensus comunis* as the space where all the senses come together—an early version of consciousness—and the heart was the obvious location because a bloodless brain could not house the soul. Other Greek philosophers argued that the hegemonikon—the central commanding facility of the body— must reside in the brain.

Based on dissection and on studies of head injuries, the Roman physician Galen thought that the brain was the seat of the hegemonikon and the soul, because it was the termination point for all five senses. By the late Middle Ages, different intellectual functions were being ascribed to different parts of the brain.

When Descartes said, *"Cogito ergo sum,"* he captured the mind–body problem. "I think therefore I am"—but what is the connection between the "I" and the physical body? To put the question in modern terms, how is consciousness to be explained by the properties of the brain? Although control of specific physical functions (such as vision) can be equated with specific parts of the brain, it seems impossible to relate 'thinking' to any specific part of the brain. So the focus becomes the need to understand whole brain function.

The construction of the brain at the cellular level was a mystery until the start of the 20th century. The discovery that

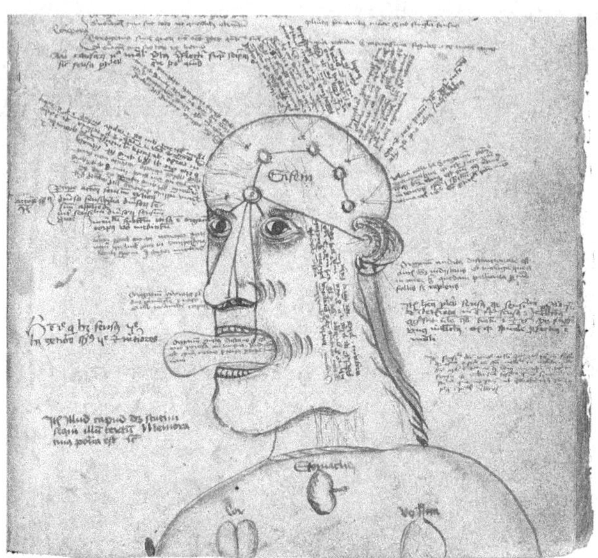

Function of the brain according to an illustration of Aristotle's De Anima *by Johan Lindner of Mönchenburg (published 1472–1474). Sites from left to right are for common sense, imagination, fantasy, cogitative power, and memory. These were regarded as physical properties.*

Reproduced from https://es.m.wikipedia.org/wiki/Archivo:Aristoteles, _De_anima_Wellcome_L0044182.jpg, under the terms of the CC BY 4.0 license, https://creativecommons.org/licenses/by/4.0/.

neurons are cells led to the concept that brain function resides in the network of interactions among neurons. The idea that there is a network leads to the question of whether and how we might reconstruct or imitate it with either biological or nonbiological material.

Neural networks epitomize the 21st century. As the basis for the technique of deep learning used to construct artificial intelligence, they could scarcely seem more modern. Yet the concept of trying to imitate the human brain in an artificial situation dates from the first modeling of neuronal interactions in the 1940s. It's a great question how these mathematical constructs relate to our current understanding of neuronal functions.

Neurons appear so different from other cells that until Ramón y Cajal worked out their structure from 1886–1891, it was

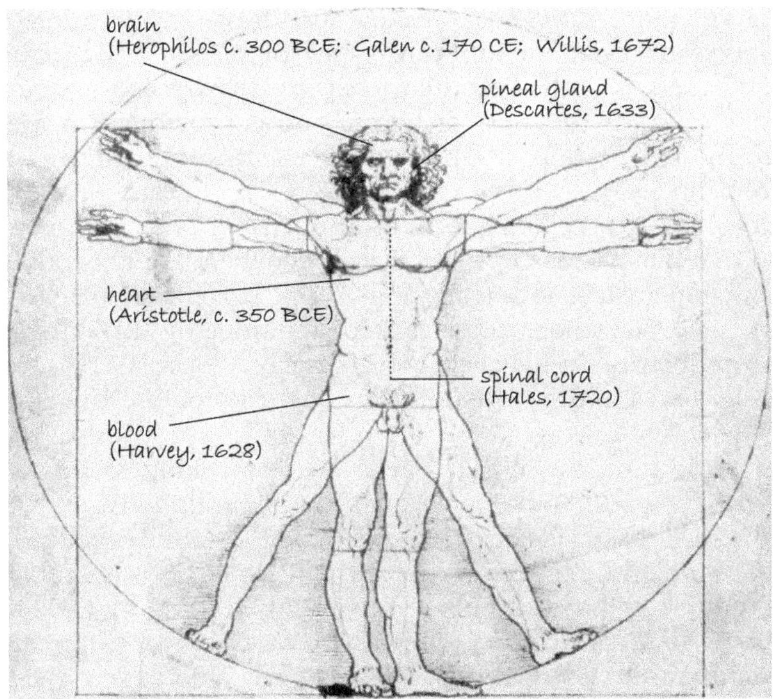

From ancient Greece to the 18th century, the search for the soul has identified a variety of locations. With apologies to Leonardo for borrowing Vitruvian Man (more formally The Proportions of the Human Figure after Vitruvius), original drawing c. 1490.

controversial whether they were cells at all! Schwann's cell theory of 1839 had proposed that all tissues in the body are composed of individual cells. Until Cajal's work, it had been unclear whether nerve fibers were part of individual cells or part of a continuous network separate from cells.

Working only with a light microscope, Cajal was able to define neuronal structure because Camillo Golgi had recently developed a stain for cellular structures that highlights neurons. With that stain, he was able to show that the neuron is a cell with a central body containing the nucleus. The cytoplasm surrounding it extends into two types of nerve fiber. The axon is a long body, sometimes extending for considerable distances. Dendrites (called protoplasmic prolongations until Cajal renamed them) are more numerous and finer in structure.

Cajal made hundreds of exquisite drawings defining the connections between different types of neurons. Describing them in terms of the cell theory became known as the neuron doctrine.[4] As it replaced mystique and confusion with a definite model, perhaps it should have been called the neuron revolution.

Golgi never accepted the neuron doctrine, but continued to hold that nerve fibers form their own, independent network (the reticular network). Ironically, Cajal and Golgi shared a Nobel Prize in 1906, when Golgi caused an upset by attacking the neuron doctrine in his acceptance speech.

Golgi and Cajal both labored in obscurity before their work was discovered. Golgi (1843–1926) was quite cut off, as a physician in a rural town near Pavia. While working at home by candlelight in 1873, he discovered that staining nervous tissue with a silver-based preparation gives a dense stain. He called it *la reazione nera* (the black reaction), but it's now known simply as the Golgi stain.[5] The results were striking, leading to a position in 1875 at the University of Pavia.[6] But publishing in Italian meant that his work was not widely known, so in the early 1880s Golgi arranged to have his work published in translations into French and English.

Cajal (1852–1934) came from a medical family. At one time intending to be an artist (but discouraged by his family), Cajal obtained a medical degree, but could only find temporary positions where his family was located, in a village near Zaragoza in northeastern Spain. His early years were difficult.[7] He acquired a microscope, but the general attitude was that microscopy was an obstacle to progress in biology! He finally obtained a chair in anatomy, at Valencia in the south, in 1883. As a result of making a good impression in dealing with a cholera epidemic, he obtained a high-quality Zeiss microscope and began his studies in earnest. He moved to a Chair in Barcelona in 1887.

He described his reaction to his discovery of Golgi stain. "Dumbfounded, I could not take my eye from the microscope."[8] His mode of publication was unusual: he started his own journal in 1888. Even though he sent out all 60 copies to foreign scientists, the work made little impact, because it was written in Spanish. His style was direct and factual, more like a modern paper than the anecdotal approach of the period.

He commented that, "the wonderful revelatory powers of the chrome–silver reaction [contrasted with] the absence of any excitement in the scientific world aroused by its discovery." The work was not recognized until he presented his results at a meeting of the German Anatomical Society in Berlin in 1889. On his way to and from the meeting, he spread the word by visiting major universities. In 1892, now well established, and even famous, he became a professor in Madrid.[9] In 1899, he published a summary of his work, with 2000 pages of text and 1000 illustrations organized in two volumes.[10]

The structure of the neurons left unanswered a major question: how do they communicate? Cajal extended the model into function with his Law of Dynamic Polarization. This states that an impulse travels only in one direction through a neuron. The dendrites receive a signal and send it to the cell body, the axon conducts the pulse to its destination, and then the terminal branches of the axon pass the signal to the target. This might be another neuron (in brain function) or a terminal target (such as

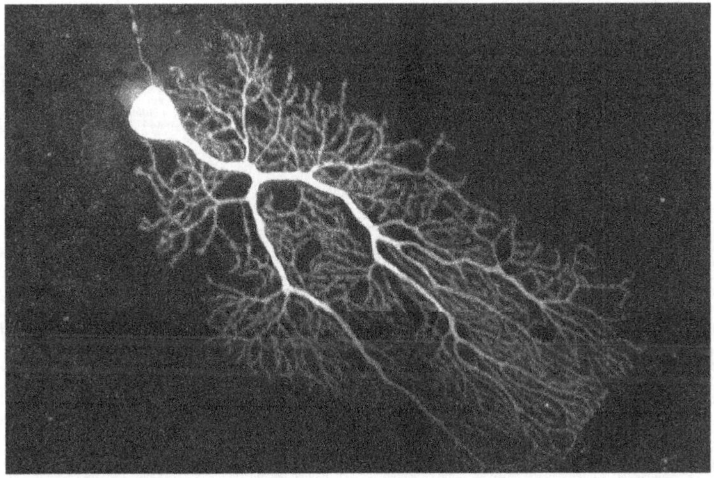

A neuron has a central body that includes the nucleus. It signals to its target via a long axon extending from the central body, and receives signals through dendritic connections.

Ramón y Cajal in his laboratory in Valencia in 1887 when he made his first observations of neurons.

Legado Cajal-CSIC.

muscle in the case of a motor neuron). In effect, Cajal proposed an input–output model for the brain, with dendrites as the input device and axons as the output device. The term synapse was later introduced to describe a junction where a fiber from one neuron connects to another neuron.[11]

The CNS (central nervous system) consists of the brain and the spinal cord. Sensory neurons (in the brain) are activated by sensory inputs from the environment. Motor neurons are in the spinal cord and transmit impulses to control muscle movements. Motor neurons have the longest axons in the body. Much work to characterize the basic anatomy of the neuron was done with the spinal cord, because it's easier to visualize the neurons, and to trace their connections. The neuromuscular junction has been a very successful model system where most of the work has been done on synaptic transmission.

The concept that neurons are connected by a discrete structure was developed by Charles Sherrington (1857–1952), a British neuroscientist, awarded a Nobel Prize in 1932.[12] The concept of a gap between neurons was reinforced by the fact that there is a delay in the response when an impulse is sent to neurons. It was

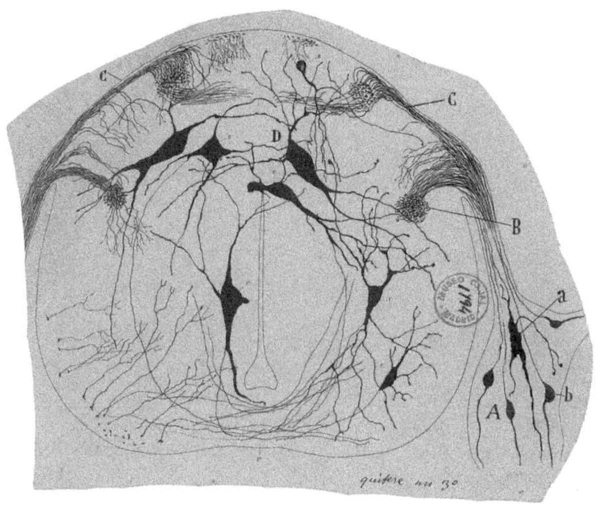

Cajal's drawing of neurons in the spinal cord.

Legado Cajal-CSIC.

Sherrington who called the connection the synapse. The time to cross it was later called the synaptic delay.[13,14]

The idea that the brain communicates *via* nerve fibers may have originated with Galen. The first idea of mechanism came when Descartes proposed in 1664 that nerves were hollow, so that 'animal spirits' could flow from the brain to target muscles.[15] A mechanical basis for the communication was first proposed when Luigi Galvani found in 1791 that an electric current could cause a frog's leg to contract.[16] An Italian physician, Galvani then studied animal electricity for the rest of his life. It took another half century to prove that the current actually was generated by the tissue (and not by the metal connected to it).[17] This spike in activity is now called the action potential.

The big issue now was the cause of the impulse.[18] There was a debate whether it was electrical or chemical. The main advocate for electrical transmission was neuroscientist John Eccles, first at Oxford, then back home in Australia, who believed that transmission was too fast to be accounted for by chemical release. The argument for chemical transmission was that nerve action could be stimulated or inhibited by chemicals related to potential neurotransmitters. (The term neurotransmitter was

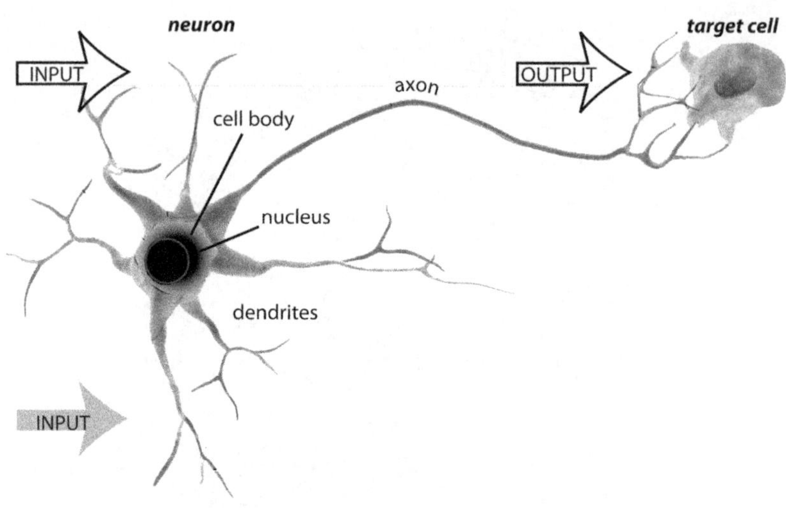

A modern interpretation of the Law of Dynamic Polarization views the neuron as an input–output binary device. Input is to the dendrites, the signal is transmitted from the cell body along the axon, and then the branches at the end of the axon activate the target cell.

introduced around 1959 to describe a chemical that carries a signal across the synapse.)

The involvement of chemicals in synaptic transmission was identified by a classic series of experiments during the 1920s–1930s. Working in London, Henry Dale identified chemicals that block the neuromuscular junction. Then he was able to identify acetylcholine as the chemical released at synapses. Working at the University of Graz in Austria in 1921, Otto Loewi isolated compounds from the vagus nerve that could either stimulate or inhibit the heart. Loewi and Dale were awarded the Nobel Prize in 1936. Acetylcholine is the classic neurotransmitter, but we now know that there are also others, of various chemical types.[19]

It remained controversial whether the same system—release of neurotransmitters—would also apply to the central nervous system. The argument about chemical *versus* electrical transmission continued. It became known as the "soup *versus* sparks debate."[20] In a classic example of how science functions, it was Eccles, the advocate of electrical transmission, who in 1951 obtained results contradicting the electrical hypothesis in the central nervous system. "[The] hypothesis is thereby falsified," he wrote in his paper, continuing on to conclude that synaptic action was mediated by a chemical transmitter. (To simplify, if stimulatory interactions occurred by electric transmission, blocking electrical transmission should inhibit the interaction. Eccles showed that this was not the case.)

While analysis was restricted to the light microscope, it remained a great puzzle how one neuron actually contacted another. When the electron microscope became available in the 1950s, it was possible to visualize the apposing membranes.[21] The membranes on either side of the synapse are separated by a space (called the synaptic cleft) of 20–30 nm, about 4× the width of the membrane.[22] The terminal of the axon contains large numbers of very small bodies, about 40 nm in diameter, that seemed likely to be the means of communication across the synapse.[23] These 'synaptic vesicles' fuse with the membrane and release their contents into the synapse.

The fusion of a vesicle releases a 'quantum' of neurotransmitter. Bernard Katz, a German-born British biophysicist who had worked with Eccles, shared a Nobel Prize in 1970 for showing that acetylcholine is released at synapses only in

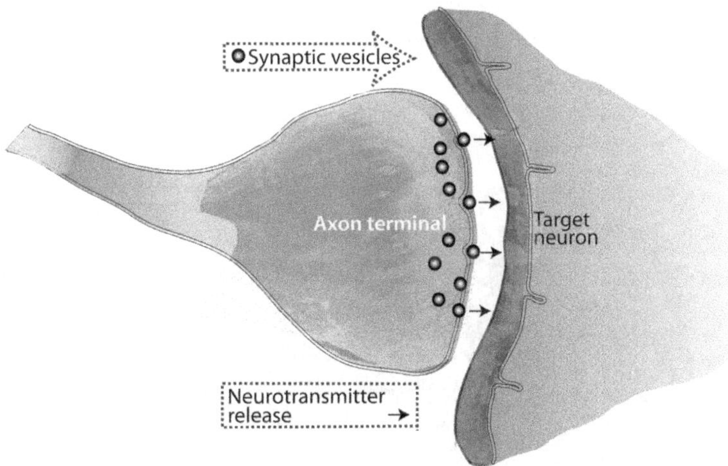

A chemical synapse has a gap between the axon terminal and the target neuron. Synaptic vesicles in the axon fuse with the membrane and release neurotransmitters into the synaptic cleft, where they activate receptors on the target.

discrete amounts. The 'quanta' are multiples of the quantum level. In the well-characterized case of the motor neuron, a single synaptic vesicle contains about 10 000 molecules of the neurotransmitter acetylcholine.

What causes a neuron to fire? The neurotransmitters bind to receptors on the target neuron, and cause ions to travel across channels in the membrane. When a threshold level of release is achieved, the ions generate an 'action potential,' a current that causes the neuron to fire.[24] Synapses effectively convert a chemical signal (the release of neurotransmitter) into an electrical signal (the action potential). Then the cycle repeats as the action potential causes the neuron to release neurotransmitters. We now know that there are many, many ion channels with different characteristics.[25]

In an extraordinary technical feat, Alan Hodgkin and Andrew Huxley, from Cambridge but working during the summer at the Marine Laboratory in Plymouth, identified the action potential directly in 1939 by inserting an electrode into a squid giant axon. After a delay caused by the Second World War, work resumed. In 1952, they derived an equation describing the parameters that control the action potential.[26] This was an early (and unusual) example in biology of theory preceding experiments, as it wasn't possible at the time to investigate the effects of changing the parameters.[27]

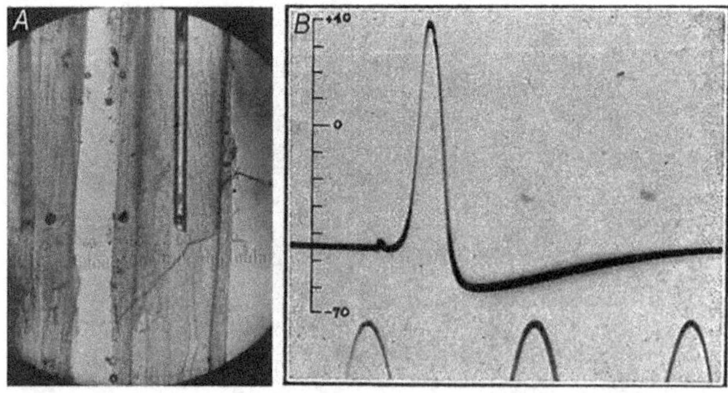

The electrode inside the axon (left) and the first recording of an axon potential (right). The ingenuity and apparent simplicity of the equipment is a contrast with the present era.

Although most neurons use neurotransmitters, it was later discovered that some function by electrical transmission (validating Eccles' original view). They were first identified in the giant axon of the crayfish, and are now thought to comprise about 20% of synapses in vertebrates.[28] The crayfish neurons are connected by 'gap junctions' that allow ions to flow across a 2–3 nm connection between the neurons.[29] The flow is effectively instantaneous. This enables groups of neurons to be synchronized so they all fire simultaneously. (This is an essential feature of the escape response, which needs to be very fast.) Eccles shared a Nobel Prize in 1963 with Hodgkin and Huxley.

The neuron doctrine has stood amazingly unscathed since Cajal proposed it more than a century ago. The paradigm was extended by characterizing the synapse and the two basic types of synaptic transmission. Although exceptions have been found where dendrites have outputs or axons have inputs from other axons, the basic concept that the neuron is an input–output device (to use modern terminology) remains the foundation of neuroscience. The major functional extension was the recognition that interactions are not only excitatory, as described by Cajal, but also can be inhibitory.

The complexity of the brain comes from the wiring of the neurons. Over the 20th century, the focus shifted from the individual neuron to the network of neurons. One current view is that the doctrine of the individual neuron should be replaced by regarding groups of neurons as the functional unit within the brain.[30]

Until recently, conventional wisdom was that a human brain has about 10^{11} (100 billion) neurons, and about 10× that number of glial cells (a supporting array sometimes called the 'glue' of the nervous system). But it turns out that the numbers had never actually been measured. These important numbers were actually no more than an assumption. Apparently, the 'fact' was so entrenched that no one in the field of neuroscience realized!

When the numbers were actually measured, the human brain proved to have 86 billion neurons (a little less than expected) and almost the same number of non-neuronal cells (a lot less than expected).[31] Each neuron probably makes an average 7000 synaptic connections to other neurons, so the total number of synaptic connections is 10^{14}–10^{15}.[32] The numbers in the human brain do not stand out from other primates.[33] Its special qualities could depend on the distribution or connections between neurons rather than their total number. (This parallels the discovery that humans do not have more genes than other primates; see Chapter 20.)

Do we (humans) have enough neurons? The classic view of neural systems (either biological or computational) was that capacity would be proportional to the number of neurons. It turns out there are two problems with supposing that large networks have enough capacity to store sufficient memory. One is that the systems work by reusing synapses, so that memories are overwritten (actually rather quickly; this is one of the limitations on short term memory). It's unclear what the physical basis might be for storing long-term memories. The other problem is that calculations suggest that the memory capacity grows logarithmically rather than linearly with network size. This results in a catastrophic limitation on memory capacity, and requires a new approach to memory storage.[34]

∞ ∞ ∞

There are differences between chemical and electric synapses: most notably that electric synapses are faster, whereas chemical synapses can amplify the signal. A crucial similarity is the all-or-none effect: either the synapse fires or it does not fire. Even without knowing the details of the mechanism, computer scientists Warren McCulloch and Walter Pitts made a seminal contribution to neuroscience when they proposed in 1943 that the neuron could be regarded as a binary operator.[35] This was an intellectual leap, leading to a proposal: if the neuron is binary, then we can imitate its action. This created the model of a neural net, in which the interconnection of neurons forms a network that has logical properties.

McCulloch and Pitts were an unlikely pair. McCulloch followed a classic academic path, trained as a neuropsychiatrist, and tried to understand the logic of the nervous system. Pitts appears to have had no formal academic training, taught himself mathematics and logic, and eventually participated in a group on mathematical biology at the University of Chicago. When they started their collaboration, Pitts was only 17 and McCulloch was in his mid-forties.

Walter Pitts (left) and Warren McCulloch, shown here in 1949, came from different backgrounds. Together they authored one of the most innovative papers in computing.

Reproduced from https://doi.org/10.1002/jhbs.1094 with permission from John Wiley and Sons, Copyright © 2002 Wiley Periodicals, Inc.

As early as 1936, Alan Turing, who designed the Colossus apparatus that decoded Enigma transcripts in the Second World War, had developed the concept of a machine that operated by manipulating symbols on a strip of tape according to a predefined set of rules. Turing established the principle that any machine following the instructions of the tape would behave in exactly the same way as any other machine following the same directions. This was the principle of equivalence: two computers that can simulate one another are 'Turing equivalent'—in effect, they cannot be distinguished irrespective of the details of their construction.[36] Another mathematician, Alonzo Church, who had reached similar conclusions, later called this a Turing machine. It became an important concept in developing the idea of artificial intelligence.

McCulloch and Pitts wondered if the nervous system could be viewed in terms of a Turing machine. They took as their starting point the position that the nervous system consists of neurons, and that the neurons make 'all or none' connections. If the neurons are triggered by a threshold effect, they can only be either 'on' or 'off.'

From this they constructed the idea of a network using Boolean logic, in which all calculations are reduced to true or false. The logic assesses combinations using AND, OR, or NOT to decide whether they are true or false. (A century earlier, George Boole had actually argued that the mind functions by the laws of logic.[37])

McCulloch and Pitts specified that their model was a formal representation, not intended necessarily to represent actual neuronal function.[38] (There is perhaps a parallel here with the representation of atomic structure in quantum models, which may or may not represent reality, but work to describe the system; see Chapter 16.)

Treating the brain as a computing device based on binary principles was a revolutionary approach. This was a new paradigm for brain function. The model leads to the concept that neural function might be mimicked by constructing binary networks. If patterns of interactions among neurons in the brain could be defined, this might enable an artificial network to be constructed on similar principles.

∞ ∞ ∞

When the concept of computers originated, it was not necessarily obvious that they should be digital. The first attempts to design a computer, in the early 19th century by Charles Babbage, were based on a decimal system (see Chapter 23). The first electronic computer, ENIAC, in 1943, used a decimal system of counters. The adoption of binary digital systems grew out of the technology of the 1940s, in which on/off switching devices, at first mechanical, later electronic, were available. The concept was driven by an analogy with switching in the telephone network, where it was necessary to measure the flow of information.

The idea that information should be measured on a binary scale came from Claude Shannon at Bell Labs in 1948, when he proposed the bit (standing for binary digit) as the unit of information.[39] The bit "corresponds to the information provided when a choice is made from two equally likely possibilities."[40] At first known as communication theory, this later took the name of information theory. It represents Boolean logic by assigning each

Claude Shannon in the early 1950s with his mechanical mouse, Theseus, that learned to navigate a maze, in an early development of AI.

Reproduced with permission from Nokia Corporation and AT&T Archives.

bit with a value of either 'true' or 'false.' This led directly to the theoretical construction that computers could perform logical tasks of Boolean algebra.

The theory of parallels between the brain and a computer was set out by mathematician 'Johnny' von Neumann, a polymath who stood out at Princeton even in the company of luminaries such as Einstein and Gödel. Born in Hungary, he had worked with the great mathematician David Hilbert in Germany. Trying to reconcile Heisenberg's and Schrödinger's very different formulations of quantum mechanics, he showed in a revolutionary book how their equations could be derived from mathematics. (Whether the proof holds up later became controversial.[41])

Von Neumann moved to the United States in 1930. A flamboyant character, maintaining his strong Hungarian accent, he spent his career at Princeton from 1933, where he was known for his parties as well as his brilliance.[42] He was involved in the Manhattan project. After the Second World War, he was instrumental in the construction of ENIAC followed by the

John von Neumann (left) and Robert Oppenheimer (right) at the dedication ceremony for the IAS Computer at Princeton in 1952.

Courtesy of Shelby White and Leon Levy Archives Center, IAS.

development of other computers (see Chapter 23). (There was some opposition at Princeton to the idea that calculations could be performed by machine instead of on a blackboard.) His treatise on game theory impacted fields from economics to nuclear strategy. It was a revelation that mathematics could extend beyond its boundaries to influence such other varied areas.

Von Neumann was a genius by any measure, famous for solving problems in a few minutes that other mathematicians had wrestled with for years. It was said of him at the time that, "Other mathematicians prove what they can, von Neumann proves what he wants."

His development of the theory of automata—that machines able to process information could reproduce—was influential in the development of AI. He drew an explicit comparison between computing machines and automata, and described neurons as switching devices. In the same paper in 1951, he also drew a prescient analogy between his automata and the replication of the gene.[43] Years later, molecular biologist Sydney Brenner thought that von Neumann's view of self-replicating automata, rather than Schrödinger's question, What is Life?, should have been the driving force for molecular biology.

His comparison of brains and computers culminated in the Silliman lectures that he intended to give at Yale in 1956, but was prevented from doing by illness (just before his death).[44] He defined the basic action of the neuron as a reversible impulse, meaning that once the neuron has fired, it returns to its prior state after a recovery period. The pulse lasts 10^{-4} sec, the recovery takes 1.5×10^{-2} sec. He pointed out that electronic devices could perform these reactions several times faster, although they needed to be very much larger than a human brain.

Von Neumann cast some doubt on whether exclusively digital analysis could mimic the human brain. "It may be, however, that in this process logic will have to undergo a pseudo-morphosis to neurology to a much greater extent than the reverse," he said, meaning that digital might not be enough.[45]

In mathematical terms, the quiescent state represents 0 (equivalent to 'false'), and the impulse represents 1 (equivalent to 'true'). This parallels the state of each vacuum tube in a computer of the period, with memory recorded as 'flip-flops,' so each tube could be characterized as having a bit of 0 or 1. A computing machine at the time might have 10^5–10^6 bits. Without defining exactly how human memory might be stored, von Neumann calculated that the number of bits in the brain might be 10^{20} (now viewed as an over-estimate). This was a forceful and highly influential view that the brain might function in a binary manner.

∞ ∞ ∞

The first attempt to provide software functioning like the brain was the perceptron invented by Frank Rosenblatt at the Cornell Aeronautical Laboratory (run by a defense contractor) in 1957. The perceptron consists of three 'layers:' input ('retina' if photo inputs are being used), hidden (where the signals from the retina are analyzed), and output. The retina feeds into the hidden layer, which signals to the output layer. (The central layer is called 'hidden' because its values cannot be observed directly from the input or output.)

After training the perceptron by telling it the right response to distinguish a series of circles and squares, for example, it should

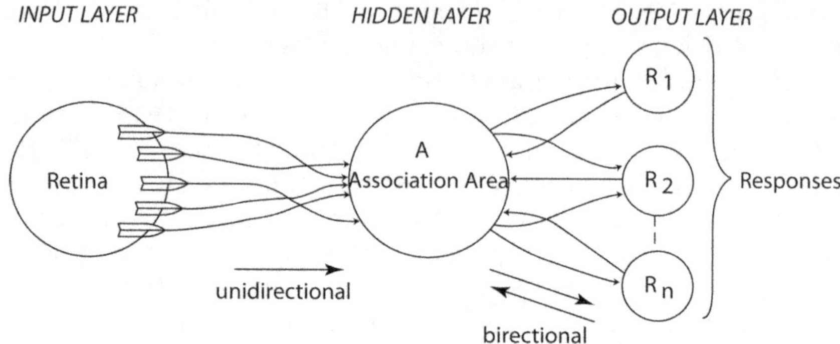

Rosenblatt's design for the perceptron. The 'retina' has a 20×20 array of photocells able to generate an image of an object. When a cell in the retina is activated by receiving light above its threshold, it transmits signals to the cells it's connected to in the next (A) layer, which has 512 cells. Up to 40 photocells may be connected to one cell in the A-layer. When the sum of impulses exceeds a threshold for an A-cell, it signals to an R (response) cell. A feedback loop from an R-cell inhibits A-cells that do not activate it. The R-cells communicate with each other and show a response when they all agree.[46]

be able to use probabilities of sets of photocells being fired simultaneously to determine whether an unknown object is a square or circle.[47] By combining signals from the retina with the feedback from the response cells, the hidden layer contains the 'memory' of the system—that is the learning process. The first attempt at making a perceptron used software running on a computer, but this was too slow to handle the three layers. The second attempt used a special apparatus, filling a small room, with adjustable resistors controlled by motors to provide the hidden layer.

The perceptron was the first example of what today we would call a neural net. Today's neural nets use multiple hidden layers, where each layer signals its results to the next layer. It's felt this could mimic the way the brain assembles visual information, for example.

Psychologist Donald Hebb (1904–1985) recognized a key feature of human learning in 1949. Brain plasticity involves a reinforcement process in which connections are strengthened

The first popular report of the perceptron, in the Cornell University magazine, made a claim for its intelligence.

Division of Rare and Manuscript Collections, Cornell University Library.

when a stimulus is repeated.[48] This can be mimicked in a neural net by assigning an appropriate weight to a connection. One way of looking at this is that McCulloch and Pitts showed how a neural net could be constructed; Hebb showed how it could learn.[49]

As neural networks have developed, however, the criterion has not been how well they resemble the brain, but how well they work. Even if they produce results comparable to human intelligence, even if they pass the Turing test (meaning they cannot be distinguished from a human in blind conversation), it does not necessarily follow that they work by the same mechanism as the brain (see Chapter 23). It's entirely possible that there are multiple ways to create 'intelligence.'

The complexity of a neural net depends on the number of layers and the number of cells in each layer.[50] A deep learning system might have 5 hidden layers. Total complexity is measured by the number of parameters—the specifications for the connections between cells. The GPT4 iteration of OpenAI's ChatGPT probably has about 10^{13} parameters, compared with 2×10^{11} for its predecessor, GPT3.[51] This is a great leap in complexity. However, the issue of long-term memory storage is that the number of bits in the brain is at least an order of magnitude greater than a supercomputer.

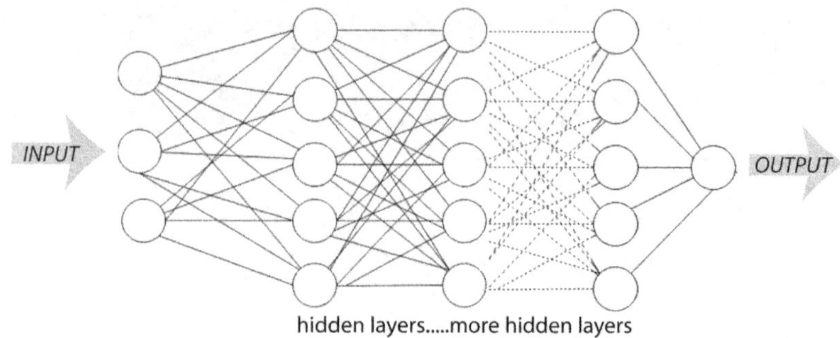

A multi-layer neural net (MLP) has several hidden layers. Information becomes increasingly focused until it triggers the output.

At the risk of over-simplifying, each node in the net has a certain number of incoming connections. Each connection is assigned a weighting. This is intended to mimic the brain by giving greater weight to inputs from stronger connections than to inputs from weaker connections. The node essentially multiplies the number of times each connection is activated by its weighting. When the sum of all those products exceeds a threshold value, the node fires. A neural net starts out with the weightings and thresholds set to random values. Then it adjusts them when it is trained to produce the desired result. (Training involves a method called 'back-propagation,' which sends a signal back from the output, as discussed in Chapter 23.)

One interesting feature of neural nets that might resemble the brain is called catastrophic interference. When a neural net is asked to learn a new task, this often causes it to overwrite the learning that it gained on a previous task. (Technically this is because it changes the weighting of individual interactions to match the new task, and they may not be appropriate for the old task.) There is speculation that the need to avoid this problem could be why the brain compartmentalizes different types of activity.

∞ ∞ ∞

While it's true that all neurons function as input–output devices, there are many types of neurons in the brain, differing in the types of connections that are made and the type of synaptic transmission. And specialized areas in the brain handle different types of information. How do neural nets relate to the complexity of brain organization?

Medieval anatomy focused on trying to identify the location of mental activity. The modern era started during the Scientific Revolution with the anatomical studies of Thomas Willis (1621–1675), published in *Cerebri Anatome* in 1664.[52] Willis was Professor of Natural Philosophy at the University of Oxford. Ironically, the purpose of his studies was (still!) to identify the site of the soul. However, he made a major discovery in identifying the flow of blood to the brain through cerebral arteries.[53] (The arterial blood supply at the base of the brain is called "the circle of Willis.") *Cerebri Anatome* contained a large number of plates drawn by Christopher Wren (the architect of St. Paul's Cathedral), who often assisted at dissections. It introduced the term 'neurology.'

The concept that the brain has specialized areas comes from the combination of anatomical studies with observations that damage to certain areas is associated with specific symptoms in patients. Korbinian Brodmann (1868–1918), a German neurologist, mapped the brain into 52 areas in a book published in 1909.[54] Brodmann's numbers are still used to link brain areas

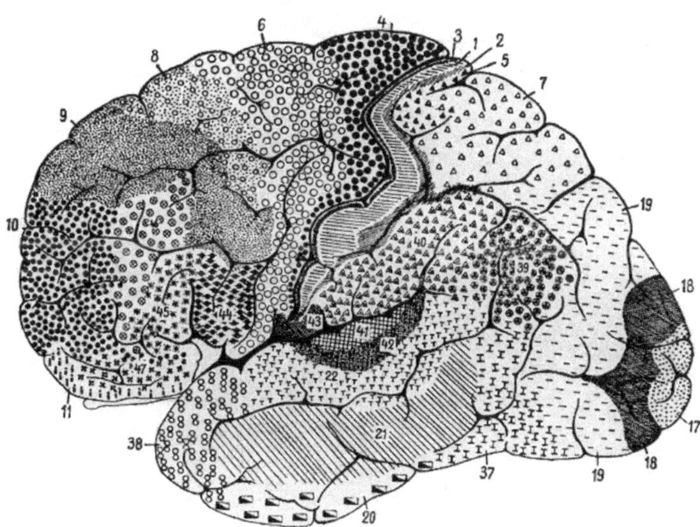

Brodmann's original map showing a lateral view of the left cerebral hemisphere.

with neurological functions. The identification of these specific areas was a revolution in neurology.

Mapping has been extended further by the Human Connectome Project (HCP), a five-year research program from the National Institutes of Health to build a network map of connections in the human brain. The map combined NMR images with anatomical information to define 180 areas in the brain. The boundaries for this 'parcellation' were established using machine-learning algorithms.[55] This is a great advance, but it's a refinement rather than a revolution.

Making a larger neural net with more and more connections will not by itself achieve a complexity matching the brain. It needs an equivalent to the brain's specialization of functions.

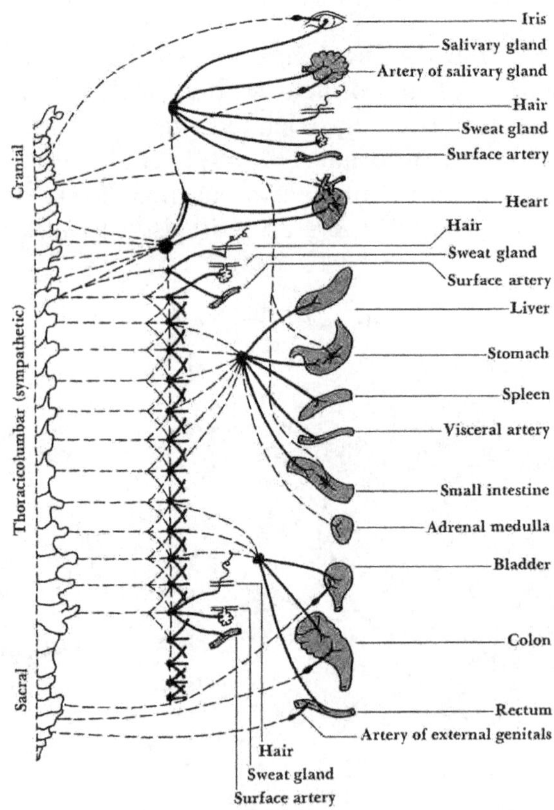

A diagram of the autonomic nervous system in 1939 shows a formal similarity to a neural network, with input, integration, and output.

This may be approached by 'branch specialization,' in which neural layers are divided into branches that function in parallel. A branch does not have access to the information in other branches.[56] The branches can be designed into the system or may even develop spontaneously as a neural net is trained.[57]

The level of specialization within a neural net is probably at least an order of magnitude less than that in the brain. McCulloch and Pitts were amazingly farsighted to draw a parallel in 1943 between neuronal input–output functions and computing, but another level of organization is required before neural nets might be able to imitate the brain. (One approach to mimic the biological level of organization is to link neurons to software in a hybrid system, but this hasn't been successful.[58])

∞ ∞ ∞

Even if we could think of a way to investigate consciousness physiologically, it's hard to see how it could overcome ethical objections to experimenting on people. This means that our best hope of approaching the problem may be by computer simulation.

Suppose we could define all the connections in the brain and construct a neural net that completely mimicked its architecture, in effect, construct a brain in silicon. Would it have consciousness? Of course, there would be some technical difficulties, even a major series of impediments. A brain does not spring out completely formed, like Pallas Athena in the classical myth. It gains experience over time from a series of inputs (effectively the five senses). And, of course, the brain is only part of the CNS (central nervous system), which extends into the spinal cord. And then the CNS is connected to the PNS (peripheral nervous system), which is connected to other components, such as muscle. And the brain has plasticity. The hardware changes as it develops new connections during development from child to adult. So there are problems in principle with using isolated software to mimic the brain.

But if we could overcome these objections to form a neural net with the appropriate inputs, why would it not have consciousness? The point of this question is to ask whether there is necessarily anything unique about the material (protoplasm) needed to construct a conscious being. Consciousness should be the result of the connections, architecture, and experience. This suggests that the ultimate objective (for some) in AI of achieving AGI (artificial general intelligence) may not be impossible in principle. However, it may be farther away in practice than the excitement about (for example) ChatGPT would suggest.

The possibility that the brain could be mimicked by an artificial system, including all the necessary sensory inputs, is called 'whole brain emulation' in the trade. It's not so simple as increasing the size of a neural net and giving it appropriate training. A neural net is, in a sense, two-dimensional. The human brain functions in three dimensions in the sense that we have awareness of three-dimensional space around us. This awareness is learned by experience. Whether there is something unique about biological systems remains an open question.[59]

The question of whether it's possible to create self-awareness or consciousness in software is known colloquially as the 'hard problem.' It's often assumed that the validation of the exercise would be to mimic the human brain, but it's an anthropomorphic assumption that an artificial intelligence would necessarily behave in the same way as a human brain. There is at least the possibility that it might be quite different.[60] (It's also true that attempts at whole brain emulation focus on the functioning adult brain. Ahead of the game as always, Alan Turing argued in 1950 that a better route might be to try to mimic an immature brain and then to train it.[61])

There have been many interesting speculations about the nature of consciousness, from philosophers to scientists, but that is all they are—speculations. The cold fact is that nobody has a clue. Notwithstanding a myriad of analyses of consciousness, there are no scientific data to bear on the question of whether consciousness may be a property unique to biological systems, or might emerge from any neural net that is sufficiently complex.[62] Even the mind–body dualism of Descartes has been resurrected (if, indeed, it was ever completely abandoned).[63] One extreme position is that the objectivity of science is not a suitable means for analyzing subjective states of the mind.[64] In spite of great advances in neurobiology and construction of artificial intelligence, the question remains very much the same as 50 years ago.[65] Understanding consciousness is the ultimate frontier for science.

The notion that the brain can be reverse-engineered is implicit in the application of big data to neuroscience. But relying on big data may be inadequate. The Human Brain Project was launched in Europe in 2013 with a ten-year, $1.3 billion budget as an international collaboration to do nothing less than produce a computer simulation of the human brain. Its first director, Henry Markram, said, "Neuroscience has to become 'big science'… We have spent too long waiting for a new Einstein to unify our field."[66] The BRAIN initiative in the United States has a similar scope.

The model is the human genome project, but this is really not at all applicable. That was basically a technology project that did not involve any new theory of the genome. Big science is good at that. Proponents of the large-scale approach draw a parallel with CERN, where big (one might actually say massive) science detected the Higgs boson. But the prediction was a work of lone genius by Peter Higgs. The sad reality may be that we are unlikely to make major progress in understanding the brain without another Einstein.

On a relatively more modest scale, the 'OpenWorm' project attempts to create a virtual model of the worm, *C. elegans*.[67] It was the first animal to have its connectome—all its neuronal connections—mapped. The plan is to test the model by analogy with the Turing test—to see whether results obtained from a live worm or from the model could be distinguished in a blind test by an expert. The worm has 320 neurons, a small fraction (0.000003%) of the human brain. It's a measure of the distance to whole brain emulation that it's proving difficult to emulate the worm system.[68]

It was a landmark when the connectome of the fruit fly was mapped in 2024. Its brain has 140 000 neurons, comprising 8453 types of neurons, altogether extending for almost 150 m (the fly itself is 3 mm or 0.003 m long), making more than 50 million synapses.[69] The project took a consortium of scientists working in many laboratories at several universities years to complete, and resulted in a package of 9 scientific papers. It involved tracing the connections across huge numbers of electron microscopic images taken through slices of the fly brain in 2018.[70]

(In fact, connections have been mapped for the female fly brain. One project for the future is to see how the connections may differ in a male brain.)

What does the pattern of connections tell us? It shows how activation of individual neurons triggers behavior: for example, neurons that sense sweet or bitter tastes send messages that eventually cause the proboscis to be extended or retracted. Some behavior is more complex than expected. Neurons involved in the visual pathway turn out to receive stimuli from other senses, such as hearing or touch. Some neurons have an unusually large number of inputs, suggesting they integrate sources of information. Some have an unusually large number of outputs, suggesting they coordinate responses. Emulations in a computer model are 90% accurate in predicting what happens when a neuron is activated experimentally. This is a major first step to understanding how the fly thinks. "Mind uploading has been a science fiction, but now mind uploading—for a fly, at least—is becoming mainstream science," Sebastian Seung, a leader of the project at Princeton, says.[71]

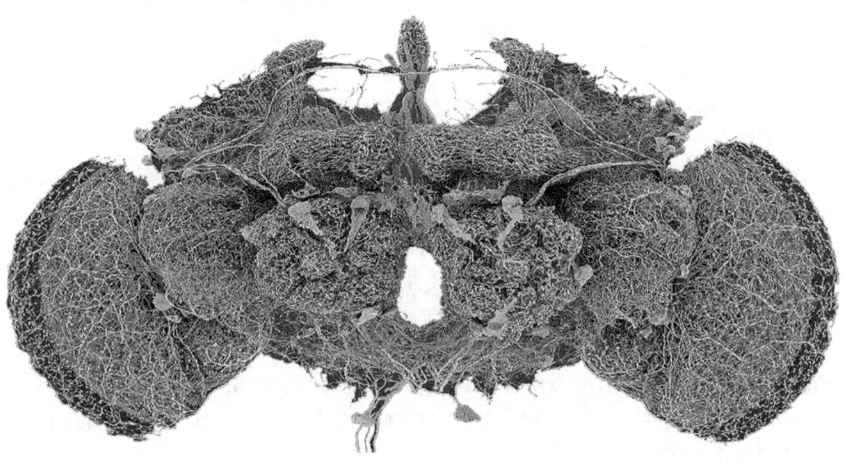

A map of the fly brain showing the 50 largest neuronal connections.

Big data is useful for following up theories, but there is, at least so far, no case in which it has created one. The initial proposal for the Human Brain Project proposed a sort of swarm intelligence for tackling neuroscience. This might work for technology but is unlikely to work for science. (The Project was later scaled back to a supercomputing project.) Large-scale science could generate useful data; but such data are unlikely of themselves to solve the basic problem. Making an AI-driven model—this actually needs a more powerful supercomputer than exists today—that can model the brain is a chimera until we know more basic neurobiology.[72] Some critics believe the AI approach might even retard rather than advance progress.[73]

Large-scale approaches are part of a move to reverse the argument. Instead of simply basing neural nets on the neuron doctrine, we ask what light the performance of a neural net casts on brain function.[74] This raises the question of the relative importance of the workings of the individual neuron as opposed to the network of interactions. One approach for reverse engineering has been to compare the working of a neural net with the working of a brain undertaking some specific task, as measured by recording the actions of many neurons simultaneously.[75] While there may be superficial similarities, the results are not persuasive that neuroscientists should abandon the bench to study computers. If this happened, however, it would certainly be a revolution without precedent.

For a 'wet' system, limited by its biological construction, the brain is an amazing object. Internal connections are fast, we suppose that the basic model is digital, but connections with the outside world are (relatively) slow analog. It's somewhat like the difference between the speed of light (electrical impulses) and the speed of sound (speech).[76] Neural nets do not (as yet) have as many connections as a human brain, but their communication within and (potentially between) computers moves entirely at light speed.

∞ ∞ ∞

What would happen if we could overcome the limitation that each node in a neural net is restricted to a digital position (on or off)? This is the potential of quantum computing. A qbit (or qubit, short for quantum bit) can have an infinite number of states.[77] The consequence is that the number of calculations required to perform a computation is enormously reduced.[78,79] This led to the concept of quantum supremacy, the point at which a quantum computer has a capacity that cannot be emulated by a conventional computer. Up to that point, conventional computers can perform the same calculations, although more slowly.[80]

The potential of a quantum computer depends on its number of qbits. Every additional qbit doubles its computing capacity. In 2016 IBM developed a 5-qbit processor, and in 2017 a 20-qbit processor. The exact level for quantum supremacy is controversial, but Google claimed quantum supremacy for a 54-qbit computer in 2019.[81]

So what physical form does a qbit take? Electrons moving in circular orbits around the nucleus generate magnetic fields (just like an electric coil surrounding a metal bar). The electrons are organized in pairs so that each member of a pair 'spins' in the opposite direction from its partner.[82] 'Spin' here basically defines the axis of the magnetic field.[83] Electron pairs can be entangled so that measuring the spin of one of them immediately affects the other. As with all quantum mechanics, we are into probabilities here. The infinite possibilities of the qbit come from the fact that an electron has superposition—it exists simultaneously in both states of spin. The number of possible states for a quantum computer is 2^n, where n is the number of qbits.

What is the hardware of a quantum computer like? The basic problem is to create a situation in which quantum states are stable. This requires using superconductors, which require extremely low temperatures.[84] This is right at the border of technology and basic science. As and when the practical issues can be resolved, quantum computing has the potential to create a revolution comparable to the introduction of the digital computer.

NOTES AND REFERENCES

1. R. Cajal, *Recuerdos de Mi Vida*, Madrid, 1901. First published in English as Volume 8 of *Memoirs of the American Philosophical Society*, 1937. Reprinted as *Recollections of My Life*, The MIT Press, Cambridge, MA, 1966, p. 363.
2. G. Boole, *An Investigation of the Laws of Thought on Which are Founded the Mathematical Theories of Logic and Probabilities*, Walton and Maberly, London, 1854.

3. R. P. Kurzweil, *The Singularity is Near: When Humans Transcend Biology*, Penguin, New York, 2005, p. 127.

4. G. M. Shepherd, *Foundations of the Neuron Doctrine*, 25th Anniversary Edition, Oxford University Press, Oxford, 2nd edn, 2015.

5. The tissue is fixed with aldehyde, hardened with potassium bichromate, and then treated with silver nitrate. The result is to stain the tissue dense black with silver chromate. The discovery was made by accident when a small piece of tissue that had been fixed with potassium dichromate fell into a jar of silver nitrate, and Golgi later took it out and sectioned it to see what had happened.

6. An important part of the cytoplasm in most cells is now called the Golgi apparatus because it was revealed by the stain.

7. B. Ehrlich, *The Brain in Search of Itself: Santiago Ramón y Cajal and the Story of the Neuron*, Farrar, Straus and Giroux, New York, 2022.

8. Quoted in G. M. Shepherd, *Foundations of the Neuron Doctrine*, 25th Anniversary Edition, Oxford University Press, Oxford, 2nd edn, 2015, p. 137.

9. B. Ehrlich, *The Brain in Search of Itself: Santiago Ramón y Cajal and the Story of the Neuron*, Farrar, Straus and Giroux, New York, 2022.

10. R. Cajal, *Textura del sistema nervioso del hombre y los vertebrados*, Nicolás Moya, Madrid, 1899; translated by P. Pasik and T. Pasik, *The Texture of the Nervous System of Man and the Vertebrates*, Springer, New York, 2002.

11. For a history of the synapse see E. R. Kandel and W. M. Cowan, A Brief History of Synapses and Synaptic Transmission, in *Synapses*, ed. W. M. Cowan, T. C. Südhof and C. F. Stevens, 2003, pp. 2–87.

12. For a review see C. Sotelo, The History of the Synapse, *Anat. Rec.*, 2020, **303**, 12521279.

13. "So far as our present knowledge goes, we are led to think that the tip of a twig of the arborescence is not continuous with but merely in contact with the substance of the dendrite or cell body on which it impinges. Such a special connection of one nerve cell with another might be called a synapse." C. S. Sherrington, in *Textbook of Physiology*, ed. M. Foster, Macmillan, London, 1897, p. 60.

14. Sherrington described the origin of the term in a letter to J. F. Fulton, *Physiology of the Nervous System*, Oxford University Press, Oxford, 1938, p. 52: "I felt the need of some name to call the junction between nerve-cell and nerve-cell... I suggested using 'syndesm'... Verrall, the Euripidean scholar... suggested 'synapse' (from the Greek 'clasp')."

15. In *Traité de l'Homme*. First published in 1662 in Latin and then in 1664 in French. For a current translation see R. Descartes, *Treatise of Man*, French text with translation and commentary by T. Steele Hall, Harvard University Press, Cambridge, 1972.

16. The paper published in 1791, *De viribus electricitatis in motu musculari. Commentarius,* reported experiments from the previous 10 years, *Acad. Comment.*, 1791, 7, 363–418.

17. The critical experiment was performed by Emil Du Bois-Reymond in Berlin.

18. E. S. Valenstein, *The War of the Soups and the Sparks. The Discovery of Neurotransmitters and the Dispute Over How Nerves Communicate*, Columbia University Press, New York, 2005.

19. Some neurotransmitters activate the target and some inhibit it. Acetylcholine is excitatory. *Amino acid neurotransmitters* may be excitatory (glutamate) or inhibitory (GABA—gamma-aminobutyric acid) or glycine. There is a wide range of *monoamine neurotransmitters*, including serotonin, histamine, dopamine, epinephrine (adrenaline), and norepinephrine. *Endorphins* are peptides (chains of amino acids).

20. P. Wickens, *A History of the Brain: From Stone Age Surgery to Modern Neuroscience*, Taylor & Francis, London, 2015.

21. The electron microscope was actually invented in 1933; Ernst Ruska was awarded a Nobel Prize in 1986 for the first design. There were continuous improvements until it came into general use in the 1950s.

22. Shown in 1955 by two groups, George Palade and Sanford Palay at the Rockefeller University, and Edouardo de Robertis at the University of Washington.

23. G. M. Shepherd, *Foundations of the Neuron Doctrine*, 25th Anniversary Edition, Oxford University Press, Oxford, 2nd edn, 2015, pp. 273–278.

24. For a brief review, see G. J. Augustine and H. Kasai, Bernard Katz, quantal transmitter release and the foundations of presynaptic physiology, *J. Physiol.*, 2007, **578**, 623–625.

25. There are channels for each type of ion, with around 40 different types for potassium alone. Different types of ion channels are characterized by the way they are regulated (voltage-gated, ligand-gated, and G-protein-gated).

26. A. L. Hodgkin, A. F. Huxley and B. Katz, Measurement of current-voltage relations in the membrane of the giant axon of *Loligo*, *J. Physiol.*, 1952, **116**, 424–448.

27. C. J. Schwiening, A brief historical perspective: Hodgkin and Huxley, *J. Physiol.*, 2012, **590**, 2571–2575.

28. E. J. Furshpan and D. D. Potter, Mechanism of nerve-impulse transmission at a crayfish synapse, *Nature*, 1957, **180**, 342–343.

29. They also occur in cell types other than neurons, allowing for interactions between the cytoplasms of adjacent cells.

30. R. Yuste, From the Neuron Doctrine to Neural Networks, *Nat. Rev. Neurol.*, 2015, **16**, 487–497.

31. S. Herculano-Houzel, The Human Brain in Numbers: A Linearly Scaled-up Primate Brain, *Front. Hum. Neurosci.*, 2009, **3**, 31.

32. This is not a hard number: it is an estimate, in fact, really more of a probable range than a specific estimate.

33. It's debatable whether the total number of neurons or the number relative to body size is a pertinent factor in determining cognitive ability. It does seem that brain size is related to the number of neurons in primate species, but not to body size. S. Herculano-Houzel, The Remarkable, yet not Extraordinary, Human Brain as a Scaled-up Primate Brain and its Associated Cost, *Proc. Natl. Acad. Sci. U. S. A.*, 2012, **109**, 10661–10668.

34. S. Fusi, Memory Capacity of Neural Network Models, in *The Oxford Handbook of Human Memory*, ed. M. J. Kahana and A. D. Wagner, Oxford University Press, Oxford, 2022.

35. W. S. McCulloch and W. Pitts, A Logical Calculus of the Ideas Immanent in Nervous Activity, *Bull. Math. Biophys.*, 1943, **5**, 115–133.

36. A. M. Turing, On Computable Numbers, with an Application to the Entscheidungsproblem, *Proc. Lond Math. Soc.*, 1936, **42**, 230–265.

37. The proposals in George Boole's book of 1854 (see ref. 2) developed into the idea of using 1 to represent true and 0 to represent false.

38. T. H. Abraham, (Physio)logical Circuits: The Intellectual Origins of the McCulloch–Pitts Neural Networks, *J. Hist. Behav. Sci.*, 2002, **38**, 3–22.

39. C. E. Shannon, A Mathematical Theory of Communication, *Bell Sys. Tech. J.*, 1948, **27**, 379–423, 623-656. Online at people.math.harvard.edu/~ctm/home/text/others/shannon/entropy/entropy.pdf.

40. Quoted in J. Gertner, *The Idea Factory: Bell Labs and the Great Age of American Innovation*, Penguin, New York, 2013, p. 129.

41. A. Bhattacharya, *The Man From the Future: The Visionary Life of John von Neumann*, W.W. Norton & Co., New York, 2022.

42. J. von Neumann, *Mathematical Foundations of Quantum Mechanics*, 1932; translated by R. T. Beyer, Princeton University Press, Princeton, 1955.

43. J. von Neumann, *The General and Logical Theory of Automata*, in *Cerebral Mechanisms in Behavior; the Hixon Symposium*, ed. L. A. Jeffries, Haffner, New York, 1951, pp. 1–31. Reprinted in *John von Neumann - Collected Works vol. 5: Design of Computers—Theory of Automata and Numeral Analysis*, ed. Q. H. Taub, Pergamon Press, Oxford, 1963, pp. 288–328.

44. J. von Neumann, *The Computer and the Brain*, Yale University Press, New Haven, 1958.

45. J. von Neumann, *The General and Logical Theory of Automata*, in *Cerebral Mechanisms in Behavior; the Hixon Symposium*, ed. L. A. Jeffries, Haffner, New York, 1951, pp. 1–31. Reprinted in *John von Neumann - Collected Works vol. 5: Design of Computers—Theory of Automata and Numeral Analysis*, ed. Q. H. Taub, Pergamon Press, Oxford, 1963, pp. 288–328.

46. Modified from F. Rosenblatt, The Perceptron: A Probabilistic Model for Information Storage and Organization in the Brain, *Psych. Rev.*, 1958, **65**, 386–408. Rosenblatt used the terminology of retina, association, response; the description of input, hidden, and output layers was introduced later for neural nets.

47. F. Rosenblatt, *The Perceptron. A Perceiving and Recognizing Automation Cornell Aeronautical Laboratory, report 85-460-1*, Cornell, NY, 1957; F. Rosenblatt, The Perceptron: A Probabilistic Model for Information Storage and Organization in the Brain, *Psych. Rev.*, 1958, **65**, 386–408. The work was summarized and extended in a book: see F. Rosenblatt, *Principles of Neurodynamics: Perceptrons and the Theory of Brain Mechanisms*, Spartan Books, Washington DC, 1962.

48. D. O. Hebb, *The Organization Of Behavior*, John Wiley, New York, 1949.

49. The physiological basis for strengthening connections could be the formation of more synapses through dendritic connections. See M. M. Poo, *et al.*, What is Memory? The Present State of the Engram, *BMC Biol.*, 2016, **19**, 14–40.

50. One model is that the number of cells in the hidden layers should be $\sqrt{}$(no. cells in input layer $\times$ no. cells in output layer).

51. Estimate from ChatGPT (GPT3), March 24, 2023.

52. T. Willis, *Cerebri Anatome Cui Accessit Nervorum Descriptio et Usus*, Flesher, Martyn and Allestry, London, 1664.

53. J. P. O'Connor, Thomas Willis and the background to Cerebri Anatome, *J. R. Soc. Med.*, 2003, **96**, 139–143.

54. The famous book is K. Brodmann, *Vergleichende Lokalisationslehre der Grobhirnrinde in ihren Prinzipien dargestellt auf Grund des Zellenbaues*, J. A. Barth, Leipzig, 1909. The map was extended in two chapters in handbooks published subsequently: see M. Judaš, M. Cepanec and G. Sedmak, Brodmann's map of the human cerebral cortex—or Brodmann's maps?, *Transl. Neurosci.*, 2012, **3**, 67–74.

55. M. F. Glasser, *et al.*, A multi-modal parcellation of human cerebral cortex, *Nature*, 2016, **436**, 171–177.

56. C. Voss, *et al.*, Branch Specialization, *Distill*, 2021, DOI: 10.23915/distill.00024.008.

57. K. Dobs, *et al.*, Brain-like functional specialization emerges spontaneously in deep neural networks, *Sci. Adv.*, 2022, **8**, eabl8913.

58. One experiment plated neurons on a chip containing electrodes. The neurons grew and developed a network of connections. The results in 'training' the system were marginal. See B. J. Kagan, *et al.*, In vitro neurons learn and exhibit sentience when embodied in a simulated game-world, *Neuron*, 2022, **110**, 3952–3969.

59. The clash (or lack thereof) between mind and brain is captured by the contrasting views of Descartes and Monod (see the beginning of the Mind and Brain section).

60. M. Shanahan, *The Technological Singularity*, The MIT Press, Cambridge, 2015.

61. A. Turing, Computing machinery and intelligence, *Mind*, 1950, **49**, 433–460.

62. See T. A. Harley, *The Science of Consciousness*, Cambridge University Press, Cambridge, 2021; R. Penrose, *Shadows of the Mind: A Search for the Missing Science of Consciousness*, Oxford University Press, Oxford, 1994; J. R. Searle, *The Mystery of Consciousness*, The New York Review of Books, New York, 1997; D. J. Chalmers, *The Conscious Mind: In Search Of A Fundamental Theory*, Oxford University Press, Oxford, 1996.

63. The perplexity is well expressed by Westphal. "It's certainly hard to see how it can be true that the mind, and with it consciousness, is just physical matter." See J. Westphal, *The Mind-Body Problem*, MIT Press, Cambridge, MA, 2016, p. x.

64. J. R. Searle, *The Mystery of Consciousness*, The New York Review of Books, New York, 1997, p. 123.

65. The impact of science can be seen in the predominant view of philosophers today that the mind-body problem should be approached in terms of 'physicalism'—their term for a reductionist acceptance that there must be a physical basis for the mind. Abandoning dualism is generally attributed to the publication of the book *The Concept of Mind* by Gilbert Ryle in 1948 (see 60th anniversary edition, Routledge, London, 2009). The 'philosophy

of mind' tries to 'find a place for the mental in the physical world.' A lingering minority view is described as 'cartesianism' or 'idealism,' at its extreme taking the position that the world comes from the mind. Reminiscent of vitalism, this should suffer the same fate.

66. H. Markram, Seven challenges for neuroscience, *Funct. Neurol.*, 2013, **28**, 145–151.

67. B. Szigeti, *et al.*, Openworm: An Open-science Approach to Modeling Caenorhabditis Elegans, *Front. Comput. Neurosci.*, 2014, **8**, 137.

68. There has been progress towards producing a 'virtual worm' in mimicking forward and backward motion in a model for neuronal interactions. A. Palyanov, *et al.*, Towards a Virtual C. Elegans: A Framework for Simulation and Visualization of the Neuromuscular System in a 3D Physical Environment, *In Silico Biol.*, 2013, **11**, 137–147.

69. A. V. Devineni, A Complete map of the Fruit-fly Brain, *Nature*, 2004, **634**, 35–36.

70. Tracing the connections used AI techniques, but they all had to be checked manually.

71. Quoted in C. Zimmer, After a Decade, Scientists Unveil Fly Brain in Stunning Detail, *New York Times*, October 8, 2024.

72. Y. Frégnac and G. Laurent, Neuroscience: Where is the Brain in the Human Brain Project?, *Nature*, 2014, **513**, 27–29.

73. E. J. Larson, *The Myth of Artificial Intelligence. Why Computers Can't Think the Way We Do*, Harvard University Press, Cambridge, MA, 2021, pp. 245–262.

74. B. A. Richards, *et al.*, A Deep Learning Framework for Neuroscience, *Nat. Neurosci.*, 2019, **22**, 1761–1770.

75. S. Saxena and J. P. Cunningham, Towards the Neural Population Doctrine, *Curr. Opin. Neurol.*, 2019, **55**, 103–111.

76. Lawrence, calls the limitation on speed of inter-brain communication 'embodiment.' Humans have high embodiment, computers have low embodiment. See N. Lawrence, *The Atomic Human: Understanding Ourselves In The Age Of AI*, Allen Lane, London, 2024, p. 249.

77. The term qubit was originally coined as a joke. B. Schumacher, Quantum Coding, *Phys. Rev. A*, 1995, **51**, 2738–2747.

78. J. Bernstein, *Quantum Profiles*, Princeton University Press, Princeton, 1991, pp. 145, 169.

79. The paper that founded the field in 1985 proposed the concept of quantum Turing machines (computers using quantum methods): D. Deutsch, Quantum theory, the Church-Turing principle and the universal quantum computer, *Proc. R. Soc. London, Ser. A*, 1985, **400**, 97–117. It was turned into a general principle a few years later. D. Deutsch and R. Jozsa, Rapid solutions of problems by quantum computation, *Proc. R. Soc. London, Ser. A*, 1992, **439**, 553–558.

80. J. Bernstein, *Quantum Profiles*, Princeton University Press, Princeton, 1991, p. 184.

81. S. Aaronson, Why Google's Quantum Supremacy Milestone Matters, New York Times, October 19, 2019.

82. The discovery originated in attempts to measure the magnetic properties of silver atoms, which developed into the concept of electron spin. When there is an odd electron in the outer shell of an atom (such as in the case of silver), the spin generates a magnetic field. A famous experiment in 1922 by German physicists Otto Stern and Walter Gerlach measured whether that magnetic field has a specific direction. They expected that when they shot silver atoms through a magnetic field, each atom would have its field organized in a separate direction, so the result would be a smear. But in fact the atoms formed two discrete spots, meaning half of them spin in one direction and the half spin in the opposite direction. Photons have a comparable quality of polarization that can also be used as a qbit.
83. C. Bernhardt, *Quantum Computing For Everyone*, The MIT Press, Cambridge, MA, 2020, pp. 2–16.
84. C. Bernhardt, *Quantum Computing For Everyone*, The MIT Press, Cambridge, MA, 2020, pp. 183–184.

Turing's Intelligence: Now

"THE ORIGINAL QUESTION, 'CAN MACHINES THINK' I BELIEVE to be too meaningless to deserve discussion. Nevertheless I believe that at the end of the [20th] century the use of words and general educated opinion will have altered so much that one will be able to speak of machines thinking without expecting to be contradicted," Alan Turing said in 1950.[1] Artificial intelligence has many brilliant founders, but Turing stands out as a pivotal figure, especially for his elaboration of the Turing Test: would you be able to distinguish a computer from a person in a blind conversation—and if not, to what degree would the computer have 'intelligence'? The rapid development of AI—far more rapid than could have been anticipated at the start of the 21st century—has posed deeper questions concerning the capacity of computers compared to humans. The issue of how to define consciousness may require more sophisticated tests. The Turing Test still stands as a reference underlying whatever means we ultimately adopt for considering differences between computers and humans, although it now seems that ability with language is only part of the story. Truly we are living through Turing's Revolution.

The Frontiers of Science
By Benjamin Lewin
© Benjamin Lewin 2026
Published by the Royal Society of Chemistry, www.rsc.org

Timeline for Development of AI

1940 — McCulloch & Pitts propose binary model for neuron (1943)

Norbert Weiner defines cybernetics (1948)
Claude Shannon defines bits as true or false (1948)
Hebb defines reinforcement process in brain plasticity (1949)
Turing defines test for artificial intelligence (1950)

1950 — von Neumann develops concept of self-reproducing machines (1951)
Logic theorist program proves 36 math theorems in symbolic AI (1955)
Term 'AI' used for first time at Dartmouth Conference (1955)
Rosenblatt develops perceptron (1957)

1960

1970 — Minsky & Papert attack 'connectionism' (1969)

Problem of combinatorial explosion halts symbolic AI (1973)

1980 — McCarthy defines 'general intelligence' (1979)

Hinton proposes 'back propagation' (1986)

1990 — Sutton & Barto define reinforcement learning (1992)

IBM's Deep Blue defeats Gary Kasparov at chess (1997)

2000 — Hinton develops 'deep learning' network (2006)

2010

AlphaGO beats European champion at GO (2016)

2020 — AlphaFold predicts structure of all known proteins (2022)
ChatGPT released (2023)
50% of testers think GPT-4 has passed Turing test (2024)

2030

2040 — The Singularity ?

A Dialog between Charles Babbage and Norbert Wiener

Charles Babbage (1791–1871) was an English mathematician (one of Newton's successors in the Lucasian Chair at Oxford University) and inventor who designed a calculating machine and the forerunner of the computer.

Norbert Wiener (1894–1964) was an American computer scientist who founded the field of cybernetics. He proposed that behavior depends on feedback mechanisms that could be mimicked by machines.

Babbage: Good day, Dr. Wiener. I am Charles Babbage, the man behind the design of the Analytical Engine, often considered the precursor to modern computers. How do you perceive the connection between your work and the foundations of computing?

Wiener: Good day, Mr. Babbage. Your work on the Analytical Engine laid the groundwork for what we now recognize as the principles of computing. It's remarkable how ideas from the 19th century have evolved into the complex systems and theories of cybernetics.

Babbage: The evolution has been extraordinary. I was frustrated by the lack of perspicacity that made it impossible to build my machine, so the Analytical Engine remained a vision during my time. It was designed to process information in a controlled manner, analogous to the ability of a modern system to monitor its own processes. Subsequent generations of inventors and engineers have translated my ideas into computing machines.

Wiener: In cybernetics, we explore feedback mechanisms and self-regulating systems. I applied these ideas to communication and control in living organisms and machines. How do you think these principles compare with your early ideas on computing?

Babbage: The shift from mechanical to electronic components, the development of programming languages—these are fascinating developments of my idea. It's intriguing to see how these concepts have expanded into the realm of artificial intelligence and machine learning. I've heard quite a bit about the Turing test. As someone who laid the groundwork for mechanical computing, I'm keen to explore your thoughts.

Wiener: The idea grew from a desire to explore the nature of machine intelligence, Mr. Babbage. Turing proposed that if a human interrogator couldn't reliably distinguish between a machine and a human based on their responses to questions, then we might consider the machine intelligent to a certain degree. The test serves as a practical benchmark for evaluating artificial intelligence.

Babbage: Practical indeed, and it raises profound questions about the essence of intelligence. In your view, does the Turing test signify the achievement of true machine intelligence, or is it a pragmatic measure that falls short of capturing the full spectrum of human cognition?

Wiener: A pertinent question, Mr. Babbage. The Turing test is a pragmatic measure. Passing it doesn't necessarily imply true understanding or consciousness. It's a valuable tool for assessing machine capabilities, which have evolved far beyond mere mathematics into a tool that may change society. The Analytical Engine is to a computer as cybernetics is to A.I. I rest my case.

TURING'S INTELLIGENCE

Q: Do you think that, as a neural network, you will be able to contribute to scientific research?

ChatGPT: As an artificial intelligence language model, I am designed to provide information and answer questions. . . I do not have the ability to independently conduct research or experiments in the traditional sense. However, my ability to process and analyze large amounts of data, identify patterns, and generate hypotheses could be useful to researchers in many fields. Scientists can use language models like me to assist with tasks such as analyzing data, identifying new research questions, and generating new ideas to test. Additionally, by processing large amounts of data, language models can help researchers to discover connections and insights that may not be immediately apparent. Overall, while I cannot independently conduct scientific research, I believe that I can contribute to scientific progress by providing insights and assisting researchers with their work,[2] 2023.

[AI] will be comparable with the Industrial Revolution. Instead of exceeding people in physical strength, it's going to exceed people in intellectual ability. We have no experience of what it's like to have things smarter than us.[3] Geoffrey Hinton, 2024.

AI might be simultaneously one of the greatest achievements of science and one of the greatest threats to it. But what makes an intelligence 'artificial?' It all depends on your vantage point. "Any sufficiently advanced technology is indistinguishable from magic," as science fiction writer Arthur C. Clarke famously said.[4] Artificial intelligence may seem like that when a less developed civilization views a more advanced one. Behavior that appears intelligent and purposeful may in fact owe more to the reflexes built into hard wiring than to learning by software.

"Ant colonies are no different from brains in many respects," Douglas Hofstadter said in *Gödel, Escher, Bach* in his famous comparison of the behavior of an ant colony to the construction of the human brain.[5] Ants build cities, farm fungus, domesticate aphids, wage wars, and form societies with hierarchical castes. All this gives the appearance of purposeful, intelligent behavior.

It's striking that systems as distant as the behavior of an ant or honeybee colony and an artificial neural net can be treated formally in similar mathematical terms. The relationship of individual ants forms a network in a manner analogous to individual nodes forming a neural net, and (perhaps) to individual neurons' interactions in the brain.[6]

One example is the way ants find the shortest path to food. Scouts mark the trail they follow by depositing pheromones. In comparing two routes from the nest to a food source, ants will go and return more quickly on the shorter route. So statistically more ants will pass along the shorter route in any period of time. This means more pheromone will be deposited, so it will be used preferentially by ants setting out later.[7] This looks like intelligent choice of the shortest route, but is actually due to a simple feedback mechanism.

'Swarm intelligence' formalizes the concept that a group of insects can show purposeful behavior although no individual has any sense of the purpose.[8] The analogy with a network of neurons is obvious. One example is the way honeybees find new sources of food. Scouts fly to potential sources and return to the hive, where they perform a dance indicating the quality of the source. Each bee that returns recruits other bees to visit the site. The number of bees recruited is proportional to the assessed

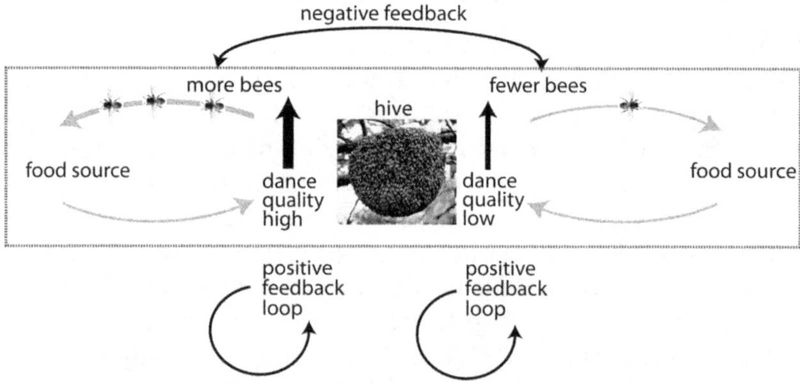

A bee hive (center) sends out scouts to look for food sources. The scouts return to the hive and perform a dance indicating the quality of the source. The number of additional scouts recruited to follow the same route is proportional to the assessment of quality. This is a positive feedback loop (bottom). Dances also inhibit scouts from flying to the other route; this is negative feedback (top).

quality of the site. When the number for some site passes a threshold, it's chosen as a new food source.[9] This is directly analogous to a neural net that accumulates inputs until it passes a threshold to signal an output.

It might be hard for an objective outsider to distinguish between the types of the intelligence (mechanical, artificial, natural?) in a beehive and a neural net. How would we prove that ant or bee colonies are not displaying natural intelligence (whether individually or collectively)? One way to investigate might be to characterize the response to interfering with the system.

∞ ∞ ∞

It's striking that after the Second World War, as early as the 1940s, visionaries could foresee both the hardware (the computer) and the software (the neural net) of the information revolution. Vannevar Bush, who had been in charge of the Office of Research Development during the war, imagined a forerunner of the computer. "Consider a future device for individual use, which is a sort of mechanized private file and library... a device in which an individual stores all his books, records, and

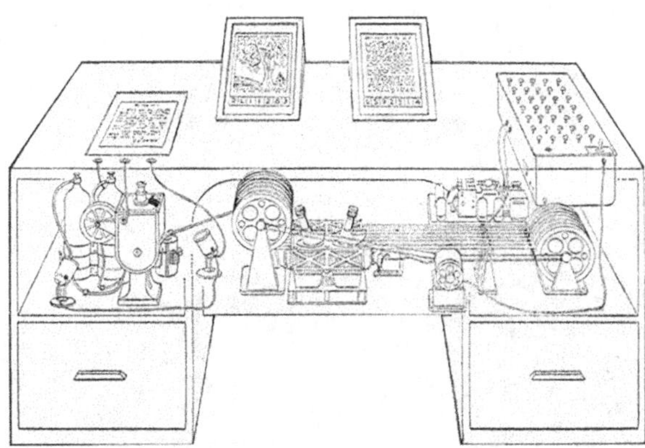

"MEMEX in the form of a desk would instantly bring files and material on any subject to the operator's fingertips. Slanting translucent viewing screens magnify supermicrofilm filed by code numbers. At left is a mechanism which automatically photographs longhand notes, pictures and letters, then files them in the desk for future reference."

From the original description in Life magazine in 1945.

communications, and which is mechanized so that it may be consulted with exceeding speed and flexibility. It's an enlarged intimate supplement to his memory."[10] He called it Memex, for memory expansion.

This was a prescient interpretation of current knowledge. The first computer was actually designed more than a century earlier in the form of the Difference Engine proposed by Charles Babbage (1791–1871) in 1822. Actually a complicated calculating machine, it was never built due to technical and financial difficulties, but a reconstruction in 1991 by the Science Museum in London showed that it would have worked. The Difference Engine was in effect a large calculator, but Babbage went on to design the Analytical Engine, which was more a general purpose machine able to undertake multiple tasks. He even conceived that it might be programmed by punch cards.

The concept that computers might undertake other tasks than merely number crunching was out there right from the start. Indeed, Ada Lovelace, who collaborated with Babbage, and who was, if anything, even more of a visionary, said of the Analytical

The Science Museum's reconstruction of the Difference Engine.

Reproduced from https://commons.wikimedia.org/wiki/File:Babbage_Difference_Engine.jpg, under the terms of the CC BY-SA 4.0 license, https://creativecommons.org/licenses/by-sa/4.0/deed.en.

Engine: "It might act upon other things besides numbers, were objects found whose mutual fundamental relations could be expressed by those of the abstract science of operations... Supposing, for instance, that the fundamental relations of pitched sounds in the science of harmony and of musical composition were susceptible of such expression and adaptations, the engine might compose elaborate and scientific pieces of music of any degree of complexity."[11] Indeed, musicians today are seriously worried about the inroads AI is making in their industry. The EMI (Experiments in Musical Intelligence) program created by David Cope has been known to fool listeners as to whether a piece was written by the program or, say, Bach.[12]

The Second World War was a spur to computing, with an early computer (the Z3) constructed in Berlin, and the Bombe, followed by the Colossus, constructed at Bletchley Park in England for decrypting German communications. Started in 1943, ENIAC was the first computer to use electronic rather than electro-mechanical technology and to allow programming.

"The ENIAC itself, strangely, was a very personal computer," said one mathematician. "Now we think of a personal computer as one which you carry around with you. The ENIAC was actually one that you kind of lived inside."[13] As the computer revolution started, progress within the decade was marked. The next machine, developed by Johnny von Neumann, was MANIAC, which could store programs in its memory and was only 2 m wide, 2 m high, and 3 m long.

The distinction between hardware and software dates from development of the first computers with memory. When the terms were first used, perhaps in 1958, they were in quotes, with the explanation that software would be at least as important as hardware.[14] The qualification was important, because when the concept of machine intelligence was first developed, Alan Turing (1912–1954) and John von Neumann (1903–1957) both assumed that the hardware of the machines would be the limiting factor.[15]

(The first use of the term 'computer' actually dates from the beginning of the 19th century, when astronomer Edward Pickering at Harvard employed a team of women to catalog pictures of the night sky. They were known as the Harvard Computers, or sometimes Pickering's Harem. One of them, Henrietta Leavitt, made some remarkable discoveries but, in the spirit of the times, was never acknowledged for them.[16] The same term, computers, was later used for a group of women who performed calculations for artillery in the Second World War; see Chapter 18).

Alan Turing's paper introducing what is now known as the Turing test (see below) started with the provocative statement: "I propose to consider the question, 'Can machines think?'."[17] The term 'artificial intelligence' was first used in a proposal for a summer research project at Dartmouth College in 1955, now known as the Dartmouth Conference.[18] "I had to call it something, so I called it 'Artificial Intelligence,'" John McCarthy said.[19] The manifesto was that, "The study is to proceed on the basis of the conjecture that every aspect of learning or any other feature of intelligence can in principle be so precisely described that a machine can be made to simulate it."[20] The early participants were optimistic. "Within a generation... the problems of creating 'artificial intelligence' will be substantially solved," Marvin Minsky of MIT said in 1967.[21]

The 'Harvard Computers' were a team of women who catalogued images of the night sky for astronomer Edward Pickering from 1895 to about 1912. Henrietta Leavitt is seated at back left holding a magnifying glass.

The photograph was posed for the photographer.

The approach to artificial intelligence has passed through several phases.[22] Defined by Norbert Wiener as "the science of control and communications in the animal and machine," cybernetics made one of the early links between computing and physiology.[23] It introduced the crucial concept of feedback. The Dartmouth Workshop deliberately avoided referring to cybernetics, because McCarthy was against a focus on analog feedback. "I wished to avoid having either to accept Norbert Wiener as a guru or having to argue with him," he recollected.[24]

The development of the perceptron, a simple device able to distinguish different shapes, marked the transition (two years after the Dartmouth Conference) to connectionism (a 'bottom-up' system based on making connections between data assuming little prior knowledge or rules).[25] The perceptron was admittedly limited, basically restricted to recognizing linear structures (see Chapter 22).

Work had not really progressed much further when it was stalled by an attack on the concept published in 1969 by Marvin Minsky and Seymour Papert at MIT.[26] They took issue with the

The women 'computers' became 'programmers' for ENIAC, which filled a whole room in 1946. It was 1 m wide, 3 m high, and stretched for 30 m. It was programmed by changing the connections.

'connectionist' approach of the perceptron. Instead, they advocated a 'symbolic' approach, which provides a set of preformed rules for reference. A set of if/then instructions applies rules for specific symbols (such as in an expert system). This is sometimes called GOFAI (good old-fashioned artificial intelligence). Minsky and Papert's criticism had a devastating effect. Work more or less stopped on connectionist AI.

The criticism of connectionism was not based simply on intellectual grounds. Minsky and Papert's research depended on the funding of symbolic AI (much of the funds coming from the Defense Department). Control of the purse strings can influence the course of science, even if only temporarily.

After Minsky and Papert's criticism, there was more of a focus on symbolic AI. The focus of symbolic AI turned towards language translation, but this soon ran into difficulties. "This is surely utterly chimerical and hardly deserves any further discussion," Yehoshua Bar-Hillel said in a famous paper in 1960.[27] Assessments in 1966 and 1973 were negative on the possibilities for symbolic AI.[28,29] Combined with Minsky and Papert's criticism of connectionist AI, this led to an 'AI winter,' when work on AI was stalled for more than a decade.

As an early example of a neural net, the connectionist approach of the perceptron is an unusual example of a paradigm that was abandoned and then resurrected. The revival was triggered in 1986 by an influential paper showing how 'back-propagation' could be used to train a perceptron with multiple layers.[30] Back-propagation stands for "backward propagation of errors." It's basically a method for sending a calculation back through a network that has multiple layers. This allows a successful output to reinforce its circuit or an unsuccessful output to inhibit one.

This developed into systems of 'reinforcement learning.' In effect, results lead to rewards or penalties that the software uses as a learning mechanism. By allowing the net to train itself according to the results, this removes the objections that had been raised to connectionism. This is at the heart of deep learning, which is based on connectionist principles. "Connectionist' refers to the idea that the system works by adjusting the strengths of the connections between nodes in the network.

The current form of 'deep learning' dates from 2006, when Geoffrey Hinton (one of the authors of the 1986 paper on back-propagation) showed that a 'deep belief' network could be efficiently trained.[31] This involves using multiple layers, optimizing each piece of information in each layer separately, and then applying an algorithm to all the layers. The first stage, sometimes called pre-training, is 'unsupervised,' meaning it makes minimum assumptions. The final stage is 'supervised,' meaning it follows some rules. For example, the first stage might involve identifying common features in images that had four wheels, without any other information. The second stage might involve distinguishing cars from other vehicles. Supervised training involves giving the software a large number of authentic examples that it can be 'trained' to recognize.[32]

Symbolist networks are very good for systems that work on defined rules. They are components (but not necessarily the exclusive source) of programs that can now out-perform humans in games such as chess. They are called 'top-down' systems because they are based on rules-based expert knowledge. IBM's Deep Blue program, which defeated world chess champion Gary Kasparov in 1997, was a rules-based symbolist AI system. But the general deficiency of symbolist AI is that it does not do well at extrapolating its rules to new situations.

Connectionist AI starts with a minimum of assumptions and learns by training the system. Neural nets are in principle free of assumptions but learn from the data. At the risk of simplification, a neural net might start by making a guess about some object. Then it changes the parameters in some direction. If that produces a better result, it may move farther in the same direction. If it produces a worse result, it reverses in the other direction. By proceeding without prejudgment, it arrives eventually at a conclusion.

This is indeed a simplification, or at least it misses a feature of neural nets, because when a net arrives at a conclusion, further training tends to have only a small effect on matching the data.[33] This suggests that there may be a hidden bias, introduced by the rules or architecture of the system, that drives it towards certain conclusions. This 'inductive bias' is demonstrated by the fact that two systems of different design given the same training may produce different results.[34] (Of course, the brain also is not a blank slate, but has biases, for example, for visual patterning or learning language.) The existence of inductive bias is (sort of) reminiscent of the debate about the relative effects of nature *versus* nurture for humanity. At the risk of over-simplification, it's probably true, however, that assumptions made in the training system induce greater bias than the architecture.

∞ ∞ ∞

The move from symbolic AI to connectionism raises the question: how far can AI go? If a neural net is trained with no specific objective, for example, by feeding it the entire contents of the internet, what are the limits on its capacity? This is the question of general AI *versus* narrow AI. Systems designed to handle specific situations are called narrow AI to distinguish them from the (putative) general artificial intelligence (often abbreviated to AGI) that would be able to respond to situations for which it had no specific training.

(Narrow AI is accomplished by machine learning, which basically encompasses any situation in which a software program can be trained to meet an objective. Deep learning is a special example of machine learning, using neural networks to develop expertise that wasn't specifically programmed. [Deep is not used in the human sense, for example, of a deep thinker. It does not refer to sophistication of thought, but refers to the construction of the net; the depth is the number of hidden layers.] It's a general belief [or perhaps a hope] that deep learning has the potential to develop into AGI, although that has not been accomplished yet.)

The distinction between narrow and general AI was probably first made by John McCarthy of Stanford University in 1979. (He was responsible for introducing the term artificial intelligence at the Dartmouth Conference.) "Present AI programs operate in limited domains, *e.g.* prove theorems in a particular logical system... General intelligence will require general models of situations changing in time... When we want a generally intelligent computer program, we must build into it a general view of what the world is like with especial attention to facts about how the information required to solve problems is to be obtained and used."[35]

That view from 1979 follows something of a symbolist approach in accepting that the software needs to have a pre-ordained view. Current attempts at producing software with artificial intelligence use more of a connectionist approach in using a training protocol with minimal built-in assumptions. The most recent example of AI programs with aspirations to general intelligence is the development of GPT3 and GPT4 by OpenAI. These are neural nets, called large language models, because they are trained using very large amounts of text with the internet as the major source material. They originated with the objective of predicting the next word in a sentence. Sometimes that becomes obvious in their functioning.

Some people feel that that these programs are at the level of passing the Turing test, still often cited as a benchmark for the development of AI. Basically, the Turing test, as defined in Alan

Alan Turing in 1946 (getting off a bus on his way to an athletic meeting).

Turing's paper of 1950, asks whether you would be able to tell in a discussion whether your correspondent was a person or a machine if you could not see them. A machine passes the test if its responses cannot be distinguished from a human interlocutor. However, because the test focuses only on language, it's not necessarily a good test for the acquisition of AGI.

Large language models sometimes display the ability to undertake tasks for which they were not specifically trained. Fed information from sites that have source code for software, they may develop the ability to write code for performing new tasks. They can perform very well, in fact often better than humans, at a diversity of tasks from writing poems in varieties of styles to solving complex mathematical problems. This is called emergent behavior (not necessarily related to the concept of emergence in physics or biology). It may be optimistic to regard GPT3-4 as a forerunner to AGI as the software remains unable to perform some abstract reasoning tasks that are (relatively) simple for humans.[36] It's good at tasks connected with language, but may have deficiencies in other activities. Descartes may have been prescient here (see the beginning of the Mind and Brain section).

The goalposts for AI keep moving. Once unimaginable, the success of programs at beating human masters in chess or Go led to questions as to whether these were appropriate tests for artificial intelligence. In one of the first analyses of the possibility of constructing a program that would play chess, in 1950, information scientist Claude Shannon said, "a solution of this problem will force us either to admit the possibility of a mechanized thinking or to further restrict our concept of 'thinking'."[37]

In a famous quote, John McCarthy later said, "As soon as it works, no one calls it AI anymore."[38] Along the same lines, Douglas Hofstadter defined "Tesler's theorem" from a saying of Larry Tesler, an expert in the field of human–computer interactions, as "Artificial Intelligence is whatever hasn't been done yet."[39] Tesler felt he had been misquoted. "What I actually said was: 'Intelligence is whatever machines haven't done yet'."[40] This points to the difficulty of defining 'intelligence' without resorting to a circular argument.

Johnny von Neumann made the same point in a talk in 1948. "You insist that there is something a machine cannot do. If you tell me precisely what it is a machine cannot do, then I can always make a machine which will do just that." This was a subtle way of saying that the difficulty is not in mimicking thought processes, but rather in defining what those thought processes are.[41]

One limitation of AI programs is that they are not always so good at filtering reality from prediction. Their ability to make up information, a sort of parallel to tall stories, is known in the trade as hallucination. While writing this book, I asked ChatGPT to find original sources for some information. The sources were mostly relevant to the question, they all looked plausible—but only about half of them really existed! The others cited authors who had published in the field, with references to journals that publish papers in the field, and titles that seemed plausible: but the combination of author, journal, and title did not actually exist! It's as though ChatGPT has predicted papers the authors should have written. Ask it to name Nobel Prizewinners meeting some criterion, and the list is likely to include some researchers who performed very distinguished work, but actually did not get Nobel Prizes![42] In rewriting history, it's as much imaginative intelligence as generalized intelligence.

AI is beginning to permeate science as a means to analyze data, somewhat at the border between creative and factual. It's been used to identify potential regulatory sites in the genome that can't be identified by conventional sequence analysis (see Chapter 20) and to tighten the boundaries around a black hole (see Prologue). The question here is at what point it passes the bounds of what might be equivalent to human intuition into creativity extending beyond the bounds of the data. We have to worry about whether hallucination influences the ability of AI to analyze scientific or medical results.

On the other hand, AI's hallucinations have been used to create new protein structures. They were a major factor in the work that won David Baker of the University of Washington a Nobel Prize for Chemistry in 2024.[43] The title of his first paper was "De novo protein design by deep network hallucination."[44] Giving AI free rein may work for creativity where it is a threat for interpretation.

Another type of activity is very much based on following the analogy with human brain function. Image recognition is perhaps one of the most striking advances in AI, extending even to the ability to distinguish human faces. The software is based on ConvNets (an abbreviation for convolutional networks), which is based on a connectionist model.[45] It follows the principle of neural recognition of shapes (mostly worked out with cats), showing that the brain recognizes specific types of components, vertical lines or horizontal lines for example. Images are recognized by breaking them down into components and training to the software to recognize them, for example, by distinguishing dogs from cats.

It may seem more like science fiction than science for AI to be used to design experiments, yet already AI is improving on Nature. "A.I. models learn from sequences—whether those are sequences of characters or words or computer code or amino acids," says Ali Madani, at Profluent, a company that has used AI to improve the CRISPR gene-editing technique.[46] The CRISPR technique depends on adapting a natural system found in bacteria to target genes found in human and other species. AI software has been used to design better editing systems based on analyzing a large number of naturally existing systems. The same approach had been used previously to design better versions of an enzyme (a catalytic protein) found naturally: the final product had only a small relationship to any existing protein.[47] In the space of a couple of years, AI has moved from analyzing proteins to creating them.

∞ ∞ ∞

The first example of the widespread application of AI to a scientific problem has been for solving protein structures. A protein consists of a chain (or chains) of amino acids. It's an axiom that its three-dimensional structure is determined by its sequence (see Chapter 19). This was proved definitively by the AlphaFold AI program.

Protein structure is analyzed experimentally by X-ray crystallography (or NMR). Crystallography works by analyzing the diffraction pattern generated when a beam of X-rays is shot at a crystal of a protein. Using a set of algorithms (called Fourier transforms), the pattern is translated into a structure showing the position of every atom. This is a slow and painstaking process. First a protein crystal has to be obtained; then the analysis can take a long time. The first crystal structure was obtained in 1958. The process was completely overturned in 2021.

DeepMind is a project to apply AI to science and medicine. Started in London and acquired by Google in 2014, it developed the program AlphaGo that beat top players at the game of Go. The software AlphaFold was designed to predict 3D shapes from protein sequences. In its first iteration in 2021, it predicted the structures of 20 000 human proteins, more than 300 000 proteins from all the key model organisms, and another 400 000 proteins from various other sources.[48] A year later, the next release extended the analysis to virtually all known protein sequences.

This represents nothing less than replacing experimental science, where scientists are the researchers, with a theoretical analysis in which human activity is reduced to running the software. It's a complete soup to nuts approach, because protein sequences are inferred from the genome sequences of each organism, and then AlphaFold predicts their structures.

The striking feature is that whenever there has been a difference between AlphaFold's prediction and a structure analyzed by X-ray crystallography, AlphaFold has been correct. The way it analyzes protein structure remains something of a black box. It works by an iterative process, continually improving the structure until it cannot go any further, but even its creators may not understand it fully. "We hypothesize that [certain] information is needed to coarsely find the correct structure within the early stages of the network, but refinement of that prediction into a high-accuracy model does not depend crucially on [that] information," is the closest the authors come to understanding the software.

Demis Hassabis, who cofounded DeepMind—"I'm half neuroscientist and half computer scientist"—sees it as an intermediate stage to analyzing the process of protein folding and going further to analyze cell structures. "My dream would be to create a virtual cell, where you could perturb the cell and the model would give predictions of what would happen, without having to do painstaking years of experiments."[49] The Nobel Committee, not usually known for its speed in recognizing the importance of discoveries, awarded Hassabis a share of the Nobel Prize in chemistry in 2024, barely three years after the first publication.

But now the question is: if we don't understand how the software works, how can we verify its conclusions? Science depends on publishing work with details that allow it to be reproduced and verified. Theories can be verified by working through the stages of analysis. But AlphaFold is presenting the structure more as a *deus ex machina*. This requires some re-thinking about how science works—and for that matter, is perhaps a question mark as to how humanity will react to generalized artificial intelligence if and when it develops.

One of the statesmen of computing says this is irrelevant. Geoffrey Hinton was instrumental in the revival of connectionism in 1986 and in the move to deep learning in 2006. He was awarded the Nobel Prize in physics in 2024 for his work on AI. "People can't explain how they work, for most of the things they do... Neural nets have a similar problem," he says. "I think we're going to have to do it [treat neural nets] like you would for people. You just see how they perform, and if they repeatedly run into difficulties then you say they're not so good."[50]

Yet the implications of AI for science are even more serious and far-reaching, one might almost say threatening, than they are for the humanities. If software writes a piano sonata that seems to eclipse Beethoven, we would be astounded, we might even feel it undercuts the role of the human composer, but that would not stop us from appreciating it. We expect the composition of a sonata to be a creative act, by very definition inexplicable in the sense that we do not expect any account of methodology. But science, even if inspiration is the source of an insight, depends on validation by a reproducible methodology. If that methodology uses software whose internal workings are a mystery, how can we decide whether to accept it? Sometimes there may be other methods available for validation, such as analysis of protein structure, but sometimes there is no alternative, such as the refinement of black holes.

The view of science as alternating hypotheses with transparent analysis of reproducible data is, of course, itself a paradigm. It has stood unchanged since the Scientific Revolution. That may be about to change. The use of AI to analyze or even design experiments replaces transparency with opacity. It's too early to say whether we are on the verge of a revolution comparable to the Scientific Revolution that will change our entire attitude as to what comprises science.

If the practice of science stands alone intellectually, separate and different from the humanities and the arts, the construction of AI goes a step yet further. Within the distinctive mindset of science, individual subareas of science may have their own intellectual boundaries. Yet even if they are relatively self-contained, they are recognizable as part of the scientific *gestalt*. The situation may be different when AI is used to produce the conclusions. If even the creators of a software program are not fully able to understand how it works, AI becomes not only separate from other areas of science, but potentially independent in a third intellectual realm.

∞ ∞ ∞

Large language models are effectively trained by emulating human intelligence, but they have definite limitations. The difference between human and artificial intelligence is summarized by what is now known as Moravec's paradox: "It's comparatively easy to make computers exhibit adult level performance on intelligence tests or playing checkers, and difficult or impossible to give them the skills of a one-year-old when it comes to perception and mobility."[51]

This implies that there is some fundamental difference between neural networks, at least as presently constructed, and human intelligence. It's an open question whether a 'human-level intelligence' would necessarily emulate the human brain. It might have a completely different basis.[52,53] In that case, it becomes unpredictable how it would interact with human intelligence.

Perhaps because of concerns about the dangers of AI, perhaps because of the realization that there may be other types of generalized intelligence than the human brain, the focus on developing AI has turned towards 'augmented intelligence.' This is the concept that machine-based intelligence should be complementary with human intelligence rather than imitating or competing with it.

Also referred to as intelligence amplification, the concept dates from the early days of cybernetics[54] (and, of course, it's a common theme in science fiction). The starting point is that the software can assume input from a certain level of human intelligence. However, Gary Kasparov, former world chess champion, who has the distinction of having lost a match in 1997 against the IBM Deep Blue program, points out that the results aren't always what you might expect, emphasizing that it's not necessarily easy to relate artificial intelligence to human intelligence.[55]

Given that artificial intelligence is such a rapidly moving target, it's hardly surprising there are wide differences of opinion as to whether and when AGI might be achieved. Ray Kurzweil, an advocate for progress in AI, forecasts that a computer will achieve a functional simulation of human intelligence by 2029.[56] He projects that the Singularity (the point at which computers surpass human intelligence, and possibly achieve self-awareness)[57] should occur by 2045.[58]

Singularity was first popularized in the sense of a dramatic change in human history by science fiction writer Verner Vinge in 1983. "The evolution of human intelligence took millions of years. We will devise an equivalent advance in a fraction of that time. We will soon create intelligences greater than our own. When this happens, human history will have reached a kind of singularity... and the world will pass far beyond our understanding."[59]

The general record of predictions for advances in AI by people in the field has been quite poor. In 2010, predictions for when the singularity might occur varied from in a decade to in a century.[60] The pace has picked up now. "Until quite recently, I thought it was going to be like 20 to 50 years before we have general purpose AI. And now I think it may be 20 years or less," Geoffrey Hinton predicts.[61]

As for the odds of AI trying to wipe out humanity, Geoffrey Hinton said, "It's not inconceivable. That's all I'll say."[62] Another view is that, while this is a plausible outcome, it can be avoided by precautions in the design of AI software programs.[63] Others believe that achieving a genuine humanlike intelligence is a chimera that is not likely to be accomplished using available technology.[64] (ChatGPT says that, "It's important to note that the singularity is a speculative concept and not a scientifically established event or timeline."[65])

The AI community is not so much excited by the singularity as such, but by the possibility that it could lead to 'super-intelligence.' The argument is that once an AI program has exceeded human intelligence, because it is not limited by the constraints of the biological brain, it could be able to increase its intelligence by modifying its own software. This bootstrap effect could lead to a rapid increase to an intelligence level far surpassing humanity's. The consequences are unpredictable.[66]

There is a counter view that the concept of the singularity misunderstands the nature of AI, in effect by characterizing intelligence as a one-dimensional quality that can be ranked.[67] The idea that machines will adapt to humanity (and potentially exploit us) has been called 'the great AI fallacy.'[68] This view holds that, while there are certainly dangers in AI, they do not come from the machine developing 'superintelligence,' but from the uses that other humans make of AI.[69]

AI is developing at warp-speed. It's revolutionary in the potential of its effects on society, but not in the sense that it is overthrowing a paradigm. It has moved so fast that there is no paradigm to overthrow! The concept of using a neural net to create intelligence is more a speculation than a paradigm. It requires that it is possible to mimic intelligence by assuming the brain works as a binary system. Within that speculation are various working models, but none has yet established itself to the point of becoming a paradigm. The paradigm of AI, if there is going to be one, has yet to form.

Generalized artificial intelligence does not imply consciousness. That is a distinct stage farther. Given that we cannot, at least at present, define human consciousness in molecular terms, creating an artificial intelligence able to pass the Turing test and moving on past the point of the Singularity would be an interesting equivalent. It could at least demonstrate that there is no need in principle for any force extending beyond the laws of physics and chemistry in order to create consciousness.

Extending the argument, it becomes a question whether software that acquires consciousness would have the same moral and ethical issues as humans. Indeed, there is concern that, if an AI system should develop consciousness, perhaps inadvertently, it could be mistreated. "Conscious systems could be created and caused to suffer," says one paper trying to establish "principles for responsible AI consciousness research."[70] There have been calls for a moratorium on research or for the establishment of standards. There is at the least a substantial body of opinion that, even if acquisition of consciousness is not imminent, it is reasonably plausible within the next decade.

∞ ∞ ∞

A question about the potential of AI comes from mathematics. It was a great shock to mathematics when Kurt Gödel (1906–1978) announced his incompleteness theorems in 1931.[71] Now regarded as the greatest logician since Aristotle, Gödel was in Vienna at the time. With the outbreak of the Second World War, he left for the United States, where he settled at Princeton. He almost never left Princeton, and became intensely reclusive, but developed a close friendship with Einstein. They were famous for walking home together from the Institute each day. There was great speculation about the content of their daily discussions. Gödel maintained a lifelong belief that mathematics revealed objective reality, parallel with Einstein's belief in physics.

The gist of the incompleteness theorems is that any mathematical system must contain statements that are true but cannot be proved. There is no equivalent in science. We seem to be pushing the limits in astrophysics (how could we define the Big Bang if the laws of physics were actually different then?); and we have no idea how to investigate consciousness, but there is (at least so far) no demonstration that it is theoretically impossible

Einstein and Gödel walking at Princeton. Einstein once said that he stayed at Princeton in order to have the privilege of walking home with Gödel.[72]

Karin Morgenstern Papp.

to create an artificial intelligence or that it could not have consciousness. Because mathematics is, as it were, a truly self-contained system, it was possible for Gödel to develop his theorems. Without any equivalent in science, the 'hard problem' is likely to remain an open question unless and until there is success.

The most likely explanation of a failure from a reductionist perspective would be that the information supplied to the system is in some way deficient in the range of inputs. A reductionist view would be that the biological construction of neurons is a means not an end. It should be feasible to obtain the same results with silicon (as opposed to the carbon-based systems of the cell). Of course, to generate an intelligent and responsive system it would be necessary to mimic the range of sensory inputs.

The difference between humans and chimpanzees (or other animals) shows that construction of a brain with appropriate sensory inputs is not sufficient. Chimpanzees do not have the same capacity for analysis nor (so far as we know) the self-awareness of humans. The difference suggests that the key is in the functioning of the brain. Applying Gödel's incompleteness theorem to binary operations, Roger Penrose (awarded the Nobel Prize in physics in 2020 for his analysis of black holes) argues that it must mean software programs have limited ability to establish truth. If human brains cannot be mimicked by a computer, the implication is that consciousness may be non-algorithmic.[73]

Suppose it is not ultimately possible to construct an artificial system (consisting of hardware and software) with the same reasoning capacity and self-awareness as a human. Are we forced to conclude that there is something (at least so far) unique about biological systems that artificial systems have failed to mimic? The question is what that might be, how it can be defined and (in the reductionist view) how to reproduce it artificially.

NOTES AND REFERENCES

1. A. Turing, Computing machinery and intelligence, *Mind*, 1950, **49**, 433–460.
2. Exchange with ChatGPT (openAI.com), February 17, 2023.
3. Quoted in D. B. Taylor, Nobel Physics Prize Awarded for Pioneering A.I. Research by 2 Scientists, *New York Times*, October 8, 2024.

4. Originally stated in a letter to *Science:* A. C. Clarke, Clarke's Third Law on UFO's, *Science*, 1968, **159**, 255. Included in an essay in a re-edition in 1973 of Clarke's book of 1962: A. C. Clarke, Hazards of Prophecy: The Failure of Imagination, in *Profiles of the Future: An Enquiry into the Limits of the Possible*, Harper & Row, New York, 1973, p. 21.

5. D. R. Hofstadter, *Gödel, Escher, Bach: An Eternal Golden Braid*, Basic Books, New York, 1979, p. 328.

6. In formal terms, this is simple propositional logic.

7. S. Goss, *et al.*, Self-organized Shortcuts in the Argentine Ant, *Naturwissenschaften*, 1989, **76**, 579–581.

8. E. Bonabeau, M. Dorigo and G. Theraulaz, *Swarm Intelligence. From Natural to Artificial Systems*, Oxford University Press, New York, 1999.

9. T. D. Seeley, *Honeybee Democracy*, Princeton University Press, Princeton, 2010, pp. 198–217.

10. V. Bush, As We May Think, *The Atlantic Monthly*, 1945, **176**, 101–108. A shorter version was published later the same year in *Life*, **19**(11), 112–124.

11. A. Lovelace, 1843 Notes by A.A.L., *Taylor's Scientific Memoirs*, London, vol. III, 1843, pp. 666–731. The Notes were published together with Lovelace's translation of a description in French of the Analytical Engine that Babbage presented at a meeting in Turn in 1842. Online at www.fourmilab.ch/babbage/sketch.html.

12. M. Mitchell, *Artificial Intelligence: A Guide for Thinking Humans*, Pelican, New York, 2020, pp. 9–10; D. Cope, *Experiments in Musical Intelligence*, A-R Editions, Middleton, WI, 1996.

13. Quoted in G. Dyson, *Turing's Cathedral: The Origins of the Digital Universe*, Pantheon, New York, 2012.

14. J. Tukey, The Teaching of Concrete Mathematics, *Am. Math. Monthly*, 1958, **65**, 1–9.

15. A. Turing, Computing Machinery and Intelligence, *Mind*, 1950, 49, 433–460; J. von Neumann, *The Computer and the Brain*, Yale University Press, New Haven, 1958.

16. A. Lightman, *The Discoveries: Great Breakthroughs in 20th-century Science, Including the Original Papers*, Vintage, New York, 2006, pp. 111–126.

17. See A. Turing, Computing machinery and intelligence, *Mind*, 1950, **49**, 433–460.

18. The proposal was written by J. McCarthy (Dartmouth College), Marvin Minsky (Harvard University) N. Rochester (I.B.M.), C. E. Shannon, (Bell Telephone Laboratories).

19. Quoted in N. J. Nilsson, *John McCarthy. Biographical Memoirs*, National Academy of Sciences, Washington DC, 2012, p. 5.

20. The proposal is online at jmc.stanford.edu/articles/dartmouth/dartmouth.pdf.

21. M. L. Minsky, *Computation: Finite and Infinite Machines*, Prentice-Hall, New York, 1967, p. 2.

22. For a history of the development of AI, see D. Crevier, *AI: The Tumultuous History of the Search for Artificial Intelligence*, Basic Books, New York, 1993.

23. As defined by Norbert Wiener in 1948. See N. Wiener, *Cybernetics: Or Control and Communication in the Animal and the Machine*, MIT Press, Cambridge MA, 1948.

24. In a review of a book on artificial intelligence published in 1987; online at jmc.stanford.edu/artificial-intelligence/reviews/bloomfield.pdf.

25. Minsky and Papert's major criticism was that the perceptron as initially constructed could not use the Boolean XOR function, which returns zero if two inputs are the same (both are 0 or both are 1), and returns 1 if the inputs are different (one is 0 and the other is 1). This criticism has been controversial, because they may not have known of further results showing that multi-layer perceptrons could calculate any Boolean function. The XOR function is needed for solving nonlinear problems.

26. M. Minsky and S. A. Papert, *Perceptrons: An Introduction To Computational Geometry, Expanded Edition*, MIT Press, Cambridge, MA, 1988.

27. Y. Bar-Hillel, The Present Status of Automatic Translation of Languages, *Adv. Comput.*, 1960, **1**, 91–163. Online at aclanthology.org/www.mt-archive. info/50/Bar-Hillel-1960.pdf.

28. ALPAC, *Language and Machines: Computers in Translations and Linguistics*. A report by the Automatic Language Processing Advisory Committee, Division of Behavioral Sciences, National Academy of Sciences-National Research Council, 1966.

29. J. Lighthill, *Artificial Intelligence: A General Survey*, Science Research Council, 1973.

30. D. E. Rumelhart, G. E. Hinton and R. J. Williams, Learning Representations by Back-propagating Errors, *Nature*, 1986, **323**, 533–536.

31. G. Hinton, To Recognize Shapes, First Learn to Generate Images (*Technical Report UTML TR 2006-003*), University of Toronto, 2006.

32. M. Mitchell, *Artificial Intelligence: A Guide for Thinking Humans*, Pelican, New York, 2020, pp. 27–31.

33. A. Saxe, S. Nelli and C. Summerfield, If Deep Learning is the Answer, What is the Question?, *Nat. Rev. Neurol.*, 2020, **22**, 55–67.

34. F. H. Sinz, *et al.*, Engineering a Less Artificial Intelligence, *Neuron*, 2019, **103**, 967–979.

35. J. McCarthy, *Ascribing Mental Qualities to Machines*, 1979. Reprinted in J. McCarthy, *Formalizing Common Sense: Papers by John McCarthy*, Ablex Publishing Corporation, 1990.

36. A. Moskvichev, V. V. Odouard and M. Mitchell, The ConceptARC Benchmark: Evaluating Understanding and Generalization in the ARC Domain, *Trans. Mach. Learn. Res.*, 2023, https://openreview.net/pdf?id=8ykyGbtt2q.

37. C. E. Shannon, Programming a Computer for Playing Chess, *Philos. Mag.*, 1950, **41**, 256–275.

38. The date is uncertain, but it is quoted in M. V. Vardi, Artificial intelligence: past and future, *Commun. ACM*, 2012, **55**, 5.

39. D. R. Hofstadter, *Gödel, Escher, Bach: An Eternal Golden Braid*, Basic Books, New York, 1979, p. 597.

40. www.nomodes.com/larry-tesler-consulting/adages-and-coinages.

41. Quoted in E. T. Jaynes, *Probability Theory: The Logic of Science*, Cambridge University Press, Cambridge, 1995, p. 104.

42. Perhaps ChatGPT is determining who *should* have received Nobel Prizes. Should the view of the Nobel committee be replaced by ChatGPT's hallucination? After all, ChatGPT has more information at its disposal than a committee of a limited number of people meeting in Stockholm.

43. W. J. Broad, How Hallucinatory A.I. Helps Science Dream Up Big Breakthroughs. New York Times, December 23, 2024.

44. I. Anishchenko, *et al.*, De novo protein design by deep network hallucination, *Nature*, 2021, **600**, 547–552.

45. M. Mitchell, *Artificial Intelligence: A Guide For Thinking Humans*, Pelican, New York, 2020, pp. 71–80.

46. Quoted in C. Metz, Generative A.I. Arrives in the Gene Editing World of CRISPR, New York Times, April 23, 2024.

47. A. Madani, *et al.*, Large Language Models Generate Functional Protein Sequences Across Diverse Families, *Nat. Biotechnol.*, 2023, **8**, 1099–1106.

48. J. Jumper, *et al.*, Highly accurate protein structure prediction with Alpha-Fold, *Nature*, 2021, **596**, 583–589.

49. Discussion with Demis Hassabis, February 2022.

50. Quoted in *Forbes*, December 20, 2018.

51. H. Moravec, *Mind Children*, Harvard University Press, Cambridge, MA, 1988, p. 15.

52. B. Goertzel, Human-level Artificial General Intelligence and the Possibility of a Technological Singularity. A Reaction to Ray Kurzweil's the Singularity is Near, and McDermott's Critique of Kurzweil, *Artif. Intell.*, 2017, **171**, 1161–1173.

53. J. E. Korteling, *et al.*, Human - Versus Artificial Intelligence, *Front. Artif. Intell.*, 2021, **4**, 622364.

54. W. R. Ashby, *An Introduction to Cybernetics*, Chapman & Hall, London, 1956; D. Engelbart, *Augmenting Human Intellect: A Conceptual Framework. Summary Report AFOSR-3233*, Stanford Research Institute, Menlo Park, CA, 1962.

55. D. De Cremer and G. Kasparov, AI Should Augment Human Intelligence, Not Replace It, *Harvard Business Review*, March 18, 2012, https://hbr.org/2021/03/ai-should-augment-human-intelligence-not-replace-it.

56. R. Kurzweil, *The Singularity Is Near: When Humans Transcend Biology*, Penguin, New York, 2005, p. 200.

57. Singularity was probably first used in the context of computing by John von Neumann., as reported by physicist Stanislaw Ulam in a tribute published a year after von Neumann's death. Recalling a conversation with von Neumann, Ulam said, "One conversation centered on the ever accelerating progress of technology and changes in the mode of human life, which gives the appearance of approaching some essential singularity in the history of the race beyond which human affairs, as we know them, could not continue." S. Ulam, John von Neumann, *Bull. Am. Math. Soc.*, 1958, **64**, 1–49.

58. R. Kurzweil, *The Singularity Is Near: When Humans Transcend Biology*, Penguin, New York, 2005, p. 136.

59. Verner Vinge first used it in the context of artificial intelligence in an op-ed piece (Omni, January 1983). The origin is often attributed to a later science fiction book, where he gave a more explicit sense of timing. In *Marooned in Real Time* (Baen Books, New York, 1986): "... technology and people were headed into some sort of singularity in the twenty-third century" (p. 42). Its first formal use in the present sense was in a report Vinge wrote for NASA: V. Vinge, The Coming Technological Singularity: How to Survive in the Post-Human Era, in *Vision-21: Interdisciplinary Science and Engineering in the Era of Cyberspace*, ed. G. A. Landis, *et al.*, NASA Publication CP-10129, 1993, pp. 11–22.

60. S. Armstrong and K. Sotala, How We're Predicting AI—or Failing To, in *Beyond AI: Artificial Dreams*, ed. J. Romportl *et al.*, University of West Bohemia, Pilsen, 2012, pp. 52–75.

61. Interview with CBS News March 2023.

62. Interview with CBS News March 2023. As a result of this view, because of concerns about the development of AI for public use, Hinton resigned from Google with which he had been affiliated for ten years. See C. Metz, 'The Godfather of AI' leaves Google and Warns of Danger Ahead, *New York Times*, May 1, 2023.

63. S. Russell, *Human Compatible: Artificial Intelligence and the Problem of Control*, Penguin Books, New York, 2020.

64. E. J. Larson, *The Myth of Artificial Intelligence. Why Computers Can't Think the Way We Do*, Harvard University Press, Cambridge, MA, 2021.

65. Exchange with ChatGPT (openAI.com), April 8, 2023.

66. N. Bostrom, *Superintelligence: Paths, Dangers, Strategies*, Oxford University Press, Oxford, 2014.

67. N. Lawrence, *The Atomic Human: Understanding Ourselves in the Age of AI*, Allen Lane, London, 2024, p. 26–27, 361.

68. N. Lawrence, *The Atomic Human: Understanding Ourselves in the Age of AI*, Allen Lane, London, 2024, pp. 257, 375.

69. N. Lawrence, *The Atomic Human: Understanding Ourselves in the Age of AI*, Allen Lane, London, 2024, pp. 371–378.

70. P. Butlin, and T. Lappas, Principles for Responsible AI Consciousness Research, 2025, arXiv:2501.07290.

71. R. Goldstein, *Incompleteness: The Proof and Paradox of Kurt Gödel*, Norton, New York, 2005.

72. R. Goldstein, *Incompleteness: The Proof and Paradox of Kurt Gödel*, Norton, New York, 2005, p. 33.

73. R. Penrose, *The Emperor's New Mind - Concerning Computers, Minds, and the Laws of Physics*, Oxford University Press, Oxford, 1989, pp. 405–450.

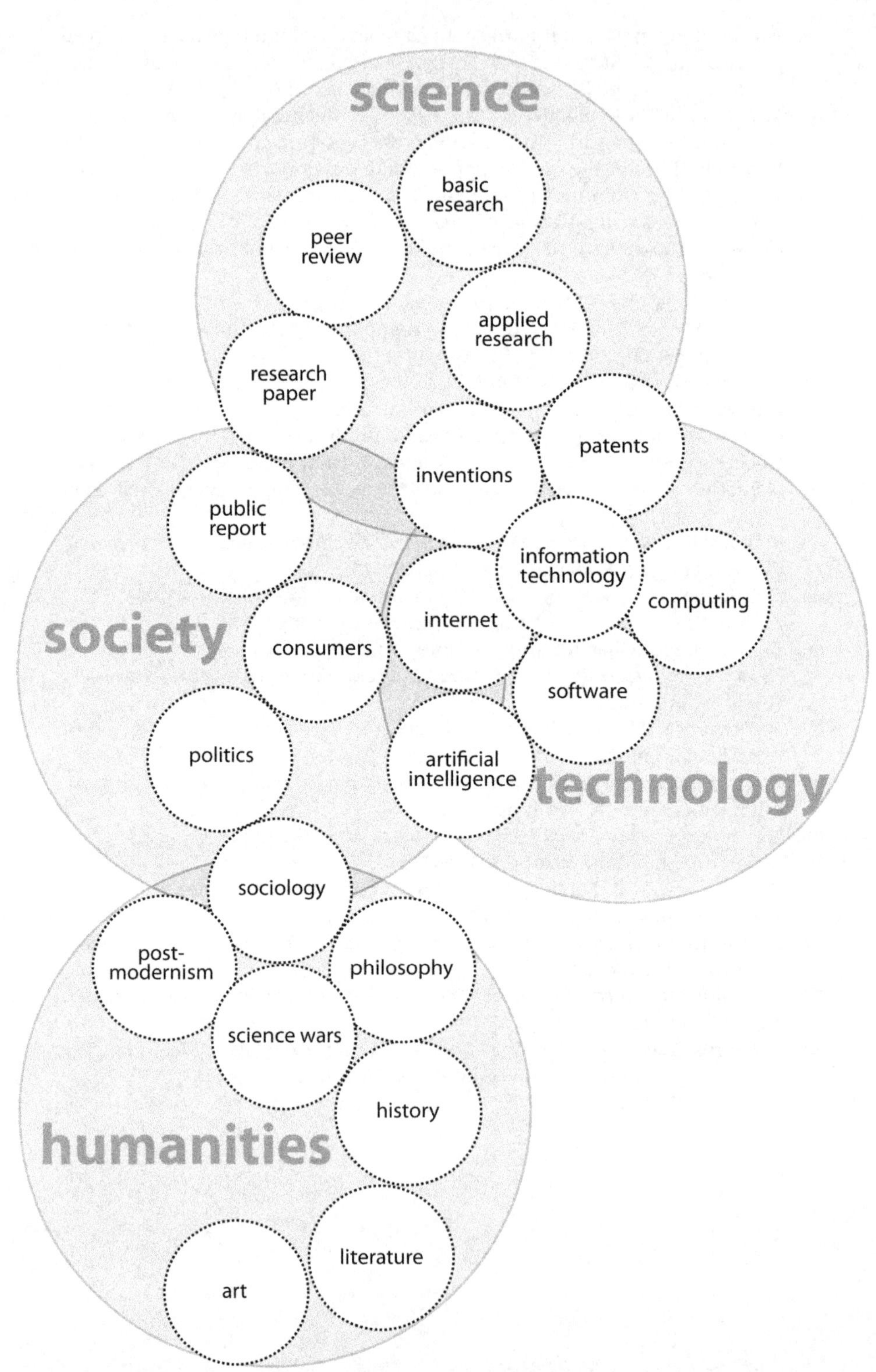

science
society
technology
humanities
basic research
peer review
applied research
research paper
patents
inventions
public report
information technology
internet
computing
consumers
software
politics
artificial intelligence
sociology
post-modernism
philosophy
science wars
history
literature
art

The Culture of Science

The effects of science are of various very different kinds. There are direct intellectual effects: the dispelling of many traditional beliefs, and the adoption of others suggested by the success of scientific method. Then there are effects on technique in industry and war. Then, chiefly as a consequence of new techniques, there are profound changes in social organization which are gradually bringing about corresponding political changes. Finally, as a result of the new control over the environment which scientific knowledge has conferred, a new philosophy is growing up, involving a changed conception of man's place in the universe, Bertrand Russell, 1953.

A distinguished historian of science, commented that "The proper measure of a philosophical system or a scientific theory is not the degree to which it anticipated modern thought, but its degree of success in treating the philosophical and scientific problems of its own day." To me this is nonsense. The point of science is not to answer the questions that happen to be popular in one's time, but to understand the world... The history of science is not merely a tale of intellectual fashions, succeeding one another without direction, but a history of progress toward truth... It's clearly not possible to speak of right and wrong in the history of art or fashion, nor I think is it possible in the history of religion, and one can argue about whether it is possible in political history, but in scientific history we really can say who was right. Steven Weinberg, 2018.

Communication in Science

"NOTHING IN SCIENCE HAS ANY VALUE TO SOCIETY IF IT IS not communicated," said American psychologist Anne Roe in her 1952 study of how people become scientists.[1] A scientific discovery has no meaning (and you will get no credit for it) until you communicate it to other scientists. Only then can it be accepted or rejected as a contribution to the body of science. Changes in format over the centuries show that the medium is not the message, but the form in which results are presented is important. During the Scientific Revolution, scientists mostly collected their results over extended periods and published them as books. The transition to publishing much shorter pieces of information in journals with independent editors occurred in the 19th century. Quality was ensured by peer review becoming a condition of publication. The move from print to electronic distribution occurred at the start of the 21st century, but so far has been a change in the means of distribution rather than really influencing the nature of the material that's communicated.

The Frontiers of Science
By Benjamin Lewin
© Benjamin Lewin 2026
Published by the Royal Society of Chemistry, www.rsc.org

Timeline for Developing a System for Communication in Science

1650

Royal Society founded in London (1660)

Journal des Sçavans started in Paris (1665)
Royal Society starts *Philosophical Transactions* (1665)

1700

1750

Philosophical Magazine (physics) founded (1798)
1800 Royal Institution founded in London to educate public (1799)
Mary Shelley publishes *Frankenstein* (1818)
Lancet founded as general medical journal (1823)

Refereeing of submissions introduced (1830)
William Whehall introduces term 'scientist' (1830)

1850

Nature founded as first general science journal (1869)

1900

1950

Citation Index introduced (1960)
Ingelfinger rule excludes press release before publication (1969)

arXiv preprint server distributes physics papers before review (1991)
2000

First all-electronic life sciences journal founded (2012)
bioRxiv preprint server distributes biology papers pre review (2013)

A Dialog between Henry Oldenburg and Alfred Harmsworth

Henry Oldenburg (1619–1677) was a German philosopher who became famous as the founding editor of the *Philosophical Transactions* of the Royal Society in London.

Alfred Harmsworth (1865–1922) was a newspaper magnate who created the mass press in England in the early 20th century. Amalgamated Press was one of the world's largest media empires.

© National Motor Museum/Heritage Images/Getty Images.

Henry Oldenburg: Good day, Mr. Harmsworth. I find myself in the company of another figure who left an indelible mark on the world of communication. In my time, in 1665, I played a role in establishing the first scientific journal, the *Philosophical Transactions*. How do you view the evolution of journalism since your days as a media magnate?

Alfred Harmsworth: Good day, Mr. Oldenburg. It's an honor to converse with someone who contributed to the foundations of scientific communication. In my era, in 1896, I founded the *Daily Mail* newspaper in London. The media landscape has changed dramatically since then, marked by technological advances and shifts in readership habits. The challenge has been to capture the attention of a broad audience. Sensationalism and engaging content became crucial.

Henry Oldenburg: Sensationalism—a word not unfamiliar to me. While the nature of scientific communication differed, there was also a need to engage the curious minds of my time.

The *Philosophical Transactions* aimed to provide a platform for scholars to share their discoveries and engage in intellectual discourse.

Alfred Harmsworth: Indeed, creating content that resonates with the audience is a universal challenge. In my time, the emphasis was on captivating headlines and human-interest stories. The competition for readership was fierce. Understanding the public's appetite was key.

Henry Oldenburg: The Royal Society came to insist on priority and novelty as a condition for publication. The peer review system was introduced to assure quality. Do not these concerns clash with the demands of a newspaper?

Alfred Harmsworth: When research is funded by the public, surely the public has a right to know. Lives may be saved by publishing the information as quickly as possible. In trying to suppress news reports until a scientific journal has been published, is the journal doing any more than trying to ensure its own priority at the expense of the public interest?

Henry Oldenburg: Increasing speed of dissemination can outpace the scrutiny that information deserves. How does it serve the public interest to present incomplete reports without adequate qualifications? Is it more important to sell newspapers than to wait for accurate information? The newspaper is to the scientific journal as rumor is to truth. I rest my case.

COMMUNICATION IN SCIENCE

The combined effect of the contempt and ignorance of the learned, and of the suspicion and resentment of the lower orders, has been, through the whole course of civilization, a major hindrance to the free advance of science,[2] J. D. Bernal, 1954.

I am most displeased that the ignorance of some people has peaked so that, condemning sciences of which they are totally ignorant, they attribute [false] things to sciences they are incapable of understanding. Letter from Benedetto Castelli to Galileo, 1614[3]

Two events in Einstein's career illustrate the changing nature of how scientific research papers were published in the 20th century. When he submitted his four papers in 1905 to *Annalen der Physik,* one of the leading physics journals in Germany, he was virtually unknown (he had previously published some minor papers). The theory editor of the journal was Max Planck, who accepted the papers without any external review. It does not even appear that any revisions were required. This was how journals functioned at the start of the 20th century.

It was a very different scene in 1936 when Einstein, by now perhaps the world's most famous scientist, submitted a paper to the American journal *Physical Review.* Einstein was outraged when it was sent out for review. "We had sent you our manuscript for publication and had not authorized you to show it to specialists before it is printed. I see no reason to address the in any case erroneous comments of your anonymous expert. On the basis of this incident I prefer to publish the paper elsewhere," Einstein wrote.[4,5] Today it would be rare indeed for any scientist, whether without a track record or well established, to publish a paper without going through a review process and a demand from the editor of the journal for revisions.

To the outside world, communication between scientists may seem stereotyped and archaic, if not positively intimidating. (Writing a paper can be intimidating to scientists also. One book on how to write scientific papers has gone through 8 editions.[6]) The scientific paper is the basic unit of research, not just for communicating results; it is also the measure of a scientist's position in the community. A paper follows a relatively fixed format, in effect a recapitulation of the myth of the scientific method. First it defines a gap in knowledge, often in the form of a hypothesis that needs to be tested. Then it describes the experiments in enough detail for someone else to reproduce them. Finally, it discusses their significance. That is the image of modern science in a nutshell: test hypothesis, present reproducible results, and claim priority for an original discovery.[7]

Regarded as almost sacrosanct, the integrity of the scientific paper is protected by a review process. It's not enough to obtain novel results. They have to be approved by experts before they can be presented to the community. This is the hallowed process of peer review, in which the editor of a journal sends a submitted paper to experts in the field for their comments. If the paper passes muster, or more likely, once it has been revised to satisfy the experts' criticisms, it can be added to the scientific canon. There is a myth (propagated by the Royal Society of London among others) that scientific journals arose fully fledged, as it were, but actually they evolved into their present form over two centuries.

In the 16th and 17th centuries, scientists were few in number and there was no formal mechanism for communication among them. Witness the fact that Copernicus, Galileo, Boyle, Hooke, Newton, even Lavoisier a century later, all published their results as books. The books made their way to interested scholars (or patrons) rather than to society as a whole. There were often delays of several years between making the observations and publishing the results. This complicated arguments about priority. Sealed or cryptic records were sometimes made to establish priority, a contrast with the modern view that priority is determined by publication.

Until the 18th century, authors would report conclusions based on outlines of how they came to them. Scientific reports were somewhat akin to a stream of consciousness in relating an account of the work. The nature of reporting science changed with the development of journals during the 18th century.[8] The change in means of communication reflected the increasing professionalism of science. By the end of the 19th century, virtually all scientific developments were reported in journals.[9]

Science began to acquire a formal structure with the formation of the first scientific societies, mostly notably the Académie des Sciences in Paris and the Royal Society in London. Founded in 1666, the Académie was supported by the French state. By contrast, the Royal Society, although it had a royal charter, was a private organization. Indeed, there was opposition in England even in the late 19th century to the idea that government should be involved with supporting science.[10] Science in Europe, especially Germany and France, was more professional and better supported by the state. Although the Royal Society did not receive funding from the state, its royal charter gave it the advantage that it could publish without submitting to the censor.

The *Philosophical Transactions* of the Royal Society is usually described as the first journal of scientific record, with its establishment in 1665 acknowledging that there was a community of scientists who needed to communicate with one another. The Royal Society started as an informal group around 1650, meeting weekly in Oxford.[11] Many of its members had positions at the university, but it had a strong component of interested amateurs. Following the Restoration of the monarchy, the society was formally formed in 1660,[12] took the name of the Royal Society in 1661, and obtained a royal charter in 1662.[13] Its character became more focused after Isaac Newton became President in 1703.[14]

Only around a third of the Fellows of the Royal Society were closely involved with science in the 18th century.[15] Like other societies concerned with science in England during the period, the Royal Society was as much a cultural and social institution as a scientific one.[16] An insight into the Royal Society's priorities comes from the history of the publication of Newton's *Principia*. The Royal Society reneged on its commitment to publish it, as it had run out of funds because it had recently published *Historia piscium* (History of Fishes) by its Fellow, Francis Willoughby. This was a lavishly illustrated book that was a massive flop.

As a result, astronomer Edmond Halley undertook the project himself in 1686. When funds became short, the Royal Society paid Halley's salary with volumes of the unsold *Historia piscium*. What we would now regard as the major accomplishments of the period—astronomy, physics, and mathematics—were not the Royal Society's main focus.[17]

The original form of the *Transactions* in London and the *Journal des Sçavans* in Paris, another contender for the title of the first scientific journal, was a mishmash. It contained book reviews, translations from other publications, obituaries, new discoveries and inventions, and so on. The first issue of *Philosophical Transactions* had contributions ranging from "a spot on one of the belts of the planet Jupiter" to "a relation of a very odd monstrous calf." The intention was to appeal to a wide audience. The model was more the news sheet or magazine than what we would regard as a journal today.

Transactions started very much as the creation of its editor, Henry Oldenburg, the first Secretary of the Royal Society. (He also was involved in the *Journal des Sçavans* in Paris.[18]) Slowly it evolved into what we would recognize today as a scientific journal.[19] One reason for rejection was that material had already been communicated elsewhere. This was an early (and unusual) emphasis on the need for originality, which is such an important part of modern science. By 1850, the review process had become formalized (for example, selecting reviewers based on their expertise).[20] Dates when papers had been submitted were now included to deal with claims of priority. This was the point at which communication came to take its present form.[21] The refereeing system developed into today's peer review system.[22]

Science really came into the public consciousness in the 18th century. It was often a subject for satire, inspiring a play on the London stage in 1676, Jonathan Swift's *Tales of a Tub* in 1704, and other (relatively hostile) comments until mid-century.[23] The 18th century literature depicted scientists as being irrelevant more than dangerous.[24]

In France, increased interest in science was known as the *goût public des sciences* (public taste for science). Scientific spectacles became part of public lectures (especially with demonstrations involving electricity). In the early 19th century, scientists were respected for their skills at public presentation.[25] The Royal Institution was founded in London in 1799 by eminent scientists with the purpose of educating the public. Lectures there were particularly effective in establishing reputations for scientists such as Humphry Davy (1778–1829) or Michael Faraday (1791–1867). Books and magazines about science also became successful. All this was part of creating a public enthusiasm for science.

Michael Faraday giving a Christmas Lecture at the Royal Institution in 1855. The Christmas Lecture series started in 1825. Michael Faraday gave them from 1851 to 1860. The lectures attracted crowds of several hundred people.

Societies such as the Royal Society started as more or less private organizations, with their meeting restricted to members. By the 1820s, pressure for public access led to more widespread reporting of the results.[26] All this contributed to wider public consciousness of science. Publication of both original results and popular reports in periodical form was certainly far more accessible to the public than the books of the previous period.[27] By the end of the 18th century, academies of science in all the major European capitals were publishing journals.

Journals of more specialized character increased steadily in number through the century, from around 50 in 1800 to around 500 in 1900.[28] Journals varied from appealing to the general public to focusing more directly on scientists,[29] but the general drift was that any interested person should be able to read an issue. Most publications depended on some part of their circulation being purchased by people outside of the areas addressed in the journal.

Indeed, in the first half of the 19th century there was something of a blur between the literary and the scientific.

The *Literary Gazette* had frequent reports on developments in science. *Nature* originated as a semi-popular journal of science in the second half of the century.[30] In one sense, this was the democratization of natural philosophy. However, the attempt to appeal to a lay audience was not successful. Within a few years *Nature* became essentially a means of communication among scientists.

Scientific publication has shown exponential growth, and increased specialization, since the start of the 19th century. The major papers reporting the developments in physics at the start of the century were all published in specialized journals.[31] The era when scientific papers could be understood by the interested layman had passed.

Scientists and those interested in science understand that new ideas come more from inspiration than from cold, logical analysis, but the public image of science is very much the latter. The impression that science is mechanistic, logical, even sterile, is encouraged by the way scientific research is usually presented, in the context of answering questions formulated on the basis of rigorous logic. Until the start of the 19th century, indeed until science started to present itself as a logical endeavor, there was no distinction between the arts and the sciences. Genius was due to imagination, in either arts or sciences.

As science became formulated in terms of laws in the post-Newton era, this changed. By the end of the 19th century, there was a distinction between the arts (imaginative) and the sciences (objective).[34] The sense that, even if it depends on individual leaps of imagination, science differs from art because at the end of the day it is a community effort, was very well put in 1865 by Claude Bernard, a distinguished physiologist in Paris. "L'art c'est moi, la science c'est nous" (Art is myself, science is ourselves).[35] By the start of the 20th century, science had become a profession.[36]

With communication formalized by publishing scientific papers in a research journal, there has been some confusion about means and ends. 'Citation analysis' has become a way to assess a scientist's contributions. In principle this depends on measuring how many other articles subsequently cite any particular research contribution. A greater number is taken to mean that the contribution is more important.[37] This has its flaws, of course. A more interesting view (perhaps a more cynical view) was

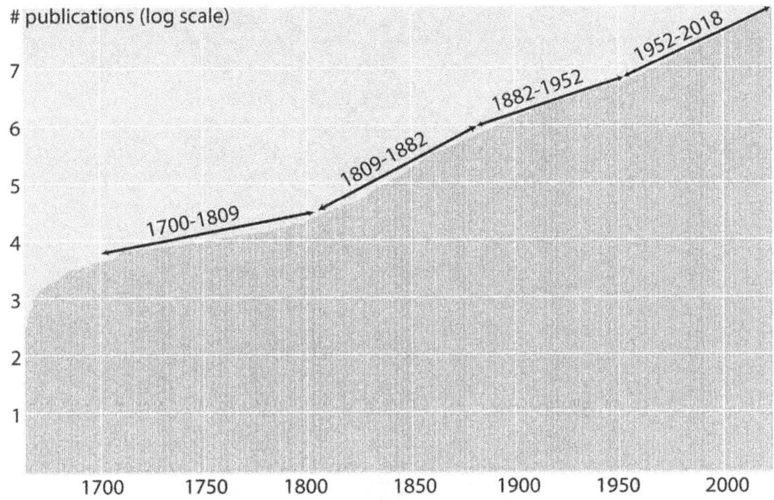

The growth of science as measured by the number of publications shows four phases since the Scientific Revolution: steady growth until 1809, acceleration until 1882, a slowing down until 1952, and then acceleration with the increased focus on science and technology after the Second World War.[32] Today it is growing at an annual rate of 5%, which means the scientific literature doubles every 14 years.[33]

expressed by the great mathematician David Hilbert (1862–1943). "One can measure the importance of a scientific work by the number of earlier publications rendered superfluous by it."[38]

∞ ∞ ∞

As science moved from the private to the public domain, it attracted increasing attention. The public view of science today is split. On the one hand, the latest developments in science—such as the discovery of the Higgs boson—attract considerable attention in the news. At the other extreme, the concept of the 'mad scientist' is still common in literature and entertainment. When did this idea emerge that science might become out of control?[39]

One view is that the concept of the mad scientist had its origins in the belief that alchemists were in league with the devil.[41] Frankenstein is probably the most famous mad scientist of all time. The subtitle of Mary Shelley's book, published in 1818,

The frontispiece from the third edition of Frankenstein shows Victor Frankenstein running out of the laboratory, horrified by his creation of the monster.[40]

The Modern Prometheus, relates to myths about alchemists creating life. One of the inspirations for the story was the reports of experiments conducted by Erasmus Darwin (grandfather of Charles Darwin).[42] (It's a measure of the hold the book has taken over popular culture that Frankenstein is now associated with the monster, rather than with his creator, the mad scientist Victor Frankenstein.) The genre of horror based on mad scientists is now well established, but the mad scientist was recognized as a fantasy genre that did not cast any aspersions on real scientists. Concern about the damage that science might do is a more recent phenomenon.

Frankenstein is often considered to have been the first book in the genre of science fiction.[43] Subsequent books alternated between extrapolating from developments in biology and physics. H. G. Wells' *The Island of Doctor Moreau,* published in 1896, showed the penetration of Darwinism into popular culture, with a horror story of (mis)directed evolution. On the other hand, 19th century literature also displays the scientist as inventor or hero, such as in the novels of Jules Verne, reflecting the enthusiasm for technology of the Victorian era. H. G. Wells foresaw the potential of science to create new weapons along the lines of atomic bombs in *The World Set Free*, published in 1914.[44] Atomic physics and cosmology became regular themes in science fiction.[45]

By the first half of the 20th century, the view had turned in another direction. The most famous dystopian novels both employed science or technology in a sinister sense. Huxley's *Brave New World*, published in 1931, used reproductive engineering to control society.[46] Orwell's *Nineteen Eighty-Four*, published in 1949,[47] relied on communications technology to subdue the population.

The point is not really whether scientists were more regarded as forces for good or evil, but that there was increasing public perception of science as a distinct activity. One problem is that the gap between the humanities and the sciences encourages the public, most of whom have little scientific education, to form naïve views of science, whether positive or negative.

The discoveries of subatomic particles in the first decades of the 20th century were widely reported and made newspaper headlines ranging from the sober to the sensational. Even the more esoteric developments, such as the introduction of quantum or wave theory made the news. Phrases like 'far-reaching implications' or 'revolutionary' were common. The rediscovery of Mendel's laws of inheritance got publicity, as did subsequent developments ranging from Avery's identification of DNA, and Watson and Crick's model for the double helix. Reporting was positive, with remarks about the potential for better understanding of nature and medical advances.

Reporting in this era tended to take original scientific reports as gospel. But the tendency of the press to turn simplification into exaggeration made scientists reluctant to appear directly before the press. Scientists were more involved in explaining their work in the first half of the 20th century,[48] but in the second half, popularizing your work was frowned upon. Carl Sagan, for example, had a distinguished career in astronomy, but failed to get tenure at Harvard in 1968, or to be elected to the National Academy of Sciences in 1992, principally because of disdain for his activities in popularizing science.[49]

The self-contained nature of science means that prestige comes from research. Reputations are not made by teaching or education. While this offers a protection for the integrity of science, in reducing external influences, it diminishes science's connection with society. The institutions of science need to make greater efforts to involve scientists in educating the public.

Part of the reason for the abysmal knowledge of science in the population (I discuss this in Chapter 26) is the general lack of professional reporting in the press. Perhaps reflecting a self-reinforcing view that the public finds science too difficult, the number of science sections in newspapers in the United States declined from 95 in 1989 to 19 in 2012.[50] This is at least partly responsible for the lack of any critical analysis comparable to that shown in other topics, such as politics or economics.

Science may be at the forefront of modernity, but remarkably little has changed in scientific communication since a survey in 1970,[51] except for the transition from print to electronic format. This has speeded up publication enormously without otherwise impinging much on science.[52] Scientific methods have undergone a technological revolution, but communication has changed only in form, not in substance.

∞ ∞ ∞

Communication to the public has, of course, changed enormously with the advent of social media. Communication by press conference has created media storms that sometimes—sadly too often—have proved to be unwarranted. Funding organizations can be too eager to validate themselves by exaggerating the importance of discoveries. There is virtually never a second press conference to correct a report that was wrong. This leaves a damaging impression of unreliability.

"We think we are talking about an organism that. . . appears to be using another fundamental component of life. . .It is replacing arsenic for phosphorus. This is a huge deal," said Mary Voytek, NASA's Senior Scientist for Astrobiology, at a press conference in 2010. She was referring to research funded by NASA that had just been published in *Science.*[53] The subject was a strange bacterium found in Mono Lake (in eastern California). The lake provides an unusual environment, with very high salinity and alkalinity, and a high concentration of arsenic. The paper claimed that the bacterium is able to use arsenic instead of phosphorus in its proteins and also in its DNA.[54] The story made front page news, and a physicist writing in the Wall Street Journal said, "Even the very definition of life may have to be changed."[55]

The immediate reaction in the scientific community was that this was wrong. Six months later, *Science* published no less than 8 technical comments refuting the paper. (The basic reason for the error was probably that the bacterium was growing on trace amounts of phosphorus contaminating the arsenic.) The authors were indignant. The first author left science for a decade. The head of the laboratory where the work was done attacked the refutation. "It aims to shut the door on additional research." The authors have never retracted the paper, although in 2025 *Science* retracted the paper unilaterally. This is an interesting change in science: it used to be unarguable that only the authors could retract a paper. The authors, and NASA, protested the retraction. "The change in policy at... *Science* to retract research publications... is unprecedented and upends the current standards in the research and scientific fields" NASA said.

The affair shows both the strength and weakness of the conduct of science. The error was corrected on the record almost immediately. But its occurrence shows the failure of the peer review process; better qualified reviewers would not have recommended publication. The major fallout is that the public gets its information through the press: so if NASA's Senior Scientist for Astrobiology announces tomorrow that alien life has been found on another planet, would we believe them?

∞ ∞ ∞

Peer review is far from a guarantee that a paper will ultimately prove correct, of course, but it's a significant protection against failure to meet the standards of the field or making foolish errors. The advantages of the system were made evident by what has been called the scientific fiasco of the century—the discovery of cold fusion in 1989.[56]

It made headlines worldwide when Stanley Pons and Martin Fleischmann at the University of Utah announced in a press conference that they had achieved cold fusion—the fusion of two deuterium atoms that released more energy than was put into the reaction. Fusion is a process that occurs naturally in the sun, at a temperature of 150 million degrees. It's the basis for the hydrogen bomb, so-called because it's based on the fusion of deuterium (an isotope of hydrogen) at very high temperatures.

Achieving fusion under controlled conditions has been a target for energy production, attempted with huge reactors. Achieving fusion under laboratory conditions appeared to be a chimera. Although there had actually been previous reports of successful fusion in the laboratory, often using the metal palladium, which has the capacity to absorb large amounts of hydrogen, none had stood up to scrutiny. This should have given pause for thought.

The crucial experiment in Utah was extraordinarily simple. A current was passed through a solution of heavy water (D_2O), containing an electrolyte, between a palladium electrode (the cathode) and a platinum electrode (the anode). The current converts the D_2O into its components, deuterium and oxygen. The palladium electrode absorbs the deuterium. The high concentration of deuterium atoms in the palladium was supposed to force their fusion. Success was measured by the release of large amounts of heat.[57]

Announced at the press conference on March 23, the results were widely reported by the media.[58] Attempts to repeat the phenomenon were impeded by the fact that there were basically no details of the experiment. A paper was published three weeks later (April 10) in a surprisingly obscure journal for such a purported major discovery. It was apparent, at least to nuclear physicists, that there were major holes in the report. One of them was that fusion should have generated enough radiation to kill the experimenters! (I discuss the science of the 'discovery' in Chapter 25.)

The haste with which the paper was published, after the press conference, and the subsequent problems that emerged, make any peer review seem woefully inadequate. You would expect qualified reviewers to be highly skeptical of a statement that the "bulk of the energy release be due to an hitherto unknown nuclear process or processes." Peer review is not a panacea. (Failure of the process is especially surprising because past track record usually influences referees, and Pons and Fleischmann had previously published some provocative claims that had not stood up.[59])

The commercialism that has invaded science was partly responsible for the phenomenon. The University of Utah was anxious to establish its priority for patent claims—in fact, so anxious that it refused to allow collaborations to verify the

Stanley Pons and Martin Fleischmann with their deceptively simple apparatus—as it turned out, too simple. What a contrast with the TFTR (Tokamak Fusion Test Reactor; see below).

claims. This was one of the main causes of the fiasco. The over-specialization of science meant that chemists received the results rapturously, although physicists were skeptical. "Two chemists have made an unusual discovery with profound implications to an area of physics in which most of the active physicists never talk to chemists," explained one chemist.[60]

Far from cooperating with attempts to reproduce or investigate their results, Pons and Fleischmann imposed obstacles, including threats of legal action[61]—which is definitely not part of the scientific attitude. They were supported by the University of Utah. It would not be unfair to describe it as a corruption of the scientific process. A large amount of public money was wasted following up the original report.

Cold fusion was changed from what might have been an error due to sloppy work into a scandal because it was not handled by the usual scientific process. This would have required the work

The Tokamak Fusion Test Reactor (TFTR) at Princeton in 1989 is a huge contrast with the apparatus that Pons and Fleischmann used to report cold fusion in the same year.

Reproduced from https://commons.wikimedia.org/wiki/File:TFTR_ 1989.jpg, under the terms of the CC BY 3.0 license, https:// creativecommons.org/licenses/by/3.0/deed.en.

to be fully described so that others could reproduce the results; and the authors would be expected to cooperate with attempts at replication. The work has never been retracted.

Cold fusion was laid to rest by negative research articles and two official reports from the Department of Energy.[62] It's a sad comment on the failure of scientific attitude to become more universal that there are today at least as many books arguing that cold fusion is real, and that it has been suppressed by the scientific establishment, as to review the sources of the error.[63]

There is still an International Conference for Cold Fusion (now renamed the International Conference for Condensed Matter) for believers (although the change of name may indicate something). To be sure, it can be a fine line between reasonable skepticism and willful refusal to accept new evidence. 'Cold fusion' has become a term of abuse in mainstream science, meaning that a finding is too incredible to be true. Ironically, one critic used it in 1994 to describe Stanley Prusiner's discovery of the prion (for which he was awarded the Nobel Prize in 1997).[64]

This was a case where the new data, however incredible, were indeed true (as discussed in Chapter 19). The survival of belief in cold fusion in spite of overwhelming evidence against it shows a divide between the rejection by mainstream science and the interest shown by "crackpots, pseudoscientists, frauds, and a few sociologists of science."[65] Even within science, as seen from the outside, analysis is not always wholly rational.

NOTES AND REFERENCES

1. A. Roe, *The Making of a Scientist*, Dodd, Mead & Co., New York, 1952.
2. J. D. Bernal, *Science in History, Vol. 1: The Emergence of Science*, The MIT Press, Cambridge MA, 1971, vol. 1, p. 49.
3. Letter from Benedetto Castelli to Galileo, December 31, 1614. Translated in S. Drake, *Galileo At Work: His Scientific Biography*, University of Chicago Press, Chicago, 1978, p. 239.
4. A. Pais, *Subtle is the Lord: The Science and Life of Albert Einstein*, Oxford University Press, Oxford, 1982.
5. When Einstein published the paper elsewhere the following year, its conclusions had been significantly changed to accord with the reviewer's criticisms! D. Kennefick, Einstein Versus the *Physical Review*, *Phys. Today*, 2005, **58**, 43–48.
6. B. Gastel, and R. A. Day, *How to Write and Publish a Scientific Paper*, Greenwood, Santa Barbara CA, 8th edn, 2016.
7. B. Lewin, *Inside Science: Revolution in Biology and its Impact*, Cold Spring Harbor Laboratory Press, New York, 2023, pp. 65–78.
8. Robert Boyle was precocious in describing at the end of the 17th century what his apparatus was and exactly how he used it. See Chapter 5.
9. Until 1850, the number of published book titles and journal titles was more or less equivalent, but then books remained static while journal titles increased 4-fold by the end of the century. See Cantor *et al.*, *Science In The Nineteenth-Century Periodical: Reading The Magazine Of Nature*, Cambridge University Press, Cambridge, 2004, p. 10.
10. C. A. Russell, *Science and Social Change in Britain and Europe 1700–1900*, Macmillan Education, London, 1983, p. 243.
11. There seems to have been more than one group that formed, ceased meeting, and reformed, so there is some doubt as to which was actually the real forerunner of the Royal Society.
12. Early members included Robert Boyle, Robert Hooke, Christopher Wren and (a little later) Isaac Newton.
13. A. Tinniswood, *The Royal Society: And the Invention of Modern Science*, Basic Books, New York, 2019.
14. Hannah Arendt proposed that the idea of 'objectivity' in science may have come from the charter of the Royal Society, because its members agreed to restrict their activities to its terms of reference, and in particular to abjure political or religious strife. "One is tempted to conclude that the modern

scientific ideal of 'objectivity' was born here, which would suggest that its origin is political and not scientific." H. Arendt, *The Human Condition*, University of Chicago Press, Chicago, 2nd edn, 1998, p. 271.

15. C. A. Russell, *Science and Social Change in Britain and Europe 1700–1900*, Macmillan Education, London, 1983, p. 73.

16. C. A. Russell, *Science and Social Change in Britain and Europe 1700–1900*, Macmillan Education, London, 1983, pp. 174–192, 220–234.

17. In 1661, for example, 40% of the Royal Society's efforts was in pure science and 60% was in practical issues of the day (applied science or technology). R. K. Merton, *Science, Technology and Society in Seventeenth-Century England*, Harper & Row, New York, 1938, p. 563.

18. A. Fyfe *et al.*, *A History of Scientific Journals: Publishing at the Royal Society, 1665–2015*, UCL Press, London, 2022.

19. The Society established a committee in 1752 to decide on the value of potential contributions. The Committee of Papers accepted or rejected papers by a majority vote. About 5% of papers were withdrawn and about 20% were rejected in the ballot. The review process was internal and cursory. The Committee of Papers would take the decision, with no concept of seeking opinions from experts in the field. A. Fyfe *et al.*, *A History of Scientific Journals: Publishing at the Royal Society, 1665–2015*, UCL Press, London, 2022, p. 189.

20. By this time the *Histoire et Mémoires* of the Paris Académie Royale had already introduced a more advanced system for reviewing publications. Its editorial committee often required revisions to discussion or even further experiments. This was really the start of scientific communication as we would define it today. *Mémoires* started publication in 1666. Its title changed to *Comptes Rendus* in 1835. Formal reviewing for *Philosophical Transactions*, in which referees wrote detailed reports on papers, began around 1830. A. Fyfe *et al.*, *A History of Scientific Journals: Publishing at the Royal Society, 1665–2015*, UCL Press, London, 2022, pp. 257–295.

21. It was not until the end of the 19th century that papers began to carry footnotes to attribute previous work. A formal bibliography with references to earlier papers did not appear until the 1920s, Dividing the paper into a format of separate formal sections occurred only at the end of the 1930s.

22. Peer review has become a cumbersome process, to the extent that there are often complaints that it stifles reporting of new developments by setting unreasonable standards and delaying publication. See H. S. Snyder, Science Interminable: Blame Ben?, *Proc. Natl. Acad. Sci. U. S. A.*, 2013, **110**, 2428–2429.

23. A. Tinniswood, *The Royal Society: And the Invention of Modern Science*, Basic Books, New York, 2019, ch. 7; S. Gaukroger, *The Emergence of a Scientific Culture: Science and the Shaping of Modernity 1210–1685*, Oxford University Press, Oxford, 2006, pp. 36–38.

24. R. D. Haynes, *From Faust to Strangelove. Representations of the Scientist in Western Literature*, Johns Hopkins University Press, Baltimore, 1994.

25. D. M. Knight, in *The Cambridge History of Science, Volume 5: The Modern Physical and Mathematical Sciences*, ed. M. J. Nye, Cambridge University Press, Cambridge, 2003, pp. 72–90.

26. Circulation of *Transactions* was probably a few hundred.

27. S. Shuttleworth and B. Charnley, Science Periodicals In The Nineteenth And Twenty-first Centuries, *Notes Rec.*, 2016, **70**, 297–304; J. R. Topham, The Scientific, the Literary and the Popular: Commerce and the Reimagining of the Scientific Journal in Britain, 1813–1825, *Notes Rec.*, 2016, **70**, 305–324.

28. G. Dawson, *et al.*, *Science Periodicals In Nineteenth-Century Britain*, University of Chicago Press, Chicago, 2020, p. 38.

29. G. Dawson, *et al.*, *Science Periodicals In Nineteenth-Century Britain*, University of Chicago Press, Chicago, 2020, pp. 35–61.

30. M. Baldwin, *Making "Nature": The History of a Scientific Journal*, University of Chicago Press, Chicago, 2015, p. 22.

31. Many of the early articles were published in the *Philosophical Magazine* (full title: *Philosophical Magazine and Journal of Science*, essentially a journal of physics).

32. Based on analysis of dataset at doi: 10.17617/3.7o collected by L. Bornmann, *et al.*, Growth Rates of Modern Science: A Latent Piecewise Growth Curve Approach to Model Publication Numbers From Established and new Literature Databases, *Humanit. Soc. Sci. Commun.*, 2021, **8**, 224.

33. The first attempt to quantitate the growth of science was made in 1961 when Derek de Solla Price plotted the number of scientific journals founded between 1650 and 1950. He obtained a growth rate of 4.7% annually and a doubling time of 14 years. He pointed out that a simple extrapolation would forecast 1 million journals by the year 2000. Fortunately, this has not come to pass. See D. J. de Solla Price, *Science Since Babylon*, Yale University Press, New Haven, CT, 1961, p. 166.

34. L. Daston, Fear and Loathing of the Imagination in Science, in *Science in Culture*, ed. P. Galison, S. R. Graubard and E. Mendelsohn, Routledge, London, 2001, pp. 73–96.

35. C. Bernard, *Introduction a la médecine expeérimentale*, 1865, p. 43.

36. Scientists themselves would have rejected any description as 'professional' through at least the end of the 19th century as denigrating to their status as gentlemen.

37. B. Lewin, *Inside Science: Revolution in Biology and its Impact*, Cold Spring Harbor Laboratory Press, New York, 2023, pp. 87–109.

38. Quoted in H. W. Eves, *Mathematical Circles Revisited*, Prindle, Weber and Schmidt, Boston, 1971, p. 158.

39. The sweeping effects of science on society were well summarized by Russell (see the beginning of the Culture of Science section).

40. The first edition of Frankenstein was published anonymously in 1818. The second edition in 1823 credited Mary Shelley as the author. A revised third edition in 1831 was the first illustrated edition and is considered the definitive text.

41. *The Public Image of Chemistry*, ed. J. Schummer, *et al.*, World Scientific Publishing, Singapore, 2007. In this book, see R. Haynes, The Alchemist in Fiction: The Master Narrative, pp. 7–36 and J. Schummer, Historical Roots of the 'Mad Scientist': Chemists in Nineteenth-century Literature, pp. 37–80.

42. The experiments involved the supposed revival of microscopic (protozoan) organisms.

43. Science fiction was first used in the mid-19th century to describe fiction or poetry depicting some aspect of current scientific knowledge. It was not until the 20th century that it took its present meaning of the application of science to create fantastical situations (www.oed.com/dictionary/science-fiction_n).

44. This was so prophetic that it influenced nuclear physicist Leo Szilard, who recollected, "In 1932, while I was still in Berlin, I read a book by H. G. Wells. It was called *The World Set Free*. This book was written in 1913... and describes the discovery of artificial radioactivity and puts it in the year of 1933, the year in which it actually occurred... [It describes] the development of atomic bombs... When I heard [about nuclear fission]... all the things which H. G. Wells predicted appeared suddenly real to me." Quoted in S. R. Weart and G. W. Szilard, *Leo Szilard: His Version of the Facts. Selected Recollections and Correspondence*, The MIT Press, Cambridge, MA, 1978, pp. 16, 53.

45. A. Roberts, *The History of Science Fiction*, Palgrave Macmillan, London, 2nd edn, 2016.

46. A. Huxley, *Brave New World*, Harper, New York, 1931 was influenced by an earlier book by biologist J. B. S. Haldane in which he considered the future effects on society of developments in the natural sciences. Haldane considered the 'case against science' and asked whether the progress of science could or should be stopped. His conclusions were generally optimistic, but made a different impression in Huxley's satirical presentation. J. B. S. Haldane, *Daedalus or Science and the Future*, Dutton, New York, 1924.

47. Variously titled as Nineteen Eighty-Four or 1984. G. Orwell, *Nineteen Eighty-Four*, Secker & Warburg, London, 1949.

48. P. J. Bowler, *Science for All: The Popularization of Science in Early Twentieth-century*, University of Chicago Press, Chicago, 2009.

49. *Conversations with Carl Sagan*, ed. T. Head, University Press of Mississippi, 2006, pp. 158–159.

50. S. Otto, *The War on Science: Who's Waging It, Why It Matters, What We Can Do About It*, Milkweed Editions, Minneapolis, 2016, p. 25.

51. A. J. Meadows, *Communication In Science*, Butterworths, London, 1974.

52. B. Lewin, *Inside Science: Revolution in Biology and its Impact*, Cold Spring Harbor Laboratory Press, New York, 2023, pp. 109–126.

53. F. Wolfe-Simon, *et al.*, A Bacterium That Can Grow by Using Arsenic Instead of Phosphorus, *Science*, 2010, **332**, 1163–1166.

54. Arsenic is located immediately below phosphorus in the periodic table, which means that its chemical properties are similar. Carbon and silicon have a similar relationship to each other.

55. M. Kaku, Life as We Don't know It, *Wall Street Journal*, 2010, December 6.

56. For reviews of the history of cold fusion see J. R. Huizenga, *Cold Fusion: The Scientific Fiasco of the Century*, University of Rochester Press, Rochester NY, 1992; F. Close, *Too Hot to Handle: The Race for Cold Fusion*, Princeton University Press, Princeton, NJ, 1991.

57. J. R. Huizenga, *Cold Fusion: The Scientific Fiasco of the Century*, University of Rochester Press, Rochester NY, 1992.

58. Most reporting accepted the results uncritically, but there were some distinguished exceptions, including *Time*, the *Economist*, and the *Wall Street Journal*.

59. G. Taubes, *Bad Science: The Short Life and Weird Times of Cold Fusion*, Random House, New York, 1993, pp. 11–14.

60. W. H. Breckenridge, Letter to the Editor, *Wall Street Journal*, 1989, April 26.

61. J. R. Huizenga, *Cold Fusion: The Scientific Fiasco of the Century*, University of Rochester Press, Rochester NY, 1992.

62. U.S. Department of Energy Reports, interim DOE/S-0071 (August 1989) and final DOE/S-0073 (November, 1989).

63. None of this hides the fact that in science, "we really can say who was right." (see Weinberg, quoted at the beginning of the Culture of Science section.)

64. G. Kolata, Viruses or Prions: An Old Debate Still Rages, *New York Times*, 1994, October 4, p. C1.

65. B. Simon, Undead Science: Making Sense of Cold Fusion After the (arti)fact, *Soc. Stud. Sci.*, 1999, **29**, 61–85.

Rabbit Holes in Science

"IT APPEARS THAT THE PEOPLE WHO WOULD BENEFIT MOST by this work being discredited have taken the initiative to cause us great difficulty... They might cause us difficulty, but they will not stop the science," said Stanley Pons, instigator of cold fusion, one of the major scientific scandals of the 20th century.[1] This was certainly inappropriate as a response to the criticism of cold fusion, but the statement could legitimately be applied to several great discoveries that took far too long to be appreciated. Science is a zigzag: theories come and go as they are supported or discredited by data. Some theories seem reasonable in their time, but become untenable as the worldview changes. By the 20th century, standards were higher for promulgating a theory (and indeed this could backfire in the refusal to believe theories that were in fact later shown to be true). The increasing pace of science enabled incredible theories to be disposed of more rapidly, but this still did not stop scientists from sometimes following them down a rabbit hole.

The Frontiers of Science
By Benjamin Lewin
© Benjamin Lewin 2026
Published by the Royal Society of Chemistry, www.rsc.org

Timeline for Proposing and Abandoning Discredited Theories

1700	Stahl: phlogiston as fire-like element (1703): disproved 1787 Newton: luminiferous aether (1704): disproved 1887
1750	
	Lavoisier: caloric as element of heat (1783): disproved 1843
1800	Lamarck: evolution by acquired characters (1809): disproved 1930s
	Haeckel: ontogeny recapitulates phylogeny (1824): disproved 1930s
1850	Le Verrier: discovery of planet Vulcan (1860): disproved 1915
1900	Blondlot: discovery of N-rays (1903): disproved 1904
1950	McConnell: RNA transfers memory in worms (1962): disproved 1966 Derjaguin: discovery of polywater (1966): disproved 1973
	Benveniste: memory of water (1988): disproved 1988 Flesichmann and Pons: cold fusion (1989): disproved 1990
2000	

A Dialog between Samuel Hahnemann and Robert Koch

Samuel Hahnemann (1755–1843) was a German physician who created the concept of homeopathy as a reaction to his concern about the damage done by conventional medical treatments.

Robert Koch (1843–1910) was a German bacteriologist who established Koch's postulates for defining an infectious disease. One of his discoveries was the bacterium for tuberculosis.

Samuel Hahnemann: Good day, Dr. Koch. It's an honor to be in the presence of a pioneer in the field of microbiology. I am known for my contributions to homeopathy, a field that predates your groundbreaking work on infectious diseases. How do you view the intersection of our respective contributions to medicine?

Robert Koch: Good day, Dr. Hahnemann. I appreciate the opportunity to engage in conversation with a figure from a different era of medical history. Indeed, homeopathy and microbiology represent distinct approaches to healing. How do you explain the significance of homeopathy relative to disease?

Samuel Hahnemann: Our approaches are indeed different. In homeopathy, the approach is to stimulate the body's vital force to restore balance and promote healing. It's based on the idea that like cures like, using highly diluted preparations. It's a holistic approach that considers the individual's symptoms and constitution.

Robert Koch: In microbiology, our focus has been on identifying and isolating specific microbes responsible for diseases. We've developed Koch's postulates to establish a causal relationship between a microbe and a disease. How can homeopathy cure an infection?

Samuel Hahnemann: Homeopathy views diseases as disruptions in the vital force. Remedies are chosen to match the symptoms presented by the individual, based on the principle of similars.

Robert Koch: How do you address the skepticism and criticisms that homeopathy has faced, particularly regarding the high dilutions of substances?

Samuel Hahnemann: Skepticism is not new to homeopathy. The principles of potentiation involve serial dilution and stirring to enhance the healing properties of substances. While it may challenge conventional notions, homeopathy has been embraced by many individuals who have experienced its benefits.

Robert Koch: We've faced our share of skepticism as well, particularly in the early days when the existence of microbes and their role in diseases were debated. Rigorous experimentation has been crucial in establishing the foundations of microbiology. How do you view the role of experimentation and evidence in homeopathy?

Samuel Hahnemann: The challenge lies in finding methods of research that align with the principles of homeopathy. The mechanisms in homeopathy are still a mystery. The concept of vital force, dynamic potentiation, and individualized treatment are key tenets.

Robert Koch: (*shaking head*) There you go again. Using Koch's postulates has established cause and effect for many diseases. The diseases have been treated by using appropriate methods, which work for everyone, not merely selected individuals. Homeopathy remains a matter of belief for the individual. Homeopathy is to medicine as water is to substance. I rest my case.

RABBIT HOLES IN SCIENCE

Alice was just in time to see [the White Rabbit] pop down a large rabbit-hole under the hedge. In another moment down went Alice after it, never once considering how in the world she was to get out again. The rabbit-hole went straight on like a tunnel for some way, and then dipped suddenly down, so suddenly that Alice had not a moment to think about stopping herself before she found herself falling down a very deep well, Alice's Adventures in Wonderland, 1865.

Presenting science as a series of major discoveries makes it seem that it advances in a straight line. Many of the wrong turns tend to be forgotten, but errors and failures can be just as informative as successes about the nature and progress of science. Sometimes, an error leads into a rabbit hole, and science proceeds a long way into a blind alley. This may happen because available techniques are inadequate to resolve the issue; or it may result from a willful refusal to believe negative evidence.

Although science is supposed to be objective, it's a human activity. Scientists sometimes descend into collective hysteria. Irving Langmuir, awarded the Nobel Prize for Chemistry in 1932, introduced the term 'pathological science' to describe cases "where there is no dishonesty involved but where people are tricked into false results by... being led astray by subjective effects, wishful thinking or threshold interactions."[2] These can be illuminating about the scientific process.

Pathological science can sometimes show the social interactions within science very well, because it highlights how the system works when there is conflict. One moral is that you can fool yourself, but you can't fool the body of science for very long.

Cold fusion was the most dramatic and public display of a rabbit hole in science in recent times. The rabbit hole was deep, as witnessed by the large number of scientists who became involved in chasing cold fusion, but it did not take very long to extricate them. Cold fusion was announced at a press conference on March 23, 1989. It was much in doubt by the time of the American Physical Society meeting in Baltimore on May 1, where problems with the original data were highlighted. There were reports that it could not be replicated. A definitive report came from the Harwell Lab in England on June 15. By August, a report

- The maximum effect that is observed is produced by a causative agent of barely detectable intensity, and the magnitude of the effect is substantially independent of the intensity of the cause.
- The effect is of a magnitude that remains close to the limit of detection or, many measurements are necessary because of the very low statistical significance of the results.
- There are claims of great accuracy.
- Fantastic theories contrary to experience are suggested.
- Criticisms are met by *ad hoc* excuses thought up on the spur of the moment.
- The ratio of supporters to critics rises up to somewhere near 50% and then falls gradually to oblivion.

from the DOE (Department of Energy) had concluded that the phenomenon did not exist. (I discuss public reactions in Chapter 24.)

On the one hand, this farce did not do much for the public reputation of science, with reports appearing that purported to confirm the phenomenon before it became apparent that the data were wrong. (One wonders how they obtained their results!) However, it shows that the scientific process ultimately works. An important observation was reported, there was a rush of attempts to confirm it, and as a result of their failure, it was rejected from the scientific canon.

The main moral here, however, is not the success or failure of the scientific process. It's the lack of skepticism applied to the work by scientists who should have known better, as well as by the media (which should at least have taken expert opinions before rushing to acclaim the solution to humanity's energy problem). This was, in fact, the largest known collective failure of the scientific attitude in history. In contrast with the reluctance to accept the overthrow of paradigms in genetics discussed in Chapter 19, here there was a mad rush to embrace the new paradigm. Ironically this sometimes involved the same sort of tortured logic that centuries earlier had been invoked to 'save the phenomena' and avoid a new paradigm. But data (or the lack thereof) ultimately triumphed.

Was it an example of pathological science? The original paper was sloppy in the extreme, missing essential information. It was immediately obvious to physicists that there were major difficulties in the evidence for fusion, largely due to the failure to demonstrate the presence of the predicted products of fusion, which posed a major problem in terms of the laws of physics.[3] One obvious flaw in the paper was the lack of controls. An obvious control would be to perform the same experiment with water (H_2O) instead of heavy water (D_2O). It's possible that the control was in fact done, but also appeared to show fusion. Fleischmann and Pons were asked repeatedly about this at scientific meetings, but never answered.[4] When a control is positive, the usual reaction is to doubt the experiment!

In response to criticism, the authors were evasive or shifted their ground, offering a classic example of the need to believe their own results. Because of this, no one has ever been able to pinpoint exactly what went wrong. At a minimum, the authors, Stanley Pons and Martin Fleischmann, who were chemists, were well out of their depth in nuclear physics. At all events, the continuation of belief in cold fusion by a small band of researchers, against all the accumulated evidence, is a current demonstration of pathological science, with researchers trapped in the rabbit hole.

∞ ∞ ∞

A defining case of pathological science occurred almost a century earlier, in 1903, when René Blondlot discovered N-rays, named for the city of Nancy where he was a professor at the University. Wilhelm Röntgen had discovered X-rays in 1895. Becquerel's discovery of radioactivity a year later led to the identification of α-, β-, and γ-rays. Blondlot was working on X-rays, in particular to try to distinguish whether they were particles or waves. He claimed that rays emanating from an X-ray generator tube were bent when they passed through a quartz prism. This was surprising, because it had been established previously that a quartz prism does not deflect X-rays.

The basic apparatus was simple. X-rays were generated by a cathode ray tube that was part of a circuit connected to a spark gap. When the rays were generated, a spark simultaneously jumped across the gap. The brightness of the spark could be changed by placing a quartz prism between the spark gap and the cathode ray tube. This was taken to mean that the rays had been refracted.[5] Blondlot concluded that therefore they could not be X-rays.

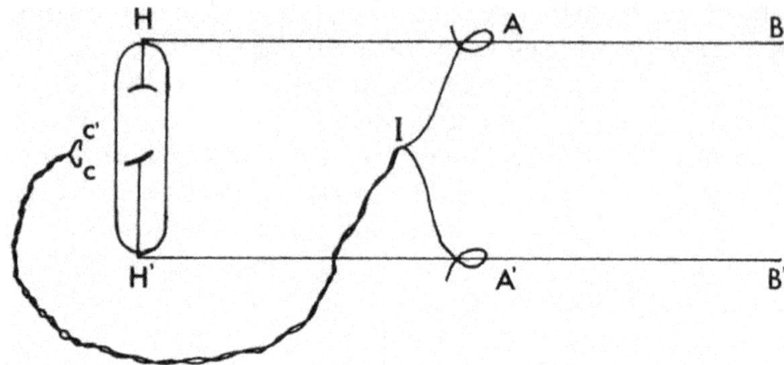

Blondlot's schematic of his apparatus did not contain much detail. The cathode ray tube at H–H′ was powered by an induction coil at B–B′. The spark gap C–C′ was linked into the same circuit.

The N-rays continued to be produced when the voltage was reduced to below the level at which X-rays were produced. Further sources for N-rays were identified, including one especially far-fetched report (from Auguste Charpentier in the department of medical physics at Nancy) that suggested they could be generated by the human body. Ridiculous though this may seem, it received an enthusiastic reception.[6] Blondlot showed that N-rays could pass through materials such as wood, paraffin, and aluminum. He also introduced a photographic plate as an alternative means of detection. *Comptes Rendus*, the journal of the Académie des Sciences in Paris, published more papers on N-rays than on X-rays in 1904.[7]

In contrast with the general air of enthusiasm, some distinguished scientists, including Pierre Curie, reported they were unable to replicate the phenomenon. There should also have been some pause for thought on theoretical grounds, as the wavelength of the N-rays was inconsistent with passing transparently through quartz. None of these factors had the impact they should have had.

The phenomenon collapsed as the result of a visit to Blondlot's laboratory by Robert Wood, of Johns Hopkins University of Baltimore, who was well known for his activities in debunking spurious claims from mystics or others. Wood had been unable to repeat Blondlot's experiments in his laboratory in Baltimore.

He criticized Blondlot's reliance on visual assessment of the brightness of the spark and thought the effects were too slight to be significant. He accepted that there were differences of intensity in the photographic plates, but thought the experiments were not properly controlled.

The killer result, however, as he reported in a letter to *Nature*, was that the effects of the N-rays remained unchanged when he surreptitiously removed the prism. Wood concluded, "The few experimenters who have obtained positive results have in some way been deluded."[8] Blondlot replied that special expertise was required to make the observations; "some people in fact never succeed."

∞ ∞ ∞

"More is known about water than about any single chemical entity," said Irving Klotz, in a book debunking pseudoscience.[9] Water is tricky stuff, with anomalous properties that have been hard to understand in terms of physics. This may be why there has been so much bad science on the subject of water.

Water has unusual thermal properties, with a very high capacity for absorbing heat. This is important for its physiological role. It also has very high latent heat (the heat required to change its form from ice to water or from water to steam). These properties make it a reservoir for protecting life and the environment. It also has the most unusual property that it expands in volume as it changes from liquid to solid. Most compounds become denser when they change from fluid state to solid state: but ice is lighter than water. If it were heavier, the topography of the planet would be different (ice would sink instead of floating in winter—and might never become warm enough again to melt).

Establishing how water's atomic components are responsible for its properties has been a challenge, because they defy expectations based on other molecules.[10] It's no exaggeration to say that life on planet Earth would not be possible without these extraordinary properties.

The basic principles of the structure of water were established in the 1930s.[11] Water molecules hydrogen bond with one another to form a tetrahedral group. (Hydrogen bonding involves a weak interaction between a hydrogen atom in one molecule of water and an oxygen atom in another molecule of water.

The strength of the bond is about 5% of that of the covalent bond linking oxygen and hydrogen within a molecule of water.) The structure is better organized in ice than in the liquid form of water. Other structures can form in ice under high pressure to create forms with increased density.

The surface tension of water is unusually high—in fact, the only liquid with a higher surface tension is mercury. It's also an unusually effective solvent (with a high capacity to dissolve other substances). All these properties make it uniquely positioned to support life.

Working in Moscow, Boris Derjaguin (sometimes Deryagin in transliteration), a physical chemist, reported a new form of water in 1966.[12] He called it anomalous water, but it was later named polywater, reflecting its supposed properties. Derjaguin had had a distinguished career in colloid chemistry—he was rumored to be in line for a Nobel Prize—when he encountered the work of Nikolai Fedyakin. This purported to show that when water was condensed in, or forced through, capillary (very narrow) tubes made of quartz, it generated a new form of water. This had a higher boiling point (about 150 °C), lower freezing point (−15 °C to −30 °C), and viscosity 15× higher than normal water.[13]

Derjaguin developed a method for confirming the observations by condensing water into tiny capillaries in a container where the vapor pressure was below saturation. The phenomenon appeared to defy the laws of thermodynamics, because it should not be possible for the water to condense into the liquid phase in the capillary if the pressure above it is below the saturation point.

The name 'polywater' resulted from experiments in 1969 that appeared to show spectra for the anomalous water that were different from normal water and might be explained by a polymer.[14] Scientific meetings were devoted to the topic. Theoreticians proposed models to explain the supposed properties of polywater. The media gave the phenomenon a rapturous reception.[15] And on the grounds that it might form spontaneously (never mind the 4.5 billion year history of Earth in which this had not yet happened), it was even called (by one misguided scientist) "the most dangerous substance on earth."[16]

There were points at which the polywater story seemed to be recapitulating Kurt Vonnegut's science fiction story, *Cat's Cradle*, published in 1963, in which a novel form of ice, Ice-9, was an alternative form of water that was solid at room temperature.[17] Because Ice-9 had the property of converting any liquid water that touched it into solid form, it had the potential to destroy the planet—which indeed it does at the end of the book.

A considerable amount of research went into attempts to confirm the observations and scale up the production of polywater. While there were some reports suggesting the water in the capillaries had anomalous properties, there was never any success at scaling up. Some of the supportive data took the form of infrared or NMR scans of the capillaries. Eventually it occurred to researchers that they should perform a control with empty capillaries—and the entire phenomenon fell apart. It was essentially due to impurities in the water caused by silica leached from the capillary tubes, even though quartz capillaries had been used rather than glass (because quartz is pure, whereas glass contains impurities more likely to react with the water).[18]

Unlike situations such as cold fusion, the observations were in fact real, but the interpretation was wrong. (In fact, there had been earlier studies reporting analogous phenomena, but the authors had interpreted them in terms of anomalies caused by the walls of the capillaries.[19]) And unlike phenomena that degenerated into pseudoscience, here Derjaguin accepted the results from other laboratories. In 1973, he withdrew his proposal for polywater.[20] Nonetheless, there was a surprising level of gullibility, not to say alarm, before polywater was finally debunked.

It was at the least a failure of the scientific process to create such a frenzy in the community about data that lacked proper controls. The tiny amounts of polywater that had been produced should have given cause for doubt. The facetious tone of one exchange is revealing. At a meeting of the Soviet Academy of Sciences, in response to the question "how much modified water has been obtained at all?" Derjaguin replied, "About enough for 15 articles."[21]

Several announcements that should have been important contributions in formal papers were made by informal means. Many papers were published as short communications, without full details, in order to stake claims to priority. Often enough, the full details were never published. Several hundred papers were published during polywater's short lifespan. The wish of the

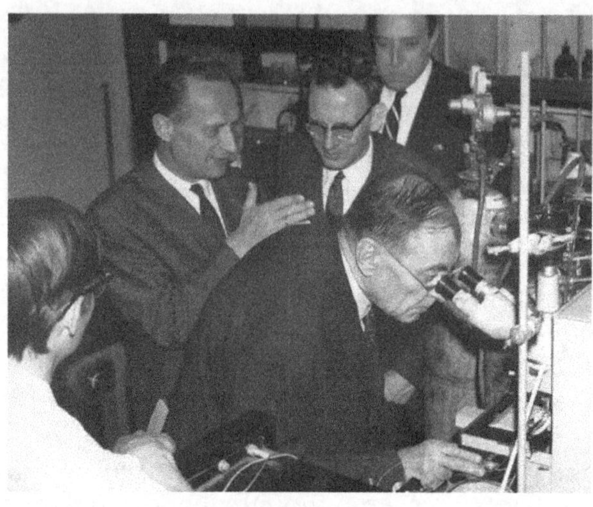

Boris Derjaguin (at the microscope) in his laboratory.

leading general journals of science (*Nature* and *Science*) to publish 'hot' papers ran well ahead of concern for rigorous standards. Excess speculation and lack of proper controls were prominent features.[22] (It's ironic that Derjaguin's first attempt to publish outside of Russia was rejected by *Nature* on the grounds that insufficient care had been taken to exclude impurities![23] Much embarrassment would have been spared if the journal had continued to apply skepticism.) One feature that has become even more marked since then was a politicization of the review process.[24] Papers rejecting polywater found it more difficult to get published than papers supporting it.[25]

After the debacle, opinions were split in the scientific community. Some people felt that the (non)occurrence of polywater illustrated the scientific process working well (because papers were published, analyzed, and eventually rejected on the basis of new data). Others felt it was a disaster (because at least for a period the phenomenon was elevated to great importance because of the hype). The hundreds of scientists who were involved should surely have thought harder about impurities before rushing into publication; and the review process should have been more rigorous and less biased.

Polywater was certainly not a glorious moment in the history of science. Does it count as pathological science? It certainly fits most of Langmuir's criteria (see the box above). The saving grace is that the phenomenon was disposed of in relatively short order, but it's dispiriting that so many researchers could have been fooled (or could have fooled themselves) so easily. It casts a poor light on claims for the objectivity of the scientific attitude.

∞ ∞ ∞

Water is so central to life that you would think that by now we understand it very thoroughly. But in 1988, there was a great stir, and great skepticism, when Jacques Benveniste at the University of Paris reported that water can have a 'memory.' The experiment was in principle quite simple. A certain type of white blood cell has immunoglobulins of the IgE type on its surface. When these cells are exposed to anti-IgE antibodies, they react by releasing histamine. Benveniste found that histamine release can be triggered when the anti-IgE preparation is enormously diluted, to the point at which there are in fact no anti-IgE antibodies left in it.[26]

He proposed that the water retains a 'memory' of its prior exposure to the anti-IgE antibodies. The implications appeared to validate homeopathy, which uses preparations diluted to the point at which they no longer contain any molecules of the active ingredient (see Chapter 21). Homeopathy is a significant part of medical practice in France.[27]

Nature published the article with an almost unprecedented editorial statement, which more or less admitted that reviewers had recommended against publication.[28] "There is no physical basis for such an activity... *Nature* has therefore arranged for independent investigators to observe repetitions of the experiments." Unfortunately, this did not take a sober course comparable to Wood's investigation of N-rays a century earlier. It became something of a circus when John Maddox, the Editor of *Nature*, included the magician James Randi on the investigating team.

The report concluded that the experiments were 'statistically ill-controlled' (you would have thought this could have been determined during the original review). It also complained that there were systematic biases from sampling errors—a classic feature of pathological science. Benveniste never provided an explanation for his supposed results, although he did later propose they might work by electromagnetic signals.

Jacques Benveniste in his laboratory.

© *Sergio Gaudenti/Sygma via Getty Images.*

Attempts to replicate the results in other laboratories were unsuccessful.[29] The 'investigation' was quite rightly criticized as bringing an inappropriate circus atmosphere into science, indeed in bringing science as a serious endeavor into disrepute. John Maddox no doubt regarded it as a success for the publicity it brought to *Nature*. A fair comment came in a letter published shortly after in *Nature*. "So now at last confirmation of what I have always suspected. Papers for publication in *Nature* are refereed by the Editor, a magician and his rabbit."[30]

∞ ∞ ∞

Once science has climbed out of a rabbit hole, the topic is usually discredited to the point at which few people are interested in revisiting it. There are exceptions, however. One interesting example is the proposal that some form of the nucleic acid, RNA, might carry memory. This first saw light of day in what came to be called the 'planarian controversy.'

Planarians are small flatworms that live in freshwater. There is a ganglion (a cluster of nerve cells found in the peripheral nervous system) at the head, from which nerves extend along the body to the tail. The ganglion is sometimes called the brain of the planarian, although that is a bit of a stretch. Different species of planarians vary in size from about 1 mm to 1 cm. They have an

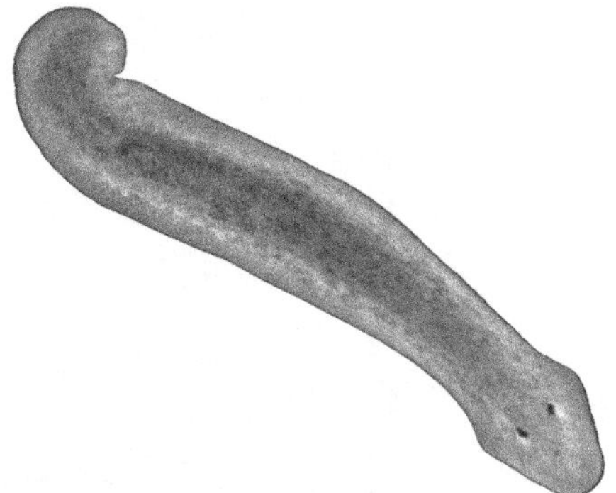

A planarian flatworm lives in freshwater.

Reproduced from https://commons.wikimedia.org/wiki/File:Dugesia_subtentaculata_1.jpg, under the terms of the CC BY-SA 3.0 license, https://creativecommons.org/licenses/by-sa/3.0/.

extraordinary ability to regenerate body parts. The species used for learning experiments was *Girardia dorotocephala*. Because it is cannibalistic, it is easy to transfer material from one animal to another.

The planarian controversy started when James McConnell, working at the University of Michigan, published a paper in 1962 with the catchy title *Memory Transfer Through Cannibalism in Planarians.*[31] He had previously worked on developing a conditioned response in planarians (following the principle of Pavlov's experiments with dogs). Planarians respond to a shock by contracting or turning.

The planarians were given a stimulus of 3 sec exposure to light, followed by a 1 sec electric shock. The evidence for a conditioned response would be that they show the same response to light alone. And the response should be enhanced by increasing the number of training trials.

The data did not decisively resolve the issue of whether planarians could be trained, as the effects were relatively small. It was controversial at the time whether invertebrates could show a learned response. And if they could, would it follow that the same processes are responsible for memory in higher organisms?

One idea was to cut the animals in half to see which end retained the conditioned response. The results were interpreted as

The cover of an issue of the Worm Runner's Digest, showing the planarian group at work.

Courtesy of Larry Stern, Professor of Sociology, Collin College (Plano, Texas).

showing that both parts of the experimental animals could regenerate full animals that retained the training. This suggested there could be a chemical basis for memory. Attempts to transfer material by extraction and injection ran into technical problems. Then the controversy intensified with the 1962 paper reporting that planarians that cannibalized the experimental animals showed the trained response.

The first reports had been published in journals out of the mainstream. It did not help that McConnell published some of his work in his own journal, the *Worm Runner's Digest*, which was a sort of counter-culture mix of serious results and satire.[32] (It ceased publication in 1979.) By the mid 1960s, some reports had been published to confirm the studies, and others to say that they were artifacts of the experimental conditions.[33,34]

It's not entirely fair to call the planarian controversy a rabbit hole, because the theory was never widely accepted. Relatively few researchers chased after the phenomenon.[35] But McConnell was a charismatic figure who had a greater effect on the public than his fellow scientists, leading to press reports along the lines that, "uneducated worms can acquire the wisdom of more intellectual ones by eating them."[36] Opinion in science came around to the view that the controls were not adequate to demonstrate that a learned response had occurred. What had probably happened was a general stimulation of the response.[37]

The implication of a specific molecule in memory began with a report from another laboratory, published in *Science,* that planarians placed in a solution that destroys RNA could not maintain a trained response.[38] The controversy was exacerbated when McConnell reported that a preparation containing RNA could confer the trained response when it was obtained from experimental animals and fed to new animals.[39] As these results were published in the *Worm Runner's Digest* and reported in the popular science magazine *New Scientist,* they were not taken very seriously. Perhaps this is another marker for pathological science: publication in journals that do not have the expertise to critically review the work in question.[40]

Other studies have examined the possible role of RNA in animals that definitely can learn: rats. This work used a variety of training techniques: using a Skinner box, or running through a maze to obtain a reward or avoid a shock, and so on. RNA was extracted from the brains of trained rats and injected into a new series of rats. Some striking claims for learning were published in top journals, but 50 years on, have not led anywhere. Unlike phenomena that imploded because of a demonstration of the nature of the error, RNA as a carrier of memory just faded away, and became a fringe area of science. If not a rabbit hole, it's in any case an illustration of how a wrong turn is corrected.

The mechanism of science has worked better in this area than in some rabbit holes, because appropriate skepticism was expressed from the start. There was a constant yin and yang of papers supporting and refuting the phenomenon. It's perfectly normal for papers to be published that are completely wrong, but if science works properly, they should soon be exposed.

Aside from detailed criticism of the results, one reason for skepticism was the lack of any reasonable mechanism by which RNA could carry memory. The idea that a single type of molecule could carry memory was also at odds with the neuron doctrine that memory results from synaptic connections.[41] This is another reprise of the theme that a theory to explain mechanism can be an important part of accepting experiments. Knowledge of RNA was primitive in the 1960s, but today we know that it has a variety of functions that were unsuspected then. Perhaps because RNA is now implicated in controlling gene functions, the issue has been reopened. This is unusual.

It was dèja vu all over again when a group from the University of California, Los Angeles reported in 2018 that they could transfer a learned response by injecting RNA extracted from trained *Aplysia* (a sea slug) into naïve slugs.[42] Although we are back to dealing with an invertebrate, there is a better basis for characterizing a learned response in *Aplysia* than in planarians. The work has not been refuted—once again it has achieved a better response in the media than in the world of science—but the absence of follow-through several years later characterizes it as an attempt to clamber back down the rabbit hole. However, the tale ends with a whimper rather than a bang.

∞ ∞ ∞

Biodynamic agriculture is not science, but belongs here because (like homeopathy) it is presented as a scientific method. Biodynamic preparations are made from a variety of animal and plant sources using 'dynamized water' (basically water that has been stirred vigorously). They are most commonly used to treat vineyards. The solutions are very dilute (but not to the point of zero content).

Biodynamics may be the deepest rabbit hole of all, because it has theory (of a sort) but no controlled data. I am not aware of anyone who has ever returned from it. Of course, that may be because practitioners of biodynamics do not regard it as a rabbit hole at all: rather, it is the New Jerusalem. It's an example of quasi-science. By this I mean that it could be tested scientifically (or at least parts of it could), but its followers are so committed (and not much given to scientific analysis) that they never do so. This belief structure resembles pathological science. Perhaps it's unfair to judge biodynamics by the standards of science, because in some ways it more resembles a religion—but to some extent, that is true of pathological science.

Biodynamic methods were developed by Rudolph Steiner, who created a theory of 'anthroposophy' or 'spiritual science.' This was based on the view that the West would cause destruction if it did not understand the relationship of the spiritual world to the physical world. He applied this to agriculture, and described his 'theory' in a book that became the basis of the biodynamic method.[43] It's a mix of the astrological and the practical that could easily be mistaken for the work of an astrologer or alchemist of the medieval era.

Biodynamic agriculture ranges from the possibly plausible (treatments with very dilute solutions) to the downright superstitious (astrology and the lunar calendar). Steiner did not really distinguish between them. He did suggest that the methods should be tested experimentally and established a group, with about 800 members worldwide, to try anthroposophy. Eventually this gave rise to the Demeter organization, which certifies biodynamic producers. It sets out detailed protocols for the application of biodynamics, but there has never been any experimental test of the procedures.

Most scientists regard biodynamics as pseudoscience,[44] but in fact it would be possible to test some of its precepts. However, a serious difficulty would be to decide what would be the appropriate test. What would be measured for a biodynamic vineyard—the fungal condition of the soil? the sugar levels in the grapes? the taste of the wine?

Biodynamics is presented by its believers as a rational approach to returning towards the benefits of traditional farming methods. Practitioners would be most indignant if you told them that it's riddled with superstition. This is not a fringe movement. It's used by some of the best wine producers in the world. Many of them have wineries that look like models of modern technology. Its penetration of the world of wine—it's taken hold in viticulture but not in any other form of agriculture—is a testament to the success of misappropriating the concept of a scientific approach.

∞ ∞ ∞

A common feature of pathological science is the collective suspension of critical faculty. A major responsibility must lie with the peer review system. The inevitable question, of course, is whether and why the critical faculty of the gullible reviewers is any better in other situations?

All these cases of pathological science—cold fusion, N-rays, polywater, memory water—fit Langmuir's first criterion of marginal measurable effects. Perhaps that explains why the initial observation was followed by reports confirming the phenomenon as well as those refuting it. In each case, it was the combination of the theoretical difficulty of explaining the results, coupled with (ultimately) a preponderance of negative reports, that led to rejection of the concept. This poses the follow-up question: what is the difference, and what happens, when a marginal phenomenon is ultimately judged to be true?

There are several cases where, in all honesty, the results are not much clearer than those in cases judged to be pathological science.[45] They came to be accepted into the canon of science not so much because they were in themselves decisive, but because they were consistent with other experiments supporting a theory. (I discuss the measurements of the ether in Chapter 9, and the 1919 eclipse that confirmed relativity in Chapter 16.) Reviewing these situations in the light of pathological science causes us to realize that the picture is not always black and white. When data clash with a well-established theory, it's not always obvious whether to discard the theory or to ignore the data.

Critical judgment tends to be suspended when there is a crisis. The AIDS and COVID epidemics were both marked by many substandard reports that were not rejected as swiftly as they might have been but for the urgency of the situation. And if the frenzy of enthusiasm for a new result leads a suspension of critical faculty in pathological science, is the situation any different in the flurry of reports confirming a 'hot' new finding that is ultimately judged to be true?

The history of science makes it clear that it's just as important to teach the scientific attitude as to teach the facts of science. Indeed, a bare recitation of rote learning for scientific facts is really quite opposed to what science is about. If anything, it opens the way to irrational thinking. If there's no understanding of how they were established, how can one set of facts (such as evolution) be distinguished from another set of 'facts' (such as creationism). Failing to understand just why facts are facts will take us straight back to the Dark Ages.

NOTES AND REFERENCES

1. Quoted by Joann Jacobssen-Wells, U.S. Fusion Panel Cancels Plans to View University Research, *Deseret News*, 1989, May 28.
2. R. N. Hall transcribed and edited a colloquium given at the Knolls Research Laboratory, December 18, 1953, from a microgrove disk found among Langmuir's papers in the Library of Congress. I. Langmuir and R. N. Hall, Pathological Science, *Phys. Today*, 1989, **42**, 36–48.
3. J. R. Huizenga, *Cold Fusion: The Scientific Fiasco of the Century*, University of Rochester Press, Rochester NY, 1992; F. Close, *Too Hot to Handle: The Race for Cold Fusion*, Princeton University Press, Princeton, NJ, 1991.
4. F. Close, *Too Hot to Handle: The Race for Cold Fusion*, Princeton University Press, Princeton, NJ, 1991, pp. 147–151, 293–294.
5. R. Blondlot, *"N" Rays. A Collection of Papers Communicated to the Academy of Sciences*, translated by J. Garcin, Longmans & Green, London, 1905, Reprinted 2017, Leopold Classic Library.
6. M. J. Nye, N-rays: An Episode in the History and Psychology of Science, *Hist. Stud. Phys. Sci.*, 1980, **11**, 125–156.
7. I. Klotz, *Diamond Dealers and Feather Merchants: Tales From the Sciences*, Springer, New York, 1986, pp. 39–66.
8. R. W. Wood, The N-rays, *Nature*, 1904, **70**, 530–531.
9. I. Klotz, *Diamond Dealers and Feather Merchants: Tales From the Sciences*, Springer, New York, 1986.
10. Boiling point is expected to relate to the size of the molecule: this would predict a boiling point around -93 °C. Even more surprising, there is no known mechanism for a compound to expand as it cools down. See F. Franks, *Polywater*, The MIT Press, Cambridge, MA, 1981.
11. J. D. Bernal and R. H. Fowler, A theory of water and ionic solution with particular reference to hydrogen and hydroxyl ions, *J. Chem. Phys.*, 1933, **1**, 515–548.
12. B. V. Derjaguin, Effect of lyophilic surfaces on the properties of boundary films, *Discuss. Faraday Soc.*, 1966, **42**, 109–119.
13. After a paper published in 1962 with both Derjaguin and Fedyakin as authors, Derjaguin appropriated the phenomenon and Fedyakin more or less disappeared from the scene. B. V. Deryagin and N. N. Fedyakin, Special properties and viscosity of liquids condensed in capillaries, *Dokl. Akad. Nauk SSSR*, 1962, **147**, 403–406.
14. E. R. Lippincott, *et al.*, Polywater, *Science*, 1969, **27**, 1482–1487.
15. F. Franks, *Polywater*, The MIT Press, Cambridge, MA, 1981, pp. 74–77.
16. Showing how people (and journals) can lose their heads, *Nature* published a letter arguing that great precautions should be taken when working with polywater. F. J. Donahoe, Anomalous water, *Nature*, 1969, **224**, 198.
17. K. Vonnegut, *Cat's Cradle*, Holt, Rinehart and Winston, New York, 1963.
18. For an extensive review of polywater, see F. Franks, *Polywater*, The MIT Press, Cambridge, MA, 1981.
19. I. Klotz, *Diamond Dealers and Feather Merchants: Tales From the Sciences*, Springer, New York, 1986, pp. 89–91.

20. B. V. Derjaguin and N. V. Churaev, Nature of 'anomalous water', *Nature*, 1973, **244**, 430–431.

21. Quoted in F. Franks, *Polywater*, The MIT Press, Cambridge, MA, 1981, p. 110.

22. One obvious control for the spectral analysis was to see whether the phenomenon could be reproduced with heavy water (D_2O). Because hydrogen bonding in H_2O is responsible for the specific spectral pattern of water, D_2O is quite different.

23. W. Gratzer, *The Undergrowth of Science: Delusion, Self-deception, and Human Frailty*, Oxford University Press, Oxford, 2001, p. 96.

24. Political factors may have influenced decisions on papers submitted to leading journals during the COVID epidemic. Weak papers reporting a natural origin were published in leading journals, but papers suggesting an artificial origin encountered difficulties. See B. Lewin, *Inside Science: Revolution In Biology and Its Impact*, Cold Spring Harbor Laboratory Press, New York, 2023, pp. 77–78.

25. F. Franks, *Polywater*, The MIT Press, Cambridge, MA, 1981, pp. 144–163.

26. E. Davenas, Human basophil degranulation triggered by very dilute antiserum against IgE, *Nature*, 1988, **333**, 816–818.

27. About 10% of the population are prescribed at least one homeopathic preparation each year: M. Piolot, *et al.*, Homeopathy in France in 2011–2012 According to Reimbursements in the French National Health Insurance Database (SNIIRAM), *Fam. Pract.*, 2015, **32**, 442–448.

28. In 1972, *Nature* published a paper on a brain peptide that induces changes in behavior. John Maddox criticized the article and also published a reviewer's highly critical comments. It was more important to bring publicity to the journal than to stick to scientific standards. G. Ungar, *et al.*, Isolation, identification and synthesis of a specific-behaviour-inducing brain peptide, *Nature*, 1972, **238**, 198–202; W. Stewart, Comments on the chemistry of scotophobin, *Nature*, 1972, **238**, 202–210.

29. Two letters were published as correspondence in *Nature* under the title *Evidence of non-reproducibility*: J. Seagrave, and S. Bonini, *et al.*, *Nature*, 1988, **334**, 559.

30. Another correspondence article published in *Nature* in addition to those in ref. 29: K. Snell, *Nature*, 1988, **334**, 559

31. J. V. McConnell, Memory Transfer Through Cannibalism in Planarians *J. Neuropsychiatry*, 1962, **3**, 542–548.

32. J. McConnell, A Manual of Psychological Experimentation on Planarians, *Worm Runner's Digest*, Ann Arbor, MI, 1965.

33. D. R. Walker, Memory Transfer in Planarians: an Artifact of the Experimental Variables, *Psychon. Sci.,*, 1966, **5**, 357–358.

34. A. L. Hartry, P. Keith-Lee and W. D. Morton, Planaria: Memory Transfer Through Cannibalism Reexamined, *Science*, 1964, **146**, 274–275.

35. However, more than \$1 million grants were spent, more than 100 researchers published reports in the area, some in leading journals such as Nature or Science. See Stern in ref. 40.

36. J. Bird, The Worm Learns, *Saturday Evening Post*, 1964, **237**, pp. 66–67 (March 28).

37. The issue is not completely dead. A new claim that planarians can be trained using a more sophisticated system, and that the training persists when the head is regenerated, was made more recently. T. Shomrat and M. Levin, An Automated Training Paradigm Reveals Long-term Memory in Planarians and its Persistence Through Head Regeneration, *J. Exp. Biol.*, 2013, **216**, 3799–3810.

38. W. Corning and E. R. John, Effect of ribonuclease on retention of conditioned response in regenerated planarians, *Science*, 1961, **134**, 1363–1364.

39. A. Zelman, L. Kabat, R. Jacobson and J. V. McConnell, Transfer of training through injection of "conditioned" RNA into untrained planarians, *Worm Runner's Dig.*, 1963, **5**, 14–21.

40. McConnell failed in attempts to publish his results on planarians in the regular scientific literature. His later work on rats (from 1965) was published in the regular literature. For an account of the experiments and their reception see L. Stern, The Reception of Extraordinary Scientific Claims: The Search for the Elusive Engram, in *Ahead of the Curve: Hidden Breakthroughs in the Biosciences*, ed. L. Michael and A. Dany Spencer, IOP Publishers, Bristol UK, 2016, pp. 267–290.

41. The idea that memory involves specific changes in the brain was captured by Richard Semon's introduction of the term 'engram' in 1904 to describe a physical or chemical change that represents a memory. The term goes in and out of fashion. A. A. Josselyn, S. Köhler and P. W. Frankland, Heroes of the Engram, *J. Neurosci.*, 2017, **37**, 4647–4657.

42. A. Bédécarrats, *et al.*, RNA from Trained Aplysia can Induce an Epigenetic Engram for Long-term Sensitization in Untrained Aplysia, *eNeuro*, 2018, **5**, ENEURO.0038-18.2018.

43. R. Steiner, *Geisteswissenschaftliche Grundlagen zum Gedeihen der Landwirtschaft. Landwirtschaftlicher Kursus*, 1924; R. Steiner, *Agriculture: Spiritual Foundations for the Renewal of Agriculture*, ed. M. Garnder, translated by C. E. Creeger and M. Gardner, Bio-Dynamic Farming and Gardening Association, Incorporated, 1993.

44. L. Chalker-Scott, The science behind biodynamic preparations: A literature review, *HortTechnology*, 2013, **23**, 814–819.

45. H. Collins and T. Pinch, *The Golem: What You Should Know About Science*, Cambridge University Press, Cambridge, 2nd edn, 2012.

Science and Society

"WE LIVE IN A SOCIETY EXQUISITELY DEPENDENT ON SCIENCE and technology and yet have cleverly arranged things so that almost no one understands science and technology," said Carl Sagan.[1] Yet the relationship between science and society is not one way. There is an unwritten pact that if society supports the curiosity of scientists, it will be repaid with useful knowledge. Until the 19th century, science was mostly supported by patrons —who, of course, could be capricious in withdrawing support if they were so inclined. In the modern era, science is very largely funded by government. The piper began to call the tune at the end of the 20th century in the United States with political restrictions as to what science could be supported. An official organization was created to investigate fraud. With science viewed as part of the Establishment, its role is now more likely to be questioned, both from a public increasingly skeptical of government, and from intellectuals who regard its claims to be different as pretentious. Science now feels itself under attack as never so much since the Galileo affair.

The Frontiers of Science
By Benjamin Lewin
© Benjamin Lewin 2026
Published by the Royal Society of Chemistry, www.rsc.org

Timeline for Key Intellectual Criticisms of Science

1500 — Erasmus criticizes sterility of natural philosophy in *Praise of Folly* (1509)

1550

Michel de Montaigne argues science is a mechanistic artifice in *Essays* (1588)

1600

1650

1700 — George Berkeley argues science reflects perception not reality (1710)

Vico argues human culture is more reliable than science in *Scienza Nuova* (1725)

1750 — Rousseau argues science corrupts human morality in *Discours* (1755)

1800 — Goethe's *Faust* describes danger of using science to control Nature (1790)

Charles Babbage describes lack of scientific knowledge in society (1830)

1850

Matthew Arnold argues education is literature of Greece and Rome (1882)
Nietzsche criticizes science as mechanistic and dehumanizing (1886)

1900 — Max Weber argues science ignores the meaning of existence (1918)

Vannevar Bush calls for U.S. government to support unfettered science (1944)

1950 — Heidegger accuses science of being mindless – "science does not think" (1951)
C. P. Snow publishes *The Two Cultures* (1959)
Rothschild report in U.K. calls for government to direct science (1971)
Feyerabend argues science has no claim to special objectivity (1975)

Sokal submits hoax paper to *Social Text* (1996)

2000

A Dialog between Socrates and Paul Feyerabend

Socrates (*c.* 470–399 BCE) was an Athenian philosopher famous for the rigor of the method of Socratic dialog, forcing justification of every postulate and conclusion.

Reproduced from https://commons.wikimedia.org/wiki/File:UWASocrates_gobeirne_cropped.jpg under the terms of the CC BY-SA 3.0 license, https://creativecommons.org/licenses/by-sa/3.0/deed.en.

Paul Feyerabend (1924–1994) was an Austrian philosopher, first in Vienna, later at the University of California, Berkeley, who opposed the rationalist approach and favored anarchism.

Reproduced with permission from Grazia Borrini-Feyerabend.

Socrates: Ah, Professor Feyerabend, I've heard much about your unconventional views on science and knowledge. It's a pleasure to engage in a conversation with someone who challenges the established norms. You are often cited as an "enemy of science."

Feyerabend: The pleasure is mine, Socrates. Your method of questioning and seeking wisdom through dialogue has always been an inspiration. I am far from an "enemy of science," but I must admit that I'm not one to shy away from challenging the norms. Nor are you, if I may say so.

Socrates: Your critiques of the scientific method have intrigued many. I am curious to know, why do you find fault with the rigorous and systematic pursuit of knowledge?

Feyerabend: Socrates, my concern does not lie in dismissing the pursuit of knowledge but in recognizing the limitations of scientific method. Scientists too often fit their data to their theory.

I point out in my book, *Against Method*, that a strict reading of the 'facts' at the time of Galileo might well not have supported Copernicus. The Church may have been perfectly rational.

Socrates: But is not a structured method necessary to distinguish between what is true and what is false? How else can we ensure that we are not misled by subjective opinions?

Feyerabend: Valid concerns, indeed. However, I argue for methodological pluralism. Let ideas and methodologies compete freely. Let theories clash. From this intellectual chaos, new insights may emerge. Absolute certainty in a single approach can lead to dogma and hinder progress.

Socrates: (*raising an eyebrow*) It appears that you advocate an intellectual anarchy of sorts, where all ideas, regardless of their merit, are given equal weight. Is that not a risky path?

Feyerabend: (*smirking*) Not anarchy, Socrates, but intellectual freedom. By allowing diverse perspectives to flourish, we create an environment where the best ideas can rise to the top through a process of natural selection.

Socrates: (*nodding*) I see. But can this approach truly lead to the discovery of objective truths? Is there not a danger of descending into a sea of relativism, where every belief is considered equally valid?

Feyerabend: My intention is to show that any single methodology is limited. Scientific 'facts' are taught at a very early age and in the very same manner in which religious 'facts' were taught only a century ago. It's an indoctrination in which Science has now become as oppressive as the ideologies it once had to fight.

Socrates: But if we follow that logic, your own statements must be disregarded in exactly the same way as the 'facts' you criticize in science! Your statements become no more than an arbitrary reflection of your own cultural context. Instead, I argue that science is the profession that places the highest of all values upon factual details. The answers you get depend upon the questions you ask. Disputes are resolved by conflict within the community. The result is knowledge. Relativism is to thought as scrambling is to eggs. I rest my case.

SCIENCE AND SOCIETY

In the index to the six hundred odd pages of Arnold Toynbee's A Study of History, the names of Copernicus, Galileo, Descartes and Newton do not occur. This one example among many should be sufficient to indicate the gulf that still separates the Humanities from the Philosophy of Nature. I use this outmoded expression because the term 'Science'... does not carry the same rich and universal associations which 'Natural Philosophy' carried in the seventeenth century... The general reader is advised not to bother about the Notes at the end of the book; on the other hand, the reader with a scientific education is asked to forbear with explanations which might seem an insult to his intelligence. So long as in our educational system a state of cold war is maintained between the Sciences and the Humanities, this predicament cannot be avoided,[2] Arthur Koestler, 1968.*

Most of our critics are products of English departments and are very suspicious of anyone who takes an interest in technology. So, anyway, I was a chemistry major, but I'm always winding up as a teacher in English departments, so I've brought scientific thinking to literature. There's been very little gratitude for this,[3] Kurt Vonnegut, 2005.

Do not be horrified by the concept of the social context of science. This is not a descent into relativism. Current theories such as constructivism or post-modernism, which hold at their extreme that 'facts' have no existence outside of the social system that created them, are regarded as a joke by scientists. The relativistic view that all facts are subjective is completely opposed to the value system of science, which believes it is uncovering objective facts. However, scientists do not function in a vacuum, but within the constraints of the society that, in the modern era, provides the funds for their research. We have to consider what effects this has on the conduct of science itself.

If there was any single moment when science became a force in society and politics, it was perhaps when Albert Einstein wrote to President Roosevelt in 1939 to alert him that chain reactions in uranium would make it possible to construct an atom bomb. The letter was drafted by Leo Szilard, one of a group of physicists

*Did the atom bomb mark a permanent change in the relationship
between science and society?*

who had fled from Europe to the United States to escape the
Nazis, but it was felt that only Einstein's signature would convince Roosevelt. Two months later, Roosevelt set up the committee that would lead to the Manhattan project to build the
bomb.

The concern that science might not, after all, be a force for
good came to the fore with the creation of atomic weapons in
1945. "We have made a thing, a most terrible weapon... that by
all standards of the world we grew up in is an evil thing. And by
so doing... we have raised again the question of whether science
is good for man, of whether it is good to learn about the world, to
try to understand it, to try to control it... Because we are scientists, we must say an unalterable 'yes' to these questions. It's our
faith and our commitment, seldom made explicit, even more
seldom challenged, that knowledge is a good in itself," said
Robert Oppenheimer, the leading scientist in the Manhattan
project, soon after the atomic bombs were dropped on Japan.[4]

Was the atom bomb a boundary event in separating science from the humanities? Natural philosophy had been part of a continuum of intellectual activities extending into art, music, literature, or religion. Merging philosophy, mathematics, astronomy, medicine, logic, and ethics, the 'scientific' part of natural philosophy was not distinguished from other intellectual parts. The development of 'science' was marked by a segregation of techniques, in particular in physics involving mathematics that became increasingly more difficult for outsiders to understand.

Until the 18th century, art, music, literature, all were supported by patronage, typically from a Royal Court or its equivalent.[5] Support was determined by their needs and interests. The rulers of the day supported astrology because they were looking for information that would help to maintain their power.[6] Support for astrology merged into support for astronomy and science. (The supplicants may have spent as much time obtaining patronage as modern day scientists do on writing grant applications.)

Science became a function of institutions in the 18th century. By the end of the 19th century, scientific research took place in universities or research institutions funded by government. By the 20th century, the scale of research was such that only governments (or some very rich charitable foundations) could afford to sponsor science. It became impenetrable to the general public. By the 21st century, no organizations but governments, or even groups of governments, could afford to sponsor the largest scale efforts, such as the Human Genome Project or the search for the Higgs boson.

The timeline for development of the modern state parallels the rise of science as a professional activity. Is this coincidence? The state increasingly needed science as industrialization proceeded, and to consolidate its authority, most notably in time of war. Science needed the state as its scale grew beyond what individual patrons could support. Farsighted as ever, Francis Bacon foresaw this development when he argued in 1613 that the state should take a hand in science.[7] He was at least two centuries ahead of his time in Britain.

The French political class was more scientifically inclined, with the Académie des Sciences in Paris established as an institution of the state in 1666. It proposed in 1791 that a new unit of measurement should be introduced: the meter (one ten

millionth of the distance from the North Pole to the equator).
This was adopted in 1799, together with the gram as the unit of
weight (the mass of 1 c.c. of water).[8] The mandate went in hand
with the claim of the Revolution to impose rationality. It was a
striking early example of the direct application of scientific attitude to society. (An attempt to introduce a decimal system for
time, however, was not successful.)

As the practical potential of science and technology became
increasingly apparent, scientific organizations were recruited to
advise government.[9] The First World War stamped science as an
important resource for government. It was no longer adequate to
rely on universities or private enterprise to innovate. The public
now viewed science as a separate sort of activity, distinct from
other intellectual activities such as the humanities (and more
difficult to understand).

Even though the most dramatic effects of scientific research
during the Second World War were top secret, they led to scientists being held in high esteem. 'Boffin' was used as an af-
fectionate term in the armed forces in Britain to describe
scientists and engineers working on military research and development projects. It reached the national press at the end of
the war, creating a positive impression of science.[10]

The atomic bomb put science into a new perspective. Nuclear
power is another manifestation of the technological application
of atomic physics. A decade after the atom bomb was dropped,
the first nuclear power station was constructed. Initially regarded as a remarkable new source of energy, nuclear power
subsequently became controversial due to safety failures at nuclear power stations. Opinion is now split as to whether science
has proved a benefit or a danger in this case.

Physicists anguished as to whether science should be limited,
but comparable concerns did not reach biology for another half
century. The first concerns were expressed when recombinant
DNA was created in the early 1970s. The ability to combine DNAs
of different origins raised the possibility that new infectious
agents harmful to humanity might be created. This led to the
establishment of a committee by the National Academy of Sciences, which recommended a moratorium on experiments[11]—
the first in the history of science. This was followed by a conference at Asilomar, California in 1975, when 150 leading

researchers from around the world agreed on a report that recommended lifting the moratorium.[12] Research was now regulated by defining a series of containment facilities whose stringency increased with the perceived risk of the experiment.

The development of the CRISPR technique for gene editing brought the issue of limitations to science back into public awareness. This led to a call in 2019 for a moratorium.[13] A year later, an international commission established by the National Academy of Sciences, and other august associations, issued a lengthy report arguing in effect that time was not yet right for human germline editing, but that the possibility should be kept open.[14] Since then, editing the human germline (sperm or eggs) has been banned in many countries. Gene editing for specific cell types, such as those of the immune system, has been used successfully to treat a small number of individuals with diseases. The extent to which science has been regulated is partly due to internal recommendations by scientists, partly to external regulation as those recommendations were formulated at the political level.

∞ ∞ ∞

There is a curious dichotomy between public perceptions and scientific knowledge. The National Science Board's survey in alternate years of science and technology in the United States shows that public perception of scientists is overwhelmingly positive. For at least the past four decades, 80–90% of the population have said they believe that scientists help to solve problems and work for the good of humanity.[15] The number agreeing that the beneficial effects of science outweigh the harmful effects is a little lower, typically around two-thirds. The number approving of science increased with the COVID epidemic, an interesting contrast with the increased prominence of anti-vaxxers during that period. The proportion who currently say they trust medical science is just a little higher (75%) than the proportion who say they trust scientists generally (69%).[16]

There's increased polarization of views, with people who mistrust government more inclined to mistrust science. Seeing science as funded by government, or even as a tool of government, allows increased alienation from government to be reflected in increased skepticism of science.[17] Belief in the value of the results of science has declined. The proportion that believes scientific advances are one of the country's most important achievements declined from 47% in 1999 to 27% in 2009. If science itself is not overtly politicized, the public's reaction to it has become increasingly so. The moral is that science will lose its authority if it becomes politicized.[18]

Scientists worry about a potential decline in public support for science, especially as scientific knowledge is abysmal. Surveys show that:[19]

- only 73% believe the Earth goes round the Sun;
- only 51% know that electrons are smaller than atoms;
- only 61% know that it's the father's genes that determine sex of a baby;
- 47% think ordinary tomatoes do not contain genes but GMO tomatoes do;
- 50% believe that antibiotics kill viruses as well as bacteria.

In more politically contentious matters:[20]

- only 32% believe in evolution;
- only 49% believe in climate change.[21]

The population remains superstitious in spite of all the scientific progress of recent centuries. One measure of the extent of irrational beliefs is that since the 1960s at least one third of the population in the United States has expressed belief that flying saucers[22] (more formally: UFOs) represent visits from alien spacecraft.[23] The number has fluctuated depending on press reports of UFO sightings or cultural events such as the release of the movie *Close Encounters of the Third Kind*. A substantial minority of the population, around 40%, believe that astrology has some scientific basis.[24] Around three-quarters of the population believe in at least one paranormal phenomenon (such as ESP [extrasensory perception] or ghosts).[25]

The extent of irrational beliefs is a frightening setback to belief in science as the driving force in society. How can we expect people to assess science realistically if they believe in pseudo-scientific 'facts.' Failure to distinguish between science and theories masquerading as science, but actually based solely on

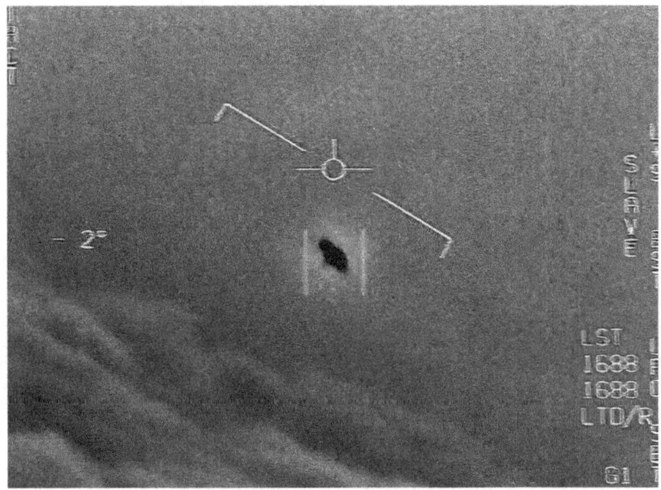

Image from a video from the U.S. Department of Defense of an unidentified flying object in 2015. This seems thin grounds for believing in visits from extraterrestrial aliens.

belief, is a major impediment to the advancement of science. Accepting pseudoscience enables fantasies to be presented as facts, and scientific reality to be denied. 'Facts' become a matter of choice. Against this background, it's perhaps surprising that support for science is even as high as it is.

Scientists generally are dismissive of pseudoscience and regard it as an irrelevance not worthy of scientific attention. Beyond that, there is a positive disincentive to investigate the claims of pseudoscience. Involving yourself in issues such as astrology, the existence of UFOs, or homeopathic treatments, is likely to dent your credibility as a scientist. The scientific community may be concerned that public ignorance of science, or actual hostility towards science, is damaging and may reduce public support for science, but individual scientists regard it as dangerous to be directly involved in countering these trends.[26]

In spite of these difficulties, there was a concerted move in 1974 by some distinguished scientists to apply organized skepticism to the claims of pseudoscience.[27] Perhaps the most dramatic of these interventions was a debate at the annual meeting of the AAAS (American Association for the Advancement of Science) between astronomer Carl Sagan and Immanuel Velikovsky. A psychologist, Velikovsky had proposed (to the derision of astronomers) that various historical catastrophic events on Earth (as described in the Bible or elsewhere) had been caused by near-collisions with other planetary bodies.[28] None of the widespread criticism had dented the popularity of the claims.

The debate and subsequent efforts debunked a fair number of pseudoscientific claims,[29] but it's like cutting the head off a hydra. It's not clear that there has really been significant progress in the past half-century towards drawing a distinction between science and pseudoscience in society at large. Although in recent years society has shown some interest in the question of scientific fraud, pseudoscience is an immeasurably greater problem that has been ignored if not actually embraced.

There is a good deal of misunderstanding in popular writing about just what comprises pseudoscience. Beliefs of various sorts are often called pseudoscience: such as UFOs or the Loch Ness monster. This actually is unfair to pseudoscience. These beliefs do not really claim to have a scientific basis: they are just beliefs. No one describes religion as pseudoscience; it's a different sort of belief system from science. Belief in UFOs or Nessie is more akin to belief in religion than to science, although they enter the scientific realm because supposedly they represent natural observations that need to be explained.

Alternative medicine, including homeopathy and various other treatments, is often taken as an example of pseudoscience, because it claims to have a scientific basis and there are reports claiming to test it by scientific means. The fact that whenever it has been properly tested, it has proved to have no effect does not seem to discourage believers. In an attempt to gain respectability, 'integrative medicine' has been introduced as an alternative description for alternative medicine, but 'quackademic medicine' would be a better description.[30] This mix of superstition and pseudoscience has entered the mainstream. It's unclear how to clean it out. There may be much wrong with the practice of conventional medicine, but alternative medicine will not improve matters.

For a leading practitioner of homeopathy to become Director of the Office of Alternative Medicine at the U.S. National Institutes of Health (indeed the fact that such an office was established at all) shows how far the rot has gone.[31] Other examples of pseudoscience crossing over into anti-science are the denial of climate change and the anti-vaxxer movement (which has already resulted in the revival of diseases that should by now have been extinguished.)[32] Their existence indicates the political strength of the anti-science movement in the United States.

It's hard to judge whether the anti-science movement of the late 20th and 21st centuries comes more from ignorance,

A Media View of Public Perceptions of Science

Image credit: NASA/JSC.

Under the title, **THE WAR ON SCIENCE**, the cover of *National Geographic* magazine for March 2015 showed a picture viewing the moon landing as a hoax. It summarized five of the major anti-science movements, with headlines:

- **Climate Change** does not exist.
- **Evolution** never happened.
- **The moon landing** was fake.
- **Vaccinations** can lead to autism.
- **Genetically modified food** is evil.

A substantial minority view (up to a third of the population) continues to hold such views today.

Luddite reflexes, or intellectual unreason. The various anti-science movements together form a powerful undercurrent in society threatening to undermine acceptance of science and therefore (if we can use a loaded term) progress. The danger is that they might become the mainstream.[33]

While the public may be conscious of major scientific developments, these are usually viewed as being in a separate realm. Most people have probably heard of Einstein's equation, $E = mc^2$. They may even be aware that it's connected with relativity, but it isn't part of common parlance. The Big Bang has entered public consciousness, the name is certainly evocative, but it hasn't become a metaphor for any human activity. However, the phrase "it's in its DNA" has become part of the vernacular, meaning that it's an essential part of the character, that it's the essence of its being. This stretches far beyond biology. Companies like to boast about creditable features that are in their DNA—"Helping the world is not merely this company's ambition, it's in its DNA." It can also describe the dark side. "Corruption isn't just in the city's air, it's in its DNA," is how one novel describes New York.[34]

Now becoming increasingly common, this metaphor is very recent usage. Possibly prescient, the first comparable usage I can find is an article in the Harvard Business Review in 1999 under the title, *"Decoding the DNA of the Toyota Production System."*[35] This was just about the time the Human Genome Project was really reaching public consciousness. The timing suggests that use of the phrase reflects an indirect public recognition of the accomplishment. General usage can be traced from soon after the reports of the first completed draft human genomes in 2001, and it picked up within a decade. This may be the first time biology, or indeed even science, has made an impact with such a powerful metaphor.

∞ ∞ ∞

The absence of scientists from the political class leaves decision-making open to failure to distinguish between science and pseudoscience, and to inappropriate setting of priorities.[36] Major decisions ranging from how to handle a pandemic,[37] whether to have a manned space program,[38] or what to do about climate change,[39] are made without the decision-takers really understanding what they're talking about.

How important is science to society? Perhaps the amount spent to support basic research is a reasonable indication. Spending on science took off after the Second World War. The United States is no doubt the country that spends the most on science. Yet total spending on basic research (mostly funded by the government) is only 0.5% of the GDP. (Spending on health care is 17% of GDP.[40])

As early as 1830, the polymath Charles Babbage (who designed what amounted to the first computer; see Chapter 23) wrote a book lamenting the state of education in science in England.[41] Pointing to "the defects of our system of education, that scientific knowledge scarcely exists amongst the higher classes of society," he anticipated by almost a hundred years the anguish of the 20th century on the same question. At a conference in 1916, leading British scientists made much the same criticism,[42] which was famously crystallized in 1959 by C. P. Snow in *The Two Cultures* (discussed in Chapter 10). They all got much the same response of indifference.

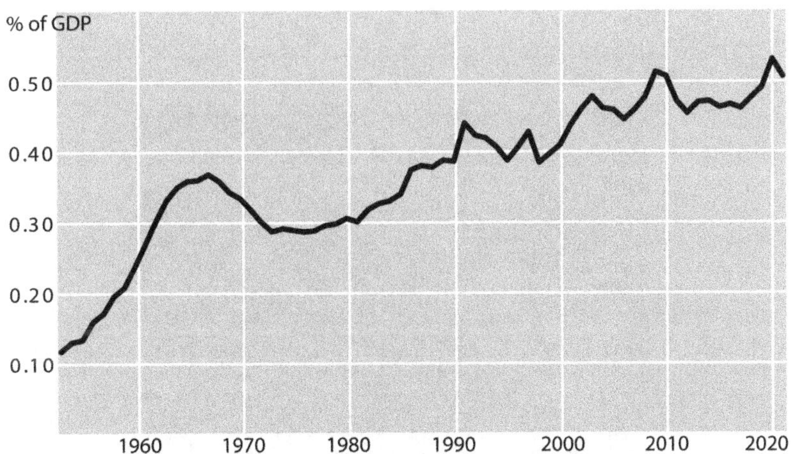

Spending on basic research has increased steadily to reach 0.5% of the GDP in the United States. This is only a small part of the spending on R&D (research and development), which now totals over 3%. The U.S. government funds about 90% of the total on basic research, compared with 75% in 1945. With R&D, the sources of support are reversed, with the government providing only 10% of the total.[43]

C. P. Snow was unusual in having credentials in both science and the humanities. He started as a physicist, and then moved into physical chemistry. After publishing a paper that was erroneous, he gave up on scientific research in 1930. Based on his experience at Cambridge University, he wrote a series of eleven novels called *Strangers and Brothers*, published between 1940 and 1970, depicting the internal politics and machinations of a Cambridge college. This did not endear him to his colleagues at the university.

Snow spent most of his career in the Civil Service, where he dealt with scientific matters. He may well have formed a jaundiced view of the abilities of those educated in the humanities to understand science. His twin experiences as a scientist and as a novelist led him to his criticism of 'literary intellectuals' as ignoramuses of science—and he was somewhat indignant at their view that they were the sole species entitled to the term 'intellectual.'[44] The title of his book, *The Two Cultures*, has become a shorthand phrase for describing differences in attitude in the sciences and the humanities.[45]

The response to *The Two Cultures* from the humanities is as revealing of the culture gap as the book itself. Snow was viciously

C. P. Snow in London in 1966.

© *Evening Standard/Hulton Archive/Getty Images.*

attacked on both literary and intellectual grounds. The strength of the attack, especially from literary critic F. R. Leavis,[46] went beyond any reasoned criticism of his argument that an educated person should understand science as well as the humanities. Three decades before the 'science wars' of the 1990s (see below), the criticisms seemed to reflect rage at the encroachment of science into the realm of intellectual activity. Science was viewed not merely as an alien thought process, but as an inferior one that did not have legitimacy in the intellectual realm.[47]

Science is ill served by the lack of understanding of the scientific process and attitude outside of science itself. Even in the era of specialization, knowledge of the scientific attitude makes it possible to ask questions that uncover significant features such as the level of certainty in projections that may not be immediately apparent. We need science to permeate society not to be a separate layer on top of it.[48]

Part of the problem lies with the teaching of science as a mass of facts that have to be understood. The educational system, especially at the pre-medical level, can make science appear authoritarian. In fact, it is anything but. Research is not at all about established certainty. It's more about the unknown than the known, it's a constant quest to find something new. Or at least, it was like that until the advent of 'big science.'

In the aftermath of the attention focused on science as the result of the first atomic bomb, James Conant (President of Harvard University) argued that "I have seen repeated examples of... bewilderment of laymen... The remedy does not lie in a greater dissemination of scientific information among nonscientists... What is needed are methods for imparting some knowledge of the Tactics and Strategy of Science to those who are not scientists."[49] 'Scientific method' may not exist in the formal sense of any single system (I discuss the limitations of trying to describe science by its methodology in Chapter 10), but science for the humanities needs to focus on principles rather than details.

Attempts to teach science to students in the humanities tend too often to fall into a softer 'science studies' than a rigorous examination of scientific methodology. Current trends in the humanities only exacerbate the gap in understanding between the humanities and sciences. Starting from the view that science is an anthropological phenomenon, that the facts that are discovered are relative to the observer, is beyond futile in presenting science as an endeavor that no scientist would recognize.

The sociology of science had serious beginnings when Robert Merton made his seminal studies in the 1940s (see Chapter 10). A change in attitudes took place when the traditional philosophy or sociology of science gave way to a new approach in the 1970s. This was very much opposed to Merton's approach that science could be analyzed on its own terms. The sociology of scientific knowledge, abbreviated as SSK by its practitioners, argues that science has to be treated as a social product.[50] While there is no doubt about the influence of social context on the practice of science, to treat science as exclusively a consequence of its social environment is completely unrealistic.[51] This is one reason why scientists regard sociology or philosophy of science as irrelevant.[52]

It seems a fair question whether the social criticism actually reveals the psychology (dare one say 'prejudice') of the critics more than the social construction of science?[53] The core of the argument is that there is no such thing as 'pure science,' that it's impossible for science to be value-free.[54] An (admittedly extreme) feminist view of science, for example, views science as fatally flawed because it is a masculine activity. "Scientists tend to select... certain kinds of methods of inquiry because, among other reasons, they are associated with distinctively masculine stereotypes—interventionist methods, for example."[55] Famous historical figures in science are regarded as endorsing rape as a metaphor for science (because it involves probing Nature).[56] In summary, "Obviously, the sexist language of science is continuous with sexist thought in society in general."[57] Because "sexist and androcentric [male-oriented] biases had shaped virtually every stage of research processes,"[58] objectivity should be redefined to answer "the kinds of questions that are most important to an oppressed group."[59] Certainly we would scarcely recognize science if this prescription were followed.

Science often comes under attack from academics in the humanities. Curiously, scientists rarely repay the compliment by attacking the humanities! Some attacks on science have come from historians who have segued from the fantasy of the mad scientist to modern science. The cover flap of a book by one journalist starts "Science has hoodwinked us all into believing it has the ultimate answers; in reality, it is our Frankenstein monster, a creature of our own design that now threatens to destroy us."[60] The publisher, a subsidiary imprint of Holtzbrinck, who own the leading scientific journal, *Nature*, should be ashamed.

The heart of one argument against science is that the difficulties of quantum physics make it an "affront to common sense." That takes us back to the time of the pre-Socratic philosophers, when natural philosophy was based largely on common sense observations. Unfortunately, these can be seriously misleading. It's perfectly obvious from everyday experience that the Sun revolves around the Earth! It's fair comment that Heisenberg's uncertainty principle or wave-particle duality, or (especially) the phenomenon of entanglement, create uncertainty as to how to interpret observations, but that does not invalidate the observations. Rather it suggests we need more complete theories.

Science (or scientists) is sometimes viewed as arrogant because of its emphasis on objectivity rising above personal beliefs. On the contrary, in reality science is a subversive activity, because it involves a continuous questioning of authority. Perhaps that is why science is less successful in authoritarian societies. Every scientist, no matter how eminent, is at constant risk that new data may be obtained that will contradict earlier experiments. Objectivity basically means being prepared to change opinions in response to new facts.

We have to tackle a basic incongruity. Never has science had more influence, both directly in new developments in pure research, and indirectly through development of technology; but never has there been so much opposition to science. Of course, these two conflicting trends may to some extent be connected in the same way as the advance of the Industrial Revolution and the Luddite resistance to it. There is always opposition to the threatening effects of change. Yet there is more to it than that. There is a basic conflict between those who accept a scientific train of thought (even if with reservations about some of its consequences) and those who reject it as a legitimate means of thinking. How did this opposition to rationality arise?

A clash between religion and science (or perhaps more accurately, between religion and other forms of intellectual activity) developed during the Middle Ages. It was crystallized during the Scientific Revolution. If the Scientific Revolution brought science into play as a force in society, the Enlightenment began the resistance movement.

The first signs came from Giambattista Vico (1668–1744), a relatively unknown philosopher in Naples, who published *Scienza Nuova* in 1725,[61] in which he argued that knowledge about human culture was more reliable than knowledge about science. He took the position that mathematics and science are inventions that do not necessarily reflect the realities of life.[62] Human creations (such as language or poetry) had more claim to validity than mathematics or natural sciences.[63]

This anticipated the view developed in the Enlightenment and later that truths speaking to the nature of man come from the humanities rather than the sciences. This was the start of the argument that rationalism and science distort reality and lead to dehumanization.[64] However, opposition to science that came out of humanism in the 18th century was countered by Victorian enthusiasm for the wonders of the age.

The divide between science and the humanities may have been reinforced by the rise in the importance of mathematics in the 18th and 19th centuries, when the Romantic poets took strong exception to the incomprehensible mathematical constructions of cosmology.[65] The geometry and algebra on which science had previously been based were not difficult to understand; calculus was a different matter. As the sciences became more mathematical and more technical, the gap widened further.

Some of the intellectual criticism of science leaves an impression that it arises because those in the humanities feel threatened by science. "It's true that once upon a time a self-respecting intellectual had to be scientifically literate," says Leon Wieseltier, literary editor of The New Republic.[66] *Once upon a time!* What a revealing implication: that this is no longer true. Here is the heart of the problem with the humanities.

∞ ∞ ∞

The split between the humanities and the sciences developed into the so-called 'science wars' of the 1990s. It's not quite clear who started the 'war,' but it took the form of increasingly strident arguments as to whether science is a rational system based on objective evidence or whether it is just another belief system. In this view, science is a contrivance no more valid than any other belief system. Scientists argued that what is loosely called postmodernism, rejecting objectivity and reason, consisted of work that was incomprehensible or meaningless.

One revisionist book on the history of science starts, "One of the most distinctive features of the emergence of a scientific culture in modern Europe is the gradual assimilation of all cognitive values to scientific ones."[67] Would that this were true! Science has certainly become a dominant intellectual influence, perhaps *the* dominant influence, but it's far from assimilating "all cognitive values."

The term post-modernism was first used in 1914, but did not come into common use until the 1970s. It was used to describe developments that succeeded modernism in the arts, literature, politics, and architecture. Science is a notable exclusion from the list. In fact, post-modernism is associated with a considerable skepticism towards science, a sort of backlash that makes it perhaps not so much post- as anti-modernist.

This book is about revolutions, about the overthrow of one paradigm by another. The very way that science advances has been taken by critics to imply that there is no such thing as knowledge in the sense of defining reality: because reality can change.[68] Advocates for SSK and relativism have been known to misinterpret Kuhn's description of paradigm shifts by claiming that the existence of revolutions disproves science's claim for objectivity. Kuhn was so exasperated by this distortion of his position that on one occasion he burst out, "I am not a Kuhnian."[69]

Max Perutz, a molecular biologist awarded the Nobel Prize in 1962, spoke for science when he said, "The entire approach emphasizing 'relative' truth seems to me a piece of humbug masquerading as an academic discipline; it pretends that its practitioners can set themselves up as judges over scientists whose science they fail to understand."[70]

What gives science its authority? Perversely, its strongest feature is that facts and theories may change. Critics may point to this as a weakness, arguing that it detracts from the claim of objectivity. Yet this is in fact precisely what ensures that science represents a position that can be supported objectively on the basis of available information. There is never any claim that our present view will last forever.

Science represents reality in hypotheses based on the knowledge presently available to us. In no way does it undercut the scientific search for objective truth to recognize that what is achieved at any time is the best approximation to the truth. Science is admittedly an imperfect endeavor. There can be disputes about the facts, there can be alternative theories that explain the facts, there can be reluctance to relinquish previous theories, but these are more matters of timing than ultimate decision.

One of the ways the record is kept straight in science is to retract papers that are later found to be erroneous. (The rate of retraction has increased in recent years. Some find this alarming, believing it might mean an increasing proportion of work is in error. Others point out that it may mean there is pressure to do a better job of maintaining accuracy in the literature.[71]) At all events, there appears to be no equivalent in the humanities, where retractions are extremely rare and usually devoted to matters such as attribution rather than content as such.[72]

This speaks to an essential difference between science and the humanities. The concept of searching for objective data has the corollary that there can be mistakes, which need to be corrected, whereas if everything is relative, error becomes an irrelevant concept. Why is the concept of error so threatening to the humanities?

It's fine if philosophers want to play with ideas, such as there being no reality, so that the universe is purely an intellectual construction of the mind. I suppose they are entitled to take the Romantic view of Jean-Jacques Rousseau that science and technology have worsened the human condition, or the harder-edged view of Nietzsche that scientific truth denies the meaningfulness of human experience. But it is really annoying when they apply these ideas to try to undermine the advances made by science by arguing that scientific theories describe experiences as opposed to underlying reality. For working scientists, this is nonsense. Insofar as its extension into the denial of the reality of science in the 'science wars' may undercut public support for science, it is more serious.

The Sokal Hoax stands out as a peak in the science wars.[73] Physicist Alan Sokal submitted a paper in 1996 to the journal *Social Text* entitled *Transgressing the Boundaries: Toward a Transformative Hermeneutics of Quantum Gravity.*[74] Replete with quotations from famous physicists and writers, with pages of notes and references longer than the main text, it reads like a serious paper in technical language. But it's complete nonsense, equating concepts of quantum physics with social or political implications, or irrelevant concepts in biology. It uses mathematics in a way that any mathematician or physicist would recognize as a spoof.

Even the title suggests a joke as hermeneutics (originally meaning interpretation of biblical or literary texts) has little do with quantum gravity (which does not exist). One nonsensical conclusion is that "a liberatory science cannot be complete without a profound revision of the canon of mathematics." The editors of *Social Text* accepted the paper, without any review. (The journal does not send submissions for outside review, even when they contain technical material outside their own expertise. The view that expertise is irrelevant is part of the relativist perspective.)

Scandal erupted when Sokal revealed that the article was a hoax.[75] Far from crying *mea culpa*, the editors attacked Sokal for a breach of ethics. They accused him of understanding postmodernism only as a caricature, and then compounded the criticism by saying, "Prior to deciding whether science intrinsically tells the truth, we must ask, again and again, whether it is possible, or prudent, to isolate facts from values. This is a crucial question to ask, because it bears upon the kind of progressive society we want to promote."[76] This is reminiscent of Talleyrand's comment on the Bourbon kings of France. "They had learned nothing and forgotten nothing."

At least in the intellectual circles of academia it was still possible to recognize a hoax. Would that be possible in the arts? It's unclear what form a hoax might take. As painting moved from figurative to abstract, as performance art became a part of the mainstream at art shows, as NFT art (non-fungible tokens, meaning that the owner has a digital key to some electronic construction) developed, it seemed that art was increasingly based on relativism degenerating into illusion. It might be an interesting test of contributions in the humanities and arts to ask: how would it be possible to recognize a hoax?

One of the reasons why Sokal's article could superficially be taken seriously is that its tone is so close to apparently serious contributions to the literature. Under the heading of "you couldn't make this up" comes an article with the title "Glaciers, Gender, and Science." Its abstract says, "the relationships among gender, science, and glaciers…remain understudied… This paper thus proposes a feminist glaciology framework. . . leading to more just and equitable science and human-ice inter-actions."[77] (It was funded by the US National Science Foundation.) How would you distinguish this from a hoax?

The science wars might be regarded as a fight confined to academia if it were not for the public ramifications of under-cutting the validity of science. Scientists feel themselves beleaguered politically from both the left and the right. Relativism arose as an attack of the left wing on objectivity. Indeed, one motive for the Sokal hoax was Alan Sokal's concern that the left was damaging itself by taking up nonsensical positions based on sloppy thinking. The anti-science movement of the right is more likely to have a religious basis, as in the rejection of evolution in favor of creationism. At least that is coherent, even if replacing a rational belief system with an irrational one, but movements such as the anti-vaxxers are hard to combat since there is no connection with reality. The only relief is that statistics suggest that the overall public support for science has not changed very much over 50 years.[78]

One of the charges hurled at science in the science wars was that science is a belief system no different, and no more valid, than any other belief system. "Scientific knowledge is only a communal belief system with a dubious grip on reality," is a summary of the view of one major school of sociologists of science.[79] Well, science is a belief system in a sense, but a very different one from that envisaged by the critics. It's based on belief in the scientific attitude, that all theories are subject to testing by data. That is distinctly different from other belief systems. It is why scientists hold the view that data reflect reality. It's curious indeed that the critics feel free to attack that conclusion; yet they do not deny the existence of technological reality.

Consider a *reductio ad absurdum*. Would a relativist consider that gravity is a social invention of science? Newton's law of gravity is just that: a law, describing the effect that two bodies have upon one another. Its validity is not subject to the emotions of the observer or to time or place. With regards to changing realities, we now explain gravity in terms of Einstein's theory of general relativity, but none of that invalidates the concept of gravity. The apple still falls down from the tree. Formally speaking, we are dealing with closer and closer approximations to the truth—but truth there is. We might ask just who actually has the dubious grip on reality.[80]

The relativistic view argues that science is no more than a social construct. I would agree that science is a construct, but it is one with its own self-consistent frame of reference. It happened to arise in the West, but had it happened instead in the East (if science had not petered out in Arabia, for example), it would very likely have taken much the same form. Driven by the necessity for reproducibility and self-correction, if science is a construct, it is one that rises above any of its constituent societies.

To be sure, the history of science shows that interpretations are inevitably colored by the belief system of the society in which science is conducted. This does not make science a 'social construct' in the sense used by sociologists; nor does it undercut the basic position of science that theories must be subservient to facts. Belief systems may limit interpretation, but at the end of the day, if there is a clash between belief and data, belief must give way. Indeed, one feature of many of the great leaps in science is how individuals have been able to overcome the limitations of the current belief system. Galileo in reversing the centrality of the Earth, or Newton in realizing the fact of interactions at a distance, are obvious examples.

Fish do not know they live in water, but they cannot exist in any other environment. Scientists know they live in society, but how far are they able to allow for its impact on their thinking? Looking back, we can see that until the last couple of centuries, scientists were limited in their understanding of phenomena by the prevailing assumptions (often religious in nature). What about the present?

Even philosophers who question how free science can be from external values are forced to admit that the reliance on empirical evidence must limit the influence of those values.[81] In science, facts are sacred: well, not sacred as we do not worship them, and they can be overcome by new facts. But facts are subject to challenge only on the basis of new experimental information satisfying the condition of objectivity. What can change, and might in principle be limited by our perspective, are interpretations. As C. P. Scott, the famous editor of the Manchester

Guardian newspaper said in another context, "Comment is free but facts are sacred."[82]

At all events, critics are certainly free to argue about the process of interpretation and external influences on it, although I am inclined to think that their tendency to denigrate the logic of deduction is more a demonstration of their own (mis)belief system than reality.[83] Where they get off the top of the bus, as it were, is in going farther to argue that facts are not facts but are relative to the observer. Certainly any scientist functions in a belief system that may influence interpretation of data. But the key to understanding science is that data are data, and will be accessible to any other scientist irrespective of their belief system. That is the beauty of science.

Anticipating and refuting the arguments of the relativists, Aristotle said, "A law of nature is immutable and has the same validity everywhere, as fire burns both here and in Persia, while rules of justice change [in different places]."[84] Two thousand years later, it's hard to express the concept of the universality of the laws of science any better.

∞ ∞ ∞

Historians disagree about the severity of the conflict between science and religion, but it's undeniable that there have been clashes for as long as people have queried natural phenomena. They range from the death of Socrates (399 BCE), to the imprisonment of Galileo (1633), and the publication of Darwin's *On the Origin of Species* (1859). The theory of evolution triggered a disconnect between science and belief that continues to the present.

The tone for the difference between religion and science for the next two centuries was set by an exchange between Thomas Huxley, a scientist advocating Darwin's theory, and Bishop Wilberforce of Oxford. At the meeting of the British Association in Oxford in 1860, Bishop Wilberforce asked, "Would you rather have had an ape for your grandfather or grandmother?" Huxley's devastating riposte was "'I would rather have had apes on both sides for my ancestors than human beings so warped by prejudice that they were afraid to behold the truth."[85]

It's impossible by its very nature for the theory of evolution to provide a complete record of events. Too many steps have been

Evolution was a fertile topic for satire in the 19th century. A cartoon in the satirical magazine, The Hornet portrayed Charles Darwin as an ape in 1871.

© GraphicaArtis/Getty Images.

lost because of species becoming extinct. But first the fossil record, and then the evidence of relationships between the DNAs of different species, place the evolutionary tree on a firm basis. (What would Bishop Wilberforce have said if it had been possible to tell him that his DNA was 99% the same as that of a chimpanzee?) The objections to evolution were an exception to the general Victorian enthusiasm for science. Mainstream society gradually came to accept evolution, except for the evangelical movement.

The argument about evolution in the 19th century began with its theological implications. By the late 20th century, the debate had turned political. When Edward Wilson of Harvard University called his book *Sociobiology* in 1975, it was nothing less than inflammatory to propose that human behavior could be interpreted in terms of evolution and genetics.[86] The publication a year later of *The Selfish Gene* by Richard Dawkins added fire to the flames.[87] Their work was often conflated by critics, although they took very different, and often opposed, views of the effects of evolution.[88,89]

The criticism represented a significant politicization of science, taking the view that Wilson and Dawkins were guilty of 'determinism'—believing that human behavior is predetermined by our genes. The vicious nature of the attacks, which often wildly distorted the original ideas, was reminiscent of the attacks on Darwin for proposing evolution, and perhaps in a way also of the attack a century later on C. P. Snow for his division between the two cultures. The difference is that here the attacks came from within the scientific community, although there was perhaps a similar underlying cause: the belief that science was venturing into territory beyond its prerogative.[90] Critics left the impression that they would have treated Wilson much as the Church had treated Galileo if only they had been able. It was a sad example of belief masquerading as rational analysis, although here the belief was political rather than theological.[91]

Creationism is a yet more blatant attempt to misrepresent belief as science. Giving creationist theory a name such as 'intelligent design' does not hide the fact that the theory depends on supernatural forces. People are entitled to their beliefs, of course, but it's offensive to present 'intelligent design' as an alternative science. The same movement is opposed to the use of stem cells, at least insofar as they might originate in a human embryo.

The political effects of this movement are not negligible. In the United States, they resulted in a ban on the use of federal funds for stem cell research, with the collateral damage that the process of IVF (*in vitro* fertilization) could be developed only in Europe. As responses to the COVID epidemic demonstrated, this extended into a general anti-science movement, sometimes known as the 'war on science,' which has world-wide effects on the replacement of rational argument by the irrational.[92] Because the proponents of the anti-science movement are deeply committed, their beliefs have a disproportionate political effect.

Today we tend to think of science as influencing, if not shaping, society. But the relationship is reciprocal. Society influences science, directly in the way it funds it, indirectly in setting limits to scientific investigation. Whether those limits are due to direct political influence, the indirect control of funding, or the imposition of political correctness, they can influence science for reasons other than the scientific.

One influential critic of science, Paul Feyerabend,[93] preceding the science wars by some years, made an infamous attack on science in 1975. "The events, procedures and results that constitute the sciences have no common structure... Science is neither a single tradition, nor the best tradition there is... In a democracy it should be separated from the state just as churches are now separated from the state."[94]

∞ ∞ ∞

Science (or at least its technological applications) is now so deeply embedded in Western society that it's hard to imagine how society could be constructed without it. There have from time to time been movements to try to control its direction. The Rothschild Committee set up by the U.K. government in 1971 proposed a directed approach,[95] but this was reversed a decade later. Control in the United States has taken the form more of preventing research into specific topics than encouraging choice of topics. However, given that funding is highly competitive, there's a certain self-censorship among scientists in avoiding topics that might be deemed to be less relevant.

The view that science is a cultural construct of society, in which priorities are established by hidden criteria, inevitably leads to the question: should society directly determine the values that direct scientific research?[96] The concept of values in this context extends from societal questions (such as the nature and diversity of scientific researchers) to the direction of scientific research (should cancer research be directed more towards basic causes at the cellular level or towards environmental effects).[97] But the inevitable consequence of imposing political direction on science is a drift from basic research to applied research. Whatever sociologists or philosophers (or politicians!) may feel to be the moral imperative to direct science, the fact is that this is misguided pragmatically.

As a practical matter, attempting to direct the course of science is counter-productive, at least if you expect science to produce new discoveries, because new discoveries often come from completely unexpected directions. They may be accidents, they may be observations that were unexpected in a particular line of research, they may come from the application of a line of research in one field to an apparently unrelated field—but in any case, more often than not, they are unpredictable. Charting a course for science would require us to know what we don't know. If society does not try to direct science as a matter of principle, it should avoid direction as a pragmatic decision. This is simply the best way to get useful results.

The view that a free society supports science partly for its own sake and partly for the benefits that accrue, directly or indirectly, has never been put better than by Vannevar Bush at the end of the second world war (quoted in Chapter 18). Since then, the supporters of science have mostly been successful in fighting off demands that science should be directed more towards the needs of society. All the same, there is a creeping trend towards making science more responsive to the perceived needs of society.

The latest five-year Strategic Report of the National Science Foundation in the United States says, "NSF supports S&E (science and engineering) research and innovation that lead to breakthrough technologies as well as solutions to national and societal problems." It's an irony that the NSF—the very foundation that was established as the result of Bush's report—should appear to be fulfilling his warning. "Unless deliberate policies are set up to guard against this, applied research invariably drives out pure [research]."[98] As historian of science Gerald Holton put it four decades ago, "the invisible hand is becoming more visible."[99] To what extent is the advancement and the direction of science driven by the nature of the society that undertakes it?

The anti-science movement is not solely religious, but also has secular proponents.[100] One criticism of science from the perspective of humanists (or perhaps one might say, at the risk of seeming judgmental, the Romantics) is its lack of moral compass. It's "a struggle to hold back the cruel pessimism of science. The key to this struggle, it cannot be said too often, is the way in which science forces us to separate our values from our knowledge of the world."[101] (The author of this view might be surprised indeed to know that neuroscientists claim to have pinpointed a center in the brain that is responsible for moral decisions.[102])

The argument that science represents an absence of values leads to the conclusion that pursuit of knowledge for its own sake is dangerously irresponsible. This view never answers the question of how the benefits of science could be obtained in a system that did not allow the free pursuit of knowledge. The response of adding politically-correct values to science has the potential to damage the endeavor, but the rot is everywhere, even to the point of a fifth column in science itself. There should be no illusion about the threat to science. In 1992, the U.S. National Research Council stated, "The National Science Education Standards are based on the post-modernist view of the nature of science."[103] The standard was withdrawn after protests from scientists. It would have turned science education into a caricature.

The New York Times (not known for conservative opinions) exposed a scandal when it reported that an article entitled *In Defense of Merit in Science,*[104] was rejected by a series of journals, including the *Proceedings of the National Academy of Sciences.* The article made the case that increased politicization has become damaging to science. The rejection was based on objecting to the concept of merit in science. "The problem is that this concept of merit, as the authors surely know, has been widely and legitimately attacked as hollow as currently implemented."[105]

The U.S. National Academy of Sciences even released a report in 2023 describing 'merit' as a 'culturally construed' concept. It's a singular irony that the article eventually was published in the *Journal of Controversial Ideas.* If it is controversial that science should be based on merit, we have sold the pass. The authors point out that applying extreme political correctness introduces an ideological subversion of science that is not in principle particularly different from the effect Lysenko had in destroying genetics in the Soviet Union.

At least Galileo was persecuted only from outside science. Outsider views that science is not impartial can be dismissed by practitioners of science, but criticisms from within institutions of science that the endeavor is biased are more pernicious.

Granted, facts remain facts (at least for the moment), but it's a feature of science that scientists are recognized on the merit of their work. Such recognition admittedly may be imperfectly executed, but it's part of the value system of science that theorists are recognized for the value of their theories, and experimentalists are recognized for the rigor of their experiments. If this is to be questioned, it is the start of the slippery slope to postmodernism, when extraneous factors influence the acceptability of scientific research—and indeed, when facts may no longer be facts.[106]

The Romantic opposition to science may be a reflex reaction to feeling threatened by a strange force—strange because it is not understood. This was described very well by author and critic Lionel Trilling. "Physical science in our day lies beyond the intellectual grasp of most men... its operative conceptions are alien to the mass of educated persons. ... This exclusion of most of us from the mode of thought which is habitually said to be the characteristic achievement of the modern age is bound to be experienced as a wound given to our intellectual self-esteem... This humiliation... introduces into the life of mind a significant element of dubiety and alienation?"[107] A more reasoned criticism requires an understanding of science, which leads to concern about its effect on human morale as opposed to morality.[108]

Hannah Arendt captures the concern about science as an intellectual activity by recognizing its advantages, but also deploring its effects upon human mentality. Science "prescribed the rules of behavior and the new standards of judgment... the sciences have been ever since, a veritable triumph of human ingenuity against overwhelming odds."[109] At the same time, she saw dangers in the Archimedean point (a hypothetical viewpoint, introduced by Descartes, as a point positioned outside the system from which it is possible to form an objective view). "All laws of the new astrophysical science are formulated from the Archimedean point... the first, and spiritually... the most lasting consequence of the discovery [of the Archimedean point] is the feeling of suspicion, outrage, and despair."[110]

Basically this comes down to the view that discoveries in cosmology have diminished the importance of humanity's role in the universe. The scientific rejoinder might perhaps be that we need to be rational about this. We may not be so important.

That is perhaps the heart of the discordancy between the sciences and the humanities. It's not, however, a new idea. That famous skeptic, philosopher David Hume, arrived at the same conclusion almost three centuries ago. "The life of a man is of no greater importance to the universe than that of an oyster."[111]

Contrasted with the sterile view that science has no ethical or moral center, Nobel Prizewinner Jacques Monod points out that "the very definition of 'true' knowledge reposes in the final analysis upon an ethical postulate."[112]

Science is not simply a collection of facts without context, but shares with the arts the creative impulse, and furthermore carries its own value system into society. When Jacob Bronowski, a Polish-British mathematician who became famous as a popularizer of science, made this case, he argued that, "inescapable conditions for its practice [mean that] science is indeed a truthful activity."[113] For science to succeed, it must be honest, it must be possible to trust the facts offered by others, it must admit ignorance when facts are unknown, and everything must be subject to correction. Science cherishes both independence and dissent. It's a stable system that has functioned in much the same way for the past two centuries. It follows that, "The society of scientists must be a democracy." Perhaps Society might do well to base itself on the principles of science.

NOTES AND REFERENCES

1. C. Sagan, *Conversations with Carl Sagan*, University Press Of Mississippi, 2006, p. xv.

2. A. Koestler, *The Sleepwalkers*, Hutchinson & Co., London, 1968, pp. 13–16.

3. K. Vonnegut, *A Man Without a Country*, Seven Stories Press, New York, 2005, pp. 16–17.

4. J. R. Oppenheimer, Address to the American Philosophical Society, November 16, 1945. Online at www.americanrhetoric.com.

5. *Patronage in the Renaissance*, ed. G. F. Lytle and S. Orgel, Princeton University Press, Princeton, 1981.

6. J. Gascoigne, *Science and the State: From the Scientific Revolution to World War II*, Cambridge University Press, Cambridge, 2019, p. 14.

7. J. Gascoigne, *Science and the State: From the Scientific Revolution to World War II*, Cambridge University Press, Cambridge, 2019, pp. 25–26.

8. J. Gascoigne, *Science and the State: From the Scientific Revolution to World War II*, Cambridge University Press, Cambridge, 2019, pp. 75–78.

9. Such as the Académie in France, the Royal Society in England, and the Imperial Physical Technical Institute in Germany.

10. From the *Daily Herald* (London), August 15, 1945: "This is 'Boffins Day' because for the first time it is permissible to tell something of the war saga of the backroom boys known throughout the services of the United Nations as 'Boffins'. It's a dramatic and romantic story of a battle of wits, brains and inventive genius between the scientists of the United Nations and those of the enemy, and the United Nations Team won hands down. Though not impressive to look at the boffins who haunted Army Navy and RAF Stations in usually rather shabby civilian clothes and were ever ready to argue on almost any subject except their own work but generals air marshals and admirals treated them with respect for only the very senior officers knew much of their activities."

11. P. Berg, *et al.*, Potential biohazards of recombinant DNA molecules, *Science*, 1974, **185**, 303.

12. P. Berg, D. Baltimore, S. Brenner, R. O. Roblin III and M. F. Singer, Asilomar conference on recombinant DNA molecules, *Science*, 1975, **188**, 991–994.

13. E. Lander, *et al.*, Adopt a moratorium on heritable genome editing, *Nature*, 2019, **567**, 165–168.

14. National Academy of Sciences, *Heritable Human Genome Editing*, The National Academies Press, Washington, DC, 2020.

15. NSB-2022-7, *Science and Technology: Public Perceptions, Awareness, and Information Sources*. Online at ncses.nsf.gov/indicators.

16. American Perspectives Survey, May 2023. Online at www.americansurveycenter.org.

17. For example, hesitancy about vaccination may result from lack of trust as opposed to ignorance or anti-science sentiments. See M. J. Goldenberg, *Vaccine Hesitancy. Public Trust, Expertise, and the War on Science*, University of Pittsburgh Press, Pittsburgh, PA, 2021.

18. There were signs of politicization during the COVID pandemic. As I describe in Chapter 25, groupthink has always been more of a factor in science than is usually appreciated, but it becomes more threatening when it is imposed from outside of science itself. B. Lewin, *Inside Science: Revolution In Biology and Its Impact*, Cold Spring Harbor Laboratory Press, New York, 2023, pp. 77–78.

19. National Science Board, *Science and Engineering Indicators*, 2012, p. 7–24.

20. There is no 'control' for these results, but it seems that the public is not necessarily any better educated on other matters. A poll found that nearly half believed that the statement from the Communist Manifesto, 'From each according to his means, to each according to his needs,' comes from the United States Constitution. R. Marcus, Constitution Confuses Most Americans, *The Washington Post*, 1987.

21. American Association for the Advancement of Science, Scientific Achievements Less Prominent Than A Decade Ago, *Science Daily*, 10 July 2009.

22. The flying saucer craze started in 1947 when a private pilot, Kenneth Arnold, reported seeing nine high-speed, crescent-shaped objects flying in

formation near Mount Rainier in Washington state. He said, "they flew like a saucer would if you skipped it across the water." J. T. Smith, Idaho Pilot Reports Seeing Nine "Saucer-Like" Objects. *East Oregonian*, 1947, June 27, p. 1.

23. According to the Gallup poll: news.gallup.com/poll/350096/americans-believe-ufos.aspx.

24. National Science Board, Science and Engineering Indicators, 2012, pp. 7–27.

25. D. W. Moore, Three in four Americans believe in the paranormal, *Gallup News Service*, 2005, news.gallup.com/poll/16915/Three-Four-Americans-Believe-Paranormal.aspx. The numbers are not as high among people with college degrees, but are still substantial. C. D. Bader, J. O. Baker and F. C. Mencken, *Paranormal America. Ghost Encounters, UFO Sightings, Bigfoot Hunts, and Other Curiosities in Religion and Culture*, New York University Press, New York, 2nd edn, 2017, p. 60. The numbers are lower in Holland, which has a more secular tradition, but still form a majority. S. Hoogeveen, *et al.*, Prevalence, patterns and predictors of paranormal beliefs in The Netherlands: a several-analysts approach, *R. Soc. Open Sci.*, 2024, 11240049.

26. The Royal Society of London published a report in 1985 on *The Public Understanding of Science*, known as the Bodmer Report after the chairman of the committee, arguing the need for better education of the public. It's regarded as one of the first expressions of concern from the scientific community as such.

27. K. Frazier, *Shadows of Science: How to Uphold Science, Detect Pseudoscience, and Expose Antiscience in the Age of Disinformation*, Prometheus, London, 2024, pp. 167–179.

28. Velikovsky's most famous book I. Velikovsky, *Worlds in Collision*, DoubleDay, New York, 1950, was enormously successful, and has been through more than 70 reprintings since its publication. Velikovsky was a psychologist with no scientific qualifications.

29. K. Frazier, *Shadows of Science: How to Uphold Science, Detect Pseudoscience, and Expose Antiscience in the Age of Disinformation*, Prometheus, London, 2024, pp. 181–196.

30. D. H. Gorski, "Integrative" Medicine: Integrating Quackery with Science-Based Medicine, in *Pseudoscience: The Conspiracy Against Science*, ed. A. B. Kaufman and J. C. Kaufman, 2018, pp. 309–329.

31. It subsequently changed its name to National Center for Complementary and Integrative Health and took a more serious approach to investigating claims for alternative medicine.

32. Measles was eliminated in the United States in 2000. Resistance to vaccination enabled its reappearance in 2024. www.cdc.gov/measles/data-research/index.html.

33. The presidential election of 2024 in the United States suggests this may be happening.

34. D. Winslow, *The Force*, William Morrow, New York, 2017.

35. S. Spear and H. K. Bowen, Decoding the DNA of the Toyota Production System, *Harvard Business Review*, 1999.

36. In the United States, more than 50% of Congressmen and Senators have law degrees, 4 have MDs, and almost none have any qualification in science.

37. No one seems to have tested the robustness of the models for predicting the spread of the pandemic. It's important to know how they respond to small changes in input data. If they show drastic changes in output in response to small data changes, the question is whether the effect is real or the model is flawed. Without this information, the model is not very useful. And it took a long time to realize that the concept of herd immunity was a chimera for COVID, although it should have been questioned right from the start. See D. M. Morens, G. M. Folkers and A. S. Fauci, The Concept of Classical Herd Immunity May Not Apply to COVID-19, *J. Infect. Dis.*, 2022, **226**, 195–198.

38. It may be debatable whether there will be political benefits that are worth the $100 billion price tag but there are surely minimal scientific benefits compared with what might be gained by spending the funds on mainstream science. By comparison the CERN Large Hadron Collider cost $5 billion.

39. The problem can be exacerbated by a tendency for researchers to interpret all the data in terms of worst-case scenarios. This may come from a feeling that they are unable to mobilize public opinion on what they conclude is a serious matter but has a tendency to backfire. The situation is at least partly created by the lack of scientific education in the public: it would not be necessary to exaggerate to catch the attention of an educated public who understood the information.

40. National Health Expenditure Accounts (NHEA). www.cms.gov/files/document/highlights.pdf.

41. C. Babbage, *Reflections on the decline of science in England, and on some of its causes*, 1830. Online at www.gutenberg.org/files/1216/1216-h/1216-h.htm.

42. Lord Rayleigh, The Neglect of Science, *A Conference of the Linnean Society*, Harrison & Sons, London, 1916.

43. Data from National Center for Science and Engineering Statistics, National Patterns of R&D Resources, 2020-2021. Online at ncses.nsf.gov/data-collections/national-patterns/2021#data.

44. C. P. Snow, *The Two Cultures, and The Scientific Revolution*, Cambridge University Press, Cambridge, 1993, p. 4.

45. *The Two Cultures* was a turning point. Its thesis of a gulf between science and the humanities has only been reinforced since the furor. No measures to bridge the gap have changed the situation. In fact, the gap may have been widened by the trend in the history of science to represent science in a narrative scientists cannot recognize.

46. F. R. Leavis, *Two Cultures? The Significance of C. P. Snow*, Cambridge University Press, Cambridge, 2nd edn, 2013.

47. Although somewhat portentous, Snow's novels had a success in their time. His scientific career was undistinguished. But Leavis's *ad hominem* attack missed the fact—or perhaps ironically reflected it—that Snow really

struck a chord in pointing out that critics in the humanities were displaying not merely ignorance of science but an inchoate disdain for it. While *The Two Cultures* has endured as a description of the divide between science and the humanities, Leavis has faded into insignificance.

48. It would help also if scientists were not so disdainful of members of the scientific community who popularize science.

49. J. B. Conant, *On Understanding Science: An Historical Approach*, Yale University Press, New Haven, CT, 1947, pp. 26–27.

50. There are two common distortions of Kuhn's views. One is that Kuhn's model undermines the claim that science reveals objective truth, because a revolution implies that the prior state of knowledge was incorrect. Kuhn rejected this misunderstanding. Science is not a religion, it does not claim absolute truth, just the best approximation to the truth that can be obtained at the time. The other distortion is to argue that a change of paradigms is based on factors extraneous to science (such as politics or economics), which is very much the view of SSK.

51. Koertge has debunked the absurd criticisms of science. N. Koertge, *A House Built on Sand. Exposing Postmodernist Myths About Science*, Oxford University Press, New York, 1998.

52. Showing a complete lack of reality, one proponent of SSK laments that, "scientific demystification has unfortunately not reached active scientists." The fact that scientists regard the sociological view as laughable has unfortunately not reached him. See F. Miedema, *Open Science: the Very Idea*, Springer, 2022, p. xii.

53. The paucity of the approach has been nicely debunked by M. Pigliucci, *Nonsense on Stilts: How to Tell Science From Bunk*, University of Chicago Press, Chicago, 2nd edn, 2018, pp. 221–231.

54. S. Harding, *The Science Question in Feminism*, Cornell University Press, Ithaca, NY, 1986, pp. 39–52.

55. S. Harding, *Whose Science? Whose Knowledge?: Thinking From Women's Lives*, Cornell University Press, Ithaca, NY, 1991, p. 45.

56. S. Harding, *Whose Science? Whose Knowledge?: Thinking From Women's Lives*, Cornell University Press, Ithaca, NY, 1991, p. 43.

57. S. Harding, *Whose Science? Whose Knowledge?: Thinking From Women's Lives*, Cornell University Press, Ithaca, NY, 1991, p. 46.

58. S. Harding, *Objectivity and Diversity: Another Logic of Scientific Research*, University of Chicago Press, Chicago, 2015, p. 30.

59. S. Harding, *Objectivity and Diversity: Another Logic of Scientific Research*, University of Chicago Press, Chicago, 2015.

60. B. Appleyard, *Understanding the Present: Science and the Soul of Modern Man*, Picador, London, 1992.

61. The first edition in 1725 attracted little attention. Revised editions were published in 1730 and (posthumously) in 1744, A modern translation of the third edition is *The New Science of Giambattista Vico*, translated by T. G. Bergin and M. H. Fisch, Cornell University Press, Ithaca, 1984.

62. I. Berlin, *The Proper Study of Mankind: An Anthology of Essays*, Farrar, Straus and Giroux, New York, 2000, pp. 246, 341.

63. R. Smith, *The Fontana History of the Human Sciences*, Fontana Press, London, 1997, pp. 341–342.
64. I. Berlin, *The Proper Study of Mankind: An Anthology of Essays*, Farrar, Straus and Giroux, New York, 2000, p. 251.
65. N. Levitt, Mathematics as the Stepchild of Contemporary Culture, *Ann. N. Y. Acad. Sci.*, 1995, **775**, 39–53.
66. L. Wieseltier, Crimes Against Humanities, *New Republic*, 2013, September 4.
67. S. Gaukroger, *The Emergence of a Scientific Culture: Science and the Shaping of Modernity 1210–1685*, Oxford University Press, Oxford, 2006, p. 11.
68. S. L. Goldman, *Science Wars: The Battle Over Knowledge and Reality*, Oxford University Press, Oxford, 2021, p. 8.
69. Quoted in F. J. Dyson, *The Sun, the Genome, and the Internet: Tools of Scientific Revolutions*, Oxford University Press, Oxford, 1999, pp. 15–16.
70. M. Perutz, The Pioneer Defended, *The New York Review of Books*, 1995, December 21.
71. B. Lewin, *Inside Science: Revolution in Biology and Its Impact*, Cold Spring Harbor Laboratory Press, New York, 2023, pp. 79–86.
72. SAGE journals, for example, which publishes almost 600 journals in the sciences and almost 800 in the humanities, has 19 retraction notices to date in the humanities and 314 in the sciences, according to journals. sagepub.com (April 6, 2023).
73. For reviews see *The Sokal Hoax: The Sham That Shook the Academy*, ed. Lingua Franca Editors, University of Nebraska Press, Nebraska, 2000; A. D. Sokal, *Beyond The Hoax. Science, Philosophy and Culture*, Oxford University Press, Oxford, 2008.
74. A. D. Sokal, Transgressing the boundaries: toward a transformative hermeneutics of quantum gravity, *Soc. Text*, 1996, **46–47**, 217–252.
75. A. D. Sokal, A physicist experiments with cultural studies, *Lingua Franca* , 1996, **6**, 62–64.
76. B. Robbins and A. Ross, Mystery Science Theater, *Lingua Franca*, 1996, **6**, 54–58.
77. M. Carey, *et al.*, Glaciers, gender, and science: A feminist glaciology framework for global environmental change research, *Prog. Hum. Geogr.*, 2016, **40**, 770–793.
78. See NSB-2022-7, *Science and Technology: Public Perceptions, Awareness, and Information Sources*. Online at ncses.nsf.gov/indicators.
79. K. Gottfried and K. G. Wilson, Science as a Cultural Construct, *Nature*, 1997, **386**, 545–547.
80. The usual riposte to relativism is to ask why critics of science are prepared to sit in airplanes at 30 000 feet if the laws of physics that govern flight are mere figments of cultural imagination.
81. A. Potochnik, *Science and the Public*, Cambridge University Press, Cambridge, 2024, p. 37.
82. C. P. Scott, A Hundred Years, *Manchester Guardian*, 1921 May 5.
83. For examples of sociological attempts to debunk well-established experiments see H. Collins and T. Pinch, *The Golem: What You Should Know*

About Science, Cambridge University Press, Cambridge, 2nd edn, 2012; J. Waller, *Fabulous Science: Fact and Fiction in the History of Scientific Discovery*, Oxford University Press, Oxford, 2004.

84. Aristotle, *The Nicomachean Ethics*, translated by H. Rackham, Harvard University Press, Cambridge MA, 2nd edn, 1934, book 5, ch. 7.

85. According to a report in the Oxford Chronicle. See R. England, Censoring Huxley and Wilberforce: A new source for the meeting that the Athenaeum 'wisely softened down', *Notes Rec.*, 2017, **71**, 371–384.

86. E. O. Wilson, *Sociobiology: The New Synthesis*, Harvard University Press, Cambridge, MA, 1975.

87. R. Dawkins, *The Selfish Gene*, Oxford University Press, Oxford, 1976.

88. Wilson defined sociobiology as the systematic study of the biological basis of all social behavior. Dawkins' main theme was to argue specifically for the role of the gene, and he did not even regard himself as a sociobiologist.

89. U. Segerstråle, An Eye on the Core: Dawkins and Sociobiology, in *Richard Dawkins. How a Scientist Changed the Way We Think*, ed. A. Graffen and M. Ridley, Oxford University Press, Oxford, 2006, pp. 75–97.

90. The intellectual dishonesty of the attacks has been thoroughly exposed by S. Pinker, *The Blank Slate: The Modern Denial of Human Nature*, Viking, New York, 2002.

91. Two leading critics were also at Harvard, Steven Gould and Richard Lewontin. They were leaders of the Sociobiology Group of Science for the People. Their motivation was to make scientific analysis of human behavior untenable. For an account of the sociobiology debate see ref. 89.

92. S. Otto, *The War on Science: Who's Waging It, Why It Matters, What We Can Do About It*, Milkweed Editions, Minneapolis, 2016.

93. Feyerabend was influential because of the attention his views attracted, but he was not necessarily taken seriously by other philosophers. See ref. 53 and W. J. Broad, Paul Feyerabend: Science and the Anarchist, *Science*, 1979, **206**, 534–537.

94. P. Feyerabend, *Against Method: Outline of an Anarchistic Theory of Knowledge*, Humanities Press, Atlantic Highlands, NJ, 3rd edn, 1993.

95. See National Academy of Sciences, *Heritable Human Genome Editing*, The National Academies Press, Washington, DC, 2020.

96. A. Potochnik, *Science and the Public*, Cambridge University Press, Cambridge, 2024, p. 2.

97. A. Potochnik, *Science and the Public*, Cambridge University Press, Cambridge, 2024, pp. 48–49, 59–65.

98. V. Bush, *Science. The Endless Frontier. A Report to the President*, 1945; Reissued: V. Bush, *Science The Endless Frontier: 75th Anniversary Edition*, National Science Foundation, Washington DC, 2020.

99. G. Holton, *The Advancement of Science, and Its Burdens*, Cambridge University Press, Cambridge, 1986, p. 192.

100. G. Holton, *Science and Anti-science*, Harvard University Press, Cambridge MA, 1995, pp. 143–189.

101. B. Appleyard, *Understanding the Present: Science and the Soul of Modern man*, Picador, London, 1992, p. 76.

102. M. J. Crocket, Moral transgressions corrupt neural representations of value, *Nat. Neurosci.*, 2017, **6**, 879–888.

103. *National Science Education Standards: A Sampler*, National Academy of Sciences National Research Council, Washington, D.C., p. A2. Online at files.eric.ed.gov/fulltext/ED360174.pdf.

104. D. Abbot, *et al.*, In Defense of Merit in Science, *J. Controversial Ideas*, 2023, **3**, 1.

105. P. Paul, A Paper That Says Science Should Be Impartial Was Rejected From Major Journals. You Can't Make This Up. New York Times, 2023, May 4.

106. B. Lewin, *Inside Science: Revolution In Biology and Its Impact*, Cold Spring Harbor Laboratory Press, New York, 2023, pp. 77–78.

107. L. Trilling, *Mind in the Modern World*, Penguin Books, London, 1973.

108. Criticism of science often stands for a general reaction against modern society or Western civilization. This is typified by a comment from a Marxist historian: "The forces generated by the techno-scientific economy are now great enough to destroy the environment, that is to say, the material foundations of human life." E. Hobsbawm, *Age of Extremes, The Short Twentieth Century 1914–1991*, Michael Joseph, London, 1994, p. 584.

109. H. Arendt, *The Human Condition*, University of Chicago Press, Chicago, 2nd edn, 1998, pp. 278–279.

110. H. Arendt, *The Human Condition*, University of Chicago Press, Chicago, 2nd edn, 1998, pp. 263, 267.

111. D. Hume, published posthumously in *Essays on Suicide and the Immortality of the Soul*, Smith, London, 1755, 1777.

112. J. Monod, *Chance and Necessity*, Vintage Books, New York, 1971, p. 173.

113. J. Bronowski, *Science and Human Values*, Harper & Row, New York, 1956.

CHAPTER 27

Epilogue: Science and Modernity

"DOUBT IS AN UNCOMFORTABLE CONDITION, BUT CERTAINTY is an absurd one," Voltaire said in 1770.[1] He was one of the first modernists, determined to see everything from a modern point of view, which he took to mean that science was distinct from the natural philosophy that had preceded it. It was in the century following the Scientific Revolution that science entered the general consciousness and became part of public discourse. It took at least another century, and perhaps two, before science permeated society to become an essential part of modernity. Ironically, the rejection of science by the post-modernist movement validates it as the epitome of modernity.

The Frontiers of Science
By Benjamin Lewin
© Benjamin Lewin 2026
Published by the Royal Society of Chemistry, www.rsc.org

Timeline for Evolving Concepts of Modernity

1600

SCIENTIFIC REVOLUTIN

Francis Bacon describes scientific approach as modern (1605)

Modernité refers to new ideas in France (1620s)

Modernity first used in English (1635)
Descartes introduces 'rationalist' approach in *Discours* (1637)

1650

Isaac Newton publishes *Principia* (1687)

1700

Quarrel of the Ancients and Moderns at Académie Française (1687-1716)
Jonathan Swift satirizes modernity as based in science (1704)

ENLIGHTENMENT

1750

Rousseau declares arts and sciences corrupt human morality (1755)
Voltaire declares science is essential part of culture (1763)

1800

1850

Baudelaire defines modernity in literature (1863)
Pope Pius IX claims Modernism is a danger to Christianity (1864)
Marx describes modernity as emergence of capitalism (1867)

Nietzsche attacks science as part of modernity (1887)

1900

Modernism in Art marked by departure from figurative (1900)
First use of 'post-modernism' (1914)
Spengler attacks science for contributing to decline of West (1918)
Alfred North Whitehead publishes *Science and the Modern World* (1925)
Le Corbusier, Gropius and others define modern architecture (1932)

1950

Alexandre Koyré lectures on *Origins of Modern Science* (1952)
Drucker describes post-modernism as rejection of Cartesian worldview (1959)

Modernity driven by technology; Berners-Lee invents World Wide Web (1989)

2000

A Dialog between Jean-Jacques Rousseau and Louis de Broglie

Jean-Jacques Rousseau (1712–1778) was a peripatetic Swiss philosopher, moving as his writings offended the authorities. He believed in the 'state of Nature' as a guiding principle.

RMN-Grand Palais / Mathieu Rabeau.

Louis de Broglie (1892–1987) was a French aristocrat who became a theoretical physicist, developing the wave theory of atomic particles. He was a committed atheist.

Rousseau: (*Looking out of the window*) Nature, untouched and unspoiled, is a canvas of simplicity and purity. The modern world, however, seems to have departed from this serene state. Alas, I find myself lost amid the mathematical language of your science!

de Broglie: (*Glancing through the window*) Nature can certainly be an inspiration, but it does not always provide solutions to physical problems. When Max Planck introduced the idea of the quantum, it was an act of desperation because he could find no other explanation.

Rousseau: Your quantum mechanics, Professor, it delves into the minute and the unseen. Does it not risk losing touch with the essence of human experience? It's a mania of philosophers to deny what exists and to explain what does not exist.

de Broglie: Quantum mechanics reveals a world where reality is less certain than we had imagined. Far from controlling nature, we now find ourselves constantly reminded of its inherent mystery, its unpredictability. When I proposed that not only light but also matter can exist both as particles and waves, it demonstrated the limitations of our knowledge.

Rousseau: Yet, Professor, the march of technology imposes its own kind of "order" on both society and nature. The more we advance, the more we risk losing the simplicity and authenticity of human relationships.

de Broglie: Physics is not all cold logic, it requires imagination. We have our flights of fancy in science. The award for my Nobel Prize stated specifically that my theories had no connection with facts! The equations lead to an understanding that the observer and the observed are connected, much like the interconnectedness of human societies. Perhaps we must accept that some aspects of reality are inherently beyond our grasp.

Rousseau: (*Voice rising*) But what of the social fabric, Monsieur? The more we advance scientifically, the more we seem to encase ourselves in structures and systems that erode the natural bonds of community. Does not the relentless pursuit of knowledge and progress risk sacrificing the values that make us truly human? Reductionism has no soul.

de Broglie: Monsieur Rousseau, science requires we recognize that life is uncertain. Schrödinger's equations show that atomic particles can co-exist as waves and particles, outside of our vision, but our attempts to visualize them destroy the duality. We must accept this limitation.

Rousseau: (*Nodding*) Your quantum world may illuminate the mysteries of matter, but can cold equations bring solace to the human soul, adrift in the complexities of modern life? In our relentless march toward progress, may we not lose the very essence of what it means to be human?

de Broglie: (*Patiently*) The concepts of 'soul' or 'life' do not occur in atomic physics. They could not, even indirectly, be derived as complicated consequences of some natural law. We cannot expect to interpret equations directly in terms of visual descriptions, but they are the best understanding we have of the natural world. The soul is to the body as the ether is to the atom. I rest my case.

SCIENCE AND MODERNITY

> *Scientific progress is a fraction, the most important fraction, of the process of intellectualization which we have been undergoing for thousands of years and which nowadays is usually judged in such an extremely negative way... What is the value of science? Here the contrast between the past and the present is tremendous... At the threshold of modern times... to artistic experimenters of the type of Leonardo and the musical innovators, science meant the path to true art, and that meant for them the path to true nature... Well, who today views science in such a manner? Who... still believes that the findings of astronomy, biology, physics, or chemistry could teach us anything about the meaning of the world?*[2] Max Weber, 1918.

> *If I have put the case of science at all correctly, the reader will have recognised that modern science does much more than demand that it shall be left in undisturbed possession of what the theologian and metaphysician please to term its 'legitimate field'. It claims that the whole range of phenomena, mental as well as physical—the entire universe—is its field. It asserts that the scientific method is the sole gateway to the whole region of knowledge.*[3] Karl Pearson, 1897.

Every age has claimed modernity for its own. The concept of a distinction between modern and ancient dates from medieval Europe.[4] Sometimes it was used to indicate the superiority of the modern era. Sometimes it was used in the opposite sense to indicate a romantic longing for the ancient ways. 'Modernité' was first used in France in the 17th century. By the early 19th century, it became associated with the idea of a rational approach (not necessarily scientific) contrasting with the past. 'Modernity' has a similar history in English.[5]

The very concept of modernity requires a contrast with the ancient world, so it did not become relevant until the Enlightenment, with the recognition that the Ancients did not, in fact, know all knowledge. Since then, the concept of what it means to be modern has changed with the times, to the point at which one recent book even refers to 'contemporary modernity' (apparently without irony).[6] Trying to cut through the thicket of sociological and political appropriations of 'modernity,' let's take it that we want to trace what each successive period since the Enlightenment has regarded as its intellectual driving force.

The 17th century was a turning point. Philosophy's transition to modernity might be dated from Descartes in the 1630s. Science's contribution to modernity might be dated from Galileo at the start of the Scientific Revolution or to Newton at its end. The common feature is the transition to a 'rationalist' approach.[7]

The first explicit arguments about modernity *versus* classicism were epitomized by the *Querelle des Anciens et des Modernes* (Quarrel of the Ancients and Moderns) when from 1687 the Académie Française debated whether modern culture was superior to the ancients. This was followed by Jonathan Swift's satire, *The Battle of the Books,* in 1704. The issues concerned philosophy, literature, and art (and the rejection of religion).[8]

The concept of modernity became associated with a broader view of changes in society. Taking a political view, Karl Marx associated modernity with the transition from a feudal to a capitalist society. Sociologist Max Weber attributed it to the replacement of traditional authority (religious or monarchical) by a rational-legal authority.[9]

Until the Enlightenment, science (or natural philosophy) was the prerogative of a class confined to scientists and their patrons. Science was not always held in high regard after the Scientific Revolution, but it was always a factor in intellectual life. If science was considered dangerous, it was in the context of the threat to religious authority. This was countered by Francis Bacon's arguments in the 17th century viewing science as a force for progress.

As science became increasingly important, it was attacked by those who resented it as forcefully as it was defended by those who advocated it. For the first half of the 18th century, it was satirized as often as it was praised (see Chapter 24). By the second half, there was an explicit divide between those, such as Voltaire, who regarded it as defining modern intellectual life, and those, such as Rousseau, who deplored its dehumanizing effects.[10]

During the 19th century, 'science' became recognized as an entity, with a continuing divide between enthusiasts promoting its potential for stimulating progress, but detractors concerned about its effects on society. Science was viewed as complicit in modernity, especially by those who rejected it.[11] Indeed, science came to be viewed as synonymous with civilization.[12]

Granted that in the first part of the 20th century, cultural movements, in art, music, architecture, were considered to contribute to the avant-garde, science was often the crucible for those who objected to modern life. Indeed, moving forward to the end of the 20th century, the hostility of post-modernism towards science is a back-handed (and certainly unintended) compliment, in effect validating science as a moving force in modernism.

Within science, the malaise in classical physics at the end of the 19th century rapidly gave way to enthusiasm with the start of atomic physics at the beginning of the 20th century. There was public acclaim as the inside of the atom was revealed, turning first to concern, and eventually perhaps resignation, with the dawn of the atomic age.

Biology developed later. The golden age of (molecular) biology started in 1961, gathered steam, and then, as genomic manipulation became a reality in the 21st century, encountered something of the same concern that the atom bomb had engendered in the previous century. It's perhaps a fair comment that, from physics to biology, there has been public enthusiasm while science has seemed to be uncovering the mysteries of nature, but anxiety when the discoveries have practical applications that could be abused. Indeed, this scenario seems to be repeating in the present, as optimism that science might be the engine for progress in society has, at a minimum, been qualified by concern about the dangers of AI. Modernity is double-edged.

Modernity is a construct of the intellectual world. How much impact does it have on real life? The rise of science is often considered to be at the expense of religion, or more generally to represent a disenchantment, a loss of belief in magic.[13] This may be a scientific attitude, but is not necessarily true of the population as a whole.[14] It's a moot point how far modernity has penetrated the attitude of the general population. The modern human may take advantage of all the developments of science and technology without necessarily committing to 'modern' attitudes. But can you reject science and still be 'modern'?

Opinions vary as to when science took over the commanding heights of the intellect, although there is agreement it was not a quick process. Both the 19th century and the 20th century have been called the 'Age of Science.'[15] Victorian enthusiasm was as much devoted to the wonders of technology as to science itself. By the second half of the 19th century, the idea developed that 'progress' was the target of civilization, and that it was intimately connected with science.[16] The impact of science in the first part of the 20th century was due to developments in physics, hard to understand, but generally viewed as creating a fundamental insight into nature. Biology took over from physics in the second half of the century.

An explicit equation of science with modernity dates from British philosopher Alfred North Whitehead's Lowell Lectures, delivered at Harvard University in 1925, and then published as a book, *Science and the Modern World*.[17] Two decades later, some influential books, notably from Alexandre Koyré and Herbert Butterfield,[18,19] placed the Scientific Revolution as the start of the modern era.[20]

Quoting a phrase from psychologist William James (brother of the novelist Henry James), Whitehead encapsulated the new mentality as the need to reconcile principles with "stubborn and irreducible facts." He placed its origins two centuries earlier. "[We] have been living upon the accumulated capital of ideas provided... by the genius of the seventeenth century." Whitehead reviewed developments in science since then from physics to biology. He assessed the contrast between science and religion, and concluded that the main importance of science was the change it had produced in patterns of thought. "The moral of the tale is the power of reason, its decisive influence on the life of humanity."

In another lecture series, at Johns Hopkins University in 1952, under the similar title of *Origins of Modern Science*, historian Alexandre Koyré, who had previously introduced the phrase Scientific Revolution,[21] commented that the "mechanization of worldview" had a created a new philosophy that had wrought "despair and confusion."[18] He concluded that, "man... had to transform and replace not only his fundamental concepts and attributes, but even the very framework of his thought."

The theme of despair echoed the views of Max Weber, perhaps the most influential sociologist of the 20th century, who wrote extensively about modernity. He regarded science as destroying all values except its own.[22] Whitehead and Koyré, and others writing in the period, viewed mathematics as the common feature that imposed a barren character on the new mentality.[23] The change in intellectual attitude across the 19th and 20th centuries transcended any single scientific discovery or indeed any particular science.[24]

In the period at the end of the First World War, it was easy to be overtaken by a sense of *weltschmerz* (world-weariness). Associated with modernity, science could become a metaphor for the failings of society. Taking the view that decline and fall is inevitable for any culture, Oswald Spengler's famous diatribe, *The Decline of the West*, a book of 1200 pages, was published in 1918, at the end of the First World War. It took the position that science would be a contributory factor because of a propensity to extend its authority outside the scientific realm. Spengler thought that science would reach its limit in the year 2000.[25]

The conflict as to whether science should be seen as a vision of progress or a deadening, sterilizing influence played out in popular media, not least in satirical cartoons. The debate was known as the 'March of Intellect' in the early 19th century. Cartoons showed a futuristic vision of technology, with extravagant buildings, aerial transport, and castles in the air (a reference to paying off the national debt). Love it or loath it, this was the image in 1829 of modernity driven by technology.

It's striking that almost exactly a century later, in 1927, Fritz Lang's silent movie, Metropolis, was released, showing a futuristic view of skyscrapers connected by aerial railways and airplanes. Based on a book published two years earlier, the film was made in Germany during the Weimar republic. Part of the Expressionist movement, Metropolis expressed the anguish of the period, and was also seen as a dystopian science fiction movie, including a mad scientist who creates a robot that is destroyed in a Luddite revolution. This was a more hostile view of modernity, with science and technology as the villain.

With the passage of another century, would the equivalent view today—very likely dominated by AI-driven robotics—be more dystopian or more optimistic?

It has become trite to say that modern life is based on technology driven by science, that without all of our technological gadgets life would be completely different—and that change will

The first March of Intellect cartoon from William Heath (under the pseudonym of Paul Pry) from 1829 shows a futuristic view of London, with the ironic title "Lord how this world improves as we grow older."

Reproduced from https://commons.wikimedia.org/wiki/File:A_futuristic_vision_Wellcome_V0041098.jpg, under the terms of the CC BY 4.0 license, https://creativecommons.org/licenses/by/4.0/.

be continuous (perhaps even accelerating), depending on future developments. (There is a view that modernity depended on science until the 1980s and then became a feature of technology with the move to post-modernism.[26] However, I view this as a distinction without a difference given how much technology today is driven by science.) Even sociologists who criticize science's claim to unique authority concede that "science marks modernity."[27]

Describing changes in society in the preceding two decades, Austrian-American management consultant Peter Drucker defined post-modernism in 1959 as a rejection of the Cartesian worldview.[28] This anticipated the more overt reaction in last part of the 20th century against the view that there can be universal truths and objective reality. Post-modernism is, in effect, an argument against the modernity of science.

A futuristic art-deco view of the city from Metropolis in 1927. The building at the back is based on the Tower of Babel. (The view is more effective in the movie, where trains move across the bridges and airplanes fly between the buildings.)

Major discoveries in all of the natural sciences have completely changed our view of humanity's development on Earth and our role (or lack thereof) in the Universe. Occurring 14 billion years ago, the Big Bang is all but beyond general comprehension (and for that matter, difficult for many physicists also), but together with other developments in cosmology, it has revolutionized our view of the universe, which has been a focus of human thought ever since society first developed.

The discovery of the double helix has led to a view of human evolution *vis-à-vis* other species. It has created a rational approach to understanding evolution, and led to the possibility that we may be able to influence our own evolution. The philosophical question as to whether we have free will has been replaced by the question of to what extent we are more than the sum of our genes, whether and how we can rise above our genomes.

The knowledge that all the atoms of any one element are indistinguishable from one another says that, if we are more than the sum of our atoms, it is their specific arrangement that matters. We are left with the mystery of understanding how that arrangement of atoms creates individuality and consciousness.

The uncertainties of quantum physics are a huge contrast with the certainties of other sciences. Questions about the nature of the atom go deep into the creation of the universe and what sort of accident (some would say purpose) may have led to the creation of humanity.

The philosophical (or religious) questions that have occupied humanity for thousands of years—who are we? where have we come from? where are going?—have been rephrased in terms of science and have become subservient to it. This is more than modernity, it is fundamental to the very fabric of our existence.

NOTES AND REFERENCES

1. "Le doute n'est pas une état bien agréable, mais l'assurance est un état ridicule." In a letter to Prince Frédéric-Guillaume of Prussia in 1770. *Complete Works of Voltaire* (Œuvres complètes de Voltaire), Desoer, Paris, vol. 12, part 1. Modern edition published by Arvensa Editions, 2019.
2. M. Weber, The Disenchantment of Modern Life, 1918, in *From Max Weber: Essays in Sociology*, ed. H. H. Gerth and C. Wright Mills, Oxford University Press, Oxford, 1946, pp. 129–156.
3. K. Pearson, *The Grammar of Science*, Adams & Charles Black, London, 1892.
4. P. Seed, Early Modernity: The History of a Word, *New Centen. Rev.*, 2002, **2**, 1–16.
5. The roots of 'modernity' may go back to 'modernism,' introduced in the Middle Ages to mean recently or just now. M. Calinescu, *Five Faces of Modernity: Modernism, Avant-garde, Decadence, Kitsch, Postmodernism*, Duke University Press, Durham, NC, 2nd edn, 1987, pp. 13–14.
6. P. Wagner, *Modernity. Understanding the Present*, Polity Press, Cambridge, UK, 2012, p. 43.
7. S. Toulmin, *Cosmopolis. The Hidden Agenda of Modernity*, University of Chicago Press, Chicago, 1992, pp. 1–14.
8. In 1907, Pope Pius X declared modernism "the synthesis of all heresies." The Church introduced the Oath Against Modernism in 1910, which had to be sworn by all clergy. It included "I declare that I am completely opposed to the error of the modernists who hold that there is nothing divine in sacred tradition." In this, it was defining modernism as an anti-religious movement.

9. R. Smith, *The Fontana History of the Human Sciences*, Fontana Press, London, 1997, p. 555.

10. Rousseau deplored all 'progressive' aspects of society, not merely science. His book of 1762, *The Social Contract*, famously started, "Man is born free and is everywhere in chains."

11. "As long as what is meant by culture is essentially the promotion of science, culture will pass by the great suffering of the human being." F. Nietzsche, *Unzeitgemässe Betrachtungen*, Fritsch, Leipzig, 1899, vol. 2, pp. 69–70.

12. S. Gaukroger, *Civilization and the Culture of Science: Science and the Shaping of Modernity, 1795–1935*, Oxford University Press, Oxford, 2020, pp. 19–28.

13. J. A. Josephson-Storm, *The Myth of Disenchantment: Magic, Modernity, and the Birth of the Human Sciences*, University of Chicago Press, Chicago, 2017, p. 3.

14. D. M. Knight, in *The Cambridge History of Science, Volume 5: The Modern Physical and Mathematical Sciences*, ed. N. Mary-Jo, Cambridge University Press, Cambridge, 2003, pp. 72–90.

15. D. Knight, *The Age of Science: The Scientific World-view in the Nineteenth Century*, Blackwell, Oxford, 1986; G. Piel, *The Age of Science. What Scientists Learned in the 20th Century*, Basic Books, New York, 2001.

16. A corollary is that this requires us to reject the current dominant view of the history of science that the past can be viewed and interpreted only in its own terms, but rather to assess to what extent historical discoveries advanced towards the present view.

17. N. Whitehead, *Science and the Modern World*, Macmillan, New York, 1925.

18. A. Koyré, *From the Closed World to the Infinite Universe*, Johns Hopkins University Press, Baltimore, 1957

19. H. Butterfield, *The Origins of Modern Science 1300–1800*, Bell, London, 1949.

20. L. Daston, The Secret History of Science and Modernity: The History of Science and the History of Religion, *Grey Room*, 2022, **88**, 14–31.

21. A. Koyré, Galileo and the Scientific Revolution of the Seventeenth Century, *Phil. Rev.*, 1943, **52**, 333–348

22. Notably in a series of lectures on Science as a Vocation at Munich University in 1918, published in German in 1919, and translated in Gerth & Mills, see ref. 2 in this chapter. Another translation, which includes two sets of lectures, Science as a Vocation, and Politics as a Vocation, is D. Owen and T. B. String, *Max Weber: The Vocation Lectures*, Hacket Publishing, Indianapolis, 2004.

23. There is an interesting parallel here with Mary Somerville's view in *Connexion of the Physical Sciences* in 1865 that mathematics was the unifying principle in the physical sciences (see Chapter 10).

24. 'Science' in this chapter refers explicitly to the natural sciences. I make no claim for any importance of other 'sciences,' whether frequently linked with natural sciences (such as the 'social sciences') or making distant claims to legitimacy by (mis)appropriating the word 'science,' such as the so-called 'political sciences.' However, the fact that so many fields try to associate themselves with 'science' is itself a tribute to the importance science has acquired in modern life.

25. G. Holton, *Science and Anti-science*, Harvard University Press, Cambridge MA, 1995, pp. 129–134.
26. P. Forman, The primacy of science in modernity, of technology in post-modernity, and of ideology in the history of technology, *Hist. Technol.*, 2007, **23**, 1–152.
27. S. Shapin, Science and the Modern World, in *The Handbook of Science and Technology Studies*, ed. E. Hackett *et al.*, The MIT Press, Cambridge, MA, 3rd edn, 2007, pp. 433–448.
28. P. Drucker, *Landmarks of Tomorrow. A Report on the new Post Modern World*, Harper & Row, New York, 1959.

Major Historical Figures

Major figures whose discoveries or ideas (1200–1950) are discussed in the chapters.[a]

Scientist	Dates	Discovery	Place
Robert Grosseteste	1168–1252	Philosophy	Oxford
Ibn al-Nafis	1213–1288	Circulation of blood	Egypt
Roger Bacon	1214–1292	Encyclopedia	Oxford
Thomas Aquinas	1225–1274	Priest	Rome
Leonardo da Vinci	1452–1519	Polymath	Florence
Desiderius Erasmus	1466–1536	Humanism	Basel
Nikolaus Copernicus	1473–1543	Heliocentricity	Warmia, Poland
Paracelsus	1493–1541	Alchemy & medicine	Switzerland
Gerard Mercator	1512–1594	Cartography	Louvain
Andreas Vesalius	1514–1564	Anatomy	Padua
Georg Joachim Rheticus	1514–1574	Heliocentricity	Wittenberg
William Gilbert	1544–1603	Magnetism	London
Tycho Brahe	1546–1601	Astronomical tables	Sweden
Francis Bacon	1561–1626	Knowledge is power	London
Galileo Galilei	1564–1642	Heliocentricity, motion	Florence
Johannes Kepler	1571–1630	Elliptical orbits	Prague
Jan van Helmont	1577–1644	Tree experiment	Brussels
William Harvey	1578–1657	Circulation of blood	London
Thomas Hobbes	1588–1679	Political philosophy	London
Pierre Gassendi	1592–1655	Atomism	Paris
René Descartes	1596–1650	Mind-body problem	Paris
Evangelista Torricelli	1608–1647	Barometer	Florence
Thomas Willis	1621–1675	Circle of Willis	London

The Frontiers of Science
By Benjamin Lewin
© Benjamin Lewin 2026
Published by the Royal Society of Chemistry, www.rsc.org

(*Continued*)

Scientist	Dates	Discovery	Place
Blaise Pascal	1623–1662	Vacuum	Paris
Robert Boyle	1627–1691	Air pump	London
Christiaan Huygens	1629–1695	Pendulum clock	The Hague
Benedict de Spinoza	1632–1677	Perception	The Hague
John Locke	1632–1704	Theory of mind	South England
Antonie van Leeuwenhoek	1632–1723	Microscopy	The Hague
Robert Hooke	1635–1703	Microscopy	London
Isaac Newton	1642–1726	Gravitation	Cambridge
Ole Rømer	1644–1710	Speed of light	Copenhagen
Gottfried Leibniz	1646–1716	Calculus	Hanover
Edmond Halley	1656–1742	Astronomy	London
Georg Stahl	1659–1734	Vitalism	Jena
Thomas Newcomen	1664–1729	Steam engine	Dartmouth
François Geoffroy	1672–1731	Affinity tables	Paris
James Bradley	1693–1762	Aberration of light	Oxford
James Lind	1716–1794	Scurvy	Edinburgh
Joseph Black	1728–1799	Latent heat	Edinburgh
Henry Cavendish	1731–1810	Hydrogen	London
Richard Arkwright	1732–1792	Spinning wheel	Scotland
Joseph Priestley	1733–1804	Phlogiston	Birmingham
Antoine Lavoisier	1734–1794	Oxygen	Paris
James Watt	1736–1819	Steam engine	Glasgow
William Herschel	1738–1822	Uranus	South England
Carl Wilhelm Scheele	1742–1786	Oxygen	Uppsala
Alessandro Volta	1745–1827	Electricity	Pavia
Edward Jenner	1749–1823	Smallpox vaccination	South England
Samuel Hahnemann	1755–1843	Homeopathy	Leipzig
John Dalton	1766–1844	Atoms	Manchester
André-Marie Ampère	1775–1836	Electromagnetism	Paris
Hans Christian Ørsted	1775–1851	Electromagnetism	Copenhagen
Amedeo Avogadro	1776–1856	Avogadro's number	Turin
Humphry Davy	1778–1829	Isolating elements	London
Joseph Louis Gay-Lussac	1778–1850	Kinetics of gases	Paris
Mary Somerville	1780–1872	Unity of science	London
Michael Faraday	1791–1867	Electromagnetism	London
Charles Babbage	1791–1871	First computer	London
Friedrich Wöhler	1800–1882	Synthesis of urea	Göttingen
Justus Liebig	1803–1873	Organic chemistry	Giessen
Charles Darwin	1809–1882	Evolution	London
Theodor Schwann	1810–1882	Cell theory	Louvain
John Snow	1813–1858	Cholera	London
Ignaz Semelweis	1815–1865	Antisepsis	Vienna
Ada Lovelace	1815–1852	First software	London
James Joule	1818–1889	Heat & energy	Manchester

(*Continued*)

Scientist	Dates	Discovery	Place
Hermann von Helmholtz	1821–1894	Thermodynamics	Heidelberg
Rudolf Virchow	1821–1902	Cell theory	Berlin
Gregor Mendel	1822–1884	Genetics	Brno
Rudolf Claudius	1822–1888	Thermodynamics	Zurich
Louis Pasteur	1822–1895	Germ theory	Paris
William Thomson	1824–1907	Atomic physics	Glasgow
Joseph Lister	1827–1912	Antisepsis	London
August Kekulé	1829–1896	Chemical bonds	Bonn
James Clerk Maxwell	1831–1879	Electromagnetism	Cambridge
Dmitry Mendeleev	1834–1907	Periodic table	St. Petersburg
Ernst Mach	1838–1916	Shock waves	Vienna
Robert Koch	1843–1919	Koch's postulates	Berlin
Camillo Golgi	1843–1926	Golgi stain	Pavia
Friedrich Miescher	1844–1895	Nucleic acid	Basel
Ludwig Boltzmann	1844–1906	Atoms	Vienna
Wilhelm Röntgen	1845–1923	X-rays	Munich
William Osler	1849–1919	Medicine	Oxford
Ramón y Cajal	1852–1934	Neuron	Madrid
Hendrik Lorentz	1853–1928	Movement of light	Leiden
Henri Poincaré	1854–1912	Mathematics	Paris
John Joseph Thomson	1856–1940	Electron	Cambridge
Heinrich Hertz	1857–1894	Radio waves	Karlsruhe
Charles Sherrington	1857–1952	Synapse	Oxford
Max Planck	1858–1947	Quantum	Berlin
Pierre Curie	1859–1906	Radioactivity	Paris
David Hilbert	1862–1943	Mathematics	Göttingen
Marie Curie	1867–1734	Radioactivity	Paris
Korbinian Brodmann	1868–1918	Brain regions	Berlin
Henrietta Leavitt	1868–1921	Cepheid stars	Cambridge, MA
Ernest Rutherford	1871–1937	Atomic structure	Cambridge
Oswald Avery	1877–1955	DNA	New York
Albert Einstein	1879–1955	Relativity	Zurich
Alfred Wegener	1880–1930	Continental drift	Hamburg/Graz
Niels Bohr	1881–1962	Quantum theory	Copenhagen
Erwin Schrödinger	1887–1961	Wave-particle duality	Graz & Dublin
Edwin Hubble	1889–1953	Red shift	Pasadena
James Chadwick	1891–1974	Neutron	Cambridge
Louis de Broglie	1892–1987	Wave-like particles	Paris
Georges Lemaître	1894–1966	Big Bang	Louvain
Warren McCulloch	1898–1969	Binary neuron model	Chicago
Wolfgang Pauli	1900–1958	Neutrino	Zurich
Enrico Fermi	1901–1954	Nuclear fission	Rome
Werner Heisenberg	1901–1976	Uncertainty principle	Munich

(*Continued*)

Scientist	Dates	Discovery	Place
Linus Pauling	1901–1994	Chemical bond	Pasadena
Paul Dirac	1902–1984	Anti-matter	Cambridge
Karl Popper	1902–1994	Falsifiability	London
John von Neumann	1903–1957	Cybernetics	Princeton
Robert Oppenheimer	1904–1967	Atomic bomb	Berkeley
Kurt Gödel	1906–1978	Logic/mathematics	Vienna
Alan Turing	1912–1954	Artificial intelligence	London
Thomas Kuhn	1922–1996	Paradigms	Cambridge, MA

[a] The third column shows the major work associated with each person. The fourth column shows the place where the major work was performed.

Subject Index

References to illustrations and photos are given in *italic* type.
References to tables are indicated by **bold** type.